The Nikola Tesla Treasury

Jaline LaPlante
LMBT #2889
By appointment only:
252-256-0364

The Nikola Tesla Treasury

By Nikola Tesla

©2007 Wilder Publications

This book is a product of its time and does not reflect the same values as it would if it were written today. Parents might wish to discuss with their children how views on race, gender, sexuality, ethnicity, and interpersonal relations have changed since this book was written before allowing them to read this classic work.

All rights reserved. Printed in the United States of America. No part of this book may be used or reproduced in any manner without written permission except for brief quotations for review purposes only.

Wilder Publications, LLC.
PO Box 3005
Radford VA 24143-3005
www.wilderpublications.com

ISBN 10: 1-934451-90-8
ISBN 13: 978-1-934451-90-8

First Edition

10 9 8 7 6 5 4 3 2 1

Table of Contents

Introduction 8
A New System of Alternating Current Motors and Transformers . 13
The Tesla Alternate Current Motor 26
Mr. Nikola Tesla on Alternating Current Motors 30
The Losses Due to Hysteresis in Transformers 33
Swinburne's "Hedgehog" Transformer 35
Tesla's New Alternating Motors 38
Phenomena of Alternating Currents of Very High Frequency 43
Experiments with Alternating Currents of High Frequency 54
Alternate Current Motors 58
Electro-Motors 59
Phenomena of Currents of High Frequency 60
Alternate Current Electrostatic Induction Apparatus 64
An Electrolytic Clock 66
Experiments with Alternate Currents of Very High Frequency and Their Application to Methods of Artificial Illumination 68
Electric Discharge in Vacuum Tubes 108
Note by Prof. J. J. Thomson 113
Reply to J. J. Thomson's Note in the Electrician 114
Notes on a Unipolar Dynamo 116
Experiments with Alternate Currents of High Potential and High Frequency 121
The "Drehstrom" Patent 196
The Ewing High-Frequency Alternator and Parsons Steam Engine 198
On the Dissipation of the Electrical Energy of the Hertz Resonator 200
On Light and Other High Frequency Phenomena 204
The Physiological and Other Effects of High Frequency Currents 263
From Nikola Tesla - He Writes about His Experiments in Electrical Healing 265
On Roentgen Rays 267
On Roentgen Rays the Latest Results 272
Tesla's Latest Results: He Now Produces Radiographs At A Distance of More Than Forty Feet 274
On Reflected Roentgen Rays 276
On Roentgen Radiations 326
Roentgen Ray Investigations 287
Electrical Progress. On Apparatus for Cathography 294
An Interesting Feature of X-Ray Radiations 297

Roentgen Rays Or Streams 299
Mr. Tesla on Thermo Electricity 305
On Electricity 306
On the Hurtful Actions of Lenard and Roentgen Tubes 316
On the Source of Roentgen Rays and the Practical Construction and Safe Operation of Lenard Tubes 321
Tesla's Latest Advances in Vacuum-tube Lighting. Application of Tubes of and Other Purposes 332
Tesla on Animal Training by Electricity 384
High Frequency Oscillators for Electro-Therapeutic and Other Purposes 337
My Submarine Destroyer 351
Letter to Editor *Electrical Engineer* 354
Tesla Describes His Efforts in Various Fields of Work 355
On Current Interrupters 358
The Problem of Increasing Human Energy, With Special References to the Harnessing of the Sun's Energy 361
Tesla's New Discovery 409
Tesla's Wireless Light 412
Talking with the Planets 414
Wind Power Should Be Used More Now 419
Electrical oscillator activity ten million Horsepower Burning atmospheric nitrogen by high frequency discharges twelve million volts 420
The Transmission of Electrical Energy Without Wires 421
Letter from Nikola Tesla *New York Sun* 426
Electric Autos 428
The Transmission of Electrical Energy Without Wires as a Means for Furthering Peace 421
Tesla on Subway Dangers 440
Tesla's Reply to Edison 442
Tesla on the Peary North Pole Expedition 444
Signaling to Mars —A Problem of Electrical Engineering 447
Tuned Lightning 449
Tesla's Wireless Torpedo 452
Wireless on Railroads 453
Nikola Tesla Objects 454
Tesla's Tidal Wave to Make War Impossible 455
Mr. Tesla on the Wireless Transmission of Power 460
Can Bridge the Gap to Mars 462
Sleep From Electricity 465
Possibilities of "Wireless" 466

Tesla on Wireless 467
My Apparatus, Says Tesla 468
The Future of the Wireless Art 469
Nikola Tesla's Forecast for 1908 471
Mr. Tesla's Vision 472
Little Airplane Progress 473
Tesla on Airplanes 475
How to Signal Mars 476
What Science May Achieve this Year: New Mechanical Principle for Conservation of Energy 480
Mr. Tesla on the Future 482
The Disturbing Influence of Solar Radiation on the Wireless Transmission of Energy 483
Nikola Tesla's Plan to Keep "Wireless Thumb" on Ships At Sea . 488
From Nikola Tesla: A Tribute to George Westinghouse 490
Tesla and Marconi 491
Science and Discovery Are the Great Forces Which Will Lead to the Consummation of the War 492
How Cosmic Forces Shape Our Destinies ("Did the War Cause the Italian Earthquake") 504
Some Personal Recollections 509
The Wonder World to Be Created by Electricity 514
Nikola Tesla Sees A Wireless Vision 523
Correction by Mr. Tesla 525
Wonders of the Future 526
Electric Drive for Battle Ships 530
Nikola Tesla Tells of Country's War Problems 538
A Speech Delivered Before the American Institute of Electrical Engineers 540
The Effect of Static on Wireless Transmission 550
Famous Scientific Illusions 553
Tesla Answers Mr. Manierre and Further Explains the Axial Rotation of the Moon 566
The Moon's Rotation 568
Tesla on High Frequency Generators 574
The True Wireless 575
Electrical Oscillators 592
Nikola Tesla Tells How We May Fly Eight Miles High at 1,000 Miles an Hour 600
Signals to Mars Based on Hope of Life on Planet 606
Developments in Practice and Art of Telephotography 610
Interplanetary Communication 615

Nikola Tesla on Electric Transmission 617
My Inventions: The Autobiography of Nikola Tesla 619
   My Early Life 619
   My First Efforts At Invention 627
   My Later Endeavors 634
   The Discovery of the Tesla Coil and Transformer 641
   The Magnifying Transmitter 647
   The Art of Telautomatics 655

Introduction

Njan poetry has so distinct a charm that Goethe is said to have learned the musical tongue in which it is written rather than lose any of its native beauty. History does not record, however, any similar instance in which the Servian language, though it be that of Boskovich, expounder of the atomic theory, has been studied for the sake of the scientific secrets that might lurk therein. The vivid imagination and ready fancy of the people have been literary in their manifestation and fruit. A great Slav orator has publicly reproached his one hundred and twenty million fellows in Eastern Europe with their utter inability to invent even a mouse-trap. They were all mere barren idealists. If this were true, to equalize matters, we might perhaps barter without loss some score of ordinary American patentees for a single singer of Illyrian love-songs. But racial conditions are hardly to be offset on any terms that do not leave genius its freedom, and once in a while Nature herself rights things by producing a man whose transcendent merit compensates his nation for the very defects to which it has long been sensitive. It does not follow that such a man shall remain in a confessedly unfavorable environment. Genius is its own passport, and has always been ready to change habitats until the natural one is found. Thus it is, perchance, that while some of our artists are impelled to set up their easels in Paris or Rome, many Europeans of mark in the fields of science and research are no less apt to adopt our nationality, of free choice. They are, indeed, Americans born in exile, and seek this country instinctively as their home, needing in reality no papers of naturalization. It was thus that we welcomed Agassiz, Ericsson, and Graham Bell. In like manner Nikola Tesla, the young Servian inventor with whose work a new age in electricity is beginning, now dwells among us in New York. Mr. Tesla's career not only touches the two extremes of European civilization, east and west, in a very interesting way, but suggests an inquiry into the essential likeness between poet and inventor. He comes of an old Servian family whose members for centuries have kept watch and ward along the Turkish frontier, and whose blood was freely shed that our western vanguard might gain time for its advance upon these shores. Yet, remote as such people and conditions are to us, it is with apparatus based on ideas and principles originating among them that the energy from Niagara Falls is to be widely distributed by electricity, in the various forms of light, heat, and power. This, in itself, would seem enough to confer fame, but Mr. Tesla has done, and will do, much else. Could he be tamed to habits of moderation in work, it would be difficult to set limit to the solutions he might give us, through ripening years, of many deep problems; but when a man springs from a people who have a hundred words for knife and only one for bread, it is a little unreasonable to urge him to be careful even of his own life. Thirty-six years make a brief span, but when an inventor believes that creative fertility is restricted to the term of youth, it is no wonder that night and day witness his anxious activity, as of a relentless volcano, and that ideas well up like hot lava till the crater be suddenly exhausted and hushed.

A Slav of the Slavs, with racial characteristics strongly stamped in look, speech, and action, Mr. Tesla is a notable exemplification of the outcropping in unwonted form of tendencies suppressed. I have never heard him speak of a picture or a piece of music, but his numerous inventions, and the noble lectures that embody his famous investigations with currents of high frequency and high potential, betray the poetic temperament throughout. One would expect the line separating fact from theory to fade at the altitudes of thought to which his later speculations reach; but this lithe, spare mountaineer is accustomed to the thin, dry air, and neither loses sharpness of sight nor breathes painfully. Has the Servian poet become inventor, or is the inventor a poet? Mr. Tesla has been held a visionary, deceived by the flash of casual shooting stars; but the growing conviction of his professional brethren is that because he saw farthest, saw first the low lights flickering on tangible new continents of science. The perceptive and imaginative qualities of the mind are not often equally marked in the same man of genius. Overplus of imagination may argue dimness of perception; an ability to dream dreams may imply a want of skill in improving reapers. Now and then the two elements combine in the creative poet of epic and drama; occasionally they give us the prolific inventor like Tesla.

Jules Breton has spoken of the history of his life as being at the same time the genesis of his art. This is true of Nikola Tesla's evolution. His bent toward invention we may surely trace to his mother, who, as the wife of an eloquent clergyman in the Greek Church, made looms and churns for a pastoral household while her husband preached. Tesla's electrical work started when, as a boy in the Polytechnic School at Gratz, he first saw a direct-current Gramme — machine, and was told that the commutator was a vital and necessary feature in all such apparatus. His intuitive judgment or latent spirit of invention at once challenged the statement of his instructor, and that moment began the process of reasoning and experiment which led him to his discovery of the rotating magnetic field, and to the practical polyphase motors, in which the commutator and brushes, fruitful and endless source of trouble, are absolutely done away with. These perfected inventions did not come at once; they never do. The conditions that surrounded this youth in the airy fastnesses of the Dinaric Alps all made against the hopes he nursed of becoming an electrician ; and not the least impediment was the fond wish of his parents at Siniljan Lika that he should maintain the priestly tradition, and benefit by the preferment likely to come through his uncle, now Metropolitan in Bosnia. But Tesla felt himself destined to serve at other altars than those of his ancient faith, with other means of approach to the invisible and unknown. He persevered in mathematical and mechanical studies, mastered incidentally half a dozen languages, and at last became an assistant in the Government Telegraph Engineering Department at BudaPest. His salary was small enough to please those who hold that the best endowment of genius is poverty, and he would make no appeals to his widowed mother for help. Experimenting, of course, went on all the time at this juncture it was on telephony that he wasted his meager substance in riotous invention. Desirous of going to a fete with some friends, and anxious not to spend on clothes the money that might buy magnets and batteries, the brilliant idea occurred to him to turn his only pair of trousers inside out and to disport in them on the morrow as new. He sat up all night tailoring, but the fete came and went before he could reappear in public. This episode is quite in keeping with his boyish efforts to fly from the steep roof of the house at Smiljan, using an old umbrella as arostat; or with the peculiar tests, stopped by the family doctor before the results could be determined, as to how long he could suspend the beating of his heart by will power.

Naturally enough for a young inventor seeking larger opportunity, Tesla soon drifted west- ward from Budapest. He made his way to Paris, where he quickly secured employment in electric lighting, then a new art, and encountered an observant associate of Mr. Edison. Almost before he knew it, he was on his glad voyage across the Atlantic to work in one of the Edison shops, and to enter upon a new stage of development. He had profound faith in the value of the principles first meditated in the silence of the sterile mountains that border the Adriatic, and he knew that in a country where every new invention in electricity has its chance, his turn would come also, for he now had demonstrated his theories in actual apparatus.

If anything were needed to confirm Mr. Tesla in his hopes and enthusiasm, it would have been the close relation that he was thus thrown into with the robust, compelling genius who has created so many new things in electricity. Emerson has said that steam is half an Englishman; may we not, in view of what such men as Edison have done, add that electricity is half an American? The fiery zeal with which this young recruit flung himself on the most exacting tasks matched that of his chief. It went the length of a daily breakfast of Welsh rabbit, for weeks Mr. Tesla accepting as true, in spite of protesting stomach, the jocular suggestion that it was thus that his hero fortified himself successfully for renewed effort after their long vigils of toil. Mr. Edison, like most other people, had some difficulty in finding anywhere within the pale of civilization, as marked by the boundaries of maps, the isolated region of Mr. Tesla's birth, and once inquired seriously of his neophyte whether he had ever tasted human flesh. It was inevitable that a really delightful intimacy and apprenticeship should end. Even the most cometic genius has its orbit, and these two men are singularly representative of different kinds of training, different methods, and different aims. Mr. Tesla must needs draw apart; and stimulated by this powerful spirit, he went on his own way for his own works sake.

Of late years a sharp controversy has raged in the electrical field as to the respective advantages of the continuous and the alternating current for light and power. In bitterness and frequent descent to personalities it has resembled the polemics of the old metaphysical schoolmen, and uninstructed, plain folk have mildly wondered whether it was really worthwhile to indulge in such terrible threatening and slaughter over purely speculative topics. There is, however, a very practical aspect to the discussion, and from the first Mr. Tesla has been an advocate of the alternating current, not because he loved the direct current less, but because he knew that with the alternating he could achieve results otherwise impossible, especially in power transmission. Furthermore, all direct current generators and motors have required commutators and brushes, but Mr. Tesla, who has himself perfected many inventions based on direct currents, has shown that with the application of the rotating-field principle these elements of complication and restriction were no longer needed. This utilization of polyphase currents was a most distinct advance, made a deep imprint on the electric arts, and has been duly signalized. In America the invention found immediate sale. From Italy came an insistent cry of priority, reminding one of the anticipations that have clustered thick around the telegraph, the telephone, and the incandescent lamp. In Germany, with money raised by popular and imperial subscription, apparatus on the polyphase principle was built by which large powers were transmitted electrically more than a hundred miles from Neckar-on-the Rhine to Frankfort-on-the-Main; and now by equivalent agency Niagara is to drive the wheels of Buffalo and beyond.

So thoroughly has Mr. Tesla worked out his discovery of the rotating magnetic field, or resultant attraction, that the record of his inventions contains no fewer than twenty-four chapters on varying forms of his polyphase- current apparatus and arrangements of circuit. But ever pursuing new researches, Mr. Tesla, after the enunciation of these fundamental ideas, next brought to notice his series of even more interesting investigations on several novel groups of phenomena produced with currents of high potential and high frequency. To familiarize the American public with some of his results, he lectured upon them at Columbia College, before the American Institute of Electrical Engineers, in May, 1891. The year following, with riper results to publish, and by special invitation, he lectured twice in England, appearing before the Institution of Electrical Engineers, a (distinguished scientific body of which Professor Crookes was then president, and, later, at the Royal Institution, where the immortal Faraday lived and labored. From England he was called to France to repeat his demonstrations before the Society Internationale des Electrifies and the Society Francaise de Physique. In Germany he received the greetings of Hertz and Von Helmholtz, and from his own country came the Order of Saint Sava, conferred by the king. Since his return to this country he has lectured before the Franklin Institute at Philadelphia and the National Electric Light Association at St. Louis. But he has an intense dislike to the platform, and has returned to his laboratory with a remorseful sense of neglected work from which long months of abandonment to unremitting research will not free him.

I can only outline the vast range of the researches that these lectures, and the apparatus connected with the demonstrations, cover. Broadly stated,, Mr. Tesla has advanced the opinion, and sustained it by brilliant experiments of startling beauty and grandeur, that light and heat are produced by electrostatic forces acting between charged molecules or atoms. Perfecting a generator that would give him currents of several thousand alternations per second, and inventing his disruptive discharge coil, he has created electrostatic conditions that have already modified not a few of the accepted notions about electricity. It has been supposed that ordinary currents of one or two thousand volts potential would surely kill, but Mr. Tesla has been seen receiving through his hands currents at a potential of more than 200,000 volts, vibrating a million times per second, and manifesting themselves in dazzling streams of light. This is not a mere tour de force, but illustrates the principle that while currents of lower frequency destroy life, these are harmless. After such a striking test, which, by the way, no one has displayed a hurried inclination to repeat, Mr. Tesla's body and clothing have continued for some time to emit fine glimmers or halos of splintered light. In fact, an actual flame is produced by this agitation of electrostatically charged molecules, and the curious spectacle can be seen of puissant, white, ethereal flames, that do not consume anything, bursting from the ends of an induction coil as though it were the bush on holy ground. With such vibrations as can be maintained by a potential of 3,000,000 volts, Mr. Tesla expects some day to envelop himself in a complete sheet of lambent fire that will leave him quite uninjured. Such currents as he now uses would, he says, keep a naked man warm at the North Pole, and their use in therapeutics is but one of the practical possibilities that has been taken up.

Utilizing similar currents and mechanism, Mr. Tesla has demonstrated the fact that electric lamps and motors can not only be made to operate on one wire, instead of using a second wire on the ground to complete the circuit, but that we can operate them even by omitting the circuit. Our Subway Boards are to find their wires and occupations gone. Electric vibrations set up at any point of the earth may by resonance at any other spot serve for the transmission of either intelligence or power. With these impulses or wave discharges, Mr. Tesla also opens up an entirely new field of electric lighting. His lamps have no filaments as ordinarily known, but contain a straight fiber, a refractory button, or nothing but a gas. Tubes or bulbs of this kind, in which the imprisoned ether or air beats the crystal walls, when carried into the area or room through which these unsuspected currents are silently vibrating, burst into sudden light. If coated inwardly with phosphorescent substances, they glow in all the splendors of the sunset and the aurora.

These are only a scant handful of ideas and discoveries from the rich mint of Mr. Tesla's laboratory, where alone, secluded, intimates or assistants shut out, he reasons from cause to effect; and with severe, patient diligence not only elaborates his theories, but tries them by the rack and thumbscrew of experiment. He is of all men most dissatisfied with things as they are in his own field of work. Recently, the high-frequency generators with which he has done so much of this advanced work have been laid aside in discontent for an oscillator, which he thinks may not only replace the steam-engine with its ponderous fly-wheels and governors, but embodies the simplest possible form of efficient mechanical generator of electricity. He may be wrong, but misdirection will only suggest new avenues to the goal.

Mr. Tesla has often been urged to assume domestic ties, settle down, and till some corner of the new domain. But shall he farm or explore? Soon enough the proprietary fences will be set up; soon enough will the dusty, beaten highway, dotted with milestones and finger-posts, run straight ahead. If we would, we cannot leash the pioneers whose yearnings are for inner Nature, whose sense is keenest to her faint voices and odors, the quest of which lures onward through the trackless woods.
—Thomas Commerford Martin

A New System of Alternating Current Motors and Transformers
Delivered before the American Institute of Electrical Engineers, May 1888.

I desire to express my thanks to Professor Anthony for the help he has given me in this matter. I would also like to express my thanks to Mr. Pope and Mr. Martin for their aid. The notice was rather short, and I have not been able to treat the subject so extensively as I could have desired, my health not being in the best condition at present. I ask your kind indulgence, and I shall be very much gratified if the little I have done meets your approval.

In the presence of the existing diversity of opinion regarding the relative merits of the alternate and continuous current systems, great importance is attached to the question whether alternate currents can be successfully utilized in the operation of motors. The transformers, with their numerous advantages, have afforded us a relatively perfect system of distribution, and although, as in all branches of the art, many improvements are desirable, comparatively little remains to be done in this direction. The transmission of power, on the contrary, has been almost entirely confined to the use of continuous currents, and notwithstanding that many efforts have been made to utilize alternate currents for this purpose, they have, up to the present, at least as far as known, failed to give the result desired. Of the various motors adapted to be used on alternate current circuits the following have been mentioned: 1. A series motor with subdivided field. 2. An alternate current generator having its field excited by continuous currents. 3. Elihu Thomson's motor. 4. A combined alternate and continuous current motor. Two more motors of this kind have suggested themselves to me.

1. A motor with one of its circuits in series with a transformer and the other in the secondary of the transformer. 2. A motor having its armature circuit connected to the generator and the field coils closed upon themselves. These, however, I mention only incidentally.

The subject which I now have the pleasure of bringing to your notice is a novel system of electric distribution and transmission of power by means of alternate currents, affording peculiar advantages, particularly in the way of motors, which I am confident will at once establish the superior adaptability of these currents to the transmission of power and will show that many results heretofore unattainable can be reached by their use; results which are very much desired in the practical operation of such systems and which cannot be accomplished by means of continuous currents.

Before going into a detailed description of this system, I think it necessary to make a few remarks with reference to certain conditions existing in continuous current generators and motors, which, although generally known, are frequently disregarded.

In our dynamo machines, it is well known, we generate alternate currents which we direct by means of a commutator, a complicated device and, it may be justly said, the source of most of the troubles experienced in the operation of the machines. Now, the currents so directed cannot be utilized in the motor, but they must—again by means of a similar unreliable device—be reconverted into their original state of alternate currents. The function of the commutator is entirely external, and in no way dues it affect the internal working of the machines. In reality, therefore, all machines are alternate current machines, the currents appearing as continuous only in the external circuit during their transit from generator to motor. In view simply of this fact, alternate currents would commend themselves as a more direct application of electrical energy, and the employment of continuous currents would only be justified if we had dynamos which would primarily generate, and motors which would be directly actuated by such currents.

But the operation of the commutator on a motor is twofold; firstly, it reverses the currents through the motor, and secondly, it effects, automatically, a progressive shifting of the poles of one of its magnetic constituents. Assuming, therefore, that both of the useless operations in the system, that is to say, the directing of the alternate currents on the generator and reversing the direct currents on the motor, be eliminated, it would still be necessary, in order to cause a rotation of the motor, to produce a progressive shifting of the poles of one of its elements, and the question presented itself,—How to perform this operation by the direct action of alternate currents? I will now proceed to show how this result was accomplished.

figures 1, 1a

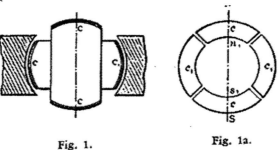

Fig. 1.   Fig. 1a.

In the first experiment a drum-armature was provided with two coils at right angles to each other, and the ends of these coils were connected to two pairs of insulated contact-rings as usual. A ring was then made of thin insulated plates of sheet-iron and wound with four coils, each two opposite coils being connected together so as to produce free poles on diametrically opposite sides of the ring. The remaining free ends of the coils were then connected to the contact-rings of the generator armature so as to form two independent circuits, as indicated in figure 9. It may now be seen what results were secured in this combination, and with this view I would refer to the diagrams, figures 1 to 8a. The field of the generator being independently excited, the rotation of the armature sets up currents in the coils C C_1, varying in strength and direction in the well-known manner. In the position shown in figure 1 the current in coil C is nil while coil C_1 is traversed by its maximum current, and the connections my be such that the ring is magnetized by the coils c_1 c_1 as indicated by the letters N S in figure 1a, the magnetizing effect of the coils c c being nil, since these coils are included in the circuit of coil C.

figures 2, 2a

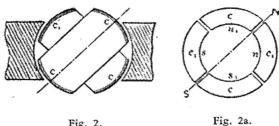

Fig. 2.   Fig. 2a.

In figure 2 the armature coils are shown in a more advanced position, one-eighth of one revolution being completed. Figure 2a illustrates the corresponding magnetic condition of the ring. At this moment the coil c1 generates a current of the same direction as previously, but weaker, producing the poles n1 s1 upon the ring; the coil c also generates a current of the same direction, and the connections may be such that the coils c c produce the poles n s, as shown in figure 2a. The resulting polarity is indicated by the letters N S, and it will be observed that the poles of the ring have been shifted one-eighth of the periphery of the same.

figures 3, 3a

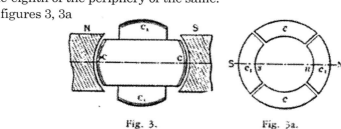

Fig. 3.   Fig. 3a.

In figure 3 the armature has completed one-quarter of one revolution. In this phase the current in coil C is maximum, and of such direction as to produce the poles N S in figure 3a, whereas the current in coil C1 is nil, this coil being at its neutral position. The poles N S in figure 3a are thus shifted one-quarter of the circumference of the ring.

figures 4, 4a

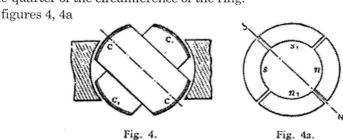

Fig. 4.   Fig. 4a.

Figure 4 shows the coils C C in a still more advanced position, the armature having completed three-eighths of one revolution. At that moment the coil C still generates a current of the same direction as before, but of less strength, producing the comparatively weaker poles n s in figure 4a, The current in the coil C1 is of the same strength, but of opposite direction. Its effect is, therefore, to produce upon the ring the poles n1 and s1 as indicated, and a polarity, N S, results, the poles now being shifted three-eighths of the periphery of the ring.

figures 5, 5a

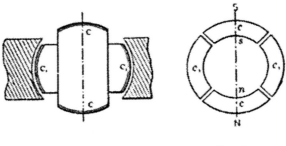

Fig. 5.   Fig. 5a.

In figure 5 one-half of one revolution of the armature is completed, and the resulting magnetic condition of the ring is indicated in figure 5a. Now, the current in coil C is nil, while the coil C1 yields its maximum current, which is of the same direction as previously; the magnetizing effect is, therefore, due to the coils Cl Cl alone, and, referring to figure 5a, it will be observed that the poles N S are shifted one-half of the circumference of the ring. During the next half revolution the operations are repeated, as represented in the figures 6 to 8a.

figures 6, 6a

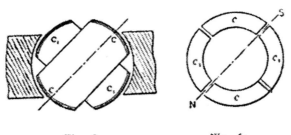

Fig. 6.   Fig. 6a.

A reference to the diagrams will make it clear that during one revolution of the armature the poles of the ring are shifted once around its periphery, and each revolution producing like effects, a rapid whirling of the poles in harmony with the rotation of the armature is the result. If the connections of either one of the circuits in the ring are reversed, the shifting of the poles is made to progress in the opposite direction, but the operation is identically the same. Instead of using four wires, with like result, three wires may be used, one forming a common return for both circuits.

figures 7, 7a

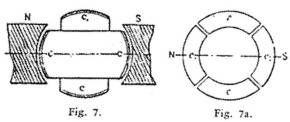

Fig. 7.   Fig. 7a.

This rotation or whirling of the poles manifests itself in a series of curious phenomena. If a delicately pivoted disc of steel or other magnetic metal is approached to the ring it is set in rapid rotation, the direction of rotation varying with the position of the disc. For instance, noting the direction outside of the ring it will be found that inside the ring it turns in an opposite direction, while it is unaffected if placed in a position symmetrical to the ring. This is easily explained. Each time that a pole approaches it induces an opposite pole in the nearest point on the disc, and an attraction is produced upon that point; owing to this, as the pole is shifted further away from the disc a tangential pull is exerted upon the same, and the action being constantly repeated, a more or less rapid rotation of the disc is the result. As the pull is exerted mainly upon that part which is nearest to the ring, the rotation outside and inside, or right and left, respectively, is in opposite directions, figure 9. When placed symmetrically to the ring, the pull on opposite sides of the disc being equal, no rotation results. The action is based on the magnetic inertia of the iron; for this reason a disc of hard steel is much more affected than a disc of soft iron, the latter being capable of very rapid variations of magnetism. Such a disc has proved to be a very useful instrument in all these investigations, as it has enabled me to detect any irregularity in the action. A curious effect is also produced upon iron filings. By placing some upon a paper and holding them externally quite close to the ring they are set in a vibrating motion, remaining in the same place, although the paper may be moved back and forth; but in lifting the paper to a certain height which seems to be dependent on the intensity of the poles and the speed of rotation, they are thrown away in a direction always opposite to the supposed movement of the poles. If a paper with filings is put flat upon the ring and the current turned on suddenly; the existence of a magnetic whirl may be easily observed.

figures 8, 8a

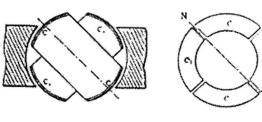

Fig. 8.   Fig. 8a.

To demonstrate the complete analogy between the ring and a revolving magnet, a strongly energized electro-magnet was rotated by mechanical power, and phenomena identical in every particular to those mentioned above were observed.

Obviously, the rotation of the poles produces corresponding inductive effects and may be utilized to generate currents in a closed conductor placed within the influence of the poles. For this purpose it is convenient to wind a ring with two sets of superimposed coils forming respectively the primary and secondary circuits, as shown in figure 10. In order to secure the most economical results the magnetic circuit should be completely closed, and with this object in view the construction may be modified at will.

The inductive effect exerted upon the secondary coils will be mainly due to the shifting or movement of the magnetic action; but there may also be currents set up in the circuits in consequence of the variations in the intensity of the poles. However, by properly designing the generator and determining the magnetizing effect of the primary coils the latter element may be made to disappear. The intensity of the poles being maintained constant, the action of the apparatus will be perfect, and the same result will be secured as though the shifting were effected by means of a commutator with an infinite number of bars. In such case the theoretical relation between the energizing effect of each set of primary coils and their resultant magnetizing effect may be expressed by the equation of a circle having its center coinciding with that of an orthogonal system of axes, and in which the radius represents the resultant and the co-ordinates both of its components. These are then respectively the sine and cosine of the angle U between the radius and one of the axes (O X). Referring to figure 11, we have $r^2 = x^2 + y^2$; where $x = r \cos a$, and $y = r \sin a$.

Assuming the magnetizing effect of each set of coils in the transformer to be proportional to the current—which may be admitted for weak degrees of magnetization—then $x = Kc$ and $y = Kc1$, where K is a constant and c and c1 the current in both sets of coils respectively. Supposing, further, the field of the generator to be uniform, we have for constant speed $c1 = K1 \sin a$ and $c = K1 \sin(90o + a) = K1 \cos a$, where K1 is a constant. See figure 12. Therefore, $x = Kc = K K1 \cos a$; $y = Kc1 = K K1 \sin a$; and $K K1 = r$.

That is, for a uniform field the disposition of the two coils at right angles will secure the theoretical result, and the intensity of the shifting poles will be constant. But from $r^2 = x^2 + y^2$ it follows that for $y = O$, $r = x$; it follows that the joint magnetizing effect of both sets of coils should be equal to the effect of one set when at its maximum action. In transformers and in a certain class of motors the fluctuation of the poles is not of great importance, but in another class of these motors it is desirable to obtain the theoretical result.

In applying this principle to the construction of motors, two typical forms of motor have been developed. First, a form having a comparatively small rotary effort at the start, but maintaining a perfectly uniform speed at all loads, which motor has been termed synchronous. Second, a form possessing a great rotary effort at the start, the speed being dependent on the load.

These motors may be operated in three different ways: 1. By the alternate currents of the source only. 2. By a combined action of these and of induced currents. 3. By the joint action of alternate and continuous currents.

figures 9,

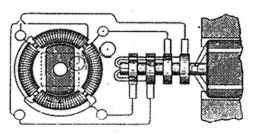

Fig. 9.

The simplest form of a synchronous motor is obtained by winding a laminated ring provided with pole projections with four coils, and connecting the same in the manner before indicated. An iron disc having a segment cut away on each side may be used as an armature. Such a motor is shown in figure 9. The disc being arranged to rotate freely within the ring in close proximity to the projections, it is evident that as the poles are shifted it will, owing to its tendency to place itself in such a position as to embrace the greatest number of the lines of force, closely follow the movement of the poles, and its motion will be synchronous with that of the armature of the generator; that is, in the peculiar disposition shown in figure 9, in which the armature produces by one revolution two current impulses in each of the circuits. It is evident that if, by one revolution of the armature, a greater number of impulses is produced, the speed of the motor will be correspondingly increased. Considering that the attraction exerted upon the disc is greatest when the same is in close proximity to the poles, it follows that such a motor will maintain exactly the same speed at all loads within the limits of its capacity.

Fifure10,

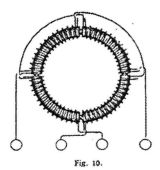

Fig. 10.

To facilitate the starting, the disc may be provided with a coil closed upon itself. The advantage secured by such a coil is evident. On the start the currents set up in the coil strongly energize the disc and increase the attraction exerted upon the same by the ring, and currents being generated in the coil as long as the speed of the armature is inferior to that of the poles, considerable work may be performed by such a motor even if the speed be below normal. The intensity of the poles being constant, no currents will be generated in the coil when the motor is turning at its normal speed.

figures 11, 12

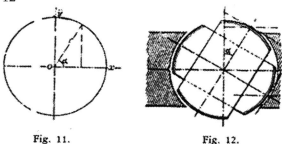

Fig. 11.   Fig. 12.

Instead of closing the coil upon itself, its ends may be connected to two insulated sliding rings, and a continuous current supplied to these from a suitable generator. The proper way to start such a motor is to close the coil upon itself until the normal speed is reached, or nearly so, and then turn on the continuous current. If the disc be very strongly energized by a continuous current the motor may not be able to start, but if it be weakly energized, or generally so that the magnetizing effect of the ring is preponderating it will start and reach the normal speed. Such a motor will maintain absolutely the same speed at all loads. It has also been found that if the motive power of the generator is not excessive, by checking the motor the speed of the generator is diminished in synchronism with that of the motor. It is characteristic of this form of motor that it cannot be reversed by reversing the continuous current through the coil.

figure 13

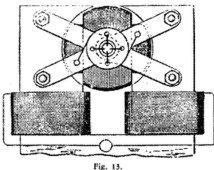

Fig. 13.

The synchronism of these motors may be demonstrated experimentally in a variety of ways. For this purpose it is best to employ a motor consisting of a stationary field magnet and an armature arranged to rotate within the same, as indicated in figure 13. In this case the shifting of the poles of the armature produces a rotation of the latter in the opposite direction. It results therefrom that when the normal speed is reached, the poles of the armature assume fixed positions relatively to the field magnet and the same is magnetized by induction, exhibiting a distinct pole on each of the pole-pieces. If a piece of soft iron is approached to the field magnet it will at the start be attracted with a rapid vibrating motion produced by the reversals of polarity of the magnet, but as the speed of the armature increases; the vibrations become less and less frequent and finally entirely cease. Then the iron is weakly but permanently attracted, showing that the synchronism is reached and the field magnet energized by induction.

The disc may also be used for the experiment. If held quite close to the armature it will turn as long as the speed of rotation of the poles exceeds that of the armature; but when the normal speed is reached, or very nearly so; it ceases to rotate and is permanently attracted.

A crude but illustrative experiment is made with an incandescent lamp. Placing the lamp in circuit with the continuous current generator, and in series with the magnet coil, rapid fluctuations are observed in the light in consequence of the induced current set up in the coil at the start; the speed increasing, the fluctuations occur at longer intervals, until they entirely disappear, showing that the motor has attained its normal speed. A telephone receiver affords a most sensitive instrument; when connected to any circuit in the motor the synchronism may be easily detected on the disappearance of the induced currents.

In motors of the synchronous type it is desirable to maintain the quantity of the shifting magnetism constant, especially if the magnets are not properly subdivided.

To obtain a rotary effort in these motors was the subject of long thought. In order to secure this result it was necessary to make such a disposition that while the poles of one element of the motor are shifted by the alternate currents of the source, the poles produced upon the other element should always be maintained in the proper relation to the former, irrespective of the speed of the motor. Such a condition exists in a continuous current motor; but in a synchronous motor, such as described, this condition is fulfilled only when the speed is normal.

figure 14

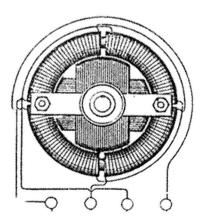

Fig. 14.

The object has been attained by placing within the ring a properly subdivided cylindrical iron core wound with several independent coils closed upon themselves. Two coils at right angles as in figure 14, are sufficient, but greater number may he advantageously employed. It results from this disposition that when the poles of the ring are shifted, currents are generated in the closed armature coils. These currents are the most intense at or near the points of the greatest density of the lines of force, and their effect is to produce poles upon the armature at right angles to those of the ring, at least theoretically so; and since action is entirely independent of the speed—that is, as far as the location of the poles is concerned—a continuous pull is exerted upon the periphery of the armature. In many respects these motors are similar to the continuous current motors. If load is put on, the speed, and also the resistance of the motor, is diminished and more current is made to pass through the energizing coils, thus increasing the effort. Upon the load being taken off, the counter-electromotive force increases and less current passes through the primary or energizing coils. Without any load the speed is very nearly equal to that of the shifting poles of the field magnet.

It will be found that the rotary effort in these motors fully equals that of the continuous current motors. The effort seems to be greatest when both armature and field magnet are without any projections; but as in such dispositions the field cannot be very concentrated, probably the best results will be obtained by leaving pole projections on one of the elements only. Generally, it may be stated that the projections diminish the torque and produce a tendency to synchronism.

A characteristic feature of motors of this kind is their capacity of being very rapidly reversed. This follows from the peculiar action of the motor. Suppose the armature to be rotating and the direction of rotation of the poles to be reversed. The apparatus then represents a dynamo machine, the power to drive this machine being the momentum stored up in the armature and its speed being the sum of the speeds of the armature and the poles.

If we now consider that the power to drive such a dynamo would be very nearly proportional to the third power of the speed, for this reason alone the armature should be quickly reversed. But simultaneously with the reversal another element is brought into action, namely, as the movement of the poles with respect to the armature is reversed, the motor acts like a transformer in which the resistance of the secondary circuit would be abnormally diminished by producing in this circuit an additional electromotive force. Owing to these causes the reversal is instantaneous.

If it is desirable to secure a constant speed, and at the same time a certain effort at the start, this result may be easily attained in a variety of ways. For instance, two armatures, one for torque and the other for synchronism, may be fastened on the same shaft, and any desired preponderance may be given to either one, or an armature may be wound for rotary effort, but a more or less pronounced tendency to synchronism may be given to it by properly constructing the iron core; and in many other ways.

As a means of obtaining the required phase of the currents in both the circuits, the disposition of the two coils at right angles is the simplest, securing the most uniform action; but the phase may be obtained in many other ways, varying with the machine employed. Any of the dynamos at present in use may be easily adapted for this purpose by making connections to proper points of the generating coils. In closed circuit armatures, such as used in the continuous current systems, it is best to make four derivations from equi-distant points or bars of the commutator, and to connect the same to four insulated sliding rings on the shaft. In this case each of the motor circuits is connected to two diametrically opposite bars of the commutator. In such a disposition the motor may also be operated at half the potential and on the three-wire plan, by connecting the motor circuits in the proper order to three of the contact rings.

In multipolar dynamo machines, such as used in the converter systems, the phase is conveniently obtained by winding upon the armature two series of coils in such a manner that while the coils of one set or series are at their maximum production of current, the coils of the other will be at their neutral position, or nearly so, whereby both sets of coils may be subjected simultaneously or successively to the inducing action of the field magnets.

figures 15, 16, 17

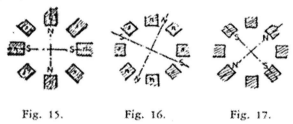

Fig. 15.   Fig. 16.   Fig. 17.

Generally the circuits in the motor will be similarly disposed, and various arrangements may be made to fulfill the requirements; but the simplest and most practicable is to arrange primary circuits on stationary parts of the motor, thereby obviating, at least in certain forms, the employment of sliding contacts. In such a case the magnet coils are connected alternately in both the circuits; that is 1, 3, 5 . . . in one, and 2, 4, 6 . . . in the other, and the coils of each set of series may be connected all in the same manner, or alternately in opposition; in the latter case a motor with half the number of poles will result, and its action will be correspondingly modified. The figures 15, 16 and 17, show three different phases, the magnet coils in each circuit being connected alternately in opposition. In this case there will be always four poles, as in figures 15 and 17, four pole projections will be neutral, and in figure 16 two adjacent pole projections will have the same polarity. If the coils are connected in the same manner there will be eight alternating poles, as indicated by the letters n' s' in fig.15.

The employment of multipolar motors secures in this system an advantage much desired and unattainable in the continuous current system, and that is, that a motor may be made to run exactly at a predetermined speed irrespective of imperfections in construction, of the load, and, within certain limits, of electromotive force and current strength.

In a general distribution system of this kind the following plan should be adopted. At the central station of supply a generator should be provided having a considerable number of poles. The motors operated from this generator should be of the synchronous type, but possessing sufficient rotary effort to insure their starting. With the observance of proper rules of construction it may be admitted that the speed of each motor will be in some inverse proportion to its size, and the number of poles should be chosen accordingly. Still exceptional demands may modify this rule. In view of this, it will be advantageous to provide each motor with a greater number of pole projections or coils, the number being preferably a multiple of two and three. By this means, by simply changing the connections of the coils, the motor may be adapted to any probable demands.

If the number of the poles in the motor is even, the action will he harmonious and the proper result will be obtained; if this is not the case the best plan to be followed is to make a motor with a double number of poles and connect the same in the manner before indicated, so that half the number of poles result. Suppose, for instance, that the generator has twelve poles, and it would be desired to obtain a speed equal to 12/7 of the speed of the generator. This would require a motor with seven pole projections or magnets, and such a motor could not be properly connected in the circuits unless fourteen armature coils would be provided, which would necessitate the employment of sliding contacts. To avoid this the motor should be provided with fourteen magnets and seven connected in each circuit, the magnets in each circuit alternating among themselves. The armature should have fourteen closed coils. The action of the motor will not be quite as perfect as in the case of an even number of poles, but the drawback will not be of a serious nature.

However, the disadvantages resulting from this unsymmetrical form will be reduced in the same proportion as the number of the poles is augmented.

If the generator has, say, n, and the motor n1 poles, the speed of the motor will be equal to that of the generator multiplied by n/n1.

figures 18, 19, 20, 21

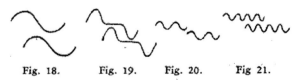

Fig. 18.   Fig. 19.   Fig. 20.   Fig 21.

The speed of the motor will generally be dependent on the number of the poles, but there may be exceptions to this rule. The speed may be modified by the phase of the currents in the circuits or by the character of the current impulses or by intervals between each or between groups of impulses. Some of the possible cases are indicated in the diagrams, figures 18, 19, 20 and 21, which are self-explanatory. Figure 18 represents the condition generally existing, and which secures the best result. In such a case, if the typical form of motor illustrated in figure 9 is employed, one complete wave in each circuit will produce one revolution of the motor. In figure 19 the same result will he effected by one wave in each circuit, the impulses being successive; in figure 20 by four, and in figure 21 by eight waves.

By such means any desired speed may be attained; that is, at least within the limits of practical demands. This system possesses this advantage besides others, resulting from simplicity. At full loads the motors show efficiency fully equal to that of the continuous current motors. The transformers present an additional advantage in their capability of operating motors. They are capable of similar modifications in construction, and will facilitate the introduction of motors and their adaptation to practical demands. Their efficiency should be higher than that of the present transformers, and I base my assertion on the following:

In a transformer as constructed at present we produce the currents in the secondary circuit by varying the strength of the primary or exciting currents. If we admit proportionality with respect to the iron core the inductive effect exerted upon the secondary coil will be proportional to the numerical sum of the variations in the strength of the exciting current per unit of time; whence it follows that for a given variation any prolongation of the primary current will result in a proportional loss. In order to obtain rapid variations in the strength of the current, essential to efficient induction, a great number of undulations are employed. From this practice various disadvantages result. These are, increased cost and diminished efficiency of the generator, more waste of energy in heating the cores, and also diminished output of the transformer, since the core is not properly utilized, the reversals being too rapid. The inductive effect is also very small in certain phases, as will be apparent from a graphic representation, and there may be periods of inaction, if there are intervals between the succeeding current impulses or waves. In producing a shifting of the poles in the transformer, and thereby inducing currents, the induction is of the ideal character, being always maintained at its maximum action. It is also reasonable to assume that by a shifting of the poles less energy will be wasted than by reversals.

The Tesla Alternate Current Motor
The Electrical Engineer - London — June 22, 1888

The interest taken in M. Tesla's contributions to electrical apparatus and to electrical literature is so great, and the subject is so important, that we do not hesitate to give further space to the subject. On May 26 a communication on the subject from Dr. Louis Duncan, of Johns Hopkins University, appeared in our American contemporary, the Electrical Review, to the effect:

"We may, for our present purposes, divide motors into two classes; Continuous, in which the armature coils are unsymmetrical with respect to the poles, and which, therefore, give a practically constant torque, and alternating motors, in which the armature coils are symmetrical with respect to the poles, and which, therefore, give a torque varying both in magnitude and sign during a period of the counter E.M.F. The Tesla motor belongs to this latter class.

"In every motor the torque is equal to the rate of change of lines of induction through the armature circuit for a small angular displacement, multiplied by the armature current, or dm/dt.

In the Tesla motor the first of these terms is greatest when the coil is opposite a pole and the field currents have their greatest amplitude. It is zero at a point about 45 deg. from this, supposing we neglect armature reactions. It depends on several things. The E.M.F. which determines it is due to changes in the number of lines of force passing through the armature circuit caused by (1) changes in the field currents; (2) the motion of the armature. The current depends on these E.M.F.'s, and on the reduced self-induction and resistance of its circuit. The motor can only do work when the first cause of E.M.F. is the greater, for a current in the direction of the ordinary counter E.M.F. would stop the motion. In some parts of a revolution the two E.M.F.'s work together, retarding the motion; in others, the induced E.M.F. produces a current causing the motor to revolve. It is impossible for me, with only a meagre description of the principles of the machine, to give an idea of the relative magnitude of these effects. Some of the results, however, are the following: Having given a definite number of reversals of the dynamo, there are a number of speeds, multiples of these reversals, at which the motor will govern itself when it is doing a certain amount of work. At one of these speeds, depending on the construction of the motor, the output will be a maximum. Now I see the statement that 'there is no difficulty whatever attendant upon starting the motor under load.' I cannot reconcile this with the above facts. That the torque for a smaller number of revolutions than ordinarily used, might be greater, one can readily see, since the counter E.M.F. is less in proportion to the induced E.M.F., but it must be remembered that for certain speeds even the induced current would tend to stop the motion; how the motor is to pass these critical speeds I do not see. Again, if the maximum load is suddenly thrown on while the motor is running at its proper speed, then, if the inertia be great, the motor will fall behind its point of maximum work, and either stop or take up some slower speed.

"What the possible efficiency and output of the motor may be, only experiment will tell. I have shown (*Inst. Elec. Engineers*, Feb., 1888.) that the output of an ordinary alternating current motor is equal to that of a continuous current motor, supplied with a corresponding E.M.F. The efficiency might be great, but is has the disadvantage that about the same current flows for no work and maximum work, so for light loads the efficiency can hardly be very high.

"With our present knowledge of alternating currents it is useless to attempt to calculate from the simple though misleading assumptions ordinarily made, the output, conditions of maximum work, &c., of this machine. Experiment alone can determine its value, and one properly conducted and interpreted set of experiments should enable us to judge both the merit of the invention and its best possible form. I cannot see, however, how, in the form described in the last issue of this journal the motor can work under conditions of a suddenly varying load as satisfactorily as continuous current motors."

To the above Mr. Tesla replied on June 2 as follows:
"I find in your issue of last week a note of Mr. Duncan referring to my system of alternate current motors.

"As I see that Dr. Duncan has not as yet been made acquainted with the real character of my invention, I cannot consider his article in the light of a serious criticism, and would think it unnecessary to respond; but desiring to express my consideration for him and the importance which I attach to his opinion, I will point out here briefly the characteristic features of my invention, inasmuch as they have a direct bearing on the article above referred to.

"The principle of action of my motor will be well understood from the following: By passing alternate currents in proper manner through independent energizing circuits in the motor, a progressive shifting or rotation of the poles of the same is effected. This shifting is more or less continuous according to the construction of the motor and the character and relative phase of the currents which should exist in order to secure the most perfect action.

"If a laminated ring be wound with four coils, and the same be connected in proper order to two independent circuits of an alternate current generator adapted for this purpose, the passage of the currents through the coils produces theoretically a rotation of the poles of the ring, and in actual practice, in a series of experiments, I have demonstrated the complete analogy between such a ring and a revolving magnet. From the application of this principle to the operation of motors, two forms of motor of a character widely differing have resulted — one designed for constant and the other for variable load. The misunderstanding of Dr. Duncan is due to the fact that the prominent features of each of these two forms have not been specifically stated. The armature of the motor is provided with two coils at right angles. As it may be believed that a symmetrical arrangement of the coils with respect to the poles is required, I will assume that the armature is provided with a great number of diametrically wound coils or conductors closed upon themselves, and forming as many independent circuits. Let it now be supposed that the ring is permanently magnetized so as to show two poles (N and S) at two points diametrically opposite, and that it is rotated by mechanical power. The armature being stationary, the rotation of the ring magnet will set up currents in the closed armature coils. These currents will be most intense at or near the points of the greatest density of the force, and they will produce poles upon the armature core at right angles to those of the ring. Of course there will be other elements entering into action which will tend to modify this, but for the present they may be left unconsidered. As far as the location of the poles upon the armature core is concerned, the currents generated in the armature coils will always act in the same manner, and will maintain continuously the poles of the core in the same position, with respect to those of the ring in any position of the latter, and independently of the speed. From the attraction between the core and the ring, a continuous rotary effort, constant in all positions, will result, the same as in a continuous current motor with a great number of armature coils. If the armature be allowed to turn, it will revolve in the direction of rotation of the ring magnet, the induced current diminishing as the speed increases, until upon the armature reaching very nearly the speed of the magnet, just enough current will flow through the coils to keep up the rotation. If, instead of rotating the ring by mechanical power, the poles of the same are shifted by the action of the alternate currents in the two circuits, the same results are obtained.

"Now compare this system with a continuous current system. In the latter we have alternate currents in the generator and motor coils, and intervening devices for commutating the currents, which on the motor besides effect automatically a progressive shifting or rotation of the poles of the armature; here we have the same elements and identically the same operation, but without the commutating devices. In view of the fact that these devices are entirely unessential to the operation, such alternate current system will — at least in many respects — show a complete similarity with a continuous current system, and the motor will act precisely like a continuous current motor. If the load is augmented, the speed is diminished and the rotary effort correspondingly increased, as more current is made to pass through the energising circuits; load being taken off, the speed increases, and the current, and consequently the effort, is lessened. The effort, of course, is greatest when the armature is in the state of rest.

"But, since the analogy is complete, how about the maximum efficiency and current passing through the circuits when the motor is running without any load? one will naturally inquire. It must be remembered that we have to deal with alternate currents. In this form the motor simply represents a transformer, in which currents are induced by a dynamic action instead of by reversals, and, as it might be expected, the efficiency will be maximum at full load. As regards the current, there will be — at least, under proper conditions — as wide a variation in its strength as in a transformer, and, by observing proper rules, it may be reduced to any desired quantity. Moreover, the current passing through the motor when running free, is no measure for the energy absorbed, since the instruments indicate only the numerical sum of the direct and induced electromotive forces and currents instead of showing their difference.

"Regarding the other class of these motors, designed for constant speed, the objections of Dr. Duncan are, in a measure applicable to certain constructions, but it should be considered that such motors are not expected to run without any, or with a very light load; and, if so, they do not, when properly constructed, present in this respect any more disadvantage than transformers under similar conditions. Besides, both features, rotary effort and tendency to constant speed, may be combined in a motor, and any desired preponderance may be given to either one, and in this manner a motor may be obtained possessing any desired character and capable of satisfying any possible demand in practice.

"In conclusion, I will remark, with all respect to Dr. Duncan, that the advantages claimed for my system are not mere assumptions, but results actually obtained, and that for this purpose experiments have been conducted through a long period, and with an assiduity such as only a deep interest in the invention could inspire; nevertheless, although my motor is the fruit of long labour and careful investigation, I do not wish to claim any other merit beyond that of having invented it, and I leave it to men more competent than myself to determine the true laws of the principle and the best mode of its application. What the result of these investigations will be the future will tell; but whatever they may be, and to whatever this principle may lead, I shall be sufficiently recompensed if later it will be admitted that I have contributed a share, however small, to the advancement of science."

Mr. Nikola Tesla on Alternating Current Motors
Electrical World — N. Y. — May 25, 1889

To the Editor of The Electrical World:

Sir: About a year ago I had the pleasure of bringing before the American Institute of Electrical Engineers the results of some of my work on alternate current motors. They were received with the interest which novel ideas never fail to excite in scientific circles, and elicited considerable comment. With truly American generosity, for which, on my part, I am ever thankful, a great deal of praise through the columns of your esteemed paper and other journals has been bestowed upon the originator of the idea, in itself insignificant. At that time it was impossible for me to bring before the Institute other results in the same line of thought. Moreover, I did not think it probable — considering the novelty of the idea — that anybody else would be likely to pursue work in the same direction. By one of the most curious coincidences, however, Professor Ferraris not only came independently to the same theoretical results, but in a manner identical almost to the smallest detail. Far from being disappointed at being prevented from calling the discovery of the principle exclusively my own, I have been excessively pleased to see my views, which I had formed and carried out long before, confirmed by this eminent man, to whom I consider myself happy to be related in spirit, and toward whom, ever since the knowledge of the facts has reached me, I have entertained feelings of the most sincere sympathy and esteem. In his able essay Prof. Ferraris omitted to mention various other ways of accomplishing similar results, some of which have later been indicated by 0. B. Shallenberger, who some time before the publication of the results obtained by Prof. Ferraris and myself had utilized the principle in the construction of his now well known alternate current meter, and at a still later period by Prof. Elihu Thomson and Mr. M. J. Wightman.

Since the original publications, for obvious reasons, little has been made known in regard to the further progress of the invention; nevertheless the work of perfecting has been carried on indefatigably with all the intelligent help and means which a corporation almost unlimited in its resources could command, and marked progress has been made in every direction. It is therefore not surprising that many unacquainted with this fact, in expressing their views as to the results obtained, have grossly erred.

In your issue of May 4, I find a communication from the electricians of Ganz & Co., of Budapest, relating to certain results observed in recent experiments with a novel form of alternate current motor. I would have nothing to say in regard to this communication unless it were to sincerely congratulate these gentlemen on any good results which they may have obtained, but for the article, seemingly inspired by them, which appeared in the London Electrical Review of April 26, wherein certain erroneous views are indorsed and some radically false assertions made, which, though they may be quite unintentional, are such as to create prejudice and affect material interests.

As to the results presented, they not only do not show anything extraordinary, but are, in fact, considerably below some figures obtained with my motors a long time ago. The main stress being laid upon the proposition between the apparent and real energy supplied, or perhaps more directly, upon the ratio of the energy apparently supplied to, and the real energy developed by, the motor, I will here submit, with your permission, to your readers, the results respectively arrived at by these gentlemen and myself.

Energy apparently supplied in watts.		Work performed in watts.		Ratio of energy apparently supplied to the real energy developed.	
Ganz & Co.	Westinghouse Co.	Ganz & Co.	Westinghouse Co.	Ganz & Co.	Westinghouse Co.
18,000	21,840	11,000	17,595	0.611	0.805
24,200	30,295	14,600	25,365	0.603	0.836
29,800	43,624	22,700	36,915	0.761	0.816
......	56,800	......	48,675	......	0.856
......	67,500	......	59,440	......	0.88
......	79,100	......	67,365	......	0.851

If we compare these figures we will find that the most favorable ratio in Ganz & Co's motor is 0.761, whereas in the Westinghouse, for about the same load, it is 0.836, while in other instances, as may be seen, it is still more favorable. Notwithstanding this, the conditions of the test were not such as to warrant the best possible results.

The factors upon which the apparent energy is mainly dependent could have been better determined by a proper construction of the motor and observance of certain conditions. In fact, with such a motor a current regulation may be obtained which, for all practical purposes, is as good as that of the direct current motors, and the only disadvantage, if it be one, is that when the motor is running without load the apparent energy cannot be reduced quite as low as might be desirable. For instance, in the case of this motor the smallest amount of apparent energy was about 3,000 watts, which is certainly not very much for a machine capable of developing 90 h. p. of work; besides, the amount could have been reduced very likely to 2,000 watts or less.

On the other hand, these motors possess the beautiful feature of maintaining an absolutely constant speed no matter how the load may vary. This feature may be illustrated best by the following experiment performed with this motor. The motor was run empty, and a load of about 200 h. p.., far exceeding the normal load, was thrown on suddenly. Both armatures of the motor and generator were seen to stop for an instant, the belts slipping over the pulleys, whereupon both came up to the normal speed with the full load, not having been thrown out of synchronism. The experiment could be repeated any number of times. In some cases, the driving power being sufficient, I have been enabled to throw on a load exceeding 8 to 9 times that which the motor was designed to carry, without affecting the speed in the least.

This will be easily understood from the manner in which the current regulation is effected. Assuming the motor to be running without any load, the poles of the armature and field have a certain relative position which is that of the highest self-induction or counter electromotive force. If load be thrown on, the poles are made to recede; the self-induction or counter electromotive force is thereby diminished and more current passed trough the stationary or movable armature-coils. This regulation is very different from that of a direct current motor. In the latter the current is varied by the motor losing a certain number of revolutions in proportion to the load, and the regulation would be impossible if the speed would be maintained constant; here the whole regulation is practically effected during a fraction of one revolution only. From this it is also apparent that it is a practical impossibility to throw such a motor out of synchronism, as the whole work must be done in an instant, it being evident that if the load is not sufficient to make a motor lose a fraction of the first revolution it will not be able to do so in the succeeding revolutions. As to the efficiency of these motors, it is perfectly practicable to obtain 94 to 95 per cent.

The results above given were obtained on a three-wire system. The same motor has been started and operated on two wires in a variety of ways, and although it was not capable of performing quite as much work as on three wires, up to about 60 h. p. it gave results practically the same as those abovementioned. In fairness to the electricians of Ganz & Co., I must state here that the speed of this motor was higher than that used in their experiments, it being about 1,500. I cannot make due allowance for this difference, as the diameter of the armature and other particulars of the Ganz & Co. motor were not given.

The motor tested had a weight of about 5,000 lbs. From this it will be seen that the performance even on two wires was quite equal to that of the best direct current motors. The motor being of a synchronous type, it might be implied that it was not capable of starting. On the contrary, however, it had a considerable torque on the start and was capable of starting under fair load.

In the article above referred to the assertion is made that the weight of such alternate current motor, for a given capacity, is "several times" larger than that of a direct current motor. In answer to this I will state here that we have motors which with a weight of about 850 pounds develop 10 h. p. with an efficiency of very nearly 90 per cent, and the spectacle of a direct current motor weighing, say 200 – 300 pounds and performing the same work, would be very gratifying for me to behold. The motor which I have just mentioned had no commutator or brushes of any kind nor did it require any direct current.

Finally, in order to refute various assertions made at random, principally in the foreign papers, I will take the liberty of calling to the attention of the critics the fact that since the discovery of the principle several types of motors have been perfected and of entirely different characteristics, each suited for a special kind of work, so that while one may be preferable on account of its ideal simplicity, another might be more efficient. It is evidently impossible to unite all imaginable advantages in one form, and it is equally unfair and unreasonable to judge all different forms according to a common standard. Which form of the existing motors is best, time will show; but even in the present state of the art we are enabled to satisfy any possible demand in practice.

Nikola Tesla
Pittsburgh, Pa.

The Losses Due to Hysteresis in Transformers
The Electrical Engineer - N. Y. — April 9, 1890

In your issue of April 2, in referring to certain remarks made by me at the recent meeting of the American Institute of Electrical Engineers on the subject of hysteresis you make the statement: "It is this constancy of relation that, as Mr. Tesla pointed out may ultimately establish the correctness of the hypothesis advanced, that in reality there is no loss due to hysteresis, and that the changes of magnetization represent a charging and discharging of molecular energy without entailing an actual expenditure of energy."

I do not recollect having made such a statement, and as I was evidently misunderstood, you will greatly oblige me in inserting the following few lines, which express the idea I meant to advance:

Up to the present no satisfactory explanation of the causes of hysteresis has been given. In the most exhaustive and competent treatise on the theory of transformers, by Fleming, static hysteresis is explained by supposing that "the magnetic molecules or molecular magnets, the arrangement of which constitutes magnetization, move stiffly, and the dissipation of energy is the work done in making the necessary magnetic displacement against a sort of magnetic friction." Commonly it is stated that this is a distinct element in the loss of energy in an iron core undergoing magnetic changes entirely independent of any currents generated therein.

Now it is difficult to reconcile these views with our present notions on the constitution of matter generally. The molecules or iron cannot be connected together by anything but elastic forces, since they are separated by an intervening elastic medium through which the forces act; and this being the case is it not reasonable to assume that if a given amount of energy is taken up to bring the molecules out of their original position an equivalent amount of energy should be restored by the molecules reassuming their original positions, as we know is the case in all molecular displacements? We cannot imagine that an appreciable amount of energy should be wasted by the elastically connected molecules swinging back and forth from their original positions, which they must constantly tend to assume, at least within the limit of elasticity, which in all probability is rarely surpassed. The losses cannot be attributed to mere displacement, as this would necessitate the supposition that the molecules are connected rigidly, which is quite unthinkable.

A current cannot act upon the particles unless it acts upon currents in the same, either previously existing or set up by it, and since the particles are held together by elastic forces the losses must be ascribed wholly to the current generated. The remarkable discovery of Ewing that the magnetization is greater on the descent than on the ascent for the same values of magnetizing force strongly points to the fact that hysteresis is intimately connected with the generation of currents either in the molecules individually or in groups of them through the space intervening. The fact observed accords perfectly with our experience on current induction, for we know that on the descent any current set up must be of the same direction with the inducing current, and, therefore, must join with the same in producing a common effect; whereas, on the ascent the contrary is the case.

Dr. Duncan stated that the ratio of increase of primary and secondary current is constant. This statement is, perhaps, not sufficiently expressive, for not only is the ratio constant but, obviously, the differential effect of primary and secondary is constant. Now any current generated - molecular or Foucault currents in the mass - must be in amount proportionate to the difference of the inductive effect of the primary and secondary, since both currents add algebraically - the ratio of windings duly considered, - and as this difference is constant the loss, if wholly accounted for in this manner, must be constant. Obviously I mean here the transformers under consideration, that is, those with a closed magnetic circuit, and I venture to say that the above will be more pronounced when the primary and secondary are wound one on top of the other than when they are wound side by side; and generally it will be the more pronounced the closer their inductive relation.

Dr. Duncan's figures also show that the loss is proportionate to the square of the electromotive force. Again this ought to be so, for an increased electromotive force causes a proportionately increased current which, in accordance with the above statements, must entail a loss in the proportion of the square.

Certainly, to account for all the phenomena of hysteresis, effects of mechanical vibration, the behavior of steel and nickel alloy, etc., a number of suppositions must be made; but can it not be assumed that, for instance, in the case of steel and nickel alloys the dissipation of energy is modified by the modified resistance; and to explain the apparent inconsistency of this view we only need to remember that the resistance of a body as a whole is not a measure of the degree of conductivity of the particles of which it is composed.

N. Tesla
New York City

Swinburne's "Hedgehog" Transformer
The Electrical Engineer - N. Y. — Sept. 24, 1890

Some time ago Mr. Swinburne advanced certain views on transformers which have elicited some comment. In *The Electrical Engineer* of Sept. 10. there are brought out further arguments on behalf of his open circuit, or, as he calls it, "hedgehog" transformer, claiming for this type a higher average efficiency than is attainable with the closed circuit forms. In regard to this, I say with Goethe, *Die Botschaft hör' ich wohl, allein mir fehlt der Glaube* — I hear the message, but I lack belief."

Many of Mr. Swinburne's arguments are in my opinion erroneous. He says: "In calculating the efficiencies of transformers, the loss in the iron has generally been left completely out of account, and the loss in copper alone considered; hence, the efficiencies of 97 and 98 per cent. claimed for closed iron circuit forms." This is a statement little complimentary to those who have made such estimates, and perhaps Mr. Swinburne would be very much embarrassed to cite names on behalf of his argument. He assumes the loss in the iron in the closed circuit forms to be 10 per cent. of the full load, and further "that in most stations the average use of lamps is less than two hours a day, including all lamps installed," and arrives at some interesting figures in regard to efficiency. Mr. Swinburne seems not to be aware of the improvements made in the iron. The loss with the best quality of iron will, I believe, not reach 8 per cent. of the full load by an intelligent use of the transformer, and there is no doubt that further improvements will be made in that direction.

As regards the second part of his assumption, I think that it is exaggerated. It must be remembered that in most central stations or large plants due care is taken that the load is favorably distributed and in many cases the wiring is such that entire circuits may be shut off at certain hours so that there is during these hours no loss whatever in the transformers.'

In his "hedgehog" form of transformer Mr. Swinburne reduces the iron considerably and comes to the conclusion that even in small transformers the iron loss is under one per cent. of the full load, while in the closed circuit forms, it is, according to him, 10 per cent. It would strengthen this argument if the iron would be dispensed with altogether. Mr. Swinburne does not appreciate fully the disadvantages which the open circuit form, operated at the usual period, entails. In order that the loss in tire iron should be reduced to one-tenth, it is necessary to reduce the weight of the iron core to one-tenth and subject every unit length of the same to the same magnoto-motive force. If a higher magneto-motive force is used the loss in the core will — within certain limits, at least — be proportionate to the square of tire magneto-motive force. The remark of Mr. Swinburne, "If the iron circuit is opened, the aides of the embracing core can be removed, so the loss by hysteresis is divided by three," is therefore not true; the loss will be divided by $-3^* \frac{F^2}{F_1^2}$ where $F_1 > F$. If the iron of the open circuit form is made up in a closed ring the advantage will be at once apparent, for, since the magnetic resistance will be much reduced, the magneto-motive force required will be correspondingly smaller. It is probable that, say, four Swinburne transformers may be joined in such a way as to form a closed magnetic circuit. In this case the amount of iron and copper would remain the same, but an advantage will be gained as as the total magnetic resistance will be diminished. The four transformers will now demand less excitation and since — under otherwise equal conditions — the gain depends on the square of the existing current, it in by no means insignificant. From the above comparison it is evident that the core of such open circuit transformer should be very short, by far shorter than it appears from the cut in *The Electrical Engineer.*

Mr. Swinburne is in error as to the motives which have caused the tendency to shorten the magnetic circuit in closed circuit transformers. It was principally on account of practical considerations and not to reduce the magnetic resistance, which has little to do with efficiency. If a ring be made of, say, 10 centimeters mean length and 10 square centimeters cross section, .and if it be wound all over with the primary and secondary wires, it will be found that it will give the best result with a certain number of alternations. If, now, a ring is made of the same quality of iron but having, say, 20 centimeters mean length and 10 square centimeters section it will give again the best result with the same number of alternations, and the efficiency will be the same as before, provided that the ring is wound all over with the primary and secondary wires. The space inside of the ring will, in the second case, be increased in proportion to the square of the diameter, and there will be no difficulty in winding on it all the wire required. So the length may be indefinitely increased and a transformer of any capacity made, as long as the ring is completely covered by the primary and secondary wires.

If the wires be wound side by side the ring of smaller diameter will give a better result, and the same will be the case if a certain fraction of the ring is not covered by the wires. It then becomes important to shorten the magnetic circuit. But, since in practice it is necessary to enclose the transformer in a casting, if such a ring be trade, it would have to be protected with a layer of laminated iron, which would increase the cost and loss. It may be inclosed in jars of some, insulating material, as Mr. Swinburne does, belt this is less practicable.

Owing to this, the constructors of the most practical forms, such, for instance, as the Westinghouse transformer, to which the Swinburne reasoning applies, have been prompted to enclose the wires as much as possible with the laminated iron, and then it became important to shorted the magnetic circuit, because in this form only a part of the magnetic circuit is surrounded by the wires, as well as for other practical considerations.

In practice it is desirable to get along with the least length of copper conductor on account of cheapness and regulation. Mr. Swinburne states that in his transformer the loss in iron is under one per cent. of the full load; all the balance of loss must, therefore, be in the copper. But since, according to him, the wires are of larger section, his transformer can hardly be an improvement in that direction. The gun-metal casting is also objectionable. There is no doubt some loss going on in the came, and beside, it increases the resistance of the wires by a factor $\sqrt{\frac{S}{S_1}}$ where S is the total cross-section of the core and S_1 the section of the iron wires. There is one important point which seems to have escaped Mr. Swinburne's attention. Whether the open circuit transformer is an improvement, or not, will depend principally on the period. The experience of most electrical engineers has resulted in the adoption of the closed-circuit transformer. I believe that I was the first to advocate in open circuit form, but to improve its efficiency I had to use a much higher period; at usual periods the, closed circuit form is preferable.

Mr. Swinburne makes some other obscure statements upon which I need not dwell, as they have no bearing on the main question.

Tesla's New Alternating Motors
The Electrical Engineer - N. Y. — Sept. 24, 1890

I hope you will allow me the privilege, to say in the columns of your esteemed journal a few words in regard to an article which appeared in *Industries* of August 22, to which my attention has been called. In this article an attempt is made to criticize some of my inventions, notably those which you have described in your issue of August 6, 1890.

The writer begins by stating: "The motor depends on a shifting of the poles under certain conditions, a principle which has been *already* employed by Mr. A. Wright in his alternating current meter." This is no surprise to me. It would rather have surprised me to learn that Mr. Wright has *not yet* employed the principle in his meter, considering what, before its appearance, was known of my work on motors, and more particularly of that of Schallenberger on meters. It has cost me years of thought to arrive at certain results, by many believed to be unattainable), for which there are now numerous claimants, and the number of these is rapidly increasing, like that of the colonels in the South after the war.

The writer then good-naturel explains the theory of action of the motive device in Wright's meter, which has greatly benefited me, for it is so long since I have arrived at this, and similar theories, that I had almost forgotten it. He then says: " Mr. Tesla has worked out some more or less complicated motors on this principle, but the curious point is that he has completely misunderstood the theory of the phenomena, and has got hold of (the old fallacy of screening." This may be *curious,* but how much *more curious* it is to find that the writer in *Industries* has *completely misunderstood* everything himself. I like nothing better than just criticism of my work, even if it be severe, but when the critic assumes a certain " l'état c'est moi" air of unquestioned Competency I want him to know what he is writing about. How little the writer in *Industries* seems to know about the matter is painfully apparent when ho connects the phenomenon in Wright's meter with the subject he has under consideration. His further remark, "He (Mr. Tesla) winds his secondary of iron instead of copper and thinks the effect is produced magnetically," is illustrative of the care with which he has perused the description of the devices contained in the issue of *The Electrical Engineer* above referred to.

I take a motor having, say eight poles, and wrap the exciting coils of four alternate cores with fine insulated iron wire. When the current is started in these coils it encounters the effect of the closed magnetic circuit and is retarded. The magnetic lines set up at the start close to the iron wire around the coils and no free poles appear at first at the ends of the four cores. As the current rises in the coils more lines are set up, which crowd more and more in the fine iron wire until finally the same becomes saturated, or nearly so, when the shielding action of the iron wire ceases and free poles appear at the ends of the four protected cores. The effect of the iron wire, as will be seen, is two-fold. First, it retards the energizing current; and second, it delays the appearance of the free poles. To produce still greater difference of phase in the magnetization of the protected and unprotected cores, I connect the iron wire surrounding the coils of the former in series with the coils of the latter, in which case, of course, the iron wire is preferably wound or connected differentially, after the fashion of the resistance, coils in a bridge, so as to have no appreciable self-induction. In other cases I obtain the desired retardation in the appearance of the free poles on one set of cores by *a* magnetic shunt, which produces a greater retardation of the current and takes up at the start a certain number of the lines set up, but becomes saturated when the current in the exciting coils reaches a predetermined strength.

In the transformer the same principle of shielding is utilized. A primary conductor is surrounded with a fine layer of laminated iron, consisting of fine iron wire or plates properly insulated and interrupted. As long as the current in the primary conductor is so small that the iron enclosure can carry all the lines of force set up by the current, there is very little action exerted upon a secondary conductor placed in vicinity to the first; but just as soon as the iron enclosure becomes saturated, or nearly so, it loses the virtue of protecting the secondary and the inducing action of the primary practically begins. What, may I ask, has all this to do with the "old fallacy of screening?"

With certain objects in view — the enumeration of which would lead me too far — an arrangement was shown in *The Electrical Engineer*, about which the writer in *Industries* says : " A ring of laminated iron is wound with a secondary. It is then encased in iron laminated in the *wrong direction* and the primary is wound outside of this. The layer of iron between the primary and secondary is supposed to screen the coil. Of course it cannot do so, such a thing Is unthinkable." This reminds me of the man who had committed some offense and engaged the services of an attorney. "They cannot commit you to prison for that," said the attorney. Finally the man *was* imprisoned. He sent for the attorney. " Sir," said the latter, " I tell you they *cannot* imprison you for that." "But, sir," retorted the prisoner, "they *have* imprisoned me." It *may not* screen, in the opinion of the writer in *Industries,* but just the same it *does*. According to the arrangement the *principal* effect of the screen may be either a retardation of the action of the primary current upon the secondary circuit or a deformation of the secondary current wave with similar results for the purposes intended. In the arrangement referred to by the writer in *Industries* be seems to be certain that the iron layer acts like a choking coil; there again he is mistaken; it does not act like a choking coil, for then its capacity for maintaining constant current would be very limited. But it acts more like a magnetic shunt in constant current transformers and dynamos, as, in my opinion, it ought to act.

There are a good many more things to be said about the remarks contained in *Industries*. In regard to the magnetic time lag the writer says: " If a bar of iron has a coil at one end, and if the core is perfectly laminated, on starting a current in the coil the induction *all along the iron* corresponds to the excitation at that instant, unless there is a microscopic time lag, *of which there is no evidence."* Yet a motor was described, the very operation of which is dependent on the time lag of magnetization of the different parts of a core. It is true the writer uses the term "perfectly laminated" (which, by the way, I would like him to explain), but if he intends to make such a "perfectly laminated" core I venture to say there is trouble in store for him. From his remarks I see that the writer completely overlooks the importance of the size of the core and of the number of the alternations pointed out; be fails to see the stress laid on the saturation of the screen, or shunt, in some of the cases described; he does not seem to recognize the fact that in the cases considered the formation of current is reduced as far as practicable in the screen, and that the same, therefore, so far as its quality of screening is concerned, has no role to perform as *a conductor.* I also see that he would want considerable information about the time lag in the magnetization of the different parts of a core, and an explanation why, in the transformer he refers to, the screen is laminated in the *wrong direction,* etc. — but the elucidation of all these points would require more time than I am able to devote to the subject. It is distressing to find all this in the columns of a leading technical journal.

In conclusion, the writer shows his true colors by making the following withering remarks: "It is questionable whether the Tesla motor will ever be a success. Such motors will go round, of course, and will give outputs, but their efficiency is doubtful; and if they need three-wire circuits and special generators there is no object in using them, as a direct current motor can be run instead with advantage."

No man of broad views will feel certain of the success of any invention, however good and original, in this period of feverish activity, when every day may bring new and unforseen developments. At the pace we are progressing the permanence of all our apparatus it its present forms becomes more and more problematical. It is impossible to foretell what type of motor will crystalize out of the united efforts of many able men; but it is my conviction that at no distant time a motor having commutator and brushes will be looked upon as an antiquated piece of mechanism. Just how much the last quoted remarks of the writer of *Industries* — considering the present state of the art — are justified, I will endeavor to show in a few lines.

First, take the transmission of power in isolated places. A case frequently occurring in practice and attracting more and more the attention of engineers is the transmission of large powers at considerable distances. In such a case the power is very likely to be cheap, and the cardinal requirements are then the reduction of the cost of the leads, cheapness of construction and maintenance of machinery and constant speed of the motors. Suppose a loss of only 25 per cent in the leads, at full load, be allowed. If a direct current motor be used, there will be, besides other difficulties, considerable variation in the speed of the motor — even if the current is supplied from a series dynamo — so much so that the motor may not be well adapted for many purposes, for instance, in cases where direct current transformation is contemplated with the object of running lights or other devices at constant potential. It is true that the condition may be bettered by employing proper regulating devices, but these will only further complicate the already complex system, and in all probability fail to secure such perfection as will be desired. In using an ordinary single-circuit alternate current motor the disadvantage is that the motor has no starting torque and that, for equal weight, its output and efficiency are more or less below that of a direct current motor. If, on the contrary, the armature of any alternator or direct current machine — large, low-speed, two-pole machines will give the best results — is wound with two circuits, a motor is at once obtained which possesses sufficient torque to start under considerable load: it runs in absolute synchronism with the generator

— an advantage much desired and hardly ever to be attained with regulating devices; it takes current in proportion to the load, and its plant efficiency within a few per cent is equal to that of a direct current motor of the same size. It will be able, however, to perform more work than a direct current motor of the same size, first, because there will be no change of speed, even if the load be doubled or tripled, within the limits of available generator power; and second, because it can be run at a higher electromotive force, the commutator and the complication and difficulties it involves in the construction and operation of the generators and motors being eliminated from the system. Such a system will, of course, require three leads, but since the plant efficiency is practically equal to that of the direct current system, it will require the same amount of copper which would be required in the latter system, and the disadvantage of the third lead will be comparatively small, if any, for three leads of smaller size may perhaps be more convenient to place than two larger leads. When more machines have to be used there may be no disadvantage whatever connected with the third wire ; however, since the simplicity of the generators and motors allows the use of higher electromotive forces, the cost of the leads may be reduced below the figure practicable with the direct current system.

Considering all the practical advantages offered by such an alternating system, I am of an opinion quite contrary to that of the author of the article in *Industries,* and think that it can quite successfully stand the competition of any direct current system, and this the more, the larger the machines built and the greater the distances.

Another case frequently occurring in practice is the transmission of small powers in numerous isolated places, such as mines, etc. In many of these cases simplicity and reliability of the apparatus are the principal objects. I believe that in many places of this kind my motor has so far proved a perfect success. In such cases a type of motor is used possessing great starting torque, requiring for its operation only alternating current and having no sliding contacts whatever on the armature, this advantage over other types of motors being highly valued in such places. The plant efficiency of this form of motor is, in the present state of perfection, inferior to that of the former form, but I am confident that improvements will be made in that direction. Be-sides, plant efficiency is in these cases of secondary importance, and in cases of transmission at considerable distances, it is no drawback, since the electromotive force may be raised as high as practicable on converters. I can not lay enough stress on this advantageous feature of my motors, and should think that it ought to be fully appreciated by engineers, for to high electromotive forces we are surely coming, and if they must be used, then the fittest apparatus will be employed. I believe that in the transmission of power with such commutatorless machines, 10,000 volts, and even more, may be used, and I would be glad to see Mr. Ferranti's enterprise succeed. His work is in the right direction, and, in my opinion, it will be of great value for the advancement of the art.

As regards the supply of power from large central stations in cities or centers of manufacture, the above arguments are applicable, and I see no reason why the three-wire motor system should not be successful. In putting up such a station, the third wire would be but a very slight drawback, and the system possesses enough advantages to over-balance this and any other disadvantage. But this question will be settled in the future, for as yet comparatively little has been done in that direction, even with the direct current system. The plant efficiency of such a three-wire system would be increased by using, in connection with the ordinary type of my motor, other types which act more like inert resistances. The plant efficiency of the whole system would, in all cases, be greater than that of each individual motor — if like motors are used — owing to the fact that they would possess different self-induction, according to the load.

The supply of power from lighting mains is, I believe, in the opinion of most engineers, limited to comparatively small powers, for obvious reasons. As the present systems are built on the two-wire plan, an efficient two-wire motor without commutator is required for this purpose, and also for traction purposes. A large number of these motors, embodying new principles, have been devised by me and are being constantly perfected. On lighting stations, however, my three-wire system may be advantageously carried out. A third wire may be run for motors and the old connections left undisturbed. The armatures of the generators may be rewound, whereby the output of the machines will be increased about 35 per cent, or even more in machines with cast iron field magnets. If the machines are worked at the same capacity, this means an increased efficiency. If power is available at the station, the gain in current may be used in motors. Those who object to the third wire, may remember that the old two-wire direct system is almost entirely superseded by the three-wire system, yet my three-wire system offers to the alternating system relatively greater advantages, than the three-wires direct possesses over the two-wire. Perhaps, if the writer in *Industries* would have taken all this in consideration, he would have been less hasty in his conclusions.

Nikola Tesla
New York
Sept. 17, 1890

Phenomena of Alternating Currents of Very High Frequency
Electrical World, Feb. 21, 1891

Electrical journals are getting to be more and more interesting. New facts are observed and new problems spring up daily which command the attention of engineers. In the last few numbers of the English journals, principally in the Electrician there have been several new matters brought up which have attracted more than usual attention. The address of Professor Crookes has revived the interest in his beautiful and skillfully performed experiments, the effect observed on the Ferranti mains has elicited the expressions of opinion of some of the leading English electricians, and Mr. Swinburne has brought out some interesting points in connection with condensers and dynamo excitation.

The writer's own experiences have induced him to venture a few remarks in regard to these and other matters, hoping that they will afford some useful information or suggestion to the reader.

Among his many experiments Professor Crookes shows some performed with tubes devoid of internal electrodes, and from his remarks it must be inferred that the results obtained with these tubes are rather unusual. If this be so, then the writer must regret that Professor Crookes, whose admirable work has been the delight of every investigator, should not have availed himself in his experiments of a properly constructed alternate current machine - namely, one capable of giving, say 10,000 to 20,000 alternations per second. His researches on this difficult but fascinating subject would then have been even more complete. It is true that when using such a machine in connection with an induction coil the distinctive character of the electrodes — which is desirable, if not essential, in many experiments — is lost, in most cases both the electrodes behaving alike; but on the other hand, the advantage is gained that the effects may be exalted at will. When using a rotating switch or commutator the rate of change obtainable in the primary current is limited. When the commutator is more rapidly revolved the primary current diminishes, and if the current be increased, the sparking, which cannot be completely overcome by the condenser, impairs considerably the virtue of the apparatus. No such limitations exist when using an alternate current machine as any desired rate of change may be produced in the primary current. It is thus; possible to obtain excessively high electromotive forces in the secondary circuit with a comparatively small primary current; moreover, the perfect regularity in the working of the apparatus may be relied upon.

The writer will incidentally mention that any one who attempts for the first time to construct such a machine will have a tale of woe to tell. He will first start out, as a matter of course, by making an armature with the required number of polar projections. He will then get the satisfaction of having produced an apparatus which is fit to accompany a thoroughly Wagnerian opera. It may besides possess the virtue of converting mechanical energy into heat in a nearly perfect manner. If there is a reversal in the polarity of the projections, he will get heat out of the machine; if there is no reversal, the heating will be less, but the output will be next to nothing. He will then abandon the iron in the armature, and he will get from the Scylla to the Charybdis. He will look for one difficulty and will find another, but, after a few trials, he may get nearly what he wanted.

Among the many experiments which may be performed with such a machine, of not the least interest are those performed with a high-tension induction coil. The character of the discharge is completely changed. The arc is established at much greater distances, and it is so easily affected by the slightest current of air that it often wriggles around in the most singular manner. It usually emits the rhythmical sound peculiar to the alternate current arcs, but the curious point is that the sound may be heard with a number of alternations far above ten thousand per second, which by many is considered to be, about the limit of audition. In many respects the coil behaves like a static machine. Points impair considerably the sparking interval, electricity escaping from them freely, and from a wire attached to one of the terminals streams of light issue, as though it were connected to a pole of a powerful Toepler machine. All these phenomena are, of course, mostly due to the enormous differences of potential obtained. As a consequence of the self-induction of the coil and the high frequency, the current is minute while there is a corresponding rise of pressure. A current impulse of some strength started in such a coil should persist to flow no less than four ten-thousandths of a second. As this time is greater than half the period, it occurs that an opposing electromotive force begins to act while the current is still flowing. As a consequence, the pressure rises as in a tube filled with liquid and vibrated rapidly around its axis. The current is so small that, in the opinion and involuntary experience of the writer, the discharge of even a very large coil cannot produce seriously injurious effects, whereas, if the same coil were operated with a current of lower frequency, though the electromotive force would be much smaller, the discharge would be most certainly injurious. This result, however, is due in part to the high frequency. The writer's experiences tend to show that the higher the frequency the greater the amount of electrical energy which may be passed through the body without serious discomfort; whence it seems certain that human tissues act as condensers.

One is not quite prepared for the behavior of the coil when connected to a Leyden jar. One, of course, anticipates that since the frequency is high the capacity of the jar should be small. He therefore takes a very small jar, about the size of a small wine glass, but he finds that even with this jar the coil is practically short-circuited. He then reduces the capacity until he comes to about the capacity of two spheres, say, ten centimetres in diameter and two to four centimetres apart. The discharge then assumes; the form of a serrated band exactly like a succession of sparks viewed in a rapidly revolving mirror; the serrations, of course, corresponding to the condenser discharges. In this case one may observe a queer phenomenon. The discharge starts at the nearest points, works gradually up, breaks somewhere near the top of the spheres, begins again at the bottom; and so on. This goes on so fast that several serrated bands are seen at once. One may be puzzled for a few minutes, but the explanation is simple enough. The discharge begins at the nearest points; the air is heated and carries the arc upward until it breaks, when it is re-established at the nearest points, etc. Since the current passes easily through a condenser of even small capacity, it will be found quite natural that connecting only one terminal to a body of the same size, no matter how well insulated, impairs considerably the striking distance of the arc.

Experiments with Geissler tubes are of special interest. An exhausted tube, devoid of electrodes of any kind, will light up at some distance from the coil. If a tube from a vacuum pump is near the coil the whole of the pump is brilliantly lighted. An incandescent lamp approached to the coil lights up and gets perceptibly hot. If a lamp have the terminals connected to one of the binding posts of the coil and the hand is approached to the bulb, a very curious and rather unpleasant discharge from the glass to the hand takes place, and the filament may become incandescent. The discharge resembles to some extent the stream issuing from the plates of a powerful Toepler machine, but is of incomparably greater quantity. The lamp in this case acts as a condenser, the rarefied gas being one coating, the operator's hand the other. By taking the globe of a lamp in the hand, and by bringing the metallic terminals near td or in contact with a conductor connected to the coil, the carbon is brought to bright incandescence and the glass is rapidly heated. With a 100-volt 10 c.p. lamp one may without great discomfort stand as much current as will bring the lamp to a considerable brilliancy; but it can be held in the hand only for a few minutes, as the glass is heated in an incredibly short time. When a tube is lighted by bringing it near to the coil it may be made to go out by interposing a metal plate on the hand between the coil and tube; but if the metal plate be fastened to a glass rod or otherwise insulated, the tube may remain lighted if the plate be interposed, or may even increase in luminosity. The effect depends on the position of the plate and tube relatively to the coil, and may be always easily foretold by assuming that conduction takes place from one terminal of the coil to the other. According to the position of the plate, it may either divert from or direct the current to the tube.

In another line of work the writer has in frequent experiments maintained incandescent lamps of 50 or 100 volts burning at any desired candle power with both the terminals of each lamp connected to a stout copper wire of no more than a few feet in length. These experiments seem interesting enough, but they are not more so than the queer experiment of Faraday, which has been revived and made much of by recent investigators, and in which a discharge is made to jump between two points of a bent copper wire. An experiment may be cited here which may seem equally interesting.

If a Geissler tube, the terminals of which are joined by a copper wire, be approached to the coil, certainly no one would be prepared to see the tube light up. Curiously enough, it does light up, and, what is more, the wire does not seem to make much difference. Now one is apt to think in the first moment that the impedance of the wire might have something to do with the phenomenon. But this is of course immediately rejected, as for this an enormous frequency would be required. This result, however, seems puzzling only at first; for upon reflection it is quite clear that the wire can make but little difference. It may be explained in more than one way, but it agrees perhaps best with observation to assume that conduction takes place from the terminals of the coil through the space. On this assumption, if the tube with the wire be held in any position, the wire can divert little more than the current which passes through the space occupied by the wire and the metallic terminals of the tube; through the adjacent space the current passes practically undisturbed. For this reason, if the tube be held in any position at right angles to the line joining the binding posts of the coil, the wire makes hardly any difference, but in a position more or less parallel with that line it impairs to a certain extent the brilliancy of the tube and its facility to light up. Numerous other phenomena may be explained on the same assumption. For instance, if the ends of the tube be provided with washers of sufficient size and held in the line joining the terminals of the coil, it will not light up, and then nearly the whole of the current, which would otherwise pass uniformly through the space between the washers, is diverted through the wire. But if the tube be inclined sufficiently to that line, it will light up in spite of the washers. Also, if a metal plate be fastened upon a glass rod and held at right angles to the line joining the binding posts, and nearer to one of them, a tube held more or less parallel with the line will light up instantly when one of the terminals touches the plate, and will go out when separated from the plate. The greater the surface of the plate, up to a certain limit, the easier the tube will light up. When a tube is placed at right angles to the straight line joining the binding posts, and then rotated, its luminosity steadily increases until it is parallel with that line. The writer must state, however, that he does not favor the idea of a leakage or current through the space any more than as a suitable explanation, for he is convinced that all these experiments could not be performed with a static machine yielding a constant difference of potential, and that condenser action is largely concerned in these phenomena.

It is well to take certain precautions when operating a Ruhmkorff coil with very rapidly alternating currents. The primary current should not be turned on too long, else the core may get so hot as to melt the guta-percha or paraffin, or otherwise injure the insulation, and this may occur in a surprisingly short time, considering the current's strength. The primary current being turned on, the fine wire terminals may be joined without great risk, the impedance being so great that it is difficult to force enough current through the fine wire so as to injure it, and in fact the coil may be on the whole much safer when the terminals of the fine wire are connected than when they are insulated; but special care should be taken when the terminals are connected to the coatings of a Leyden jar, for with anywhere near the critical capacity, which just counteracts the self-induction at the existing frequency, the coil might meet the fate of St. Polycarpus. If an expensive vacuum pump is lighted up by being near to the coil or touched with a wire connected to one of the terminals, the current should be left on no more than a few moments, else the glass will be cracked by the heating of the rarefied gas in one of the narrow passages - in the writer's own experience quod erat demonstrandum.

There are a good many other points of interest which may be observed in connection with such a machine. Experiments with the telephone, a conductor in a strong field or with a condenser or arc, seem to afford certain proof that sounds far above the usual accepted limit of hearing would be perceived. A telephone will emit notes of twelve to thirteen thousand vibrations per second; then the inability of the core to follow such rapid alternations begins to tell. If, however, the magnet and core be replaced by a condenser and the terminals connected to the high-tension secondary of a transformer, higher notes may still be heard. If the current be sent around a finely laminated core and a small piece of thin sheet iron be held gently against the core, a sound may be still heard with thirteen to fourteen thousand alternations per second, provided the current is sufficiently strong. A small coil, however, tightly packed between the poles of a powerful magnet, will emit a sound with the above number of alternations, and arcs may be audible with a still higher frequency. The limit of audition is variously estimated. In Sir William Thomson's writings it is stated somewhere that ten thousand per second, or nearly so, is the limit. Other, but less reliable, sources give it as high as twenty-four thousand per second. The above experiments have convinced the writer that notes of an incomparably higher number of vibrations per second would be perceived provided they could be produced with sufficient power. There is no reason why it should not be so. The condensations and rarefactions of the air would necessarily set the diaphragm in a corresponding vibration and some sensation would be produced, whatever — within certain limits — the velocity, of transmission to their nerve centres, though it is probable that for want of exercise the ear would not be able to distinguish any such high note. With the eye it is different; if the sense of vision is based upon some resonance effect, as many believe, no amount of increase in the intensity of the ethereal vibration could extend our range of vision on either side of the visible spectrum.

The limit of audition of an arc depends on its size. The greater the surface by a given heating effect in the arc, the higher the limit of audition. The highest notes are emitted by the high-tension discharges of an induction coil in which the arc is, so to speak, all surface. If R be the resistance of an arc, and C the current, and the linear dimensions be n times increased, then the resistance is R/n, and with the same current density the current would be $n^2 C$; hence the heating effect is n^3 times greater, while the surface is only n^2 times as, great. For this reason very large arcs would not emit any rhythmical sound even with a very low frequency. It must be observed, however, that the sound emitted depends to some extent also on the composition of the carbon. If the carbon contain highly refractory material, this, when heated, tends to maintain the` temperature' of the arc uniform and the sound is lessened; for this reason it would seem that an alternating arc requires such carbons:

With currents of such high frequencies it is possible to obtain noiseless arcs, but the regulation of the lamp is rendered extremely difficult on account of the excessively small attractions or repulsions between conductors conveying these currents:

An interesting feature of the arc produced by these rapidly alternating currents is its persistency. There are two causes for it, one of which is always present, the other sometimes only. One is due to the character of the current and the other to a property of the machine. The first cause is the more important one, and is due directly to the rapidity of the alternations. When an arc is formed by a periodically undulating current, there, is, a corresponding undulation in the temperature of the gaseous column, and, therefore, a corresponding undulation in the resistance of the arc. But the resistance of the arc varies enormously with the temperature of the gaseous column, being, practically infinite when the gas between the electrodes is cold. The persistence of the arc, therefore, depends on the inability of the column to cool. It is for this reason impossible to maintain an arc with the current alternating only a few times a second. On the other hand, with a practically continuous current, the arc is easily maintained, the column being constantly, kept at a high temperature and low resistance. The higher the frequency the smaller the time interval during which the arc may cool' and increase considerably in resistance. With a frequency of 10,000 per second or more in any arc of equal, size excessively small variations of temperature are superimposed upon a steady temperature, like ripples on the surface of a deep sea. The heating effect is practically continuous and the arc behaves like one produced, by a continuous current, with the exception, however, that it may not be quite as easily started, and that the electrodes are equally consumed; though the writer has observed 'some irregularities in this respect. The second cause alluded to, which possibly may not be present, is due to the tendency of a, machine of such high frequency to maintain a practically constant current. When the arc is lengthened, the electromotive force rises in proportion and the arc appears to be more persistent.

Such a machine is eminently adapted to maintain a constant current, but it is very unfit for a constant potential. As a matter of fact, in certain types of such machines a nearly constant current is an almost unavoidable result. As the number of poles or polar projections is greatly increased, the clearance becomes of great importance. One has really to do with' a great number of very small machines. Then there is the impedance in the armature, enormously augmented by the high frequency. Then, again, the magnetic leakage is facilitated. If there are' three or four hundred alternate poles, the leakage is so great that it is virtually the same as connecting, in a two-pole machine, the poles by a piece of iron. This disadvantage,, it is true, may be obviated more or less by using a field throughout of the same polarity, but then one encounters difficulties, of a different nature: All these things tend to maintain a constant' current in the armature circuit.

In this connection it is interesting to notice that even to-day engineers are astonished at the performance of a constant current machine, just as, some years ago, they used to consider it an extraordinary performance if a machine was capable of maintaining a constant, potential difference between the terminals. Yet one result is just as easily secured as the other. It must only be remembered that in an inductive apparatus of any kind, if constant potential is required, the inductive relation between the primary or exciting and secondary or armature circuit must be the closest possible; whereas, in an apparatus for constant current just the opposite is required. Furthermore, the opposition to the current's flow in the induced circuit must be as small as possible in the former and as great as possible in the latter case. But opposition to a current's flow may be caused in more than one way. It may be caused by ohmic resistance of self-induction. One may make the induced circuit of a dynamo machine or transformer of such high resistance that when operating devices of considerably smaller resistance within very wide limits a nearly constant current is maintained. But such high resistance involves a great loss in power, hence it is not practicable. Not so self-induction. Self-induction does not necessarily mean loss of power. The moral is, use self-induction instead of resistance. There is, however, a circumstance which favors the adoption of this plan, and this is, that a very high self-induction may be obtained cheaply by surrounding a comparatively small length of wire more or less completely with iron, and, furthermore, the effect may be exalted at will by causing a rapid undulation of the current. To sum up, the requirements for constant current are: Weak magnetic connection between the induced and inducing circuits, greatest possible self-induction with the least resistance, greatest practicable rate of change of the current. Constant potential, on the other hand, requires: Closest magnetic connection between the circuits, steady induced current, and, if possible, no reaction. If the latter conditions could be fully satisfied in a constant potential machine, its output would surpass many times that of a machine primarily designed to give constant current. Unfortunately, the type of machine in which these conditions may be satisfied is of little practical value, owing to the small electromotive force obtainable and the difficulties in taking off the current.

With their keen inventor's instinct, the now successful arc-light men have early recognized the desiderata of a constant current machine. Their arc light machines have weak fields, large armatures, with a great length of copper wire and few commutator segments to produce great variations in the current's strength and to bring self-induction into play. Such machines may maintain within considerable limits of variation in the resistance of the circuit a practically constant current. Their output is of course correspondingly diminished, and, perhaps with the object in view not to Cut down the output too much, a simple device compensating exceptional variations is employed. The undulation of the current is almost essential to the commercial success of an arc-light system. It introduces in the circuit a steadying element taking the place of a large ohmic resistance, without involving a great loss in power, and, what is more important, it allows the use of simple clutch lamps, which with a current of a certain number of impulses per second, best suitable for each particular lamp, will, if properly attended to, regulate even better than the finest clock-work lamps. This discovery has been made by the writer - several years too late.

It has been asserted by competent English electricians that in a constant-current machine or transformer the regulation is effected by varying the phase of the secondary current. That this view is erroneous may be easily proved by using, instead of lamps, devices each possessing self-induction and capacity or self-induction and resistance - that is, retarding and accelerating components - in such proportions as to not affect materially the phase of the secondary current. Any number of such devices may be inserted or cut out, still it will be found that the regulation occurs, a constant current being maintained, while the electromotive force is varied with the number of the devices. The change of phase of the secondary current is simply a result following from the changes in resistance, and, though secondary reaction is always of more or less importance, yet the real cause of the regulation lies in the existence of the conditions above enumerated. It should be stated, however, that in the case of a machine the above remarks are to be restricted to the cases in which the machine is independently excited. If the excitation be effected by commutating the armature current, then the fixed position of the brushes makes any shifting of the neutral line of the utmost importance, and it may not be thought immodest of the writer to mention that, as far as records go, he seems to have been the first who has successfully regulated machines by providing a bridge connection between a point of the external circuit and the commutator by means of a third brush. The armature and field being properly proportioned, and the brushes placed in their determined positions, a constant current or constant potential resulted from the shifting of the diameter of commutation by the varying loads.

In connection with machines of such high frequencies, the condenser affords an especially interesting study. It is easy to raise the electromotive force of such a machine to four or five times the value by simply connecting the condenser to the circuit, and the writer has continually used the condenser for the purposes of regulation, as suggested by Blakesley in his book on alternate currents, in which he has treated the most frequently occurring condenser problems with exquisite simplicity and clearness. The high frequency allows the use of small capacities and renders investigation easy. But; although in most of the experiments the result may be foretold, some phenomena observed seem at first curious. One experiment performed three or four months ago with such a machine and a condenser may serve as an illustration. A machine was used giving about 20,000 alternations per second. Two bare wires about twenty feet long and two millimeters in diameter, in close proximity to each other, were connected to the terminals of the machine at the one end, and to a condenser at the other. A small transformer without an iron core, of course, was used to bring the reading within range of a Cardew voltmeter by connecting the voltmeter to the secondary. On the terminals of the condenser the electromotive force was about 120 volts, and from there inch by inch it gradually fell until at the terminals of the machine it was about 65 volts. It was virtually as though the condenser were a generator, and the line and armature circuit simply a resistance connected to it. The writer looked for a case of resonance, but he was unable to augment the effect by varying the capacity very carefully and gradually or by changing the speed of the machine. A case of pure resonance he was unable to obtain. When a condenser was connected to the terminals of the machine — the self-induction of the armature being first determined in the maximum and minimum position and the mean value taken — the capacity which gave the highest electromotive force corresponded most nearly to that which just counteracted the self-induction with the existing frequency. If the capacity was increased or diminished, the electromotive force fell as expected.

With frequencies as high as the above mentioned, the condenser effects are of enormous importance. The condenser becomes a highly efficient apparatus capable of transferring considerable energy.

The writer has thought machines of high frequencies may find use at least in cases when transmission at great distances is not contemplated. The increase of the resistance may be reduced in the conductors and exalted in the devices when heating effects are wanted, transformers may be made of higher efficiency and greater outputs and valuable results may be secured by means of condensers. In using machines of high frequency the writer has been able to observe condenser effects which would have otherwise escaped his notice. He has been very much interested in the phenomenon observed on the Ferranti main which has been so much spoken of. Opinions have been expressed by competent electricians, but up to the present all still seems: to be conjecture. Undoubtedly in the views expressed the truth must be contained, but as the opinions differ some must be erroneous. Upon seeing the diagram of M. Ferranti in the Electrician of Dec. 19 the writer has formed his opinion of the effect. In the absence of all the necessary data he must content himself to express in words the process which, in his opinion, must undoubtedly occur. The condenser brings about two effects: (1) It changes the phases of the currents in the branches; (2) it changes the strength of the currents. As regards the change in phase, the effect of the condenser is to accelerate the current in the secondary at Deptford and to retard it in the primary at London. The former has the effect diminishing the self-induction in the Deptford primary, and this means lower electromotive force on the dynamo. The retardation of the primary at London, as far as merely the phase is concerned, has little or no effect since the phase of the current in the secondary in London is not arbitrarily kept.

Now, the second effect of the condenser is to increase the current in both the branches. It is immaterial whether there is equality between the currents or not; but it is necessary to point out, in order . to see the importance of the Deptford step-up transformer, that an increase of the current in both the branches produces opposite effects. At Deptford it means further lowering of the electromotive force at the primary, and at London it means increase of the electromotive force. at the secondary., Therefore, all the things co-act to bring about the phenomenon observed. Such actions, at least, have been formed to take place under similar conditions. When the dynamo is connected directly to the main, one can see that no such action can happen.

The writer has been particularly interested in, the suggestions and views expressed by Mr. Swinburne. Mr. Swinburne has frequently honored him by disagreeing with his views. Three years ago, when the writer, against the prevailing opinion of engineers, advanced' an open circuit transformer, Mr. Swinburne was the first to condemn it by stating in the Electrician: "The (Tesla) transformer must be inefficient; it has magnetic poles revolving, and has thus an open magnetic circuit." Two years later Mr. Swinburne becomes the champion of the open circuit transformer, and offers to convert him. But, *tempora mutantur, et nos mutamur in illis.*

The writer cannot believe in the armature reaction theory as expressed in Industries, though undoubtedly there is some truth in it. Mr. Swinburne's interpretation, however, is so broad that it may mean anything.

Mr. Swinburne seems to have been the first who has called attention to the heating of the condensers. The astonishment expressed at that by the ablest electrician is a striking illustration of 'the desirability to execute experiments on a large scale. To the scientific investigator, who deals with the minutest quantities, who observes the faintest effects, far more credit is due .than to one who experiments with apparatus on an industrial scale; and indeed history of science has recorded examples of marvelous skill, patience and keenness of observation. But however great the skill, and however keen the observer's perception, it can only be of advantage to magnify an effect and thus facilitate its study. Had Faraday carried out but one of his experiments on dynamic induction on a large scale it would have resulted in an incalculable benefit.

In 'the opinion of the writer, the heating of the condensers is due to three distinct causes: first, leakage or conduction; second, imperfect elasticity in the dielectric, and, third, surging of the charges in the conductor.

In many experiments he has been confronted with the problem of transferring the greatest possible amount of energy across a dielectric. For instance, he has made incandescent lamps the ends of the filaments being completely sealed in' glass, but attached to interior condenser coatings so that all the energy required had to be transferred across the glass with a condenser surface of no more than a few centimeters square. Such lamps would be a practical success with sufficiently high frequencies. With alternations as high as 15,000 per second it was easy to bring the filaments to incandescence. With lower frequencies this could also be effected, but the potential difference had, of course, to be increased. The writer has then found that the glass gets, after a while, perforated and the vacuum is impaired. The higher the frequency the longer the lamp can withstand. Such a deterioration of the dielectric always takes place when the amount of energy transferred across a dielectric of definite dimensions and by a given frequency is too great. Glass withstands best, but even glass is deteriorated. In this case the potential difference on the plates is of course too great and losses by conduction and imperfect elasticity result. If it is desirable to produce condensers capable to stand differences of potential, then the only dielectric which will involve no losses is a gas under pressure. The writer has worked with air under enormous pressures, but there are a great many practical difficulties in that direction. He thinks that in order to make the condensers of considerable practical utility, higher frequencies should be used: though such a plan has besides others the great disadvantage that the system would become very unfit for the operation of motors.

If the writer does not err Mr. Swinburne has suggested a way of exciting an alternator by means of a condenser. For a number of years past the writer has carried on experiments with the object in view of producing a practical self-exciting alternator: He has in a ,variety of ways succeeded in producing some excitation of the magnets by means of alternating currents, which were not commutated by mechanical devices. Nevertheless, his experiments have revealed a fact which stands as solid, as the rock of Gibraltar. No practical excitation can be obtained with a single periodically varying and not commutated current. The reason is that the changes in the strength of the exciting current produce corresponding changes in the field strength, with the result of inducing currents in the armature; and these currents interfere with these produced by the motion of the armature through the field, the former being a quarter phase in advance of the latter. If the field be laminated, no excitation can be produced; if it be not laminated, some excitation is produced, but .the magnets are heated. By combining two exciting currents - displaced by a quarter phase, excitation may be produced in both cases, and if the magnet be not laminated the heating effect is comparatively small, as a uniformity in the field strength is maintained, and, were it possible to produce a perfectly uniform field, excitation on this plan would give quite practical results. If such results are to be secured by the use of a condenser, as suggested by Mr. Swinburne, it is necessary to combine two circuits separated by a quarter phase; that is to say, the armature coils must be wound in two sets and connected to one or two independent condensers. The writer has done some work in that direction, but must defer the description of the devices for some future time.

Experiments with Alternating Currents of High Frequency
The Electrical Engineer - N. Y. — March. 18, 1891

In *The Electrical Engineer* issue of 11th inst., I find a note of Prof. Elihu Thomson relating to some of my experiments with alternating currents of very high frequency.

Prof. Thomson calls the attention of your readers to the interesting fact that he has performed some experiments in the same line. I was not quite unprepared to hear this, as a letter from him has appeared in the *Electrician* a few months ago, in which he mentions a small alternate current machine which was capable of giving, I believe, 5,000 alternations per second, from which letter it likewise appears that his investigations on that subject are of a more recent date.

Prof. Thomson describes an experiment with a bulb enclosing a carbon filament which was brought to incandescence by the bombardment of the molecules of the residual gas when the bulb was immersed in water. "rendered slightly conducting by salt dissolved therein," (?) and a potential of 1,000 volts alternating 5,000 time a second applied to the carbon strip. Similar experiments have, of course, been performed by many experimenters, the only distinctive feature in Prof. Thomson's experiment being the comparatively high rate of alternation. These experiments can also be performed with a steady difference of potential between the water and the carbon strip in which case, of course, conduction through the glass takes place, the difference of potential required being in proportion to the thickness of the glass. With 5,000 alternations per second, conduction still takes place, but the condenser effect is preponderating. It goes, of course, without saying that the healing of the glass in such a case is principally due to the bombardment of the molecules, partly also to leakage or conduction, but it is an undeniable fact that the glass may also be heated merely by the molecular displacement. The interesting feature in ray experiments was that a lamp would light up when brought near to an induction coil, and that it could be held in the hand and the filament brought to incandescence.

Experiments of the kind described I have followed up for a long time with some practical objects in view. In connection with the experiment described by Prof. Thomson, if may be of interest to mention a very pretty phenomenon which may be observed with an incandescent lamp. If a lamp be immersed in water as far as practicable and the filament and the vessel connected to the terminals of an induction coil operated from a machine such as I have used in my experiments, one may see the dull red filament surrounded by a very luminous globe around which there is a less luminous space. The effect is probably due to reflection, as the globe is sharply defined, but may also be due to a "dark space;" at any rate it is so pretty that it must be seen to be appreciated.

Prof. Thomson has misunderstood my statement about the limit of audition. I was perfectly well aware of the fact that opinions differ widely on this point. Nor was I surprised to find that arcs of about 10,000 impulses per second, emit a sound. My statement " the *curious* point is," etc. was only made in deference to an opinion expressed by Sir William Thomson. There was absolutely no stress laid on the precise number. The popular belief was that something like 10,000 to 20,000 per second, or 20,000 to 40,000, at the utmost was the limit. For my argument this was immaterial. I contended that sounds of an incomparably greater number, that is, many times even the highest number, could be heard if they could be produced with sufficient power. My statement was only speculative, but I have devised means which I think may allow me to learn something definite on that point. I have not the least doubt that it is simply a question of power. A very short arc may be silent with 10,000 per second, but just as soon as it is lengthened it begins to emit a sound. The vibrations are the same in number, but more powerful.

Prof. Thomson states that I am taking as the limit of " audition sounds from 5,000 to 10,000 complete waves per second." There is nothing in my statements from which the above could be inferred, but Prof. Thomson has perhaps not thought that there are two sound vibrations for each complete current wave, the former being independent of the direction of the current.

I am glad to learn that Prof. Thomson agrees with me as to the causes of the persistence of the arc. Theoretical considerations considerable time since have led me to the belief that arcs produced by currents of such high frequency would possess this and other desirable features. One of my objects in this direction has been to produce a practicable small arc. With these, currents, for many reasons, much smaller arcs are practicable.

The interpretation by Prof. Thomson of my statements about the arc system leads me *now*, he will pardon me for saying so, to believe that what is most essential to the success of an arc system is a good management. Nevertheless I feel confident of the correctness of the views expressed. The conditions in practice are so manifold that it is impossible for any type of machine to prove best in all the different conditions.

In one case, where the circuit is many miles long, it is desirable to employ the most efficient machine with the least internal resistance ; in another case such a machine would not be the best to employ. It will certainly be admitted that a machine of any type must have a greater resistance if intended to operate arc lights than if it is designed to supply incandescent lamps in series. When arc lights are operated and the resistance is small, the lamps are unsteady, unless a type of lamp is employed in which the carbons are separated by a mechanism which has no further influence upon the feed, the feeding being effected by an independent mechanism ; but even in this case the resistance must be considerably greater to allow a quiet working of the lamps. Now, if the machine be such as to yield a steady current, there is no way of attaining the desired result except by putting the required resistance somewhere either inside or outside of the machine. The latter is hardly practicable, for the customer may stand a hot machine, but he looks with suspicion upon a hot resistance box. A good automatic regulator of course improves the machine and allows us to reduce the internal resistance to some extent, but not as far as would be desirable. Now, since resistance is loss, we can advantageously replace resistance in the machine by an equivalent impedance. But to produce a great impedance with small ohmic resistance, it is necessary to have self-induction and variation of current, and the greater the self-induction and the rate of change of the current, the greater the impedance may be made, while the ohmic resistance may be very small. It may also be remarked that the impedance of the circuit external to the machine is likewise increased. As regards the increase in ohmic resistance in consequence of the variation of the current, the same is, in the commercial machines now in use, very small. Clearly then a great advantage is gained by providing self induction in the machine circuit and undulating the current, for it is possible to replace a machine which has a resistance of, say, 16 ohms by one which has no more than 2 or 3 ohms, and the lights will work even steadier. It seems to me therefore, that my saying that self-induction is essential to the commercial success of an arc system is justified. What is still more important, such a machine will cost considerably less. But to realize fully the benefits, it is preferably to employ an alternate current machine, as in this case a greater rate of change in the current is obtainable. Just what the ratio of resistance to impedance is in the Brush and Thomson machines is nowhere stated, but I think that it is smaller in the Brush machine, judging from its construction.

As regards the better working of clutch lamps with undulating currents, there is, according to my experience, not the least doubt about it. I have proved it on a variety of lamps to the complete satisfaction not only of myself, but of many others. To see the improvement in the feed and to the jar of the clutch at its best it is desirable to employ a lamp in which an independent clutch mechanism effects the feed, and the release of the rod is independent of the up and down movement. In such a lamp the clutch has a small inertia and is very sensitive to vibration, whereas, if the feed is effected by the up and down movement of the lever carrying the rod, the inertia of the system is so great that it is not affected as much by vibration, especially if, as in many cases, a dash pot is employed. During the year 1885 I perfected such a lamp which wan calculated to be operated with undulating currents. With about 1,500 to 1,800 current impulses per minute the feed of this lamp is such that absolutely no movement of the rod can be observed, even if the arc be magnified fifty-fold by means of a lens; whereas, if a steady current is employed, the lamp feeds by small steps. I have, however, demonstrated this feature on other types of lamps, among them being a derived circuit lamp such as Prof. Thomson refers to. I conceived the idea of such a lamp early in 1884, and when my first company was started, this was the first lamp I perfected. It was not until the lamp was ready for manufacture that, on receiving copies of applications from the Patent Office, I learned for the first time, not having had any knowledge of the state of the art in America, that Prof. Thomson had anticipated me and had obtained many patents on this principle, which, of course, greatly disappointed and embarrassed me at that time. I observed the improvement of the feed with undulating currents on that lamp, but I recognized the advantage of providing a light and independent clutch unhampered in its movements. Circumstances did not allow me to carry out at that time some designs of machines I had in mind, and with the existing machines the lamp has worked at a great disadvantage. I cannot agree with Prof. Thomson that small vibrations would benefit a clockwork lamp as much as a clutch lamp; in fact, I think that they do not at all benefit a clockwork lamp.

It would be interesting to learn the opinion of Mr. Charles F. Brush on these points.

Prof. Thomson states that he has run with perfect success clutch lamps " in circuit with coils of such large self-induction that any but very slight fluctuations were wiped out." Surely Prof. Thomson does not mean to say that self-induction wiped out the periodical fluctuations of the current. For this, just the opposite quality, namely, capacity, is required. The self-induction of the coils in this case simply augmented the impedance and prevented the great variations occurring at large time intervals, which take place when the resistance in circuit with the lamps is too small, or even with larger resistance in circuit when the dash pots either in the lamps or elsewhere are too loose.

Prof. Thomson further states that in a lamp in which the feed mechanism is under the control of the derived circuit magnet only, the fluctuations pass through the arc without affecting the magnet to a perceptible degree. It is true that the variations of the resistance of the arc, in consequence of the variations in the current strength, are such as to dampen the fluctuation. Nevertheless, the periodical fluctuations are transmitted through the derived circuit, as one may convince himself easily of, by holding a thin plate of iron against the magnet.

In regard to the physiological effects of the currents I may state that upon reading the memorable lecture of Sir William Thomson, in which he advanced his views on the propagation of the alternate currents through conductors, it instantly occurred to me that currents of high frequencies would be less injurious. I have been looking for a proof that the mode of distribution through the body is the cause of the smaller physiological effects. At times I have thought to have been able to locate the pain in the outer portions of the body, but it is very uncertain. It is most certain, however, that the feeling with currents of very high frequencies is somewhat different from that with low frequencies. I have also noted the enormous importance of one being prepared for the shock or not. If one is prepared, the effect upon the nerves is not nearly as great as when unprepared. With alternations as high as 10,000 per second and upwards, one feels but little pain in the central portion of the body. A remarkable feature of such currents of high tension is that one receives a burn instantly he touches the wire, but beyond that the pain is hardly noticeable.

But since the potential difference across the body by a given current through it is very small, the effects can not be very well ascribed to the surface distribution of the current, and the excessively low resistance of the body to such rapidly varying currents speak rather for a condenser action.

In regard to the suggestion of Dr. Tatum, which Prof. Thomson mentions in another article in the same issue, I would state that I have constructed machines up to 480 poles, from which it is possible to obtain about 30,000 alternations per second, and perhaps more. I have also designed types of machines in which the field would revolve in an opposite direction to the armature, by which means it would be possible to obtain from a similar machine 60,000 alternations per second or more.

I highly value the appreciation of Prof. Thomson of my work, but I must confess that in his conclusion he makes a most astounding statement as to the motives of his critical remarks. I have never for a moment thought that his remarks would be dictated by anything but friendly motives. Often we are forced in daily life to represent opposing interests or opinions, but surely in the higher aims the feelings of friendship and mutual consideration should not be affected by such things as these.

Alternate Current Motors

The Electrical Engineer - April 3, 1891

Sir — In your issue of March 6 I find the passage: "Mr. Kapp described the position as it exists. He showed how Ferraris first of all pointed out the right way to get an alternating-current motor that was self-starting, and how Tesla and others had worked in the direction indicated by Ferraris," etc.

I would be very glad to learn how Mr. Kapp succeeded in showing this. I may call his attention to the fact that the date of filing of my American patent anticipates the publication of the results of Prof. Ferraris in Italy by something like six months. The date of filing of my application is, therefore, the first public record of the invention. Considering this fact, it seems to me that it would be desireable that Mr. Kapp should modify his statement.
—Yours, etc.
Nikola Tesla
New York, 17th March, 1891

Electro-Motors
Electrical Review - London — April 3, 1891

Fifteen or sixteen years ago, when I was pursuing my course at the college, I was told by an eminent physicist that a motor could not he operated without the use of brushes and commutators, or mechanical means of some kind for commutating the current. It was then I determined to solve the problem.

After years of persistent thought I finally arrived at a solution. I worked out the theory to the last detail, and confirmed all of my theoretical conclusions by experiments. Recognizing the value of the invention, I applied myself to the work of perfecting it, and after long continued labor I produced several types of practical motors.

Now all this I did long before anything whatever transpired in the whole scientific literature — as far as it could be ascertained — which would have even pointed at the possibility of obtaining such a result, but quite contrary at a time when scientific and practical men alike considered this result unattainable. In all civilized countries patents have been obtained almost without a single reference to anything which would have in the least degree rendered questionable the novelty of the invention. The first published essay — an account of some laboratory experiments by Prof. Ferraris — was published in Italy six or seven months after the date of filing of my applications for the foundation patents. The date of filing of my patents is thus the first public record of the invention. Yet in your issue of March 6th I read the passage: " For several years past, from the days of Prof. Ferraris's investigations, which were followed by those of Tesla, Zipernowsky and a host of imitators," etc.

No one can say that I have not been free in acknowledging the merit of Prof. Ferraris, and I hope that my statement of facts will not be misinterpreted. Even if Prof. Ferraris's essay would have anticipated the date of filing of my application, yet, in the opinion of all fairminded men, I would have been entitled to the credit of having been the first to produce a practical motor ; for Prof. Ferraris himself denies in his essay the value of the invention for the transmission of power, and only points out the possibility of using a properly-constructed generator, which is the only way of obtaining the required difference of phase without losses; for even with condensers — by means of which it is possible to obtain a quarter phase — there are considerable losses, the cost of the condensers not considered.

Thus, in the most essential features of the system — the generators with the two or three circuits of differing phase, the three-wire system, the closed coil armature, the motors with direct current in the field, etc. — I would stand alone, even had Prof. Ferraris's essay been published many years ago.

As regards the most practicable form of two-wire motor, namely, one with a single energising circuit and induced circuits, of which there are now thousands in use, I likewise stand alone.

Most of these facts, if not all, are perfectly well known in England; yet, according to some papers, one of the leading English electricians does not hesitate to say that I have worked in the direction indicated by Prof. Ferraris, and in your issue above referred to it seems I am called an imitator. Now, I ask you where is that world-known English fairness ? I am a pioneer, and I am called an imitator. I am not an imitator. I produce original work or none at all.

Nikola Tesla

Phenomena of Currents of High Frequency
The Electrical Engineer - N. Y. — April. 8, 1891

I cannot pass without comment the note of Prof. Thomson in your issue of April 1, although I dislike very much to engage in a prolonged controversy. I would gladly let Prof. Thomson have the last word, were it not that some of his statements render a reply from me necessary.

I did not mean to imply that, whatever work Prof. Thomson has done in alternating currents of very high frequency, he had done subsequent to his letter published in the *Electrician*. I thought it possible, and even probable, that he had made his experiments some time before, and my statement in regard to this was meant in this general way. It is more than probable that quite a number of experimenters have built such machines and observed effects similar to those described by Prof. Thomson. It is doubtful, however, whether, in the absence of any publication on this subject, the luminous phenomena described by me have been observed by others, the more so, as very few would be likely to go to the trouble I did, and I would myself not have done so had I not had beforehand the firm conviction, gained from the study of the works of the most advanced thinkers, that I would obtain the results sought for. Now, that I have indicated the direction, many will probably follow, and for this very purpose I have shown some of the results I have reached.

Prof. Thomson states decisively in regard to experiments with the incandescent lamp bulb and the filament mounted on a single wire, that he cannot agree with me at all that conduction through the glass has anything to do with the phenomenon observed. He mentions the well-known fact that an incandescent lamp acts as a Leyden jar and says that " if conduction through the glass were a possibility this action could not occur." I think I may confidently assert that very few electricians will share this view. For the possibility of the condenser effect taking place it is only necessary that the rate at which the charges can equalize through the glass by conduction should be somewhat below the rate at which they are stored.

Prof. Thomson seems to think that conduction through the glass is an impossibility. Has he then never measured insulation resistance, and has he then not measured it by means of a conduction current? Does he think that there *is* such a thing as a perfect non-conductor among the bodies we are able to perceive ? Does he *not* think that as regards conductivity there can be question only of degree? If glass were a perfect non-conductor, how could we account for the leakage of a glass condenser when subjected to steady differences of potential?

While not directly connected with the present controversy, I would here point out that there exists a popular error in regard to the properties of dielectric bodies. Many electricians frequently confound the theoretical dielectric of Maxwell with the dielectric bodies in use. They do not stop to think that the only perfect dielectric is ether, and that all other bodies, the existence of which is known to us, must be conductors, judging from their physical properties.

My statement that conduction is concerned to some, although perhaps negligible, extent in the experiment above described was, however, made not only on account of the fact that all bodies conduct more or less, but principally on account of the heating of the glass during the experiment. Prof. Thomson seems to overlook the fact that the insulating power of glass diminishes enormously with the increase in temperature, so much so, that melted glass is comparatively an excellent conductor. I have, moreover, stated in my first reply to Prof. Thomson in your issue of March 18, that the same experiment can be performed by means of an unvarying difference of potential. In this case it must be assumed that some such process as conduction through the glass takes place, and all the more as it is possible to show by experiment, that with a sufficiently high steady difference of potential, enough current can be passed through the glass of a condenser with mercury coating to light up a Geissler tube joined in series with the condenser. When the potential is alternating, the condenser action comes in and conduction becomes insignificant, and the more so, the greater the rate of alternation or change per unit of time. Nevertheless, in my opinion, conduction must always exist, especially if the glass is hot, though it may be negligible with very high frequencies.

Prof. Thomson states, further, that from his point of view I have misunderstood his statement about the limit of audition. He says that 10,000 to 20,000 alternations correspond to 5,000 to 10,000 complete waves of sound. In my first reply to Prof. Thomson's remarks (in your issue of March 18,) I avoided pointing out directly that Prof. Thomson was mistaken, but now I see no way out of it. Prof. Thomson will pardon me if I call his attention to the fact he seems to disregard, namely, that 10,000 to 20,000 alternations of current in an arc — which was the subject under discussion — do not mean 5,000 to 10,000, but 10,000 to 20,000 *complete waves of sound.*

He says that I have adopted or suggested as the limit of audition 10,000 waves per second, but I have neither adopted nor suggested it. Prof. Thomson states that I have been working with 5,000 to 10,000 complete waves, while I have nowhere made any such statement. He says that this would be working below the limit of audition, and cites as an argument that at the Central High School, in Philadelphia, he has heard 20,000 waves per second ; but he wholly overlooks a point on which I have dwelt at some length, namely, that the limit of audition of an arc is something entirely different from the limit of audition in general.

Prof. Thomson further states, in reply to some of my views expressed in regard to the constant current machines that five or six years ago it occurred to him to try the construction of a dynamo for constant current, in which "the armature coils were of a highly efficient type, that is, of comparatively short wire length for the voltage and moving in a dense magnetic field." Exteriorly to the coils and to the field he had placed in the circuit of each coil an impedance coil which consisted of an iron core wound with a considerable length of wire and connected directly in circuit with the armature coil. He thus obtained, he thought, "the property of considerable self-induction along with efficient current generation." Prof. Thomson says he expected "that possibly the effects would be very much the same as those obtainable from the regularly constructed apparatus." But he was disappointed, he adds. With all the consideration due to Prof. Thomson, I would say that, to expect a good result from such a combination, was rather sanguine. Earth is not farther from Heaven than this arrangement is from one, in which there would be a length of wire, sufficient to give the same self-induction, wound on the armature and utilized to produce useful E. M. F., instead of doing just the opposite, let alone the loss in the iron cores. But it is, of course, only fair to remember that this experiment was performed five or six years ago, when even the foremost electricians lacked the necessary information in these and other matters.

Prof. Thomson seems to think that self-induction wipes out the periodical undulations of current. Now self-induction does not produce any such effect, but, if anything, it renders the undulation more pronounced. This is self-evident. Let us insert a self-induction coil in a circuit traversed by an undulating current and see what happens. During the period of the greatest rate of change, when the current has a small value, the self-induction opposes more than during the time of the small rate of change when the current is at, or near, its maximum value. The consequence is, that with the same frequency the maximum value of the current becomes the greater, the greater the self-induction. As the sound in a telephone depends only on the maximum value, it is clear that self-induction is the very thing required in a telephone circuit. The larger the self-induction, the louder and clearer the speech, provided the same current is passed through the circuit. I have had ample opportunity to study this subject during my telephone experience of several years. As regard the fact that a self-induction coil in series with a telephone diminishes the loudness of the sound, Prof. Thomson seems to overlook the fact that this effect is wholly due to the impedance of the coil, *i. e.,* to its property of diminishing the *current strength*. But while the current strength is diminished the undulation is rendered only more pronounced. Obviously, when comparisons are made they must be made with the same current.

In an arc machine, such as that of Prof. Thomson's, the effect is different. There, one has to deal with a make and break. There are then two induced currents, one in the opposite, the other in the same direction with the main current. If the function of the mechanism be the same whether a self-induction coil be present or not, the undulations could not possibly be wiped out. But Prof. Thomson seems, likewise, to forget that the effect is wholly due to the defect of the commutator ; namely, the induced current of the break, which is of the same direction with the main current and of great intensity, when large self-induction is present, simply bridges the adjacent commutator segments, or, if not entirely so, at least shortens the interval during which the circuit is open and thus reduces the undulation.

In regard to the improvement in the feeding of the lamps by vibrations or undulations, Prof. Thomson expresses a decisive opinion. He now says that the vibrations *must* improve the feeding of a clock-work lamp. He says that I "contented myself by simply saying," that I cannot agree with him on that point.

Now, s*aying it,* is not the only thing I did. I have passed many a night watching a lamp feed, and I leave it to any skilled experimenter to investigate whether my statements are correct. My opinion is, that a clock-work lamp; that is, a lamp in which the descent of the carbon is regulated not by a clutch or friction mechanism, but by an escapement, cannot feed any more perfectly than tooth by tooth, which may be a movement of, say, 1/15 of an inch or less. Such a lamp will feed in nearly the same manner whether the current be perfectly smooth or undulating, providing the conditions of the circuit are otherwise stable. If there is any advantage, I think it would be in the use of a smooth current, for, with an undulating current, the lamp is likely to miss some time and feed by more than one tooth. But in a lamp in which the descent of the carbon is regulated by friction mechanism, an indulating current of the proper number of undulations per second will always give a better result. Of course, to realize fully the benefits of the undulating current the release ought to be effected independently of the up-and-down movement I have pointed out before.

In regard to the physiological effects, Prof. Thomson says, that in such a comparatively poor conductive material as animal tissue the distribution of current cannot be governed by self-induction to any appreciable extent, but he does not consider the two-fold effect of the large cross-section, pointed out by Sir William Thomson. As the resistance of the body to such currents is low, we must assume either condenser action or induction of currents in the body.
Nikola Tesla
New York, April 4, 1891

Alternate Current Electrostatic Induction Apparatus
The Electrical Engineer - N.Y. — May 6, 1891

About a year and a half ago while engaged in the study of alternate currents of short period, it occurred to me that such currents could be obtained by rotating charged surfaces in close proximity to conductors. Accordingly I devised various forms of experimental apparatus of which two are illustrated in the accompanying engravings.

Figure 1.

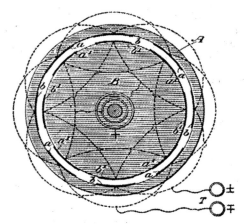

In the apparatus shown in Fig. 1, A is a ring of dry shellacked hard wood provided on its inside with two sets of tin-foil coatings, a and b, all the a coatings and all the b coatings being connected together, respectively, but independent from each other. These two sets of coatings are connected to two terminals, T. For the sake of clearness only a few coatings are shown. Inside of the ring A, and in close proximity to it there is arranged to rotate a cylinder B, likewise of dry, shellacked hard wood, and provided with two similar sets of coatings, a_1 and b_1, all the coatings a1 being connected to one ring and all the others, b_1, to another marked + and –. These two sets, a_1 and b_1 are charged to a high potential by a Holtz or a Wimshurst machine, and may be connected to a jar of some capacity. The inside of ring A is coated with mica in order to increase the induction and also to allow higher potentials to be used.

Figure 2,

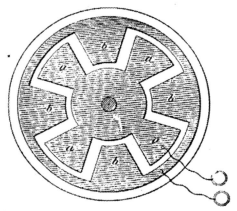

When the cylinder B with the charged coatings is rotated, a circuit connected to the terminals T is traversed by alternating currents. Another form of apparatus is illustrated in Fig. 2. In this apparatus the two sets of tinfoil coatings are glued on a plate of ebonite, and a similar plate which is rotated, and the coatings of which are charged as in Fig. 1, is provided.

The output of such an apparatus is very small, but some of the effects peculiar to alternating currents of short periods may be observed. The effects, however, cannot be compared with those obtainable with an induction coil which is operated by an alternate current machine of high frequency, some of which were described by me a short while ago.

An Electrolytic Clock
May 6, 1891 The Electrical Engineer

If a delicately and pivoted and well-balanced metal disc or cylinder be placed in a proper plating solution midway between the anode and cathode, one half of the disc becomes electro positive and the other half electro negative. Owing to this fact metal is deposited on one, and taken off from the other half, and the disc is caused to rotate under the action of gravity. As the amount of metal deposited and taken off is proportionate to the current strength, the speed of rotation, if it be small, is proportionate to the current.

The first device of this kind was operated by me early in 1888, in the endeavor to construct an electric meter. Upon learning, however, that I had been anticipated by others, as far as the principle is concerned, I devised the apparatus illustrated in the accompanying engraving. Here F is a rectangular frame of hard rubber which is fastened upon a wooden base. This frame is about — inch thick, 6 inches long and 5 inches high.

On both of its upright sides are fastened thick metal plates which serve as the electrodes. These plates are held firmly against the rubber frame by the binding posts T T and T1 T1 On the lateral sides of the frame are fastened the brass plates B and B1, respectively, of the same shape as the rubber frame F. These brass plates serve to keep in place two plates of polished glass, and the vessel is hermetically sealed by placing a soft rubber washer under and above each of the glass plates. In this manner the plates may be screwed on tight without fear of breaking them.

The plating solution, which in this case is a concentrated solution of sulphate of copper, is poured in through an opening on the top of the rubber frame, which is closed by a plug R.

In the center of the vessel is placed a light and delicately balanced copper disc D, the axis of which is supported by a capillary glass tube which is fixed to one of the glass plates by means of sealing wax, or other material not attacked by the liquid. To diminish the friction as much as possible, the capillary tube which serves as a bearing contains a drop of oil. The center of disc should be equi-distant from both the electrodes. To one side of the axis of the disc is fastened a very light indicator or pointer consisting preferably of a thin glass thread. The glass plate next to this pointer has a circle with the usual hour divisions engraved upon it, as on a clock dial. This circle may be movable so that it can be put in any position relatively to the pointer. If the dial is not movable then a thin wire of annealed iron may be used as on a pointer. The wire should then be so placed that it is exactly in the center of the solution. By means of a horse-shoe magnet the disc may then be rotated and set in proper position.

The copper solution being carefully poured in, and the plug R replaced, the terminals of a constant current battery are connected to the binding-posts TT1, and from time to time the rotation of the disc is observed. A shunt is connected to the other two binding-posts TT1, and by varying the resistance of this shunt, or other disc, the speed of rotation is regulated until it is made to correspond to the division of the dial; that is, until, for instance, one turn is made in 12 hours.

Obviously this instrument was not devised for a practical purpose. Neither will it be quite exact in its indications. There are certain errors, unavoidable from the principle; for instance, the friction, which cannot be completely overcome. But the device is interesting as a means of indicating time in a novel manner. It will, however, be found that by a careful construction, constant current, and a temperature compensator, it may be made to rotate with almost perfect uniformity. The current density should, of course, be very small to secure the best results, and the disc of about 3 inches diameter should turn once in 6 hours. It is probable that with a silver solution and a silver plate better results would be obtained.

It is very interesting to note the appearance of the solution and disc in such a narrow transparent vessel. The solution appears a clear blue, one side of the disc seems to be silver white in a certain position, and the other half is dark like tarnished silver. There is no line of demarcation, but the shades melt beautifully together.

Experiments with Alternate Currents of Very High Frequency and Their Application to Methods of Artificial Illumination

Delivered before the American Institute of Electrical Engineers at Columbia College May 20th 1891

There is no subject more captivating, more worthy of study, than nature. To understand this great mechanism, to discover the forces which are active, and the lams which govern them, is the highest aim of the intellect of man.

Nature has stored up in the universe infinite energy. The eternal recipient and transmitter of this infinite energy is the ether. The recognition of the existence of ether, and of the functions it performs, is one of the most important results of modern scientific research. The mere abandoning of the idea of action at a distance, the assumption of a medium pervading all space and connecting all gross matter, has freed the minds of thinkers of an ever present doubt, and, by opening a new horizon — new and unforeseen possibilities — has given fresh interest to phenomena with which we are familiar of old. It has been a great step towards the understanding of the forces of nature and their multifold manifestations to our senses. It has been for the enlightened student of physics what the understanding of the mechanism of the firearm or of the steam engine is for the barbarian. Phenomena upon which we used to look as wonders baffling explanation, we now see in a different light. The spark of an induction coil, the glow of an incandescent lamp, the manifestations of the mechanical forces of currents and magnets are no longer beyond our grasp; instead of the incomprehensible, as before, their observation suggests now in our minds a simple mechanism, and although as to its precise nature all is still conjecture, yet we know .that the truth cannot be much longer hidden, and instinctively we feel that the understanding is dawning upon us. We still admire these beautiful phenomena, these strange forces, but we are helpless no longer; we can in a certain measure explain them, account for them, and we are hopeful of finally succeeding in unraveling the mystery which surrounds them.

In how far we can understand the world around us is the ultimate thought of every student of nature. The coarseness of our senses prevents us from recognizing the ulterior construction of matter, and astronomy, this grandest and most positive of natural sciences, can only teach us something that happens, as it were, in our immediate neighborhood; of the remoter portions of the boundless universe, with its numberless stars and suns, we know nothing, But far beyond the limit of perception of our senses the spirit still can guide us, and so we may hope that even these unknown worlds — infinitely small and great — may in a measure became known to us. Still, even if this knowledge should reach us, the searching mind will find a barrier, perhaps forever unsurpassable, to the *true* recognition of that which *seems* to be, the mere *appearance* of which is the only and slender basis of all our philosophy.

Of all the forms of nature's immeasurable, all-pervading energy, which ever and ever changing and moving; like a soul animates the inert universe, electricity and magnetism are perhaps the most fascinating. The effects of gravitation, of heat and light we observe daily, and soon we get accustomed to them, and soon they lose for us the character of the marvelous and wonderful; but electricity and magnetism, with their singular relationship, with their seemingly dual character, unique among the forces in nature, with their phenomena of attractions, repulsions and rotations, strange manifestations of mysterious agents; stimulate and excite the mind to thought and research. What is electricity, and what is magnetism? These questions have been asked again and again. The most able intellects have ceaselessly wrestled with the problem; still the question has not as yet been fully answered. But while we cannot even to-day state what these singular forces are, we have made good headway towards the solution of the problem. We are now confident that electric and magnetic phenomena are attributable to ether, and we are perhaps justified in saying that the effects of static electricity are effects of ether under strain, and those of dynamic electricity and electro-magnetism effects of ether in motion. But this still leaves the question, as to what electricity and magnetism are, unanswered.

First, we naturally inquire, What is electricity, and is there such a thing as electricity ? In interpreting electric phenomena: we may speak of electricity or of an electric condition, state or effect. If we speak of electric effects we must distinguish two such effects, opposite in character and neutralizing each other, as observation shows that two such opposite effects exist. This is unavoidable, for in a medium of the properties of ether, we cannot possibly exert a strain, or produce a displacement or motion of any kind, without causing in the surrounding medium an equivalent and opposite effect. But if we speak of electricity, meaning *a thing,* we must, I think, abandon the idea of two electricities, as tie existence of two such things is highly improbable. For how can we imagine that there should be two things, equivalent in amount, alike in their properties, but of opposite character, both clinging to matter, both attracting and completely neutralizing each other? Such an assumption, though suggested by many phenomena, though most convenient for explaining them, has little to commend it. If there *is* such a thing as electricity, there can be only *one* such thing, and; excess and want of that one thin, possibly; but more probably its condition determines the positive and negative character. The old theory of Franklin, though falling short in some respects; is, from a certain point of view, after all, the most plausible one. Still, in spite of this, the theory of the two electricities is generally accepted, as it apparently explains electric phenomena in a more satisfactory manner. But a theory which better explains the facts is not necessarily true. Ingenious minds will invent theories to suit observation, and almost every independent thinker has his own views on the subject.

It is not with the, object of advancing an opinion; but with the desire of acquainting you better with some of the results, which I will describe, to show you the reasoning I have followed, the departures I have made — that I venture to express, in a few words, the views and convictions which have led me to these results.

I adhere to the idea that there is a thing which we have been in the habit of calling electricity. The question is, What is that thing? or, What, of all things, the existence of which we know, have we the best reason to call electricity? We know that it acts like an incompressible fluid; that there must be a constant quantity of it in nature; that it can be neither produced nor destroyed; and, what is more important, the electro-magnetic theory of light and all facts observed teach us that electric and ether phenomena are identical. The idea at once suggests itself, therefore, that electricity might be called ether. In fact, this view has in a certain sense been advanced by Dr. Lodge. His interesting work has been read by everyone and many have been convinced by his arguments. Isis great ability and the interesting nature of the subject, keep the reader spellbound; but when the impressions fade, one realizes that he has to deal only with ingenious explanations. I must confess, that I cannot believe in two electricities, much less in a doubly-constituted ether. The puzzling behavior of tile ether as a solid waves of light anti heat, and as a fluid to the motion of bodies through it, is certainly explained in the most natural and satisfactory manner by assuming it to be in motion, as Sir William Thomson has suggested; but regardless of this, there is nothing which would enable us to conclude with certainty that, while a fluid is not capable of transmitting transverse vibrations of a few hundred or thousand per second, it might not be capable of transmitting such vibrations when they range into hundreds of million millions per second. Nor can anyone prove that there are transverse ether waves emitted from an alternate current machine, giving a small number of alternations per second; to such slow disturbances, the ether, if at rest, may behave as a true fluid.

Returning to the subject, and bearing in mind that the existence of two electricities is, to say the least, highly improbable, we must remember, that we have no evidence of electricity, nor can we hope to get it, unless gross matter is present. Electricity, therefore, cannot be called ether in the broad sense of the term; but nothing would seem to stand in the *way of* calling electricity ether associated with matter, or bound other; or, in other words, that the so-called static charge of the molecule is ether associated in some way with the molecule. Looking at it in that light, we would be justified in saying, that electricity is concerned in all molecular actions.

Now, precisely what the ether surrounding tine molecules is, wherein it differs from ether in general, can only be conjectured. It cannot differ in density, ether being incompressible; it must, therefore, be under some strain or is motion, and the latter is the` most probable: To understand its functions, it would be necessary to have an exact idea of the physical construction of matter, of which, of course, we can only form a mental picture.

But of all the views on nature, the one which assumes one matter and one force, and a perfect uniformity throughout, is the most scientific, and most likely to be true. An infinitesimal world, with the molecules and their atoms spinning and moving in orbits, in much the same manner as celestial bodies, carrying with them and probably spinning with them ether, or in other words; carrying with them static charges, seems to my mind the most probable view, and one which; in a plausible manner, accounts for most of the phenomena observed. The spinning of the molecules and their ether sets up the ether tensions or electrostatic strains; the equalization of ether tensions sets up ether motions or electric currents, and the orbital movements produce the effects of electro and permanent magnetism.

About fifteen, years ago, Prof. Rowland demonstrated a most interesting and important fact; namely, that a static charge carried around produces the effects of an electric current. Leaving out of consideration the precise nature of the mechanism, which produces the attraction and repulsion of currents, and conceiving the electrostatically charged molecules in motion, this experimental fact gives us a fair idea of magnetism. We can conceive lines or tubes of force which physically exist, being formed of rows of directed moving molecules; we can see that these lines must be closed, that they must tend to shorten and expand, etc. It likewise explains in a reasonable way, the most puzzling phenomenon. of all, permanent magnetism, and, in general, has all the beauties of the Ampere *theory* without possessing the vital defect of the same, namely, the assumption of molecular currents. Without enlarging further upon the subject, I would say, that I look upon all electrostatic, current and magnetic phenomena as being due to electrostatic molecular forces.

The preceding remarks I have deemed necessary to a full understanding; of the subject a s it presents itself to my mind.

Of all these phenomena the most important to study' are the current phenomena, on account of the already extensive and evergrowing use of currents for industrial purposes. It is now a century since the first practical source of current was produced, and, ever since, the phenomena which accompany the flow of currents have been diligently studied, and through the untiring efforts of scientific men the simple laws which govern them have been discovered. But these laws are found to hold good only when the currents are of a steady character. When the currents are rapidly varying in strength, quite different phenomena, often unexpected, present themselves, and quite different laws hold good, which even now have not been determined as fully as is desirable, though through the work, principally, of English scientists, enough knowledge has been gained on the subject to enable us to treat simple cases which now present themselves in daily practice.

The phenomena which are peculiar to the changing character of the currents are greatly exalted when the rate of change is increased, hence the study of these currents is considerably facilitated by the employment of properly constructed apparatus. It was with this and other objects in view that I constructed alternate current machines capable of giving more than two million reversals of current per minute, and to this circumstance it is principally due, that I am able to bring to your attention some of the results thus far reached; which I hope will prove to be a step in advance on account of their direct bearing upon one of the most important problems, namely, the production of a practical and efficient source of light.

The study of such rapidly alternating currents is very interesting. Nearly every experiment discloses something new. Many results may, of course, be predicted, but many more are unforeseen. The experimenter makes many interesting observations. For instance, we take a piece of iron and hold it against a magnet. Starting from low alternations and running up higher and higher we feel the impulses succeed each other faster and faster, get weaker and weaker, and finally disappear. We then observe a continuous pull; the pull, of course, is not continuous; it only appears so to us; our sense of touch is imperfect.

We may next establish an arc between the electrodes and observe, as the alternations rise, that the note which accompanies alternating arcs gets shriller and shriller, gradually weakens, and finally ceases. The air vibrations, of course, continue, but they are too weak to be perceived; our sense of hearing fails us.

We observe the small physiological effects, the rapid heating of the iron cores and conductors, curious inductive effects, interesting condenser phenomena, and still more interesting light phenomena with a high tension induction coil. All these experiments and observations would be of the greatest interest to the student, but their description would lead me too far from the principal subject. Partly for this reason, and partly on account of their vastly greater importance, I will confine myself to the description of the light effects produced by these currents.

In the experiments to this end a high tension induction coil or equivalent apparatus for converting currents of comparatively low into currents of high tension is used.

If you will be sufficiently interested in the results I shall describe as to enter into an experimental study of this subject; if you will be convinced of the truth of the arguments I shall advance — your aim will be to produce high frequencies and high potentials; in other words, powerful electrostatic effects. You will then encounter many difficulties, which, if completely overcome, would allow us to produce truly wonderful results.

First will be met the difficulty of obtaining the required frequencies by means of mechanical apparatus, and, if they be obtained otherwise, obstacles of a different nature will present themselves. Next it will be found difficult to provide the requisite insulation without considerably increasing the size of the apparatus, for the potentials required are high, and, owing to the rapidity of the alternations, the insulation presents peculiar difficulties. So, for instance, when a gas is present, the discharge may work, by the molecular bombardment of the gas and consequent heating, through as much as an inch of the best solid insulating material, such as glass, hard rubber, porcelain, sealing wax, etc.; in fact, through any known insulating substance. The chief requisite in the insulation of the apparatus is, therefore, the exclusion of any gaseous matter.

In general my experience tends to show that bodies which possess the highest specific inductive capacity, such as glass, afford a rather inferior insulation to others, which, while they are good insulators, have a much smaller specific inductive capacity, such as oils, for instance, the dielectric losses being no doubt greater in the former. The difficulty of insulating, of course, only exists when the potentials are excessively high, for with potentials such as a few thousand volts there is no particular difficulty encountered in conveying currents from a machine giving, say, 20,000 alternations per second, to quite a distance. This number of alternations, however, is by far too small for many purposes, though quite sufficient for some practical applications. This difficulty of insulating is fortunately not a vital drawback; it affects mostly the size of the apparatus, for, when excessively high potentials would be used, the light-giving devices would be located not far from the apparatus, and often they would be quite close to it. As the air-bombardment of the insulated wire is dependent on condenser action, the loss may be reduced to a trifle by using excessively thin wires heavily insulated.

Another difficulty will be encountered in the capacity and self-induction necessarily possessed by the coil. If the toil be large, that is, if it contain a great length of wire, it will be generally unsuited for excessively high frequencies; if it be small, it may be well adapted for such frequencies, but the potential might then not be as high as desired. A good insulator, and preferably one possessing a small specific inductive capacity, would afford a two-fold advantage. First, it would enable us to construct a very small coil capable of withstanding enormous differences of potential; and secondly, such a small coil, by reason of its smaller capacity and self-induction, would be capable of a quicker and more vigorous vibration. The problem then of constructing a coil or induction apparatus of any kind possessing the requisite qualities I regard as one of no small importance, and it has occupied me for a considerable time.

The investigator who desires to repeat the experiments which I will describe, with an alternate current machine, capable of supplying currents of the desired frequency, and an induction coil, will do well to take the primary coil out and mount the secondary in such a manner as to be able to look through the tube upon which the secondary is wound. He will then be able to observe the streams which pass from the primary to the insulating tube, and from their intensity he will know how far he can strain the coil. Without this precaution he is sure to injure the insulation. This arrangement permits, however, an easy exchange of the primaries, which is desirable in these experiments.

The selection of the type of machine best suited for the purpose must be left to the judgment of the experimenter. There are here illustrated three distinct types of machines, which, besides others, I have used in my experiments.

Fig. 1 represents the machine used in my experiments before this Institute. The field magnet consists of a ring of wrought iron with 384 pole projections. The armature comprises a steel disc to which is fastened a thin, carefully welded rim of wrought iron. Upon the rim are wound several layers of fine, well annealed iron wire, which, when wound, is passed through shellac. The armature wires are wound around brass pins, wrapped with silk thread: The diameter of the armature wire in this type of machine should not be more than 1/8. of the thickness of the pole projections, else the local action will be considerable.

FIG. 1.—High Frequency Alternator with Drum Armature.

FIG. 2.—High Frequency Alternator with Revolving Disc Armature.

Fig. 2 represents a larger machine of a different type. The field magnet of this machine consists of two like parts which either enclose an exciting coil, or else are independently wound. Each part has 480 pole projections, the projections of one facing those of the other. The armature consists of a wheel of hard bronze, carrying the conductors which revolve between the projections of the field magnet. To wind the armature conductors, I have found it most convenient to proceed in the following manner. I construct a ring of hard bronze of the required size. This ring and the rim a the wheel are provided with the proper number of pins, and both fastened upon a plate. The armature conductors being wound, the pins are cut off and the ends of the conductors fastened by two rings which screw to the bronze ring and the rim of the wheel, respectively. The whole may then be taken off and forms a solid structure. The conductors in such a type of machine should consist of sheet copper, the thickness of which, of course, depends on the thickness of the pale projections; or else twisted thin wires should be employed.

Fig. 3 is a smaller machine, in many respects similar to the former, only here the armature conductors and the exciting coil are kept stationary, while only a block of wrought iron is revolved.

It would be uselessly lengthening this description were I to dwell more on the details of construction of these machines. Besides, they have been described somewhat more elaborately in *The Electrical Engineer,* of March 18,

FIG. 8.—High Frequency Alternator with Stationary Disc Armature and Stationary Exciting Coil.

Experiments with Alternate Currents of Very High Frequency and Their Application to Methods of Artificial Illumination 1891. I deem it well, however, to call the attention of the investigator to two
things, the importance of which, though self evident, he is nevertheless apt to underestimate; namely, to the local action in the conductors which must be carefully avoided, and to the clearance, which must be small. I may add, that since it is desirable to use very high peripheral speeds, the armature should he of very large diameter in order to avoid impracticable belt speeds. Of the several types of these machines which have been constructed by me, I have found that the type illustrated in Fig. 1 caused me the least trouble in construction, as well as in maintenance, and on the whole, it has been a good experimental machine.

In operating an induction coil with very rapidly alternating currents, among the first luminous phenomena noticed are naturally those, presented by the high-tension discharge. As the number of alternations per second is increased, or as — the number being high — the current through the primary is varied, the discharge gradually changes in appearance. It would be difficult to describe the minor changes which occur, and the conditions which bring them about, but one may note five distinct forms of the discharge.

First, one may observe a weak, sensitive discharge in the form of a thin, feeble-colored thread (Fig. 4). It always occurs when, the number of alternations per second being high, the current through the primary is very small. In spite of the excessively small current, the rate of change is great, and the difference of potential at the terminals of the secondary is therefore considerable, so that the arc is established at great distances; but the quantity of "electricity" set in motion is insignificant, barely sufficient to maintain a thin, threadlike arc. It is excessively, sensitive and may be made so to such a degree that the mere act of breathing near the coil will affect it, and unless it is perfectly well protected from currents of air, it wriggles around constantly. Nevertheless, it is in this form excessively persistent, and when the terminals are approached to, say, one-third of the striking distance, it can be blown out only with difficulty. This exceptional persistency, when short, is largely due to the arc being excessively thin; presenting, therefore, a very small surface to the blast. Its great sensitiveness, when very long, is probably due to the motion of the particles of dust suspended in the air.

When the current through the primary is increased, the discharge gets broader and stronger, and the effect of the capacity of the coil becomes visible until, finally, under proper conditions, a white flaming arc, Fig. 5, often as thick as one's finger, and striking across the whole coil, is produce. It develops remarkable heat, and may be further characterized by the absence of the high note which accompanies the less powerful discharges. To take a shock from .the coil under these conditions would not be advisable, although under different conditions the potential being much higher; a shock from the coil may be taken with impunity. To produce this kind of discharge the number of alternations per second must not be too great for the coil used; and, generally speaking, certain relations between capacity, self-induction and frequency must be observed.

Fig. 4.—Sensitive Thread Discharge.   Fig. 5.—Flaming Discharge.

The importance of these elements in an alternate current circuit is now well-known, and under ordinary conditions, the general rules are applicable. But in an induction coil exceptional conditions prevail. First, the self-induction is of little importance before the arc is established, when it asserts itself, but perhaps never as prominently as in ordinary alternate current circuits, because the capacity is distributed all along the coil, and by reason of the fact that the coil usually discharges through very great remittances; hence the currents are exceptionally small. Secondly, the capacity goes on increasing continually as the potential rises, in consequence of absorption which takes place to a considerable extent. Owing to this there exists no critical relationship between these quantities, and ordinary rules would not seem: to be applicable: As the potential is increased either in consequence of the increased frequency or of the increased current through the primary, the amount of the energy stored becomes greater and greater, and the capacity gains more and more in importance. Up to a certain point the capacity is beneficial, but after that it begins to be an enormous drawback. It follows from this that each coil gives the best result with a given frequency and primary current. A very large coil, when operated with currents of very high frequency, may not give as much as 1/8 inch spark. By adding capacity to the terminals, the condition may be improved, but what the coil really wants is a lower frequency.

When the flaming discharge occurs, the conditions are evidently such that the greatest current is made to flow through the circuit. These conditions may be attained by varying the frequency within wide limits, but the highest frequency at which the flaming arc can still be produced, determines, for a given primary current, the maximum striking distance of the coil. In the flaming discharge the *eclat* effect of the capacity is not perceptible; the rate at which the energy is being stored then just equals the rate at which it can be disposed of through the circuit. This kind of discharge is the severest test for a coil; the break, when it occurs, is of the nature of that in an overcharged Leyden jar. To give a rough approximation I would state that, with an ordinary coil of, say, 10,000 ohms resistance, the most powerful arc would be produced with about 12,000 alternations per second.

When the frequency is increased beyond that rate, the potential, of course, rises, but the striking distance may, nevertheless, diminish, paradoxical as it may seem. As the potential rises the coil attains more and more the properties of a static machine until, finally, one may observe the beautiful phenomenon of the streaming discharge, Fig. 5, which may be produced across the whole length of the coil. At that stage streams begin to issue freely from all points and projections. These streams will also be seen to pass in abundance in the space between the primary and the insulating tube. When the potential is excessively high they will always appear; even if the frequency be low, and even if the primary be surrounded by as much as an inch of wax, hard rubber, glass, or any other insulating substance. This limits greatly the output of the coil, but I will later show how I have been able to overcome to a considerable extent this disadvantage in the ordinary coil.

Besides the potential, the intensity of the streams depends on the frequency; but if the coil be very large they show themselves, no matter how low the frequencies used. For instance, in a very large coil of a resistance of 67,000 ohms, constructed by me some time ago, they appear with as low as 100 alternations per second and less, the insulation of the secondary being 3/4 inch of ebonite. When very intense they produce a noise similar to that produced by the charging of a Holtz machine, but much more powerful, and they emit a strong smell of ozone. The lower the frequency, the more apt they are to suddenly injure the coil. With excessively high frequencies they may pass freely without producing any other effect than to heat the insulation slowly and uniformly.

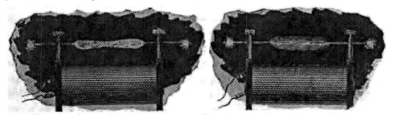

FIG. 6.— Streaming Discharge.   FIG. 7.—Brush and Spray Discharge.

The existence of these streams shows the importance of constructing an expensive coil so as to permit of one's seeing through the tube surrounding the primary, and the latter should be easily exchangeable; or else the space between the primary and secondary should be completely filled up with insulating material so as to exclude all air. The non-observance of this simple rule in the construction of commercial coils is responsible for the destruction of many an expensive coil.

At the stage when the streaming discharge occurs, or with somewhat higher frequencies, one may, by approaching the terminals quite nearly, and regulating properly the effect of capacity, produce a veritable spray of small silver-white sparks, or a bunch of excessively thin silvery threads (Fig. 7) amidst a powerful brush — each spark or thread possibly corresponding to one alternation. ibis, when produced under proper conditions, is probably the most beautiful discharge, and when an air blast is directed against it, it presents a singular appearance. The spray of sparks, when received through the body, causes some inconvenience, whereas, when the discharge simply streams, nothing at all is likely to be felt if large conducting objects are held in the hands to protect them from receiving small burns.

If the frequency is still more increased, then the coil refuses to give any spark unless at comparatively small distances, and the fifth typical form of discharge may be observed (Fig. 8). The tendency to stream out and dissipate is then so great that when the brush is produced at one terminal no sparking occurs; even if, as I have repeatedly tried, the hand, or any conducting object, is held within the stream; and. what is mere singular, the luminous stream is not at all easily deflected by the approach of a conducting body.

Fig. 8.—Fifth Typical Form of Discharge.

Fig. 9.—Luminous Discharge with Interposed Insulators.

At this stage the streams seemingly pass with the greatest freedom through considerable thicknesses of insulators, and it is particularly interesting to study their behavior. For ibis purpose it is convenient to connect to the terminals of the coil two metallic spheres which may be placed at any desired distance, Fig. 9. Spheres arc preferable to plates, as the discharge can be better observed. By inserting dielectric bodies between the spheres, beautiful discharge phenomena tray be observed. If the spheres be quite close and the spark be playing between them, by interposing a thin plate of ebonite between the spheres the span: instantly ceases and the discharge spread; into an intensely luminous circle several inches in diameter, provided the spheres are sufficiently large. The passage of the streams heats, and; after a while, softens, the rubber so much that two plates may be made to stick together in this manner. If the spheres are so far apart that no spark occurs, even if they are far beyond the striking distance, by inserting a thick plate of mass the discharge is instantly induced to pass from the spheres to the glass is the form of luminous streams. It appears almost as though these streams pass *through* the dielectric. In reality this is not the case, as the streams are due to the molecules of the air which are violently agitated in the space between the oppositely charged surfaces of the spheres. When no dielectric other than air is present, the bombardment goes on, but is too weak to be visible; by inserting, a dielectric the inductive effect is much increased, and besides, the projected air molecules find an obstacle and the bombardment becomes so intense that the streams become luminous. If *by any* mechanical means we could effect such a violent agitation of the molecules we could produce the same phenomenon. A jet of air escaping through a small hole under enormous pressure and striking against an insulating substance, such as glass, may be luminous in the dark, and it might be possible to produce a phosphorescence of the gloss or other insulators in this manner.

The greater the specific inductive capacity of the interposed dielectric, the more powerful the effect produced. Owing to this, the streams show themselves with excessively high potentials even if the glass be as much as one and one-half to two inches thick. But besides the heating due to bombardment, some heating goes on undoubtedly in the dielectric, being apparently greater in glass than in ebonite. I attribute this to the greater specific inductive capacity of the glass; in consequence of which, with the same potential difference, a greater amount of energy is taken up in it than in rubber. It is like connecting to a battery a copper and a brass wire of the same dimensions. The copper wire, though a more perfect conductor, would heat more by reason of its taking more current. Thus what is otherwise considered a virtue of the glass is here a defect. Glass usually gives way much quicker than ebonite; when it is heated to a certain degree, the discharge suddenly breaks through at one point, assuming then the ordinary form of an arc.

The heating effect produced by molecular bombardment of the dielectric would, of course, diminish as the pressure of tile air is increased, and at enormous pressure it would be negligible, unless the frequency would increase correspondingly.

It will be often observed in these experiments that when the spheres are beyond the striking distance, the approach of a glass plate, for instance, may induce the spark to jump between the spheres. This occurs when the capacity of the spheres is somewhat below the critical value which gives the greatest difference of potential at the terminals of the coil. By approaching a dielectric, the specific inductive capacity of the space between the spheres is increased, producing the same effect as if the capacity of the spheres were increased. The potential at. the terminals may then rise so high that the air space is cracked. The experiment is best performed with dense glass or mica.

Another interesting observation is that a plate of insulating material, when the discharge is passing through it, is strongly attracted by either of the spheres, that is by the nearer one, this being obviously due to the smaller mechanical effect of the bombardment on that side, and perhaps also to the greater electrification.

From the behavior of the dielectrics in these experiments; we may conclude that the best insulator for these rapidly alternating currents would be the one possessing the smallest specific inductive capacity and at the same time one capable of withstanding the greatest differences of potential; and thus two diametrically opposite ways of securing the required insulation are indicated, namely, to use either a. perfect vacuum or a gas under great pressure; but the former would be preferable. Unfortunately neither of these two ways is easily carried out in practice.

It is especially interesting to note the behavior of an excessively high vacuum in these experiments. If a test tube, provided with external electrodes and exhausted to the highest possible degree, be convected to the terminals of the coil, Fig. 10, the electrodes of the tube are instantly brought to a high temperature and the glass at each end of the tube is rendered intensely phosphorescent, but the middle appears comparatively dark, and for a while remains cool.

When the frequency is so high that the discharge shown in Fig. 8 is, observed, considerable dissipation no doubt occurs in the coil. Nevertheless the coil may be worked for a long time, as the heating is gradual.

In spite of the fact that the difference of potential may be enormous, little is felt when the discharge is passed through the body, provided the hands are armed. This is to some extent due to the higher frequency, but principally to the fact that less energy is available externally, when the difference of potential reaches an enormous value, owing to the circumstance that, with the rise of potential, the energy absorbed in the coil increases as the square of the potential. Up to a certain point the energy available externally increases with the rise of potential, then it begins to fall off rapidly. Thus, with the ordinary high tension induction coil, the curious paradox exists, that, while with a given current through the primary the shock might be fatal, with many times that current it might be perfectly harmless, even if the frequency be the same. With high frequencies and excessively high potentials when the terminals are not connected to bodies of some size, practically all the energy supplied to the primary is taken up by the coil. There is no breaking through, no local injury, but all the material, insulating and conducting, is uniformly heated.

Fig. 10.—Discharge Through Highest Vacuum.

Fig. 11.—Pinwheel driven by a Powerful Brush.

To avoid misunderstanding in regard to the physiological effect of alternating currents of very high frequency, I think it necessary to state that, while it is an undeniable fact that they are incomparably less dangerous than currents of low frequencies; it should not be thought that they are altogether harmless. What has just been said refers only to currents from an ordinary high tension induction coil, which currents are necessarily very small; if received directly from a machine or from a secondary of low resistance, they produce more or less powerful effects, and may cause serious injury, especially when used in conjunction with condensers.

FIG. 12.—Luminous Streams Escaping From a Cotton-Covered Wire.

FIG. 13.—Aspect Presented by a Very Thin Wire Attached to a Terminal of the Coil.

The streaming discharge of a high tension induction coil differs in many respects from that of a powerful static machine. In color it has neither the violet of the positive, nor the brightness of the negative, static discharge, but lies somewhere between, being, of course, alternatively positive and negative. But since the streaming is more powerful when the point or terminal is electrified positively, than when electrified negatively, it follows that the point of the brush is more like the positive, and the root more like the negative, static discharge. In the dark, when the brush is very powerful, the root may appear almost white. The wind produced by the escaping streams, though it may be very strong — often indeed to such a degree that it may be felt quite a distance from the coil — is, nevertheless, considering the quantity of the discharge, smaller than that produced by the positive brush of a static machine, and it affects the flame much less powerfully: From the nature of the phenomenon we can conclude that the higher the frequency, the smaller must, of course, be the wind produced by the streams, and with sufficiently high frequencies no wind at all would be produced at the ordinary atmospheric pressures. With frequencies obtainable by means of a machine, the mechanical effect is sufficiently great to revolve, with considerable speed, large pin-wheels, which in the dark present beautiful appearance owing to the abundance of the streams (Fig. 11).

In general, most of the experiments usually performed with a static machine can be performed with an induction coil when operated with very rapidly alternating currents. The effects produced, however, are much more striking; being of incomparably greater power. When a small length of ordinary cotton covered wire, Fig. 11, is attached to one terminal of the coil, the streams issuing from all points of the wire may be so intense as to produce a considerable light effect. When the potentials and frequencies are very high, a wire insulated with gutta percha or rubber and attached to one of the terminals, appears to be covered with a luminous film A very thin bare wire when attached to a terminal emits powerful streams and vibrates continually to and fro or spins in a circle, producing a singular effect (Fig. 13). Some of these experiments have been described by me in *The Electrical World,* of February 21, 1891.

Another peculiarity of the rapidly alternating discharge of the induction coil is its radically different behavior with respect to points and rounded surfaces.

If a thick wire, provided with a ball at one end and with a point at the other, be attached to the positive terminal of a static machine, practically all the charge will be lost through the point, on account of the enormously greater tension, dependent on the radius of curvature. But if such a wire is attached to one of the terminals of the induction coil, it, will be observed that with very high frequencies streams issue from the ball almost as copiously as from the point (Fig. 14).

FIG. 14.—Effect of Ball and Point.   FIG. 15.— Aspect of Coil under Powerful Brush Discharge.

It is hardly conceivable that we could produce such a condition to an equal degree in a static machine, for the simple reason, that the tension increases as the square of the density, which in turn is proportional to the radius of curvature; hence, with a steady potential an enormous charge would be required to make streams issue from a polished ball while it is connected with a point. But with. an induction coil the discharge of which alternates with great rapidity it is different: Here we have to deal with two distinct tendencies. First, there is the tendency to escape which exists in a condition of rest, and which depends on the radius of curvature; second, there is the tendency to dissipate into the surrounding air by condenser action, which depends on the surface. When one of these tendencies is at a maximum, the other is at a minimum. At the point the luminous stream is principally due to the air molecules coming bodily in contact with the point; they are attracted and repelled, charged and discharged, and, their atomic charges being thus disturbed; vibrate and emit light waves. At the ball, on the contrary, there is no doubt that the effect is to a great extent produced inductively, the air molecules not *necessarily* coming in contact with the ball, though they undoubtedly do so. To convince ourselves of this we only need to exalt the condenser action, for instance, by enveloping the ball, at some distance, by a better conductor than the surrounding medium, the conductor being, of course, insulated; or else by surrounding it with a better dielectric and approaching an insulated conductor; in both cases the streams will break forth more copiously. Also, the larger the ball with a given frequency, or the higher the frequency, the more will the ball have the advantage over the point. But, since a certain intensity of action is required to render the streams visible, it is obvious that in the experiment described the ball should not be taken too large.

In consequence of this two-fold tendency, it is possible to produce by means of points, effects identical to those produced by capacity. Thus, for instance, by attaching to one terminal of the coil a small length of soiled wire, presenting many points and offering great facility to escape, the potential of the coil may be raised to the same value as by attaching to the terminal a polished ball of a surface many times greater than that of the wire.

An interesting experiment, showing the effect of the points, may be performed in the following manner: Attach to one of the terminals of the coil a cotton covered wire about two feet in length, and adjust the conditions so that streams issue from the wire. In this experiment the primary coil should be preferably placed so that it extends only about half *way* into the secondary coil. Now touch the free terminal of the secondary with a conducting object held in the hand, or else connect it to an insulated body of some size. In this manner the potential on the wire may be enormously raised. The effect of this will be either to increase, or to diminish, the streams: If they increase, the wire is too short; if they diminish, it is too long. By adjusting the length of the wire, a point is found where the touching of the other terminal does not at all affect the streams. In this case the rise of potential is exactly counteracted by the drop through the coil. It will be observed that small lengths of wire produce considerable difference in the magnitude and luminosity of the streams. The primary coil is placed sidewise for two reasons: First, to increase the potential at the wire: and, second, to increase the drop through the coil. The sensitiveness is thus augmented.

There is still another and far more striking peculiarity of the brush discharge produced by very rapidly alternating currents. To observe this it is best to replace the usual terminals of the coil by two metal columns insulated with a good thickness of ebonite. It is also well to close all fissures and cracks with wax so that the brushes cannot form anywhere except at the tops of the columns. If the conditions are carefully adjusted — which, of course, must be left to the skill of the experimenter — so that the potential rises to an enormous value, one may produce two powerful brushes several inches long, nearly white at their roots, which in the dart: bear a striking resemblance two flames of a gas escaping under pressure (Fig. 15). But they do not only *resemble,* they *are* veritable flames, for they are hot. Certainly they are not as hot as a gas burner, *but they would be so if the frequency and the potential would be sufficiently high.* Produced with, say, twenty thousand alternations per second, the heat is easily perceptible even if the potential is not excessively high. The heat developed is, of course, due to the impact of the air molecules against the terminals and against each other. As, at the ordinary pressures, the mean free path is excessively small, it is possible that in spite of the enormous initial speed imparted to each molecule upon coming in contact with the terminal, its progress — by collision with other molecules — is retarded to such an extent, that it does not get away far from the terminal, but may strike the same many times in succession. The higher the frequency, the less the molecule is able to get away, and this the more so, as for a given effect the potential required is smaller; and a frequency is conceivable — perhaps even obtainable — at which practically the same molecules would strike the terminal. Under such conditions the exchange of the molecules would be very slow, and the heat produced at, and very near, the terminal would be excessive. But if the frequency would go on increasing constantly, the heat produced would begin to diminish for obvious reasons. In the positive brush of a static machine the exchange of the molecules is very rapid, the stream is constantly of one direction, and there are fewer collisions; hence the heating effect must be very small. Anything that impairs the facility of exchange tends to increase the local heat produced. Thus, if a bulb be held over the terminal of the coil so as to enclose the brush, the air contained in the bulb is very quickly brought to a high temperature. If a, glass tube be held over the brush so as to allow the drought to carry the brush upwards, scorching hot air escapes at the top of the tube. Anything held within the brush is, of course, rapidly heated, and the possibility of using such heating effects for some purpose or other suggests itself.

When contemplating this singular phenomenon of the hot brush, we cannot help being convinced that a similar process must take place in the ordinary flame, and it seems strange that after all these centuries past of familiarity with the flame, now, in this era of electric lighting and heating; we are finally led to recognize, that since time immemorial we have, after all, always had "electric light and: heat" at our disposal. It is also of no little interest to contemplate, that we have a possible way of producing — by other than chemical means — a veritable flame; which would give light and heat without any material being consumed, without any chemical process taking place, and to accomplish this, we only need to perfect methods of producing enormous frequencies and potentials. I have no doubt that if the potential could be made to alternate with sufficient rapidity and power, the brush formed at the end of a wire would lose its electrical characteristics and would become flame like. The flame must be due to electrostatic molecular action.

This phenomenon now explains in a manner which can hardly be doubted the frequent accidents occurring in storms. It is well known that objects are often set on fire without the lightning striking them. We shall presently see how this can happen. On a nail in a roof, for instance, or on a projection of any kind, more or less conducting, or rendered so by dampness, a powerful brush may appear. If the lightning strikes somewhere in .the neighborhood the enormous potential may be made to alternate or fluctuate perhaps many million times a second. The air molecules are violently attracted and repelled, and by their impact produce such a powerful heating effect that a fire is started. It is conceivable that a ship at sea may, in this manner, catch fire at many points at once. When we consider, that even with the comparatively low frequencies obtained from a dynamo machine, and with potentials of no more than one or two hundred thousand volts, the heating effects are considerable, we may imagine how much more powerful they must be with frequencies and potentials many times greater: and the above explanation seems, to say the least, very probable. Similar explanations may have been suggested, but I am not aware that, up to the present; the heating effects of a brush produced by a rapidly alternating potential have been experimentally demonstrated, at least not to such a remarkable degree.

By preventing completely the exchange of the air molecules, the local heating effect may be so exalted as to bring a body to incandescence. Thus, for instance, if a small button, or preferably a very thin wire or filament be enclosed in an unexhausted globe and connected with the terminal of the coil, it may be rendered incandescent. The phenomenon is made much more interesting by the rapid spinning round in a circle of the top of the filament, thus presenting the appearance of a luminous funnel, [Fig. 16], which widens when the potential is increased. When the potential is small the end of the filament may perform irregular motions, suddenly changing from one to the other, or it may describe an ellipse; but when the potential is very high it always spins in a circle; and so does generally a thin straight wire attached freely to the terminal of the coil. These motions are, of course, due to the impact of the molecules, and the irregularity. in the distribution of the potential, owing to the roughness and dissymmetry of the wire or filament. With a perfectly symmetrical and polished wire such motions would probably not occur. That the motion is not likely to be due to other causes is evident from the fact that it is not of a definite direction, and that in a very highly exhausted globe it ceases altogether. The possibility of bringing a body to incandescence in an exhausted globe, or even when not at all enclosed, would seem to afford a possible way of obtaining light effects, which, in perfecting methods of producing rapidly alternating potentials,. might be rendered available for useful purposes.

FIG. 16.—Incandescent Wire or Filament Spinning in an Unexhausted Globe.

In employing a commercial coil; the production of very powerful brush effects is attended with considerable difficulties, for when these high frequencies and enormous potentials are used, the best insulation is apt to give way. Usually the coil is insulated well enough to stand the strain from convolution to convolution, since two double silk covered paraffined wires will withstand a pressure of several thousand volts; the difficulty lies principally in preventing the breaking through from the secondary to the primary, which is ,greatly facilitated by the streams issuing from the latter. In. the coil, of course, the strain is greatest from section to section, but usually in a larger coil there are so many sections that the danger of a sudden giving *way* is not very great. No difficulty will generally be encountered in that direction, and besides, the liability of injuring the coil internally is very much reduced by the fact that the effect most likely to be produced is simply a gradual heating, which, when far enough advanced, could not fail to be observed. The principal necessity is then to prevent the streams between he primary and the tube, not only on account of the heating and possible injury, but also because the streams may diminish very considerably the potential difference available at the terminals. A few hints as to how this may be accomplished will probably be found useful in most of these experiments with the ordinary induction coil.

FIG. 17B.—Coil Arranged for Powerful Brush Effects.— St. Elmo's Hot Fire.

One of the ways is to wind a short primary, Fig. 17a, so that the difference of potential is not at that length great enough to cause the breaking forth of the streams through the insulating tube. The length of the primary should be determined by experiment. Both the ends of the coil should be brought out on one end through a plug of insulating material fitting in the tube as illustrated. In such a disposition one terminal of the secondary is attached to a body, the surface of which is determined with the greatest care so as to produce the greatest rise in the potential. At the other terminal a powerful brush appears, which may be experimented upon.

FIG. 17A.—Coil Arranged for Powerful Brush Effects.

The above plan necessitates the employment of a primary of comparatively small size, and it is apt to heat when powerful effects are desirable for a certain length of time. In such a case it is better to employ a larger coil [Fig. 17b], and introduce it from one side of the tube, until the streams begin to appear. In this case the nearest terminal of the secondary may be connected to the primary or to the ground, which is practically the same thing, if the primary is connected directly to the machine. In the case of ground connections it is well to determine experimentally the frequency which is best suited under the conditions of the test. Another way of obviating the streams, more or less, is to make the primary in sections and supply it from separate, well insulated sources.

In many of these experiments, when powerful effects are wanted for a short time, it is advantageous to use iron cores with the primaries. In such case a very large primary coil may be wound and placed side by side with the secondary, and, the nearest terminal of the latter being connected to the primary, a laminated iron core is introduced through the primary into the secondary as far as the streams will permit. Under these conditions an excessively powerful brush, several inches long, which may be appropriately called "St. Elmo's hot fire," may be caused to appear at the other terminal of the secondary, producing striking effects. It is a most powerful ozonizer, so powerful indeed, that only a few minutes are sufficient to fill the whole room with the smell of ozone, and it undoubtedly possesses the quality of exciting chemical affinities.

For the production of ozone, alternating currents of very high frequency are eminently suited, not only on account of the advantages they offer in the way of conversion but also because of the fact, that the ozonizing action of a discharge is dependent on the frequency as well as on the potential, this being undoubtedly confirmed by observation.

In these experiments if an iron core is used it should be carefully watched, as it is apt to get excessively hot in an incredibly short time. To give an idea of the rapidity of the heating, I will state, that by passing a powerful current through a coil with many turns, the inserting within the same of a thin iron wire for no more than one seconds time is sufficient to heat the wire to something like 100° C.

But this rapid heating need not discourage us in the use of iron cores in connection with rapidly alternating currents. I have for a long time been convinced that in tile industrial distribution by means of transformers, some such plan as the following might be practicable. We may use a comparatively small iron core, subdivided, or perhaps not even subdivided. We may surround this core with a considerable thickness of material which is fire-proof and conducts the heat poorly, and on top of that we may place the primary and secondary windings. By using either higher frequencies or greater magnetizing forces, we may by hysteresis and eddy currents heat the iron core so far as to bring it nearly to its maximum permeability, which, as Hopkinson has shown, may be as much as sixteen times greater than that at ordinary temperatures. If the iron core were perfectly enclosed, it would not be deteriorated by the heat, and, if the enclosure of fire-proof material would be sufficiently thick, only a limited amount of energy could be radiated in spite of the high temperature. Transformers have been constructed by me on that plan, but for lack of time, no thorough tests have as yet been made.

Another way of adapting the iron core to rapid alternations, or, generally speaking, reducing the frictional losses, is to produce by continuous magnetization a flow of something like seven thousand or eight thousand lines per square centimeter through the core, and then work with weak magnetizing forces and preferably high frequencies around the point of greatest permeability. A higher efficiency of conversion and greater output are obtainable in this manner. I have also employed this principle in connection .with machines in which there is no reversal of polarity. In these types of machines, as long as there are only few pole projections, there is no great gain; as the maxima and minima of magnetization are far from the point of maximum permeability; but when the number of the pole projections is very great, the required rate of change may be obtained, without the magnetization varying so far as to depart greatly from the point of maximum permeability, and the gain is considerable.

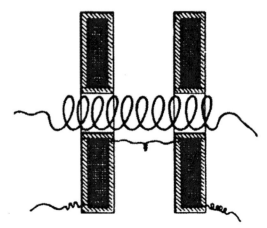

FIG. 18.—Coil for Producing Very High Difference of Potential.

The above described arrangements refer only to the use of commercial coils as ordinarily constructed. If it is desired to construct a coil for the express purpose of performing with it such experiments as I have described, or, generally, rendering it capable of withstanding the greatest possible difference of potential, then a construction as indicated in Fig. 18 will be found of advantage. The coil in this case is formed of two independent parts which are wound oppositely, the connection between both being made near the primary. The potential in the middle being zero, there is not much tendency to jump to the primary and not much insulation is required. In some cases the middle point may, however, be connected to the primary or to the ground. In such a coil the places of greatest difference of potential are far apart and the coil is capable of withstanding an enormous strain. The two parts may be movable so as to allow a slight adjustment of the capacity effect.

As to the manner of insulating the coil, it will be found convenient to proceed in the following way: First, the wire should be boiled in paraffine until all the air is out; then the coil is wound by running the wire through melted paraffine, merely for the purpose of fixing the wire. The coil is then taken off from the spool, immersed in a cylindrical vessel filled with pure melted wax and boiled for a long time until the bubbles cease to appear. The whole is then left to cool down thoroughly, and then the mass is taken out of the vessel and turned up in a lathe. A coil made in this manner arid with care is capable of withstanding enormous potential differences.

It may be found convenient to immerse the coil in paraffine oil or some other hind of oil; it is a most effective way of insulating, principally on account of the perfect exclusion of air, but it may be found that, after all, a vessel filled with oil is not a very convenient thing to handle in a laboratory.

If an ordinary coil can be dismounted, the primary may be taken out of the tube and the latter plugged up at one end, filled with oil, and the primary reinserted. This affords an excellent insulation and prevents the formation of the streams.

Of all the experiments which may be performed with rapidly alternating currents the most interesting are those which concern the production of a practical illuminant. It cannot be denied that the present methods, though they were brilliant advances, are very wasteful. Some better methods must be invented, some more perfect apparatus devised. Modern research has opened new possibilities for the production of an efficient source of light, and the attention of all has been turned in the direction indicated by able pioneers. Many have been carried away by the enthusiasm and passion to discover, but in their zeal to reach results, some have been misled. Starting with the idea of producing electro-magnetic waves, they turned their attention, perhaps, too much to the study of electro-magnetic effects, and neglected the study of electrostatic phenomena. Naturally, nearly every investigator availed himself of an apparatus similar to that used in earlier experiments. But in those forms of apparatus, while the electro-magnetic inductive effects are enormous, the electrostatic effects are excessively small.

In the Hertz experiments, for instance, a high tension induction coil is short circuited by an arc, the resistance of which is very small, the smaller, the more capacity is attached to the terminals; and the difference of potential at these is enormously diminished: On the other hand, when the discharge is not passing between the terminals, the static effects may be considerable, but only qualitatively so, not quantitatively, since their rise and fall is very sudden, and since their frequency is small. In neither case, therefore, are powerful electrostatic effects perceivable. Similar conditions exist when, as in some interesting experiments of Dr. Lodge, Leyden jars are discharged disruptively. It has been thought — and I believe asserted — that in such cases most of the energy is radiated into space. In the light of the experiments which I have described above, it will now not be thought so. I feel safe in asserting that in such cases most of the energy is partly taken up and converted into heat. in the arc of the discharge and in the conducting and insulating material of the jar, some energy being, of course, given off by electrification of the air; but the amount of the directly radiated energy is very small.

When a high tension induction coil, operated by currents alternating only 20,000 times a second, has its terminals closed through even a very small jar, practically all the energy passes through the dielectric of the jar, which is heated, and the electrostatic effects manifest themselves outwardly only to a very weak degree. Now the external circuit of a Leyden jar, that is, the arc and the connections of the coatings, may be looked upon as a circuit generating alternating currents of excessively high frequency and fairly high potential, which is closed through the coatings and the dielectric between them, and from the above it is evident that the external electrostatic effects must be very small, even if a recoil circuit be used. These conditions make it appear that with the apparatus usually at hand, the observation of powerful electrostatic effects was impossible, and what experience has been gained in that direction is only due to the great ability of the investigators.

But powerful electrostatic effects are *a sine qua non* of light production on the lines indicated by theory. Electro-magnetic effects are primarily unavailable, for the reason that to produce the required effects we would have to pass current impulses through a conductor; which, long before the required frequency of the impulses could be reached, would cease to transmit them. On the other hand, electro-magnetic waves many times longer than those of light, and producible by sudden discharge of a condenser, could not be utilized, it would seem, except we avail ourselves of their effect upon conductors as in the present methods, which are wasteful. We could not affect by means of such waves the static molecular or atomic charges of a gas, cause them to vibrate and to emit light. Long transverse waves cannot, apparently, produce such effects, since excessively small electro-magnetic disturbances may pass readily through miles of air. Such dark waves, unless they are of the length of true light waves, cannot, it would seem, excite luminous radiation in a Geissler tube; and the luminous effects, which are producible by induction in a tube devoid of electrodes, I am inclined to consider as being of an electrostatic nature.

To produce such luminous effects, straight electrostatic thrusts are required; these, whatever be their frequency, may disturb the molecular charges and produce light. Since current impulses of the required frequency cannot pass through a conductor of measurable dimensions, we must work with a gas, and then the production of powerful electrostatic effects becomes an imperative necessity.

It has occurred to me, however, that electrostatic effects are in many ways available for the production of light. For instance, we may place a body of some refractory material in a closed; and preferably more or less exhausted, globe, connect it to a source of high, rapidly alternating potential, causing the molecules of the gas to strike it many times a second at enormous speeds, and in this manner, with trillions of invisible hammers, pound it until it, gets incandescent: or we may place a body in a very highly exhausted globe, in a non-striking vacuum, and, by employing very high frequencies and potentials, transfer sufficient energy from it to other bodies in the vicinity, or in general to the surroundings, to maintain it at any degree of incandescence; or we may, by means of such rapidly alternating high potentials, disturb the ether carried by the molecules of a gas or their static charges, causing them to vibrate and to emit light.

But, electrostatic effects being dependent upon the potential and frequency, to produce the most powerful action it is desirable to increase both as far as practicable. It may be possible to obtain quite fair results by keeping either of these factors small, provided the other is sufficiently great; but we are limited in both directions. My experience demonstrates that we cannot go below a certain frequency, for, first, the potential then becomes so great that it is dangerous; and, secondly, the light production is less efficient.

I have found that, by using the ordinary low frequencies, the physiological effect of the current required to maintain at a certain degree of brightness a tube four feet long, provided at the ends with outside and inside condenser coatings, is so powerful that, I think, it might produce serious injury to those not accustomed to such shocks: whereas, with twenty thousand alternations per second, the tube may be maintained at the same degree of brightness without any effect being felt. This is due principally to the fact that a much smaller potential is required to produce the same light effect, and also to the higher efficiency in the light production. It is evident that the efficiency in such cases is the greater, the higher the frequency, for the quicker the process of charging and discharging the molecules, the less energy will be lost in the form of dark radiation. But, unfortunately, we cannot go beyond a certain frequency on account of the difficulty of producing and conveying the effects.

I have stated above that a body inclosed in an unexhausted bulb may be intensely heated by simply connecting it with a source of rapidly alternating potential. The heating in such a case is, in all probability, due mostly to the bombardment of the molecules of the gas contained in the bulb. When the bulb is exhausted, the heating of the body is much more rapid, and there is no difficulty whatever in bringing a wire or filament to any degree of incandescence by simply connecting it to one terminal of a coil of the proper dimensions. Thus, if the well-known apparatus of Prof. Crookes, consisting of a bent platinum wire with vanes mounted over it (Fig. 18), be connected to one terminal of the coil — either one or both ends of the platinum wire being connected — the wire is rendered almost instantly incandescent, and the mica vanes are rotated as though a current from a battery were used: A thin carbon filament, or, preferably, a button of some refractory material [Fig. 20], even if it be a comparatively poor conductor, inclosed in an exhausted globe, may be rendered highly incandescent; and in this manner a simple lamp capable of giving any desired candle power is provided.

Fig. 19.—The Crookes Experiment on Open Circuit.

Fig. 20.—Lamp with Single Block of Refractory Material.

The success of lamps of this kind would depend largely on the selection of the light-giving bodies contained within the bulb. Since, under the conditions described, refractory bodies — which are very poor conductors and capable of withstanding for a long time excessively high degrees of temperature — may be used, such illuminating devices may be rendered successful.

It might be thought at first that if the bulb, containing the filament or button of refractory material, be perfectly well exhausted — that is, as far as it can be done by the use of the best apparatus — the heating would be much less intense, and that in a perfect vacuum it could not occur at all. This is not confirmed *by my* experience; quite the contrary, the better the vacuum the more easily the bodies are brought to incandescence. This result is interesting for many reasons.

At the outset of this work the idea presented itself to me, whether two bodies of refractory material enclosed in a bulb exhausted to such a degree that the discharge of a large induction coil, operated in the usual manner, cannot pass through, could be rendered incandescent by mere condenser action. Obviously, to reach this result enormous potential differences and very high frequencies are required, as is evident from a simple calculation.

But such a lamp would possess a vast advantage over an ordinary incandescent lamp in regard to efficiency. It is well-known that the efficiency of a lamp is to some extent a function of the degree of incandescence, and that, could we but work a filament at many times higher degrees of incandescence, the efficiency would be much greater. In an ordinary lamp this is impracticable on account of the destruction of the filament, and it has been determined by experience how far it is advisable to push the incandescence. It is impossible to tell how much higher efficiency could be obtained if the filament could withstand indefinitely, as the investigation to this end obviously cannot be carried beyond a certain stage; but there are reasons for believing that it would be very considerably higher. An improvement might be made in the ordinary lamp by employing a short and thick carbon; but then the leading-in wires would have to be thick, and, besides, there ace many other considerations which render such a modification entirely impracticable. But in a lamp as above described, the leading-in wires may be very small, the incandescent refractory material may be in the shape of blocks offering a very small radiating surface, so that less energy would be required to keep them at the desired incandescence; and in addition to this, the refractory material need not be carbon, but may be manufactured from mixtures of oxides, for instance, with carbon or other material, or may be selected from bodies which are practically non-conductors, and capable of withstanding enormous degrees of temperature.

All this would point to the possibility of obtaining a much higher efficiency with such a lamp than is obtainable in ordinary lamps. In my experience it has been demonstrated that the blocks are brought to high degrees of incandescence with much lower potentials than those determined by calculation, and the blocks may be set at greater distances from each other. We may freely assume, and it is probable, that the molecular bombardment is an important element in the heating, even if the globe be exhausted with the utmost care, as I have done; for although the number of the molecules is, comparatively speaking, insignificant, yet on account of the mean free path being very great, there are fewer collisions, and the molecules may reach much higher speeds, so that the heating effect due to this cause may be considerable, as in the Crookes experiments with radiant matter.

But it is likewise possible that we have to deal here with an increased facility of losing the charge in very high vacuum, when the potential is rapidly alternating, in which case most of the heating would be directly due to the surging of the charges in the heated bodies. Or else the observed fact may be largely attributable to the effect of the points which I have mentioned above, in consequence of which the blocks or filaments contained in the vacuum are equivalent to condensers of many times greater surface than that calculated from their geometrical dimensions. Scientific men still differ in opinion as to whether a charge should, or should not, be lost in a perfect vacuum, or. in other words, whether ether is, or is not, a conductor. If the former were the case, then a thin filament enclosed in a perfectly exhausted globe, and connected to a source of enormous, steady potential, would be brought to incandescence.

FIG. 21.—Lamp with Two Filaments in Highest Vacuum with Leading-in Wires.

Various forms of lamps on the above described principle, with the refractory bodies in the form of filaments, Fig. 21, or blocks, Fig. 22, have been constructed and operated by me, and investigations are being carried on in this line. There is no difficulty in reaching such high degrees of incandescence that ordinary carbon is to all appearance melted and volatilized. If the vacuum could be made absolutely perfect, such a lamp, although inoperative with apparatus ordinarily used, would, if operated with currents of the required character, afford an illuminant which would never be destroyed, and which would be far more efficient than an ordinary incandescent lamp. This perfection can, of course, never be reached; and a very slow destruction and gradual diminution in size always occurs, as in incandescent filaments; but there is no possibility of a sudden and premature disabling which occurs in the latter by the breaking of the filament, especially when the incandescent bodies are in the shape of blocks.

With these rapidly alternating potentials there is, however, no necessity of enclosing two blocks in a globe, but a single block, as in Fig. 20, or filament, Fig. 23, may be used. The potential in this case must of course be higher, but is easily obtainable, and besides it is not necessarily dangerous.

FIG. 23.—Lamp with Two Refractory Blocks in Highest Vacuum.

FIG. 23.—Lamp with Single Straight Filament and one Leading-in Wire.

The facility with which the button or filament in such a lamp is brought to incandescence, other things being equal, depends on the size of the globe. If a perfect ,vacuum could be obtained, the size of the globe would not be of importance, for then the heating would be wholly due to the surging of the charges, and all the energy would be given off to the surroundings by radiation. But this can never occur in practice. There is always some gas left in the globe, and although the exhaustion may be carried to the highest degree, still the space inside of the bulb must be considered as conducting when such high potentials are used, and I assume that, in estimating the energy that may be given off from the filament to the surroundings, we may consider the inside surface of the bulb as one coating of a condenser, the air and other objects surrounding the bulb forming the other coating. When the alternations are very low there is no doubt that a considerable portion of the energy is given off by the electrification of the surrounding air.

In order to study this subject better, I carried on some experiments with excessively high potentials and low frequencies. I then observed that when the hand is approached to the bulb, — the filament being connected with one terminal of the coil, — a powerful vibration is felt, being due to the attraction and repulsion of the molecules of the air which are electrified by induction through the glass. In some cases when the action is very intense I have been able to hear a sound, which must be due to the same cause.

FIG. 24.—Lamps wtih One Leading-in Wire Rendered Incandescent.

FIG. 25.—Lamp with two Blocks or Filaments and a Pair of Independent Inside and Outside Condenser Coatings.

When the alternations are low, one is apt to get an excessively powerful shock from the bulb. In general, when one attaches bulbs or objects of some size to the terminals of the coil, one should look out for the rise of potential, for it may happen that by merely connecting a bulb or plate to the terminal, the potential may rise to many times its original value. When lamps are attached to the terminals, as illustrated in Fig. 24, then the capacity of the bulbs should be such as to give the maximum rise of potential under the existing conditions. In this manner one may obtain the required potential with fewer turns of wire.

The life of such lamps as described above depends, of course, largely on the degree of exhaustion, but to some extent also on the shape of the block of refractory material. Theoretically it would seem that a small sphere of carbon enclosed in a sphere of glass would not suffer deterioration from molecular bombardment, for, the matter in the globe being radiant, the molecules would move in straight lines, and would seldom strike the sphere obliquely. An interesting thought in connection with such a lamp is, that in it "electricity" and electrical energy apparently must move in the same lines.

The use of alternating currents of very high frequency makes it possible to transfer, by electrostatic or electromagnetic induction through the glass of a lamp, sufficient energy to keep a filament at incandescence and so do away with the leading-in wires. Such lamps have been proposed, but for want of proper apparatus they have not been successfully operated. Many forms of lamps on this principle with continuous and broken filaments have been constructed by me and experimented upon. When using a secondary enclosed within the lamp, a condenser is advantageously combined with the secondary. When the transference is effected by electrostatic induction, the potentials used are, of course, very high with frequencies obtainable from a machine. For instance, with a condenser surface of forty square centimeters, which is not

Fig. 26a.   Fig. 26b.
Lamp with One Filament, One Inside and One Outside Condenser Coating.

impracticably large, and with glass of good quality I mm. thick, using currents alternating twenty thousand times a second, the potential required is approximately 9,000 volts. This may seem large, but since each lamp may be included in the secondary of a transformer of very small dimensions, it would not be inconvenient, and, moreover, it would not produce fatal injury. The transformers would all be preferably in series. The regulation would offer no difficulties, as with currents of such frequencies it is very easy to maintain a constant current.

In the accompanying engravings some of the types of lamps of this kind are shown. Fig. 25 is such a lamp with a broken filament, and Figs. 26a and 26b one with a single outside and inside coating and a single filament. I have also made lamps with two outside and inside coatings and a continuous loop connecting the latter. Such lamps have been operated by me with current impulses of the enormous frequencies obtainable by the disruptive discharge of condensers.

Fig. 27.—Lamp with One Filament and Leading-in Wire and External Condenser Coating.

Fig. 28.—Lamp with One Filament, One Inside and One Outside Condenser Coating, and Auxiliary Coating.

The disruptive discharge of a condenser is especially suited for operating such lamps — with no outward electrical connections — by means of electromagnetic induction, the electromagnetic inductive effects being excessively high; and I have been able to produce the desired incandescence with only a few short turns of wire. Incandescence may also be produced in this manner in a simple closed filament.

Leaving now out of consideration the practicability of such lamps, I would only say that they possess a beautiful and desirable feature, namely, that they can be rendered, at will, more or less brilliant simply by altering the relative position of the outside and inside condenser coatings, or inducing and induced circuits.

When a lamp is lighted by connecting it to one terminal only of the source, this may be facilitated by providing the globe with an outside condenser coating, which serves at the same time as a reflector, and connecting this to an insulated body of some size. Lamps of this kind are illustrated in Fig. 27 and Fig. 28. Fig. 29 shows the plan of connection. The brilliancy of the lamp may, in this case, be regulated within wide limits by varying the size of the insulated metal plate to which the coating is connected.

It is likewise practicable to light with one leading wire lamps such as illustrated in Fig. 21 and Fig. 22, by connecting one terminal of the lamp to one terminal of the source, and the other to an insulated body of the required size. In all cases the insulated body serves to give off the energy into the surrounding space, and is equivalent to a return wire. Obviously, in the two last-named cases, instead of connecting the wires to an insulated body, connections may be made to the ground.

FIG. 29.—Increasing the Brilliancy of Lamp on One Wire.

The experiments which will prove most suggestive and of most interest to the investigator are probably those performed with exhausted tubes. As might be anticipated, a source of such rapidly alternating potentials is capable of exciting the tubes at a considerable distance, and the light effects produced are remarkable.

During my investigations in this line I endeavored to excite tubes, devoid of any electrodes, by electromagnetic induction, snaking the tube the secondary of the induction device, and passing through the primary the discharges of a Leyden jar. These tubes were made of many shapes, and I was able to obtain luminous effects which I then thought were due wholly to electromagnetic induction. But on carefully investigating the phenomena I found that the effects produced were more of an electrostatic nature. It may be attributed to this circumstance that this mode of exciting tubes is very wasteful, namely, the primary circuit being closed, the potential, and consequently the electrostatic inductive effect, is much diminished.

When an induction coil, operated as above described, is used, there is no doubt that the tubes are excited by electrostatic induction, and that electromagnetic induction has little, if anything, to do with the phenomena.

This is evident from many experiments. For instance, if a tube be taken in one hand, the observer being near the coil, it is brilliantly lighted and remains so no matter in what position it is held relatively to the observer's body. Were the action electromagnetic, the tube could not be lighted when the observer's body is interposed between it and the coil, or at least its luminosity should be considerably diminished. When the tube is held exactly over the center of the coil — the latter being wound in sections and the primary placed symmetrically to the secondary — it may remain completely dark, whereas it is rendered intensely luminous by moving it slightly to the right or left from the center of the coil. It does not light because in the middle both halves of the coil neutralize each other, and the electric potential is zero. If the action were electromagnetic, the tube should light best in the plane through the center of the toil, since the electromagnetic effect there should be a maximum. When an arc is established between the terminals, the tubes and lamps in the vicinity of the coil go out, out light up again when the arc is broken, on account of the rise of potential. Yet the electromagnetic effect should be practically the same in both cases.

By placing a tube at some distance from the coil, and nearer to one terminal — preferably at a point on the axis of the coil — one may light it by touching the remote terminal with an insulated body of some size or with the hand, thereby raising the potential at that terminal nearer to, the tube. If the tube is shifted nearer to the coil so that it is lighted by the action of the nearer terminal, it may be made to go out by holding, on an insulated support, the end of a wire connected to the remote terminal, in the vicinity of the nearer terminal, by this means counteracting the action of the latter upon the tube. These effects are evidently electrostatic. Likewise, when a tube is placed it a considerable distance from the coil, the observer may, standing upon an insulated support between coil and tube, light the latter by approaching the hand to it; or he may even render it luminous by simply stepping between it and the coil. This would be impossible with electro-magnetic induction, for the body of the observer would act as a screen.

When the coil is energized by excessively weak currents, the experimenter may, by touching one terminal of the coil with the tube, extinguish the latter, and may again light it by bringing it out of contact with the terminal and allowing a small arc to form. This is clearly due to the respective lowering and raising of the potential at that terminal. In the above experiment, when the tube is lighted through a small arc, it may go out when the arc is broken, because the electrostatic inductive effect alone is too weak, though the potential may be much higher; but when the arc is established, the electrification of the end of the tube is much greater, and it consequently lights.

If a tube is lighted by holding it near to the coil, and in the hand which is remote. by grasping the tube anywhere with the other hand, the part between the hands is rendered dark, and the singular effect of wiping out the light of the tube may he produced by passing the hand quickly along the tube and at the same time withdrawing it gently from the coil, judging properly tile distance so that the tube remains dark afterwards.

If the primary coil is placed sidewise, as in Fig. 16b for instance, and an exhausted tube be introduced from the other side in the hollow space, the tube is lighted most intensely because of the increased condenser action, and in this position the striae are most sharply defined. In all these experiments described, and in many others, the action is clearly electrostatic.

The effects of screening also indicate the electrostatic nature of the phenomena and show something of the nature of electrification through the air. For instance, if a tube is placed in the direction of the axis of the coil, and an insulated metal plate be interposed, the tube will generally increase in brilliancy, or if it be too far from the coil to light, it may even be rendered luminous by interposing an insulated metal plate. The magnitude of the effects depends to some extent on the size of the plate. But if the metal plate be connected by a wire to the ground, its interposition will always make the tube go put even if it be very near the coil. In general, the interposition of a body between the coil and tube, increases or diminishes the brilliancy of the tube, or its facility to light up, according to whether it increases or diminishes the electrification. When experimenting with an insulated plate, the plate should not be taken too large, else it will generally produce a weakening effect by reason of its great facility for giving off energy to the surroundings.

If a tube be lighted at some distance from the coil, and a plate of hard rubber or other insulating substance be interposed, the tube may be made to go .out. The interposition of the dielectric in this case *only* slightly increases the inductive effect, but diminishes considerably the electrification through the air.

In all cases, then, when we excite luminosity in exhausted tubes by means of such a coil, the effect is due to the rapidly alternating electrostatic' potential; and, furthermore, it must be attributed to the harmonic alternation produced directly by the machine, and not to any superimposed vibration which might be thought to exist. Such superimposed vibrations are impossible when we work with an alternate current machine. If a spring be gradually tightened and released, it does not perform independent vibrations; for this a sudden release is necessary. So with the alternate currents from a dynamo machine; the medium is harmonically strained and released, this giving rise to only one kind of waves; a sudden contact or break, or a sudden giving way of the dielectric, as in the disruptive discharge of a Leyden jar, are essential for the production of superimposed waves.

FIG. 30.—Ideal Method of Lighting a Room —Tubes Devoid of Any Electrodes Rendered Brilliant in an Alternating Electrostatic Field.

In all the last described experiments, tubes devoid of any electrodes may be used, and there is no difficulty in producing by their means sufficient light to read by. The light effect is, however, considerably increased by the use of phosphorescent bodies such as yttria, uranium glass, etc. A difficulty will be found when the phosphorescent material is used, for with these powerful effects, it is carried gradually away, and it is preferable to use material in the form of a solid.

Instead of depending on induction at a distance to light the tube, the same may be provided with an external — and, if desired, also with an internal — condenser coating, and it may then be suspended anywhere in the room from a conductor connected to one terminal of the coil, and in this manner a soft illumination may be provided.

The ideal way of lighting a hall or room would, however, be to produce such a condition in it that an illuminating device could be moved and put anywhere, and that it is lighted, no matter where it is put and without being electrically connected to anything. I have been able to produce such a condition by creating in the room a powerful, rapidly alternating electrostatic field. For this purpose I suspend a sheet of metal a distance from the ceiling on insulating cords and connect it to one terminal of the induction coil, the other terminal being preferably connected to the ground. Or else I suspend two sheets as illustrated in Fig. 30, each sheet being connected with on;. of the terminals of the coil, and their size being carefully determined. An exhausted tube may then be carried in the hand anywhere between the sheets or placed anywhere, even a certain distance beyond them; it remains always luminous.

In such an electrostatic field interesting phenomena may be observed, especially if the alternations are kept low and the potentials excessively high. In addition to the luminous phenomena mentioned, one may observe that any insulated conductor gives .,parks when the hand or another object is approached to it, and the sparks may often be powerful. When a large conducting object is fastened on an insulating support, and the hand approached to it, a vibration, due to the rhythmical motion of the air molecules is felt, and luminous streams may be perceived when the hand is held rear a pointed projection. When a telephone receiver is made to touch with one or both of its terminals art insulated conductor of some size, the telephone emits a loud sound; it also emits a sound when a length of wire is attached to one or both terminals, and with very powerful fields a sound may be perceived even without any wire.

How far this principle is capable of practical application, the future will tell. It might be thought that electrostatic effects are unsuited for such action at a distance. Electromagnetic inductive effects, if available for the production of light, might be thought better suited. It is true the electrostatic effects diminish nearly with the cube of the distance from the coil, whereas the electromagnetic inductive effects diminish simply with the distance. But when we establish an electrostatic field of force, the condition is very different, for then, instead of the differential effect of both the terminals, we get their conjoint effect. Besides, I would call attention to the effect that in an alternating electrostatic field, a conductor, such as an exhausted tube, for instance, tends to take up most of the energy, whereas in an electromagnetic alternating field the conductor tends to take up the least energy, the waves being reflected with but little loss. This is one reason why it is difficult to excite an exhausted tube, at a distance, by electromagnetic induction. I have wound coils of very large diameter and of many turns of wire, and connected a Geissler tube to the ends of the coil with the object of exciting the tube at a distance; but even with the powerful inductive effects producible by Leyden jar discharges, the tube could not be excited unless at a very small distance, although some judgment was used as to the dimensions of the coil. I have also found that even the most powerful Leyden jar discharges are capable of exciting only feeble luminous effects in a closed exhausted tube, and even these effects upon thorough examination I have been forced to consider of an electrostatic nature.

How then can we hope to produce the required effects at a distance by means of electromagnetic action, when even in the closest proximity to the source of disturbance, under the most advantageous conditions, we can excite but faint luminosity? It is true that when acting at a distance we have the resonance to help us out. We can connect an exhausted tube, or whatever the illuminating device may be, with an insulated system of the proper capacity, and so it may be possible to increase the effect qualitatively, and only qualitatively, for we would not get *snore* energy through the device. So we may, by resonance effect, obtain the required electromotive force in an exhausted tube, and excite faint luminous effects, but we cannot get enough energy to render the light practically available, and a simple calculation, based on experimental results, shows that even if all the energy which a tube would receive at a certain distance from the source should be wholly converted into light, it would hardly satisfy the practical requirements. Hence the necessity of directing, by means of a conducting circuit, the energy to the place of transformation. But in so doing we cannot very sensibly depart from present methods, and all we could do would be to improve the apparatus.

From these considerations it would seem that if this ideal way of lighting is to rendered practicable it will be only by the use of electrostatic effects. In such a case the most powerful electrostatic inductive effects are needed; the apparatus employed must, therefore, be capable of producing high electrostatic potentials changing in value with extreme rapidity. High frequencies are especially wanted, for practical considerations make it desirable to keep down the potential. By the employment of machines, or, generally speaking, of any mechanical apparatus, but low frequencies can be reached; recourse must, therefore, be had to some other means. The discharge of a condenser affords us a means of obtaining frequencies by far higher than are obtainable mechanically, and I have accordingly employed condensers in the experiments to the above end.

When the terminals of a high tension induction coil, [Fig. 31] are connected to a Leyden jar, and the latter is discharging disruptively into a circuit, we may look upon the arc playing between the knobs as being a source of alternating, or generally speaking, undulating currents, and then we have to deal with the familiar system of a generator of such currents, a circuit

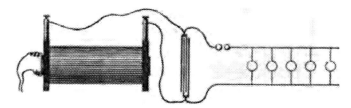

Fig. 31—Diagram of Connections for Converting from High to Low Tension by Means of the Disruptive Discharge.

connected to it, and a condenser bridging the circuit. The condenser in such case is a veritable transformer, and since the frequency is excessive, almost any ratio in the strength of the currents in both the branches may be obtained.. In reality the analogy is not quite complete, for in the disruptive discharge we have most generally a fundamental instantaneous variation of comparatively low frequency, and a superimposed harmonic vibration, and the laws governing the flow of currents are not the: same for both.

In converting in this manner, the ratio of conversion should not be too great, for the loss in the arc between the knobs increases with the square of the current, and if the jar be discharged through very thick and short conductors, with the view of obtaining a very rapid oscillation, a very considerable portion of the energy stored is lost. On the other hand, too small ratios are not practicable for many obvious reasons.

As the converted currents flow in a practically closed circuit, the electrostatic effects are necessarily small, and I therefore convert them into currents or effects of the required character. I have effected such conversions in several ways. The preferred plan of connections is illustrated in Fig. 32. The manner of operating renders it easy to obtain by means of a small and inexpensive apparatus enormous differences of potential which have been usually obtained by means of large and expensive coils. For this it is only necessary to take an ordinary small coil, adjust to it a condenser and discharging circuit, forming the primary of an auxiliary small coil, and convert upward. As the inductive effect of the primary currents is excessively great, the second coil need have comparatively but very few turns. By properly adjusting the elements, remarkable results may be secured.

In endeavoring to obtain the required electrostatic effects in this manner, I have, as might be expected, encountered many difficulties which I have been gradually overcoming, but I am not as yet prepared to dwell upon my experiences in this direction.

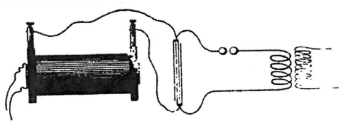

Fig. 32.—Manner of Operating an Induction Coil

I believe that the disruptive discharge of a condenser will play an important part in the future, for it offers vast possibilities, not only in the way of producing light in a more efficient manner and in the line indicated by theory, but also in many other respects.

For years the efforts of inventors have been directed towards obtaining electrical energy from heat by means of the thermopile. It might seem invidious to remark that but few know what is the real trouble with the thermopile. It is not the inefficiency or small output — though these are great drawbacks — but the fact that the thermopile has its phylloxera, that is, that by constant use it is deteriorated, which has thus far prevented its introduction on an industrial scale. Now that all modern research seems to point with certainty to the use of electricity of excessively high tension, the question must present itself to many whether it is not possible to obtain in a practicable manner this form of energy from heat. We have been used to look upon an electrostatic machine as a plaything, and somehow we couple with it the idea of the inefficient and impractical. But now we must think differently, for now we know that everywhere we have to deal with the same forces, and that it is a mere question of inventing proper methods or apparatus for rendering them available.

In the present systems of electrical distribution, the employment of the iron with its wonderful magnetic properties allows us to reduce considerably the size of the apparatus; but, in spite of this, it is still very cumbersome. The more we progress in the study of electric and magnetic phenomena, the more we become convinced that the present methods will be short-lived. For the production of light, at least, such heavy machinery would seem to be unnecessary. The energy required is very small, and if light can be obtained as efficiently as, theoretically, it appears possible, the apparatus need have but a very small output. There being a strong probability that the illuminating methods of the future will involve the use of very high potentials, it seems very desirable to perfect a contrivance capable of converting the energy of heat into energy of the requisite form. Nothing to speak of has been done towards this end, for the thought that electricity of some 50,000 or 100,000 volts pressure or more, even if obtained, would be unavailable for practical purposes, has deterred inventors from working in this direction.

In Fig. 31 a plan of connections is shown for converting currents of high, into currents of low, tension by means of the disruptive discharge of a condenser. This plan has been used by me frequently for operating a few incandescent lamps required in the laboratory. Some difficulties have been encountered in the arc of the discharge which I have been able to overcome to a great extent; besides this, and the adjustment necessary for the proper working, no other difficulties have been met with, and it was easy to operate ordinary lamps; and even motors, in this manner. The line being connected to the ground, all the wires could be handled with perfect impunity, no matter how high the potential at the terminals of the condenser. In these experiments a high tension induction coil, operated from a battery or from an alternate current machine, was employed to charge the condenser; but the induction coil might be replaced by an apparatus of a different kind, capable of giving electricity of such high tension. In this manner, direct or alternating currents may be converted, and in both cases the current-impulses may be of any desired frequency. When the currents charging the condenser are of the same direction, and it is desired that the converted currents should also be of one direction, the resistance of the discharging circuit should, of course, be so chosen that there are no oscillations.

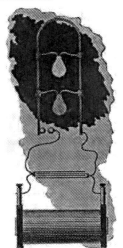

Fig. 33.—Lamp Kept at Incandescence across a Thick Copper Bar—Showing Nodes.

In operating devices on the above plan I have observed curious phenomena of impedance which are of interest. For instance if a thick copper bar be bent, as indicated in Fig. 33, and shunted by ordinary incandescent lamps, then, by passing the discharge between the knobs, the lamps may be brought to incandescence although they are short-circuited. When a large induction coil is employed it is easy to obtain nodes on the bar, which are rendered evident by the different degree of brilliancy of the lamps, as shown roughly in Fig. 33. The nodes are never clearly defined, but they are simply maxima and minima of potentials along the bar. This is probably due to the irregularity of the arc between the knobs. In general when the above-described plan of conversion from high to low tension is used, the behavior of the disruptive discharge may be closely studied. The nodes may also be investigated by means of an ordinary Cardew voltmeter which should be well insulated. Geissler tubes may also be lighted across the points of the bent bar; in this case, of course, it is better to employ smaller capacities. I have found it practicable to light up in this manner a lamp, and even a Geissler tube, shunted *by* a short, heavy block of metal, and this result seems at first very curious. In fact, the thicker the copper bar in Fig. 33; the better it is for the success of the experiments, as they appear more striking. When lamps with long slender filaments are used it will be often noted that the filaments are from time to time violently vibrated, the vibration being smallest at the nodal points. This vibration seems to be due to an electrostatic action between the filament and the glass of the bulb.

Fig. 34.—Phenomenon of Impedance in an Incandescent Lamp.

In some of the above experiments it is preferable to use special lamps having a straight filament as shown in Fig. 34. When such a lamp is used a still more curious phenomenon than those described may be observed. The lamp may be placed across the copper bar and lighted, and by using somewhat larger capacities, or, in other words, smaller frequencies or smaller impulsive impedances, the filament may be brought to any desired degree of incandescence. But when the impedance is increased, a point is reached when comparatively little current passes through the carbon, and most of it through the rarefied gas; or perhaps it may be more correct to state that the current divides nearly evenly through both, its spite of the enormous difference in the resistance, and this would be true unless the Las and the filament behave differently. It is then noted that the whole bulb is brilliantly illuminated, and the ends of the leading-in wires become incandescent and often throw off sparks in consequence of the violent bombardment, but the carbon filament remains dark. This is illustrated in Fig. 34. Instead of the filament a single wire extending through the whole bulb may be used, and in this case the phenomenon would seen to be still more interesting.

From the above experiment it will be evident, that when ordinary lamps are operated by the converted currents, those should be preferably taken in which the platinum wires are far apart, and the frequencies used should not be too great, else the discharge will occur at the ends of the filament or in the base of the lamp between the leading-in wires, and the lamp might then be damaged.

In presenting to you these results of my investigation on the subject under consideration, I have paid only a passing notice to facts upon which I could have dwelt at length, and among many observations I have selected only those which I thought most likely to interest you. The field is wide and completely unexplored, and at every step a new truth is gleaned, a novel fact observed.

How far the results here borne out are capable of practical applications will be decided in the future. As regards the production of light, some results already reached are encouraging and make me confident in asserting that the practical solution of the problem lies in the direction I have endeavored to indicate. Still, whatever may be the immediate outcome of these experiments I am hopeful that they will only prove a step in further developments towards the ideal and final perfection. The possibilities which are opened by modern research are so vast that even the most reserved must feel sanguine of the future. Eminent scientists consider the problem of utilizing one kind of radiation without the others a rational one. In an apparatus designed for the production of light by conversion from any form of energy into that of light, such a result can never be reached, for no matter what the process of producing the required vibrations, be it electrical, chemical or any other, it will not be possible to obtain the higher light vibrations without going through the lower heat vibrations. It is the problem of imparting to a body a certain velocity without passing through all lower velocities. But there is a possibility of obtaining energy not only in the form of light, but motive power, and energy of any other form, in some more direct way from the medium. The time will be when this will be accomplished, and the time has come when one may utter such words before an enlightened audience without being considered a visionary. We are whirling through endless space with an inconceivable speed, all around us everything is spinning, everything is moving, everywhere is energy. There *mart* be some way of availing ourselves of this energy more directly. Then; with the light obtained from the medium, with the power derived from it, with every form of energy obtained without effort, from the store forever inexhaustible, humanity will advance with giant strides. The mere contemplation of these magnificent possibilities expands our minds, strengthen our hopes and fills our hearts with supreme delight.

Electric Discharge in Vacuum Tubes
The Electrical Engineer - N.Y. — July 1, 1891

In *The Electrical Engineer* of June 10 I have noted the description of some experiments of Prof. J. J. Thomson, on the "Electric Discharge in Vacuum Tubes," and in your issue of June 24 Prof. Elihu Thomson describes an experiment of the same kind. The fundamental idea in these experiments is to set up an electromotive force in a vacuum tube — preferably devoid of any electrodes — by means of electromagnetic induction, and to excite the tube in this manner.

As I view the subject I should think that to any experimenter who had carefully studied the problem confronting us and who attempted to find a solution of it, this idea must present itself as naturally as, for instance, the idea of replacing the tinfoil coatings of a Leyden jar by rarefied gas and exciting luminosity in the condenser thus obtained by repeatedly charging and discharging it. The idea being obvious, whatever merit there is in this line of investigation must depend upon the completeness of the study of the subject and the correctness of the observations. The following lines are not penned with any desire on my part to put myself on record as one who has performed `similar experiments, but with a desire to assist other experimenters by pointing out certain peculiarities of the phenomena observed, which, to all appearances, have not been noted by Prof. J. J. Thomson, who, however, seems to have gone about systematically in his investigations, and who has been the first to make his results known. These peculiarities noted by me would seem to be at variance with the views of Prof. J. J. Thomson, and present the phenomena in a different light.

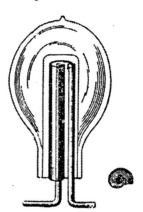

My investigations in this line occupied me principally during the winter and spring of the past year. During this time many different experiments were performed, and in my exchanges of ideas on this subject with Mr. Alfred S. Brown, of the Western Union Telegraph Company, various different dispositions were suggested which were carried out by me in practice. Fig. 1 may serve as an example of one of the many forms of apparatus used. This consisted of a large glass tube sealed at one end and projecting into an ordinary incandescent lamp bulb. The primary, usually consisting of a few turns of thick, well-insulated copper sheet was inserted within the tube, the inside space of the bulb furnishing the secondary. This form of apparatus was arrived at after some experimenting, and was used principally with the view of enabling me to place a polished reflecting surface on the inside of the tube, and for this purpose the last turn of the primary was covered with a thin silver sheet. In all forms of apparatus used there was no special difficulty in exciting a luminous circle or cylinder in proximity to the primary. ,

As to the number of turns, I cannot quite understand why Prof. J. J. Thomson should think that a few turns were "quite sufficient," but lest I should impute to him an opinion he may not have, I will add that I have gained this impression from the reading of the published abstracts of his lecture. Clearly, the number of turns which gives the best result in any case, is dependent on the dimensions of the apparatus, and, were it not for various considerations, one turn would always give the best result.

I have found that it is preferable to use. in these experiments an, alternate current machine giving a moderate number of alternations per second to excite the induction coil for charging the Leyden jar which discharges through the primary — shown diagrammatically in Fig. 2, — as in such case, before the disruptive discharge takes place, the tube or bulb is slightly excited and the formation of the luminous circle is decidedly facilitated. But I have also used a Wimshurst machine in some experiments.

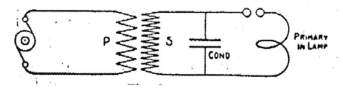

Prof. J. J. Thomson's view of the phenomena under consideration seems to be that they are wholly due to electro-magnetic action. I was, at one time, of the same opinion, but upon carefully investigating the subject I was led to the conviction that they are more of an electrostatic nature. It must be remembered that in these experiments we have to deal with primary currents of an enormous frequency or rate of change and of high potential, and that the secondary conductor consists of a rarefied gas, and that under such conditions electrostatic effects must play an important part.

In support of my view I will describe a few experiments made by me. To excite luminosity in the tube it is not absolutely necessary that the conductor should be closed. For instance, if an ordinary exhausted tube (preferably of large diameter) be surrounded by a spiral of thick copper wire serving as the primary, a feebly luminous spiral may be induced in the tube, roughly shown in Fig. 3. In one of these experiments a curious phenomenon was observed; namely, two intensely luminous circles, each of them close to a turn of the primary spiral, were formed inside of the tube, and I attributed this phenomenon to the .existence of nodes on the primary. The circles were connected by a faint luminous spiral parallel to the primary and in close proximity to it. To produce this effect I have found it necessary to strain the jar to the utmost. The turns of the spiral tend to close and form circles, but this, of course, would be expected, and does not necessarily indicate an electromagnetic effect; whereas the fact that a glow can be produced along the primary in the form of an open spiral argues for an electrostatic effect.

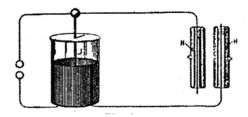

In using Dr. Lodge's recoil circuit, the electrostatic action is likewise apparent. The arrangement is illustrated in Fig. 4. In his experiments two hollow exhausted tubes H H were slipped over the wires of the recoil circuit and upon discharging the jar in the usual manner luminosity was excited in the tubes.

Another experiment performed is illustrated in Fig. 5. In this case an ordinary lamp-bulb was surrounded by one or two turns of thick copper wire P and the luminous circle L excited in the bulb by discharging the jar through the primary. The lamp-bulb was provided with a tinfoil coating on the side opposite to the primary and each time the tinfoil coating was connected to the ground or to a large object the luminosity of the circle was considerably increased. This was evidently due to electrostatic action.

In other experiments I have noted that when the primary touches the glass the luminous circle is easier produced and is more sharply defined; but I have not noted that, generally speaking, the circles induced were very sharply defined, as Prof. J. J. Thomson has observed; on the contrary, in my experiments they were broad and often the whole of the bulb or tube was illuminated; and in one ease I have observed an intensely purplish glow, to which Prof. J. J. Thomson refers. But the circles were always in close proximity to the primary and were considerably easier produced when the latter was very close to the glass, much more so than would be expected assuming the action to be electromagnetic and considering the distance; and these facts speak for an electrostatic effect.

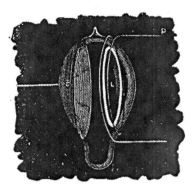

Furthermore I have observed that there is a molecular bombardment in the plane of the luminous circle at right angles to the glass — supposing the circle to be in the plane of the primary — this bombardment being evident from the rapid heating of the glass near the primary. Were the bombardment not at right angles to the glass the hating could not be so rapid. If there is a circumferential movement of the molecules constituting the luminous circle, I have thought that it might be rendered manifest by placing within the tube or bulb, radially to the circle, a thin plate of mica coated with some phosphorescent material and another such plate tangentially to the circle. If the molecules would move circumferentially, the former plate would be rendered more intensely phosphorescent. For want of time I have, however, not been able to perform the experiment.

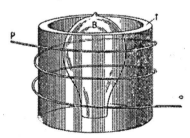

Another observation made by me was that when the specific inductive capacity of the medium between the primary and secondary is increased, the inductive effect is augmented. This is roughly illustrated in Fig. G. In this case luminosity was excited in an exhausted tube or bulb B and a glass tube T slipped between the primary and the bulb, when the effect pointed out was noted. Were the action wholly electromagnetic no change could possibly have been observed.

I have likewise noted that when a bulb is surrounded by a wire closed upon itself and in the plane of the primary, the formation of the luminous circle within the bulb is not prevented. But if instead of the wire a broad strip of tinfoil is glued upon the bulb, the formation of the luminous band was prevented, because then the action was. 'distributed over a greater surface. The effect of the closed tinfoil was no doubt of an electrostatic nature, for it presented a much greater resistance than the closed wire and produced therefore a much smaller electromagnetic effect.

Some of the experiments of Prof. J. J. Thomson also would seem to show some electrostatic action. For instance, in the experiment with the bulb enclosed in a bell jar, I should think that when the latter is exhausted so far that the gas enclosed reaches the maximum conductivity, the formation of the circle in the bulb and jar is prevented because of the space surrounding the primary being highly conducting; when the jar is further exhausted, the conductivity of the space around the primary diminishes and the circles appear necessarily first in the bell jar, as the rarefied gas is nearer to the primary. But were the inductive effect very powerful, they would probably appear in the bulb also. If, however, the bell Jar were exhausted to the highest degree they would very likely show themselves in the bulb only, that is, supposing the vacuous space to be non-conducting. On the assumption that in these phenomena electrostatic actions are concerned we find it easily explicable why the introduction of mercury or the heating of the bulb prevents the formation of the luminous band or shortens the after-glow; and also why in some cases a platinum wire may prevent the excitation of the tube. Nevertheless some of the experiments of Prof. J. J. Thomson would seem to indicate an electromagnetic effect. I may add that in one of my experiments in which a vacuum was produced by the Torricellian method, I was unable to produce the luminous band, but this may have been due to the weak exciting current employed.

My principal argument is the following: I have experimentally proved that if the same discharge which is barely sufficient to excite a luminous band in the bulb when passed through the primary circuit be so directed as to exalt the electrostatic inductive effect — namely, by converting upwards — an exhausted tube, devoid of electrodes, may be excited at a distance of several feet.

Note by Prof. J. J. Thomson
The London Electrician, July 24, 1891

"Mr. Tesla seems to ascribe the effects he observed to electrostatic action, and I have no doubt, from the description he gives of his method of conducting his experiments, that in them electrostatic action plays a very important part. He .seems, however, to have misunderstood my position with respect to the cause of these discharges, which is not, as he implies, that luminosity in tubes without electrodes cannot be produced by electrostatic action, but that it can also be produced when this action is excluded. As a matter of fact, it is very much easier to get the luminosity when these electrostatic effects are operative than when they are not. As an illustration of this I may mention that the first experiment I tried with the discharge of a Leyden jar produced luminosity in the tube, but it was not until after six weeks' continuous experimenting that I was able to get a discharge in the exhausted tube which I was satisfied was due to what is ordinarily called electrodynamic action. It is advisable to have a clear idea of what we mean by electrostatic action. If, previous to the discharge of the jar, the primary coil is raised to a high potential, it will induce over the glass of the tube a distribution of electricity. When the potential of the primary suddenly falls, this electrification will redistribute itself, and may pass through, the rarefied gas and produce luminosity in doing so. Whilst the discharge of the jar is going on, it is difficult, and, from a theoretical point of view, undesirable, to separate the effect into parts, one of which is called electrostatic, the other electromagnetic; what we can prove is that in this case the discharge is not such as would be produced by electromotive forces derived from a potential function. In my experiments the primary coil was connected to earth, and, as a further precaution, the primary was separated from the discharge tube by a screen of blotting paper, moistened with dilute sulphuric acid, and connected to earth. Wet blotting paper is a sufficiently good conductor to screen off a stationary electrostatic effect, though it is not a good enough one to stop waves of alternating electromotive intensity. When showing the experiments to the Physical Society I could not, of course, keep the tubes covered up, but, unless my memory deceives me, I stated the precautions which had been taken against the electrostatic effect. To correct misapprehension I may state that I did not read a formal paper to the Society, my object being to exhibit a few of the most typical experiments. The account of the experiments in the *Electrician* was from a reporter's note, and was not written, or even read, by me. I have now almost finished writing out, and hope very shortly to publish, an account of these and a large number of allied experiments, including some analogous to those mentioned by Mr. Tesla on the effect of conductors placed near the discharge tube, which I find, in some cases, to produce a diminution, in others an increase, in the brightness of the discharge, as well as some on the effect of the presence of substances of large specific inductive capacity. These seem to me to admit of a satisfactory explanation, for which, however, I must refer to my paper."

Reply to J. J. Thomson's Note in the *Electrician,* July 24, 1891
The Electrical Engineer - N.Y. — *August 26, 1891*

In *The Electrical Engineer* of August 12, I find some remarks of Prof. J. J. Thomson, which appeared originally in the London *Electrician* and which have a bearing upon some experiments described by me in your issue of July 1.

I did not, as Prof. J. J. Thomson seems to believe, misunderstand his position in regard to the cause of the phenomena considered, but I thought that in his experiments, as well as in my own, electrostatic effects were of great importance. It did not appear, *from* the meagre description of his experiments, that all possible precautions had been taken to exclude these effects. I did not doubt that luminosity could be excited in a closed tube when electrostatic action is completely excluded. In fact, at the outset, I myself looked for a purely electrodynamic effect and believed that I had obtained it. But many experiments performed at that time proved to me that the electrostatic effects were generally of far greater importance, and admitted of a more satisfactory explanation of most of the phenomena observed.

In using the term *electrostatic* I had reference rather to the nature of the action than to a stationary condition, which is the usual acceptance of the term. To express myself more clearly, I will suppose that near a closed exhausted tube be placed a small sphere charged to a very high potential. The sphere would act inductively upon the tube, and by distributing electricity over the same would undoubtedly produce luminosity (if the potential be sufficiently high), until a permanent condition would be reached. Assuming the tube to be perfectly well insulated, there would be only one instantaneous flash during the act of distribution. This would be due to the electrostatic action simply.

But now, suppose the charged sphere to be moved at short intervals with great speed along the exhausted tube. The tube would now be permanently excited, as the moving sphere would cause a constant redistribution of electricity and collisions of the molecules of the rarefied gas. We would still have to deal with an electrostatic effect, and in addition an electrodynamic effect would be observed. But if it were found that, for instance, the effect produced depended more on the specific inductive capacity than on the magnetic permeability of the medium — which would certainly be the case for speeds incomparably lower than that of light — then I believe I would be justified in saying that the effect produced was more of an electrostatic nature. I do not mean to say, however, that any similar condition prevails in the case of the discharge of a Leyden jar through the primary, but I think that such an action would be desirable.

It is in the spirit of the above example that I used the terms "more of an electrostatic nature," and have investigated the influence of bodies of high specific inductive capacity, and observed, for instance, the importance of the quality of glass of which the tube is made. I also endeavored to ascertain the influence of a medium of high permeability by using oxygen. It appeared from rough estimation that an oxygen tube when excited under similar conditions — that is, as far as could be determined — gives more light; but this, of course, may be due to many causes.

Without doubting in the least that, with the care and precautions taken by Prof. J. J. Thomson, the luminosity excited was due solely to electrodynamic action, I would say that in many experiments I have observed curious instances of the ineffectiveness of the screening, and I have also found that the electrification through the air is often of very great importance, and may, in some cases, determine the excitation of the tube.

In his original communication to the *Electrician,* Prof. J. J. Thomson refers to the fact that the luminosity in a tube near a wire through which a Leyden jar was discharged was noted by Hittorf. I think that the feeble luminous effect referred to has been noted by many experimenters, but in my experiments the effects were much more powerful than those usually noted.

Notes on a Unipolar Dynamo
Electrical Engineer, N.Y., Sept 1891

It is characteristic of fundamental discoveries, of great achievements of intellect, that they retain an undiminished power upon the imagination of the thinker. The memorable experiment of Faraday with a disc rotating between the two poles of a magnet, which has borne such magnificent fruit, has long passed into every-day experience; yet there are certain features about this embryo of the present dynamos and motors which even to-day appear to us striking, and are worthy of the most careful study.

Consider, for instance, the case of a disc of iron or other metal revolving between the two opposite poles of a magnet, and the polar surfaces completely covering both sides of the disc, and assume the current to be taken off or conveyed to the same by contacts uniformly from all points of the periphery of the disc. Take first the case of a motor. In all ordinary motors the operation is dependent upon some shifting or change of the resultant of the magnetic attraction exerted upon the armature, this process being effected either by some mechanical contrivance on the motor or by the action of currents of the proper character. We may explain the operation of such a motor just as we can that of a water-wheel. But if the above example of the disc surrounded completely by the polar surfaces, there is no shifting of the magnetic action, no change whatever, as far as we know, and yet rotation ensues. Here, then, ordinary considerations do not apply; we cannot even give a superficial explanation, as in ordinary motors, and the operation will be clear to us only when we shall have recognized the very nature of the forces concerned, and fathomed the mystery of the invisible connecting mechanism.

Considered as a dynamo machine, the disc is an equally interesting object of study. In addition to its peculiarity of giving currents of one direction without the employment of commutating devices, such a machine differs from ordinary dynamos in that there is no reaction between armature and field. The armature current tends to set up a magnetization at right angles to that of the field current, but since the current is taken off uniformly from all points of the periphery, and since, to he exalt, the external circuit may also be arranged perfectly symmetrical to the field magnet, no reaction can occur. This, however, is true only as long as the magnets are weakly energized, for when the magnets are more or less saturated, both magnetizations at right angles seemingly interfere with each other.

For the above reason alone it would appear that the output of such a machine; should, for the same weight, be much greater than that of any other machine in which the armature current tends to demagnetize the field. The extraordinary output of the Forbes unipolar dynamo and the experience of the writer confirm this view.

Again, the facility with which such a machine may be made to excite itself is striking, but this may be due – besides to the absence of armature reaction – to the defect smoothness of the current and non-existence of self-induction.

If the poles do not cover the disc completely on both sides, then, of course, unless the disc be properly subdivided, the machine will be very inefficient. Again, in this case there are points worthy of notice. If the disc tie rotated and the field current interrupted, the current through the armature will continue to flow and the field magnets will lose their strength comparatively slowly.

The reason for this will at once appear when we consider the direction of the currents set up in the disc.

Referring to the diagram Fig. 1, d represents the disc with the sliding contacts B B' on the shaft and periphery, N and S represent the two poles of a magnet.

[Fig. 1]

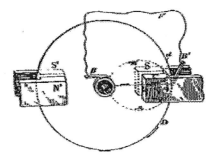

If the pole N be above, as indicated in the diagram, the disc being supposed to be in the plane of the paper, and rotating in the direction of the arrow D, the current set up in the disc will flow from the center to the periphery, as indicated by the arrow A. Since the magnetic action is more or less confined to the space between the poles N S, the other portions of the disc may be considered inactive. The current set up will therefore not wholly pass through the external circuit F, but will close through the disc itself, and generally, if the disposition be in any way similar to the one illustrated, by far the greater portion of the current generated will not appear externally, as the circuit F is practically short-circuited by the inactive portions of the disc. The direction of the resulting currents in the latter may be assumed to be as indicated by the dotted lines and arrows m and n; and the direction of the energizing field current being indicated by the arrows a b c d, an inspection of the figure shows that one of the two branches of the eddy current, that is, A B' m B, will tend to demagnetize the field, while the other branch, that is, A B' n B, will have the opposite effect. Therefore, the branch A B' m B, that is, the one which is approaching the field, will repel the lines of the same, while branch A B' n B, that is, the one leaving the field, will gather the lines of force upon itself.

In consequence of this there will be a constant tendency to reduce the current flow in the path AB' m B, while on the other hand no such opposition will exist in path A B' n B, and the effect of the latter branch or path will be more or less preponderating over that of the former. The joint effect of both the assumed branch currents might be represented by that of one single current of the same direction as that energizing the field. In other words, the eddy currents circulating in the disc will energize the field magnet. This is a result quite contrary to what we might be led to suppose at first, for we would naturally expect that the resulting effect of the armature currents would be such as to oppose the field current, as generally occurs when a primary and secondary conductor are placed in inductive relations to each Other, But it must be remembered that this result from the peculiar disposition in this case, namely, two paths being afforded to the current, and the latter selecting that path which offers the least opposition to its flow. From this we see that the eddy currents flowing in the disc partly energize the field, and for this reason when the field current is interrupted the currents in the disc will continue to flow, and the field magnet will lose its strength with comparative slowness and may even retain a certain strength as long as the rotation of the disc is continued.

[Fig. 2]   [Fig. 3]

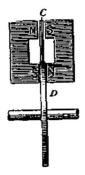
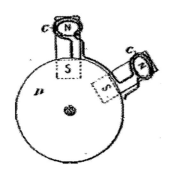

The result will, of course, largely depend on the resistance and geometrical dimensions of the path of the resulting eddy current and on the speed of rotation; these elements, namely, determine the retardation of this current and its position relative to the field. For a certain speed there would be a maximum energizing action: then at higher speeds, it would gradually fall off to zero and finally reverse, that is, the resultant eddy current effect would be to weaken the field. The reaction would be best demonstrated experimentally by arranging the fields N S, N' S', freely movable on an axis concentric with the shaft of the disc. If the latter were rotated as before in the direction of the arrow D, the field would be dragged in the same direction with a torque, which, up to a certain point, would go on increasing with the speed of rotation, then fall off, and, passing through zero, finally become negative; that is, the field would begin to rotate in opposite direction to the disc. In experiments with alternate current motors in which the field was shifted by currents of differing phase, this interesting result was observed. For very low speeds of rotation of the field the motor would show a torque of 900 lbs. or more, measured on a, pulley 12 inches in diameter. When the speed of rotation of the poles was increased, the torque would diminish, would finally go down to zero, become negative, and then the armature would begin to rotate in opposite direction to the field-To return to the principal subject; assume the conditions to be such that the eddy currents generated by the rotation of the disc strengthen the field, and suppose the latter gradually removed while the disc is kept rotating at an increased rate. The current, once started, may then be sufficient to maintain itself and even increase in strength, and then we have the case of Sir William Thomson's "current accumulator."

But from the above considerations it would seem that for the success of the experiment the employment of a disc not subdivided would be essential, for if there should be a radial subdivision, the eddy currents could not form and the self-exciting action would cease. If such a radially subdivided disc were used it would be necessary to connect the spokes by a conducting rim or in any proper manner so as to form a symmetrical system of closed circuits.

The action of the eddy currents may be utilized to excite a machine of any construction. For instance, in Figs. 2 and 3 an arrangement is shown by which a machine with a disc armature might be excited. Here a number of magnets, N S, N S, are placed radially on each side of a metal disc D carrying on its rim a. set of insulated coils, C C The magnets form two separate fields, an internal and an external one, the solid disc rotating in the field nearest the axis, and the coils in the field further from it-Assume the magnets slightly energized at the start; they could be strengthened by the action of the eddy currents in the solid disc so as to afford a stronger field for the peripheral coils. Although there is no doubt that under proper conditions a machine might be excited in this or a similar manner, there being sufficient experimental evidence to warrant such an assertion, such a mode of excitation would be wasteful.

But an unipolar dynamo or motor, such as shown in Fig, 1 may be excited in an efficient manner by simply properly subdividing the disc or cylinder in which the currents are set up, and it is practicable to do away with the field coils which are usually employed. Such a plan is illustrated in Fig. 4. The disc or cylinder D is supposed to be arranged to rotate between the two poles N and S of a magnet, which completely cover it on both sides, the contours of the disc and poles being represented by the circles d and d' respectively, the upper pole being omitted for the sake of clearness.

[Fig. 4]   [Fig. 5]

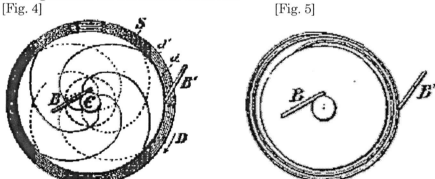

The cores of the magnet are supposed to be hollow, the shaft C of the disc passing through them. If the unmarked pole be below, and the disc be rotated screw fashion, the current will be, as before, from the center to the periphery, and may be taken off by suitable sliding contacts, B B', on the shaft and periphery respectively. In this arrangement the current flowing through the disc and external circuit will have no appreciable effect on the field magnet.

But let us now suppose the disc to be subdivided, spirally, as indicated by the full or dotted lines. Fig. 4. The difference of potential between a point on the shaft and a point on the periphery wilt remain unchanged, in sign as well as in amount. The only difference will be that the resistance of the disc will be augmented and that there will be a greater fall of potential from a point on the shaft to a point on the periphery when the same current is traversing the external circuit. But since the current is forced to follow the lines of subdivision, we see that it will tend either to energize or de-energize the field, and this will depend, other things being equal, upon the direction of the Lines of subdivision. If the subdivision be as indicated by the full lines in Fig, 4, it is evident that if the current is of the same direction as before, that is, from center to periphery, its effect will be to strengthen the field magnet; whereas, if the subdivision be as indicated by the dotted lines, the current generated will tend to weaken the magnet. In the former case the machine will be capable of exciting itself when the disc is rotated in the direction of arrow D; in the latter case the direction of rotation must be reversed. Two such discs may be combined, however, as indicated, the two discs rotating in opposite fields, and in the same or opposite direction.

Similar disposition may, of course, be made in a type of machine in which, instead of a disc, a cylinder is rotated. In such unipolar machines, in the manner indicated, the usual field coils and poles may be omitted and the machine may be made to consist only of a cylinder or of two discs enveloped by a metal casting.

Instead of subdividing the disc or cylinder spirally, as indicated in Fig. 4, it is more convenient to interpose one or more turns between the disc and the contact ring on the periphery, as illustrated in Fig. 5.

A Forbes dynamo may, for instance, be excited in such a manner. In the experience of the writer it has been found that instead of taking the current from two such discs by sliding contacts, as usual, a flexible conducting belt may be employed to advantage. The discs are in such case provided with large flanges, affording a very great contact surface. The belt should be made to beat on the flanges with spring pressure to take up the expansion. Several machines with belt contact were constructed by the writer two years ago, and worked satisfactorily; but for want of time the work in that direction has been temporarily suspended. A number of features, pointed out above have also been used by the writer in connection with some types of alternating current motors.

Experiments with Alternate Currents of High Potential and High Frequency
Delivered before the Institution of Electrical Engineers, London, February 1892

I cannot find words to express how deeply I feel the honor of addressing some of the foremost thinkers of the present time, and so many able scientific men, engineers and electricians, of the country greatest in scientific achievements.

The results which I have the honor to present before .such a gathering I cannot call my own. There are among you not a few who can lay better claim than myself on *any* feature of merit which this work may contain. I need not mention many names which are world-known — names of those among you who are recognized as the leaders in this enchanting science; but one, at least, I must mention — a name which could not be omitted in a demonstration of this kind. It is a name associated with the most beautiful invention ever made: it is Crookes !

When I was at college, a good time ago, I read, in a translation (for then I was not familiar with you magnificent language), the description of his experiments on radiant matter. I read it only once in my life — that time — yet every detail about that charming work I can remember this day. Few are the books, let me say, which can make such an impression upon the mind of a student.

But if, on the present occasion, I mention this name as one of many your institution can boast of, it is because I have more than one reason to do so. For what I have to tell you and to show you this evening concerns, in a large measure, that same vague world which Professor Crookes has so ably explored; and, more than this, when I trace back the mental process which led me to these advances — which even by myself cannot be considered trifling, since they are so appreciated by you — I believe that their real origin, that which started me to work in this direction, and brought me to them, after a long period of constant thought, was that fascinating little book which I read many years ago.

And now that I have made a feeble effort to express my homage and acknowledge my indebtedness to him and others among you, I will make a second effort, which I hope you will not find so feeble as the first, to entertain you.

Give me leave to introduce the subject in a few words.

A short time ago I had the honor to bring before our American Institute of Electrical Engineers some results then arrived at by me in a novel line of work. I need not assure you that the many evidences which I have received that English scientific men and engineers were interested in this work have been for me a great reward and encouragement. I will not dwell upon the experiments already described, except with the view of completing, or more clearly expressing, some ideas advanced by me before, and also with the view of rendering the study here presented self-contained, and my remarks on the subject of this evening's lecture consistent.

This investigation, then, it goes without saying, deals with alternating currents, and, to be more precise, with alternating currents of high potential and high frequency. Just in how much a very high frequency is essential for the production of the results presented is a question which, even with my present experience, would embarrass me to answer. Some of the experiments may be performed with low frequencies; but very high frequencies are desirable, not only on account of the many effects secured by their use, but also as a convenient means of obtaining, in the induction apparatus employed, the high potentials, which in their turn are necessary to the demonstration of most of the experiments here contemplated.

Of the various branches of electrical investigation, perhaps the most interesting and immediately the most promising is that dealing with alternating currents. The progress in this branch of applied science has been so great in recent years that it justifies the most sanguine hopes. Hardly have we become familiar with one fact, when novel experiences are met with and new avenues of research are opened. Even at this hour possibilities not dreamed of before are, by the use of these currents, partly realized. As in nature al! is ebb and tide, all is wave motion, so it seems that in all branches of industry alternating currents — electric wave motion — will have the sway.

One reason, perhaps, why this branch of science is being so rapidly developed is to be found in the interest which is attached to its experimental study. We wind a simple ring of iron with coils; we establish the connections to the generator, and witch wonder and delight we note the effects of strange forces which we bring into play, which allow us to transform, to transmit and direct energy at will. We arrange the circuits properly, and we see the mass of iron and wires behave as though it were endowed with life, spinning a heavy armature, through invisible connections, with great speed and power — with the energy possibly conveyed from a great distance. We observe how the energy of an alternating current traversing the wire manifests itself — not so much in the wire as in the surrounding space — in the most surprising manner, taking the forms of heat, light, mechanical energy, and, most surprising of all, even chemical affinity. All these observations fascinate us, and fill us with an intense desire to know more about the nature of these phenomena. Each day we go to our work in the hope of discovering, — in the hope that some one, on matter who, may find a solution of one of the pending great problems, — and each succeeding day we return to our task with renewed ardor; and even if we *are* unsuccessful, our work has not been in vain, for in these strivings, in these efforts, we have found hours of untold pleasure, and we have directed our energies to the benefit of mankind.

We may take — at random, if you choose — any of the many experiments which *may* be performed with alternating currents; a few of which only, and by no means the most striking, form the subject of this evening's demonstration; they are all equally interesting, equally inciting to thought.

Here is a simple glass tube from which the air has been partially exhausted. I take hold of it; I bring my body in contact with a wire conveying alternating currents of high potential, and the tube in my hand is brilliantly lighted. In whatever position I may put it, wherever I may move it in space, as far as I can reach, its soft, ,pleasing light persists with undiminished brightness.

Here is an exhausted bulb suspended from a single wire. Standing on an insulated support, I grasp it, and a platinum button mounted in it is brought to vivid incandescence.

Here, attached to a leading wire, is another bulb, which, as I touch its metallic socket, is filled with magnificent colors of phosphorescent light.

Here still another, which by my fingers' touch casts a shadow — the Crookes shadow, of the stem inside of it.

Here, again, insulated as I stand on this platform, I bring my body in contact with one of the terminals of the secondary of this induction coil — with the end of a wire many miles long — and you see streams of light break forth from its distant end, which is set in violent vibration.

Here, once more, I attach these two plates of wire gauze to the terminals of the coil, I set them a distance apart, and I set the coil to work. You may see a small spar'.; pass between the plates. I insert a thick plate of one of the best dielectrics between them, and instead of rendering altogether impossible, as we are used to expect, *I aid* the passage of the discharge, which, as I insert the plate, merely changes in appearance and assumes the form of luminous streams.

Is there, I ask, can there be, a more interesting study than that of alternating currents ?

In all these investigations, in all these experiments, which are so very, very interesting for many years past — ever since the greatest experimenter who lectured in this hall discovered its principle — we have had a steady companion, an appliance familiar to every one, a plaything once, a thing of momentous importance now — the induction coil. There is no dearer appliance to the electrician. From the ablest among you, I dare say, down to the inexperienced student, to your lecturer, we all have passed many delightful hours in experimenting with the induction coil. We have watched its play, and thought and pondered over the beautiful phenomena which it disclosed to our ravished eyes. So well known is this apparatus, so familiar are these phenomena to every one, that my courage nearly fails me when I think that I have ventured to address so able an audience, that I have ventured to entertain you with that same old subject. Here in reality is the same apparatus, and here are the same phenomena, only the apparatus is operated somewhat differently; the phenomena are presented in a different aspect. Some of the results we find as expected, others surprise us, but all captivate our attention, for in scientific investigation each novel result achieved may be the center of a new departure, each novel fact learned may lead to important developments.

Usually in operating an induction coil we have set up a vibration of moderate frequency in the primary, either by means of an interrupter or break, or by the use of an alternator. Earlier English investigators, to mention only Spottiswoode and J. E. H. Gordon, have used a rapid break in connection with the coil. Our knowledge and experience of to-day enables us to see clearly why these coils under the conditions of the tests did not disclose any remarkable phenomena, and why able experimenters failed to perceive many of the curious effects which have since been observed.

In the experiments such as performed this evening, we operate the coil either from a specially constructed alternator capable of giving many thousands of reversals of current per second, or, by disruptively discharging a condenser through the primary, we set up a vibration in the secondary circuit of a frequency of many hundred thousand or millions per second, if we so desire; and in using either of these means we enter a field as yet unexplored.

It is impossible to pursue an investigation in any novel line without finally making some interesting observation or learning some useful fact. That this statement is applicable to the subject of this lecture the many curious and unexpected phenomena which we observe afford a convincing proof. *By way of illustration, take for instance the most obvious phenomena, those of the discharge of the induction coil.*

Here is a coil which is operated by currents vibrating with extreme rapidity, obtained by disruptively discharging a Leyden jar. It would not surprise a student were the lecturer to say that the secondary of this coil consists of a small length of comparatively stout wire; it would not surprise him were the lecturer to state that, in spite of this, the coil is capable of giving any potential which the best insulation of the turns is able to withstand; but although he may be prepared, and even be indifferent as to the anticipated result, yet the aspect of the discharge of the coil will surprise and interest him. Every one is familiar with the discharge of an ordinary coil; it need not be reproduced here. But, by way of contrast, here is a form of discharge of a coil, the primary current of which is vibrating several hundred thousand times per second. The discharge of an ordinary coil appears as a simple line or band of light. The discharge of this coil appears in the form of powerful brushes and luminous streams issuing from all points of the two straight wires attached to the terminals of the secondary (Fig. 1.)

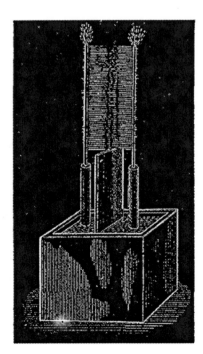
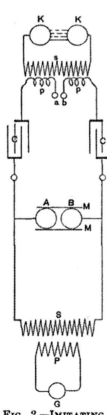

FIG. 1.—DISCHARGE BETWEEN TWO WIRES WITH FREQUENCIES OF A FEW HUNDRED THOUSAND PER SECOND.

FIG. 2.—IMITATING THE SPARK OF A HOLTZ MACHINE.

Now compare this phenomenon which you have just witnessed with the discharge of a Holtz or Wimshurst machine — that other interesting appliance so dear to the experimenter. What a difference there is between these phenomena! And yet, had I made the necessary arrangements — which could have been made easily, were it not that they would interfere with other experiments — I could have produced with this coil sparks which, had I the coil hidden from your view and only two knobs exposed, even the keenest observer among you would find it difficult, if not impossible, to distinguish from those of an influence or friction machine. This may be done in many ways — for instance, by operating the induction coil which charges the condenser from an alternating-current machine of very low frequency, and preferably adjusting the discharge circuit so that there are no oscillations set up in it. We then obtain in the secondary circuit, if the knobs are of the required size and properly set, a more or less rapid succession of sparks of great intensity and small quantity, which possess the same brilliancy, and are accompanied by the same sharp crackling sound, as those obtained from a friction or influence machine.

Another way is to pass through two primary circuits, having a common secondary, two currents of a slightly different period, which produce in the secondary circuit sparks occurring at comparatively long intervals. But, even with the means at hand this evening, I may succeed in imitating the spark of a Holtz machine. For this purpose I establish between the terminals of the coil which charges the condenser a long, unsteady arc, which is periodically interrupted by the upward current of air produced by it. To increase the current of air I place on each side of the arc, and close to it, a large plate of mica: The condenser charged from this coil discharges into the primary circuit of a second coil through a small air gap, which is necessary to produce a sudden rush of current through the primary. The scheme of connections in the present experiment is indicated in Fig. 2.

G is an ordinarily constructed alternator, supplying the primary P of an induction coil, the secondary S of which charges the condensers or jars $C\ C$: The terminals of the secondary are connected to the inside coatings of the jars, the outer coatings being connected to the ends of the primary p of a second induction coil. This primary $p\ p$ has a small air gap a,b.

The secondary s of this coil is provided with knobs or spheres $K\ K$ of the proper size and set at a distance suitable for the experiment.

A long arc is established between the terminals $A\ B$ of the first induction coil. $M\ M$ arc the mica plates.

Each time the arc is broken between A and B the jars are quickly charged and discharged through the primary $p\ p,$ producing a snapping spark between the knobs $K K.$ Upon the arc forming between A and B the potential falls, and the jars cannot be charged to such hih potential as to break through the air gap $a\ b$ until the arc is again broken by the drought.

In this manner sudden impulses, at long intervals, are produced in the primary $p\ p$, which in the secondary s give a corresponding number of impulses of great intensity. If the secondary knobs or spheres, $K K$, are of the proper size, the sparks show much resemblance to those of a Holtz machine.

But these two effects, which to the eye appear so very different, are only two of the many discharge phenomena. We only need to change the conditions of the test, and again we make other observations of interest.

When, instead of operating the induction coil as in the last two experiments, we operate it from a high frequency alternator, as in the next experiment, a systematic study of the phenomena is rendered much more easy. In such case, in varying the strength and frequency of the currents through the primary, we may observe five distinct forms of discharge, which I have described in my former paper on the subject before the American Institute of Electrical Engineers, May 20, 1891.

It would take too much time, and it would lead us too far from the subject presented this evening, to reproduce all these forms, but it seems to me desirable to show you one of them. It is a brush discharge, which is interesting in more than one respect. Viewed from a near position it resembles much a jet of gas escaping under great pressure. We know that the phenomenon is due to the agitation of the molecules near the terminal, and we anticipate that some heat must be developed by the impact of the molecules against the terminal or against each other. Indeed, we find that the brush is hot, and only a little thought leads us to the conclusion that, could we but reach sufficiently high frequencies, we could produce a brush which would give intense light and heat, and which would resemble in every particular an ordinary flame, save, perhaps, that both phenomena might not be due to the same agent — save, perhaps, that chemical affinity might not be *electrical* in its nature.

As the production of heat and light is here due to the impact of the molecules, or atoms of air, or something else besides, and, as we can augment the energy simply by raising the potential, we might, even with frequencies obtained from a dynamo machine, intensify the action to such a degree as to bring the terminal to melting heat. Bud with such low frequencies we would have to deal always with something of the nature of an electric current. If I approach a conducting object to the brush, a thin little spark passes, yet, even with the frequencies used this evening, the tendency to spark is not very great. So, for instance, if I hold a metallic sphere at some distance above the terminal you may see the whole space between the terminal and sphere illuminated by the streams without the spark passing; and with the much higher frequencies obtainable by the disruptive discharge of a condenser, were it not for the sudden impulses, which are comparatively few in number, sparking would not occur even at very small distances. However, with incomparably higher frequencies, which we may yet find means to produce efficiently, and provided that electric impulses of such high frequencies could be transmitted through a conductor, the electrical characteristics of the brush discharge would completely vanish — no spark would pass, no shock would be felt — yet we would still have to deal with an *electric* phenomenon, but in the broad, modern interpretation of the word. In my first paper before referred to I have pointed out the curious properties of the brush, and described the best manner of producing it, but I have thought it worth while to endeavor to express myself more clearly in regard to this phenomenon, because of its absorbing interest.

When a coil is operated with currents of very high frequency. beautiful brush effects may be produced, even if the coil be of comparatively small dimensions. The experimenter may vary them in many ways, and, if it were nothing else, they afford a pleasing sight. What adds to their interest is that they may be produced with one single terminal as well as with two — in fact, often better with one than with two.

But of all the discharge phenomena observed, the most pleasing to the eye, and the most instructive, are those observed with a coil which is operated by means of the disruptive discharge of a condenser. The power of the brushes, the abundance of the sparks, when the conditions are patiently adjusted, is often amazing. With even a very small coil, if it be so well insulated as to stand a difference of potential of several thousand volts per turn, the sparks may be so abundant that the whole coil may appear a complete mass of fire.

Curiously enough the sparks, when the terminals of the coil are set at a considerable distance, seem to dart in every possible direction as though the terminals were perfectly independent of each other. As the sparks would soon destroy the insulation it is necessary to prevent them. This is best done by immersing the coil in a good liquid insulator, such as boiled-out oil. Immersion in a liquid may be considered almost an absolute necessity for the continued and successful working of such a coil.

It is of course out of the question, in an experimental lecture, with only a few minutes at disposal for the performance of each experiment, to show these discharge phenomena to advantage, as to produce each phenomenon at its best a very careful adjustment is required. But even if imperfectly produced, as they are likely to be this evening, they are sufficiently striking to interest an intelligent audience.

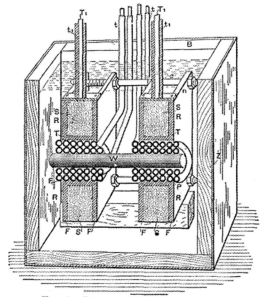

FIG. 3.—DISRUPTIVE DISCHARGE COIL.

Before showing some of these curious effects I must, for the sake of completeness, give a short description of the coil and other apparatus used in the experiments with the disruptive discharge this evening.

It is contained in a box B (Fig. 3) of thick boards of hard wood, covered on the outside with zinc sheet Z, which is carefully soldered all around. It might be advisable, in a strictly scientific investigation, when accuracy is of great importance, to do away with the metal cover, as it might introduce many errors, principally on account of its complex action upon the coil, as a condenser of very small capacity and as an electrostatic and electromagnetic screen. When the coil is used for such experiments as are here contemplated, the employment of the metal cover offers some practical advantages, but these are not of sufficient importance to be dwelt upon.

The coil should be placed symmetrically to the metal cover, and the space between should, of course, not be too small, certainly not less than, say, five centimeters, but much more if possible; especially the two sides of the zinc box, which are at right angles to the axis of the coil, should be sufficiently remote from the latter, as otherwise they might impair its action and be a source of loss.

The coil consists of two spools of hard rubber $R\,R$, held apart at a distance of 10 centimeters by bolts c and nuts n, likewise of hard rubber. Each spool comprises a tube T of approximately 8 centimeters inside diameter, and 3 millimeters thick, upon which are screwed two flanges $F\,F$, 24 centimeters square, the space between the flanges being about 3 centimeters. The secondary, $S\,S$, of the best gutta percha-covered wire, has 26 layers, 10 turns in each, giving for each half a total of 260 turns. The two halves are wound oppositely and connected in series, the connection between both being made over the primary. This disposition, besides being convenient, has the advantage that when the coil is well balanced — that is, when both of its terminals $T_1\,T_1$ are connected to bodies or devices of equal capacity — there is not much danger of breaking through to the primary, and the insulation between the primary and the secondary need not be thick. In using the coil it is advisable to attach to *both* terminals devices of nearly equal capacity, as, when the capacity of the terminals is not equal, sparks will be ant to pass to the primary. To avoid this, the middle point of the secondary may be connected to the primary, but this is not always practicable.

The primary $P\,P$ is wound in two parts, and oppositely, upon a wooden spool W, and the four ends are led out of the oil through hard rubber tubes $t\,t$. The ends of the secondary T_1, T_1 are also led out of the oil through rubber tubes $t_1\,t_1$ of great thickness. The primary and secondary layers are insulated by cotton cloth, the thickness of the insulation, of course, bearing some proportion to the difference of potential between the turns of the different layers. Each half of the primary has four layers, 24 turns in each, this giving a total of 96 turns. When both the parts are connected in series, this gives a ratio of conversion of about 1:2.7, and with the primaries in multiple, 1:5,4; but in operating with very rapidly alternating currents this ratio does not convey even an approximate idea of the ratio of the E.M.Fs. in the primary and secondary circuits. The coil is held in position in the oil on wooden supports, there being about 5 centimeters thickness of oil all round. Where the oil is not specially needed, the space is filled with pieces of wood, and for this purpose principally the wooden box B surrounding the whole is used.

The construction here shown is, of course, not the best on general principles, but I believe it is a good and convenient one for the production of effects in which an excessive potential and a very small current are needed.

In connection with the coil I use either the ordinary form of discharger or a modified form. In the former I have introduced two changes which secure some advantages, and which are obvious. If they are mentioned, it is only in the hope that some experimenter may find them of use.

One of the changes is that the adjustable knobs A and B (Fig. 4), of the discharger are held in jaws of brass, *J J*, by spring pressure, this allowing of turning them successively into different positions, and so doing away with the tedious process or frequent polishing up.

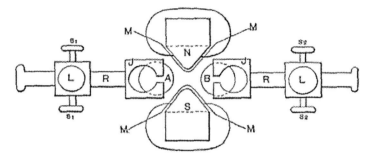

FIG. 4.—ARRANGEMENT OF IMPROVED DISCHARGER AND MAGNET.

The other change consists in the employment of a strong electromagnet *N S*, which is placed with its axis at right angles to the line joining the knobs A and B. and produces a strong magnetic field between them. The pole pieces of the magnet are movable and properly formed so as to protrude between the brass knobs, in order to make the field as intense as possible; but to prevent the discharge from jumping to the magnet the pole pieces are protected by a layer of mica, *M M,* of sufficient thickness. s_1 s_1 and s_2 s_2 are screws for fastening the wires. On each side one of the screws is for large and the other for small wires. *L L* are screws for fixing in position the rods *R R*, which support the knobs.

In another arrangement with the magnet I take the discharge between the rounded pole pieces themselves, which in such case are insulated and preferably provided with polished brass caps.

The employment of an intense magnetic field is of advantage principally when the induction coil or transformer which charges the condenser is operated by currents of very low frequency. In such a case the number of the fundamental discharges between the knobs may be so small as to render the currents produced in the secondary unsuitable for many experiments. The intense magnetic field than serves to blow out the arc between the knobs as soon as it is formed, and the fundamental discharges occur in quicker succession.

Instead of the magnet, a drought or blast of air may be employed with some advantage. In this case the arc is preferably established between the knobs A B, in Fig. 2 (the knobs a b being generally joined, or entirely done away with), as in this disposition the arc is long and unsteady, and is easily affected by the drought.

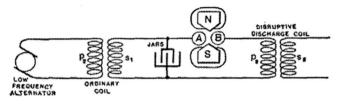

FIG. 5.—ARRANGEMENT WITH LOW-FREQUENCY ALTERNATOR AND IMPROVED DISCHARGER.

When a magnet is employed to break the arc, it is better to choose the connection indicated diagrammatically in Fig 5, as in this case the currents forming the arc are much more powerful, and the magnetic field exercises a greater influence. The use of the magnet permits, however, of the arc being replaced by a vacuum tube, but I have encountered great difficulties in working with an exhausted tube.

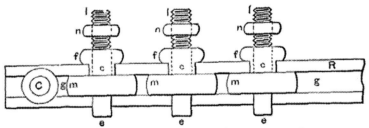

FIG. 6.—DISCHARGER WITH MULTIPLE GAPS.

The other form of discharger used in these and similar experiments is indicated in Figs. 6 and 7. It consists of a number of brass pieces c c (Fig. 6), each of which comprises a spherical middle portion m with an extension a below — which is merely used to fasten the piece in a lathe when polishing up the discharging surface — and a column above, which consists of a knurled flange f surmounted by a threaded stem l carrying a nut n, by means of which a wire is fastened to the column. The flange f conveniently serves for holding the brass piece when fastening the wire. and also for turning it in any position when it becomes necessary to present a fresh discharging surface. Two stout strips of hard rubber R R, with planed grooves g g (Fig. 7) to fit the middle portion of the pieces c c, serve to clamp the latter and hold them firmly in position by means of two bolts C C (of which only one is shown) passing through the ends of the strips.

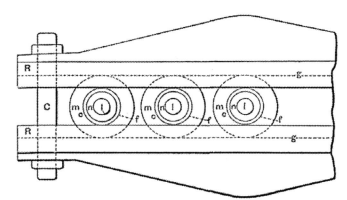

FIG. 7.—DISCHARGER WITH MULTIPLE GAPS.

In the use of this kind of discharger I have found three principal advantages over the ordinary form. First, the dielectric strength of a given total width of air space is greater when a great many small air gaps are used instead of one, which permits of working with a smaller length of air gap, and that means smaller loss and less deterioration of the metal; secondly by reason of splitting the arc up into smaller arcs, the polished surfaces are made to last much longer; and, thirdly, the apparatus affords some gauge in the experiments. I usually set the pieces by putting between them sheets of uniform thickness at a certain very small distance which is known from the experiments of Sir William Thomson to require a certain electromotive force to be bridged by the spark.

It should, of course, be remembered that the sparking distance is much diminished as the frequency is increased. By taking any number of spaces the experimenter has a rough idea of the electromotive force, and he finds it easier to repeat an experiment, as he has not the trouble of setting the knobs again and again. With this kind of discharger I have been able to maintain an oscillating motion without any spark being visible with the naked eye between the knobs, and they would not show a very appreciable rise in temperature. This form of discharge also lends itself to many arrangements of condensers and circuits which are often very convenient and time-saving. I have used it preferably in a disposition similar to that indicated in Fig. 2, when the currents forming the arc are small.

I may here mention that I have also used dischargers with single or multiple air gaps, in which the discharge surfaces were rotated with great speed. No particular advantage was, however, gained by this method, except in cases where the currents from the condenser were large and the keeping cool of the surfaces was necessary, and in cases when, the discharge not being oscillating of itself, the arc as soon as established was broken by the air current, thus starting the vibration at intervals in rapid succession. I have also used mechanical interrupters in many ways. To avoid the difficulties with frictional contacts, the preferred plan adopted was to establish the arc and rotate through it at great speed a rim of mica provided with many holes and fastened to a steel plate.

It is understood, of course, that the employment of a magnet, air current, or other interrupter, produces an effect worth noticing, unless the self-induction, capacity and resistance are so related that there are oscillations set up upon each interruption.

I will now endeavor to show you some of the most noteworthy of these discharge phenomena.

I have stretched across the room two ordinary cotton covered wires, each about 7 meters in length. They are supported on insulating cords at a distance of about 30 centimeters. I attach now to each of the terminals of the coil one of the wires and set the coil in action. Upon turning the lights off in the room you see the wires strongly illuminated by the streams issuing abundantly from their whole surface in spite of the cotton covering, which may even be very thick. When the experiment is performed under good conditions, the light from the wires is sufficiently intense to allow distinguishing the objects in a room. To produce the best result it is, of course, necessary to adjust carefully the capacity of the jars, the arc between the knobs and the length of the wires. My experience is that calculation of the length of the wires leads, in such a case, to no result whatever. The experimenter will do best to take the wires at the start very long, and then adjust by cutting off first long pieces, and then smaller and smaller ones as he approaches the right length.

A convenient way is to use an oil condenser of very small capacity, consisting of two small adjustable metal plates, in connection with this and similar experiments. In such case I take wires rather short and set at the beginning the condenser plates at maximum distance. If the streams for the wires increase by approach of the plates, the length of the wires is about right; if they diminish the wires are too long for that frequency and potential. When a condenser is used in connection with experiments with such a coil, it should be an oil condenser by all means, as in using an air condenser considerable energy might be wasted. The wires leading to the plates in the oil should be very thin, heavily coated with some insulating compound, and provided with a conducting covering — this preferably extending under the surface of the oil. The conducting cover should not be too near the terminals, or ends, of the wire, as a spark would be apt to jump from the wire to it. The conducting coating is used to diminish the air losses, in virtue of its action as an electrostatic screen. As to the size of the vessel containing the oil, and the size of the plates, the experimenter gains at once an idea from a rough trial. The size of the plates *in oil* is, however, calculable, as the dielectric losses are very small.

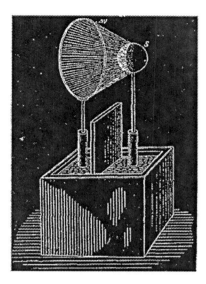

FIG. 8.—EFFECT PRODUCED BY CONCENTRATING STREAMS.

In the preceding experiment it is of considerable interest to know what relation the quantity of the light emitted bears to the frequency and potential of the electric impulses. My opinion is that the heat as well as light effects produced should be proportionate, under otherwise equal conditions of test, to the product of frequency and square of potential, but the experimental verification of the law, whatever it may be, would be exceedingly difficult. One thing is certain, at any rate, and that is, that in augmenting the potential and frequency we rapidly intensify the streams; and, though it may be *very* sanguine, it is surely not altogether hopeless to expect that we may succeed in producing a practical illuminant on these lines. We would then be simply using burners or flames, in which there would be no chemical process, no consumption of material, but merely a transfer of energy, and which would, in all probability emit more light and less heat than ordinary flames.

The luminous intensity of the streams is, of course, considerably increased when they are focused upon a small surface. This may be shown by the following experiment:

I attach to one of the terminals of the coil a wire w (Fig. 8), bent in a circle of about 30 centimeters in diameter, and to the other terminal I fasten a small brass spheres, the surface of the wire being preferably equal to the surface of the sphere, and the center of the latter being in a line at right angles to the plane of the wire circle and passing through its center. When the discharge is established under proper conditions, a luminous hollow cone is formed, and in the dark one-half of the brass sphere is strongly illuminated, as shown in the cut.

By some artifice or other, it is easy to concentrate the streams upon small surfaces and to produce very strong light effects. Two thin wires may thus be rendered intensely luminous.

In order to intensify the streams the wires should be very thin and short; but as in this case their capacity would be generally too small for the coil — at least, for such a one as the present — it is necessary to augment the capacity to the required value, while, at the same time, the surface of the wires remains very small. This may be done in many ways.

Here, for instance, I have two plates $R\ R$, of hard rubber (Fig. 9), upon which I have glued two very thin wires $w\ w$, so as to form a name. The wires may be bare or covered with the best insulation — it is immaterial for the success of the experiment. Well insulated wires, if anything, are preferable: On the back of each plate, indicated by the shaded portion, is a tinfoil coating $t\ t$. The plates are placed in line at a sufficient distance to prevent a spark passing from one to the other wire: The two tinfoil coatings I have joined by a conductor C, and the two wires I presently connect to the terminals of the coil. It is now easy, by varying the strength and frequency of the currents trough the primary, to find a paint at which the capacity of the system is best suited to the conditions, and the wires become so strongly luminous that, when the light in the room is turned off the name formed by them appears in brilliant letters.

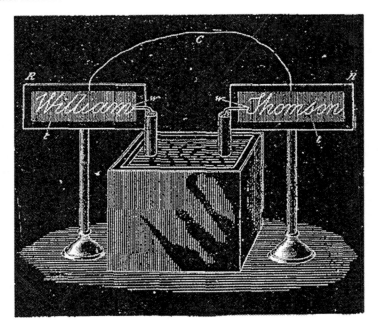

FIG. 9.—WIRES RENDERED INTENSELY LUMINOUS.

It is perhaps preferable to perform this experiment with a roil operated from an alternator of high frequency, as then, owing to the harmonic rise and fall, the streams are very uniform, though they are less abundant than when produced with such a coil as the present. This experiment; however, may be performed with low frequencies, but much less satisfactorily.

When two wires, attached to the terminals of the coil, are set at the proper distance, the streams between them may be so intense as to produce a continuous luminous sheet. To show this phenomenon I have here two circles, C and c (Fig. 10), of rather stout wire, one being about 80 centimeters and the other 30 centimeters in diameter. To each of the terminals of the coil I attach one of the circles. The supporting wires are so bent that the circles may be placed in the same plane, coinciding as nearly as possible. When the light in the room is turned off and the coil set to work, you see the whole space between the wires uniformly filled with streams, forming a luminous disc, which could be seen from a considerable distance, such is the intensity of the streams. The outer circle could have been much larger than the present one; in fact, with this coil I have used much larger circles, and I have been able to produce a strongly luminous sheet, covering an area of more than one square meter which is a remarkable effect with this very small coil. To avoid uncertainty, the circle has been taken smaller, and the area is now about 0,43 square meter.

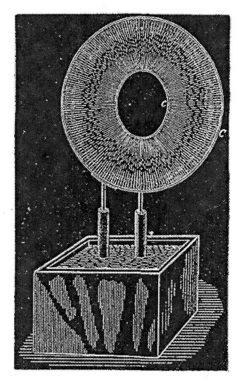

FIG. 10.—LUMINOUS DISCS.

The frequency of the vibration, and the quickness of succession of the sparks between the knobs, affect to a marked degree the appearance of the streams. When the frequency is very low, the air gives way in more or less the same manner, as by a steady difference of potential, and the streams consist of distinct threads, generally mingled with thin sparks, which probably correspond to the successive discharges occurring between the knobs. But when the frequency is extremely high, and the arc of the discharge produces a very *loud* but *smooth* sound — showing both that oscillation takes place and that the sparks succeed each other with great rapidity — then the luminous streams formed are perfectly uniform. To reach this result very small coils and jars of small capacity should be used. I take two tubes of thick Bohemian glass, about 5 centimeters in diameter and 20 centimeters long. In each of the tubes I slip a primary of very thick copper wire. On the top of each tube I wind a secondary of much thinner gutta-percha covered wire. The two secondaries I connect in series, the primaries preferably in multiple arc. The tubes are then placed in a large glass vessel, at a distance of 10 to 15 centimeters from each other, on insulating supports, and the vessel is filled with boiled out oil, the oil reaching about an inch above the tubes. The free ends of the secondary are lifted out of the oil and placed parallel to each other at a distance of about 10 centimeters. The ends which are scraped should be dipped in the oil. Two four-pint jars joined in series may be used to discharge through the primary. When the necessary adjustments in the length and distance of the wires above the oil and in the arc of discharge are made, a luminous sheet is produced between the wires which is perfectly smooth and textureless, like the ordinary discharge through a moderately exhausted tube.

I have purposely dwelt upon this apparently insignificant experiment. In trials of this kind the experimenter arrives at the startling conclusion that, to pass ordinary luminous discharges through gases, no particular degree of exhaustion is needed, but that the gas may be at ordinary or even greater pressure. To accomplish this, a very high frequency is essential; a high potential is likewise required, but this is a merely incidental necessity. These experiments teach us that, in endeavoring to discover novel methods of producing light by the agitation of atoms, or molecules, of a gas, we need not limit our research to the vacuum tube, but may look forward quite seriously to the possibility of obtaining the light effects without the use of any vessel whatever, with air at ordinary pressure.

Such discharges of very high frequency, which render luminous the air at ordinary pressures, we have probably often occasion to witness in Nature. I have no doubt that if, as many believe, the aurora borealis is produced by sudden cosmic disturbances, such as eruptions at the sun's surface, which set the electrostatic charge of the earth in an extremely rapid vibration. the red glow observed is not confined to the upper rarefied strata of the air, but the discharge traverses, by reason of its very high frequency, also the dense atmosphere in the form of *a glow,* such as we ordinarily produce in a slightly exhausted tube. If the frequency were very low, or even more so, if the charge were not at all vibrating, the dense air would break down as in a lightning discharge. Indications of such breaking down of the lower dense strata of the air have been repeatedly observed at the occurrence of this marvelous phenomenon; but if it does occur, it can only be attributed to the fundamental disturbances, which are few in number, for the vibration produced by them would be far too rapid to allow a disruptive break. It is the original and irregular impulses which affect the instruments; the superimposed vibrations probably pass unnoticed.

When an ordinary low frequency discharge is passed through moderately rarefied air, the air assumes a purplish hue. If by some means or other we increase the intensity of the molecular, or atomic, vibration, the gas changes to a white color. A similar change occurs at ordinary pressures with electric impulses of very high frequency. If the molecules of the air around a wire are moderately agitated, the brush formed is reddish or violet; if the vibration is rendered sufficiently intense, the streams become white. We may accomplish this in various ways. In the experiment before shown with the two wires across the room, I have endeavored to secure the result by pushing to a high value both the frequency and potential; in the experiment with the thin wires glued on the rubber plate I have concentrated the action upon a very small surface — in other words, I have worked with a great electric density.

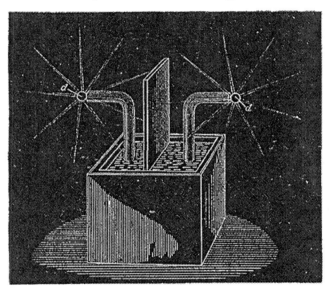

FIG. 11.—PHANTOM STREAMS.

A most curious form of discharge is observed with such a coil when the frequency and potential are pushed to the extreme limit. To perform the experiment, every part of the coil should be heavily insulated, and only two small spheres — or, better still, two sharp-edged metal discs (d d, Fig. 11) of no more than a few centimeters in diameter — should be exposed to the air. The coil here used is immersed in oil, and the ends of the secondary reaching out of the oil are covered with an air-tight cover of hard rubber of great thickness. All cracks, if there are any, should be carefully stopped up, so that the brush discharge cannot form anywhere except on the small spheres or plates which are exposed to the air. In this case, since there are no large plates or other bodies of capacity attached to the terminals, the coil is capable of an extremely rapid vibration. The potential may be raised by increasing, as far as the experimenter judges proper, the rate of change of the primary current. With a coil not widely differing from the present, it is best to connect the two primaries in multiple arc; but if the secondary should have a much greater number of turns the primaries should preferably be used in series, as otherwise the vibration might be too fast for the secondary. It occurs under these conditions that misty white streams break forth from the edges of the discs and spread out phantom-like into space. With this coil, when fairly well produced, they are about 25 to 30 centimeters long. When the hand is held against them no sensation is produced, and a spark, causing a shock, jumps from the terminal only upon the hand being brought much nearer. If the oscillation of the primary current is rendered intermittent by some means or other, there is a corresponding throbbing of the streams, and now the hand or other conducting object may he brought in still greater proximity to the terminal without a spark being caused to jump.

Among the many beautiful phenomena which may be produced with such a coil I have here selected only those which appear to possess some features of novelty, and lead us to some conclusions of interest. One will not find it at all difficult to produce in the laboratory, by means of it, many other phenomena which appeal to the eye even more than these here shown, but present no particular feature of novelty.

Early experimenters describe the display of sparks produced by an ordinary large induction coil upon an insulating plate separating the terminals. Quite recently Siemens performed some experiments in which fine effects were obtained, which were seen by many with interest. No doubt large coils, even if operated with currents of low frequencies, are capable of producing beautiful effects. But the largest coil ever made could not, by far, equal the magnificent display of streams and sparks obtained from such a disruptive discharge coil when properly adjusted. To give an idea, a coil such as the present one will cover easily a plate of 1 meter in diameter completely with the streams. The best *way* to perform such experiments is to take a very thin rubber or a glass plate and glue on one side of it a narrow ring of tinfoil of very large diameter, and on the other a circular washer, the center of the latter coinciding with that of the ring, and the surfaces of both being preferably equal, so as to keep the coil well balanced. The washer and ring should be connected to the terminals by heavily insulated thin wires. It is easy in observing the effect of the capacity to produce a sheet of uniform streams, or a fine network of thin silvery threads, or a mass of loud brilliant sparks, which completely cover the plate.

Since I have advanced the idea of the conversion by means of the disruptive discharge, in my paper before the American Institute of Electrical Engineers at the beginning of the past year, the interest excited in it has been considerable. It afford:, us a means for producing any potentials by the aid of inexpensive coils operated from ordinary systems of distribution, and — what is perhaps more appreciated — it enables us to convert currents of any frequency into currents of any other lower or higher frequency. But its chief value will perhaps be found in the help which it will afford us in the investigations of the phenomena of phosphorescence, which a disruptive discharge coil is capable of exciting in innumerable cases where ordinary coils, even the largest, would utterly fail.

Considering its probable uses for many practical purposes, and its possible introduction into laboratories for scientific research, a few additional remarks as to the construction of such a coil will perhaps not be found superfluous.

It is, of course, absolutely necessary to employ in such a coil wires provided with the best insulation.

Good coils may be produced by employing wires covered with several layers of cotton, boiling the coil a long time in pure wax, and cooling under moderate pressure. The advantage of such a coil is that it can be easily handled, but it cannot probably give as satisfactory results as a coil immersed in pure oil. Besides, it seems that the presence of a large body of wax affects the coil disadvantageously, whereas this does not seem to be the case with oil. Perhaps it is because the dielectric losses in the liquid are smaller.

I have tried at first silk and cotton covered wires with oil immersion, but I have been gradually led to use gutta-percha covered wires, which proved most satisfactory. Gutta-percha insulation adds, of course, to the capacity of the coil, and this, especially if the coil be large, is a great disadvantage when extreme frequencies are desired; but, on the other hand, gutta-percha will withstand much more than an equal thickness of oil, and this advantage should be secured at any price. Once the coil has been immersed, it should never be taken out of the oil for more than a few hours, else the gutta-percha will crack up and the coil will not be worth half as much as before. Gutta-percha is probably slowly attacked by the oil, but after an immersion of eight to nine months I have found no ill effects.

I have obtained in commerce two kinds of gutta-percha wire: in one the insulation sticks tightly to the metal, in the other it does not. Unless a special method is followed to expel all air, it is much safer to use the first kind. I wind the coil within an oil tank so that all interstices are filled up with the oil. Between the layers I use cloth boiled out thoroughly in oil, calculating the thickness according to the difference of potential between the turns. There seems not to be a very great difference whatever kind of oil is used; I use paraffine or linseed oil.

To exclude more perfectly the air, an excellent way to proceed, and easily practicable with small coils, is the following: Construct a box of hard wood of very thick boards which have been for a long time boiled in oil. The boards should be so joined as to safely withstand the external air pressure. The coil being placed and fastened in position within the box, the latter is closed with a strong lid, and covered with closely fitting metal sheets, the joints of which are soldered very carefully. On the top two small holes are drilled, passing through the metal sheet and the wood, and in these holes two small glass tubes are inserted and the joints made air-tight. One of the tubes is connected to a vacuum pump, and the other with a vessel containing a sufficient quantity of boiled-out oil. The latter tube has a very small hole at the bottom, and is provided with a stopcock. When a fairly good vacuum has been obtained, the stopcock is opened and the oil slowly fed in. Proceeding in this manner, it is impossible that any big bubbles, which are the principal danger, should remain between the turns. The air is most completely excluded, probably better than by boiling out, which, however, when gutta-percha coated wires are used, is not practicable.

For the primaries I use ordinary line wire with a thick cotton coating. Strands *of* very thin insulated wires properly interlaced would, of course, be the best to employ for the primaries, but they are not to be had.

In an experimental coil the size of the wires is not of great importance. In the coil here used the primary is No. 12 and the secondary No. 24 Brown & Sharpe gauge wire; but the sections may be varied considerably. I would only imply different adjustments; the results aimed at would not be materially affected.

I have dwelt at some length upon the various forms of brush discharge because., in studying them, we not only observe phenomena which please our eye, but also afford us food for thought, and lead us to conclusions of practical importance. In the use of alternating currents of very high tension, too much precaution cannot be taken to prevent ft! brush discharge. In a main conveying such currents, in an induction coil or transformer, or in a condenser, the brush discharge is a source of great danger to the insulation. In a condenser especially the gaseous matter must be most carefully expelled, for in it the charged surfaces are near each other, and if the potentials are high, just as sure as a weight will fall if let go, so the insulation will give way if a single gaseous bubble of some size be present, whereas, if all gaseous matter were carefully excluded, the condenser would safely withstand a much higher difference of potential. A main conveying alternating currents of very high tension may be injured merely by a blow-hole or small crack in the insulation, the more so as a blowhole is apt to contain gas at low pressure; and as it appears almost impossible to completely obviate such little imperfections, I am led to believe that in our future distribution of electrical energy by currents of very high tension liquid insulation will be used. The cost is a great drawback, but if we employ an oil as an insulator the distribution of electrical energy with something like 100,000 volts, and even more, become, at least with higher frequencies, so easy that they could be hardly called engineering feats. With oil insulation and alternate current motors transmissions of power can be effected with safety and upon an industrial basis at distances of as much as a thousand miles.

A peculiar property of oils, and liquid insulation in, general, when subjected to rapidly changing electric stresses, is to disperse any gaseous bubbles which may be present, and diffuse them through its mass, generally long before any injurious break can occur. This feature may be easily observed with an ordinary induction coil by taking the primary out, plugging up the end of the tube upon which the secondary is wound, and filling it with some fairly transparent insulator, such as paraffine oil. A primary of a diameter something like six millimeters smaller than the inside of the tube may be inserted in the oil. When the coil is set to work one may see, looking from the top through the oil, many luminous points — air bubbles which are caught by inserting the primary, and which are rendered luminous in consequence of the violent bombardment. The occluded air, by its impact against the oil, beats it; the oil begins to circulate, carrying some of the air along with it, until the bubbles are dispersed and the luminous points disappear. In this manner, unless large bubbles are occluded in such way that circulation is rendered impossible, a damaging break is averted, the only effect being a moderate warming up of the oil. If, instead of the liquid, a solid insulation, no matter how thick, were used, a breaking through and injury of the apparatus would be inevitable.

The exclusion of gaseous matter from any apparatus in which the dielectric is subjected to more or less rapidly changing electric forces is, however, not only desirable in order to avoid a possible injury of the apparatus, but also on account of economy. In a condenser, for instance, as long as only a solid or only a liquid dielectric is used, the loss is small; but if a gas under ordinary or small pressure be present the loss may be very great. Whatever the nature of the force acting in the dielectric may be, it seems that in a solid or liquid the molecular displacement produced by the force is small : hence the product of force and displacement is insignificant, unless the force be very great; but in a gas the displacement, and therefore this product, is considerable; the molecules are free to move, they reach high speeds, and the energy of their impact is lost in heat or otherwise. If the gas be strongly compressed, the displacement due to the force is made smaller, and the losses are reduced.

In most of the succeeding experiments I prefer, chiefly on account of the regular and positive action, to employ the alternator before referred to. This is one of the several machines constructed by me for the purposes of these investigations. It has 384 pole projections, and is capable of giving currents of a frequency of about 10,000 per second. This machine has been illustrated and briefly described in my first paper before the American Institute of Electrical Engineers, May 20, 1891, to which I have already referred. A more detailed description, sufficient to enable any engineer to build a similar machine, will be found in several electrical journals of that period.

The induction coils operated from the machine are rather small, containing from 5,000 to 15,000 turns in the secondary. They are immersed in boiled-out linseed oil, contained in wooden boxes covered with zinc sheet.

I have found it advantageous to reverse the usual position of the wires, and to wind, in these coils, the primaries on the top; this allowing the use of a much bigger primary, which, of course, reduces the danger of overheating and increases the output of the coil. I make the primary on each side at least one centimeter shorter than the secondary, to prevent the breaking through on the ends, which would surely occur unless the insulation on the top of the secondary be very thick, and this, of course, would be disadvantageous.

When the primary is made movable, which is necessary in some experiments, and many times convenient for the purposes of adjustment, I cover the secondary with wax, and turn it off in a lathe to a diameter slightly smaller than the inside of the primary coil. The latter I provide with a handle reaching out of the oil, which serves to shift it in any position along the secondary.

I will now venture to make, in regard to the general manipulation of induction coils, a few observations bearing upon points which have not been fully appreciated in earlier experiments with such coils, and are even now often overlooked.

The secondary of the coil possesses usually such a high self-induction that the current through the wire is inappreciable, and may be so even when the terminals are joined by a conductor of small resistance. If capacity is added to the terminals, the self-induction is counteracted, and a stronger current is made to flow through the secondary, though its terminals are insulated from each other. To one entirely unacquainted with the properties of alternating currents nothing will look more puzzling. This feature was illustrated in the experiment performed at the beginning with the top plates of wire gauze attached to the terminals and the rubber plate. When the plates of wire gauze were close together, and a small arc passed between them, the arc *prevented* a strong current from passing through the secondary, because it did away with, the capacity on the terminals; when the rubber plate was inserted between, the capacity of the condenser formed counteracted the self-induction of the secondary, a stronger current passed now, the coil performed more work, and the discharge was by far more powerful.

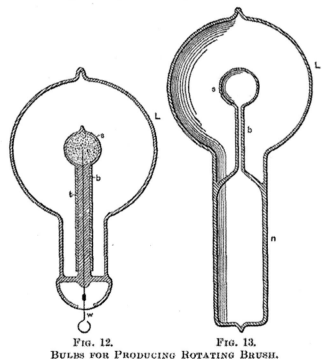

FIG. 12.   FIG. 13.
BULBS FOR PRODUCING ROTATING BRUSH.

The first thing, then, in operating the induction coil is to ,combine capacity with the secondary to overcome the self-induction. If the frequencies and potentials are very high gaseous matter should be carefully kept away from the charged surfaces. If Leyden jars are used, they should be immersed in oil, as otherwise considerable dissipation may occur if the jars are greatly strained. When high frequencies are used, it is of equal importance to combine a condenser with the primary. One may use a condenser connected to the ends of the primary or to the terminals of the alternator, but the latter is not to be recommended, as the machine might be injured. The best way is undoubtedly to use the condenser in series with the primary and with the alternator, and to adjust its capacity so as to annul the self-induction of both the latter. The condenser should be adjustable by very small steps, and for a finer adjustment a small oil condenser with movable plates may be used conveniently.

I think it best at this juncture to bring before you a phenomenon, observed by me some time ago, which to the purely scientific investigator may perhaps appear more interesting than any of the results which I have the privilege to present to you this evening.

It may be quite properly ranked among the brush phenomena — in fact, it is a brush, formed at, or near, a single terminal in high vacuum.

In bulbs provided with a conducting terminal, though it be of aluminum, the brush has but an ephemeral existence, and cannot, unfortunately, be indefinitely preserved in its most sensitive state, even in a bulb devoid of *any* conducting electrode. In studying one phenomenon, by all means a bulb having no leading-in wire should be used. I have found it best to use bulbs constructed as indicated in Figs. 12 and 13.

In Fig. 12 the bulb comprises an incandescent lamp globe L, in the neck of which is sealed a barometer tube b, the end of which is blown out to form a small sphere s. This sphere should be sealed as closely as possible in the center of the large globe. Before sealing, a thin tube t, of aluminum sheet, may be slipped in the barometer tube, but it is not important to employ it.

The small hollow sphere s is filled with some conducting powder, and a wire w, is cemented in the neck for the purpose of connecting the conducting powder with the generator.

The construction shown in Fig. 13 was chosen in order to remove from the brush any conducting body which might possibly affect it. The bulb consists in this case of a lamp globe L, which has a neck n, provided with a tube b and small sphere s, sealed to it, so that two entirely independent compartments are formed, as indicated in the drawing. When the bulb is in use, the neck n is provided with a tinfoil coating, which is connected to the generator and acts inductively upon the moderately rarefied and highly conducting gas inclosed in the neck. From there the current passes through the tube b into the small spheres, to act by induction upon the gas contained in the globe L.

It is of advantage to make the tube t very thick, the hole through it very small, and to blow the sphere s very thin, It is of the greatest importance that the sphere s be placed in the center of the globe L.

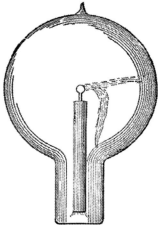

FIG. 14.—FORMS AND PHASES OF THE ROTATING BRUSH.

Figs. 14, 15 and '16 indicate different forms, or stages, of the brush. Fig. 14 shows the brush as it first appears in a bulb provided with a conducting terminal: but, as in such a bulb it very soon disappears — often after a few minutes — I will confine myself to the description of the phenomenon as seen in a bulb without conducting electrode. It is observed under the following conditions:

When the globe L (Figs. 12 and 13) is exhausted to a very high degree, generally the bulb is not excited upon connecting the wire w (Fig. 12) or the tinfoil coating of the bulb (Fig. 13) to the terminal of the induction coil. To excite it, it is usually sufficient to gasp the globe L with the hand. An intense phosphorescence then spreads at first over the globe, but soon gives place to a white, misty light. Shortly afterward one may notice that the luminosity is unevenly distributed in the globe, and after passing the current for some time the bulb appears as in Fig. 15. From this stage the phenomenon will gradually pass to that indicated in Fig. 16, after some minutes, hours, days or weeks, according as the bulb is worked. Warming the bulb or increasing the potential hastens the transit.

When the brush assumes the form indicated in Fig. 16, it may be brought to a state of extreme sensitiveness to electrostatic and magnetic influence. The bulb hanging straight down from a wire, and all objects being remote from it, the approach of the observer at a few paces from the bulb will cause the brush to fly to the opposite side, and if he walks around the bulb it will always keep on the opposite side. It may begin to spin around the terminal long before it reaches that sensitive stage. When it begins to turn around principally, but also before, it is affected by a magnet, and at a certain stage it is susceptible to magnetic influence to an astonishing degree. A small permanent magnet, with its poles at a distance of no more than two centimeters, will affect it visibly at a distance of two meters, slowing down or accelerating the rotation according to how it is held relatively to the brush. I think I have observed that at the stage when it is most sensitive to magnetic, it is not most sensitive to electrostatic, influence. My explanation is, that the electrostatic attraction between the brush and the glass of the bulb, which retards the rotation, grows much quicker than the magnetic influence when the intensity of the stream is increased.

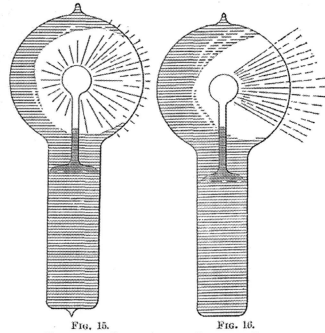

FIG. 15. FIG. 16.
FORMS AND PHASES OF THE ROTATING BRUSH.

When the bulb hangs with the globe L down, the rotation is always clockwise. In the southern hemisphere it would occur in the opposite direction and on the equator the brush should not turn at all. The rotation may be reversed by a magnet kept at some distance. The brush rotates best, seemingly, when it is at right angles to the lines of force of the earth. It very likely rotates, when at its maximum speed, in synchronism with the alternations, say 10,000 times a second. The rotation can be slowed down or accelerated by the approach or receding *of* the observer, or any conducting *body,* but it cannot be reversed by putting the bulb in any position. When it is in the state of the highest sensitiveness and the potential or frequency be varied the sensitiveness is rapidly diminished. Changing either of these but little will generally stop the rotation. The sensitiveness is likewise affected by the variations of temperature. To attain great sensitiveness it is necessary to have the small sphere s in the center of the globe L, as otherwise the electrostatic action of the glass of the globe will tend to stop the rotation. The sphere *s* should be small and of uniform thickness; any dissymmetry of course has the effect to diminish the sensitiveness.

The fact that the brush rotates in a definite direction in a permanent magnetic field seems to show that in alternating currents of very high frequency the positive and negative impulses are not equal, but that one always preponderates over the other.

Of course, this rotation in one direction may be due to the action of two elements of the same current upon each other, or to the action of the field produced by one of the elements upon the other, as in a series motor, without necessarily one impulse being stronger than the other. The fact that the brush turns, as far as I could observe, in any position, would speak for this view. In such case it would turn at any point of the earth's surface. But, on the other hand, it is then hard to explain why a permanent magnet should reverse the rotation, and one must assume the preponderance of impulses of one kind.

As to the causes of the formation of the brush or stream, I think it is due to the electrostatic action of the globe and the dissymmetry of the parts. If the small bulb s and the globe L were perfect concentric spheres, and the glass throughout of the same thickness and quality, I think the brush would not form, as the tendency to pass would, be equal on all sides. That the formation of the stream is due to an irregularity is apparent from the fact that it has the tendency to remain in one position, and rotation occurs most generally only when it is brought out of this position by electrostatic or magnetic influence. When in an extremely sensitive state it rests in one position, most curious experiments may be performed with it. For instance, the experimenter may, by selecting a proper position, approach the hand at a certain considerable distance to the bulb, and he may cause the brush to pass off 'by merely stiffening the muscles of the arm. When it begins to rotate slowly, and the hands are held at a proper distance, it is impossible to make even the slightest motion without producing a visible effect upon the brush. A metal plate connected to the other terminal of the coil affects it at a great distance, slowing down the rotation often to one turn a second.

I am firmly convinced that such a brush, when we learn how to produce it properly, will prove a valuable aid in the investigation of the nature of the forces acting in 1n electrostatic or magnetic field. If there is any motion which is measurable going on in the space, such a brush ought to reveal it. It is, so to speak, a beam of light, frictionless, devoid of inertia.

I think that it may find practical applications in telegraphy. With such a brush it would be possible to send dispatches across the Atlantic, for instance, with any speed, since its sensitiveness may be so great that the slightest changes will affect it. If it were possible to make the stream more intense and very narrow, its deflections could be easily photographed.

I have been interested to find whether there is a rotation of the stream itself, or whether there is simply a stress traveling around in the bulb. For this purpose I mounted a light mica fan so that its vanes were in the path of the brush. If the stream itself was rotating the fan would be spun around. I could produce no distinct rotation of the fan, although I tried the experiment repeatedly; but as the fan exerted a noticeable influence on the stream, and the apparent rotation of the latter was, in this case, never quite satisfactory, the experiment did not appear to be conclusive.

I have been unable to produce the phenomenon with the disruptive discharge coil, although every other of these phenomena can be well produced by it — many, in fact, much better than with coils operated from an alternator.

It may be possible to produce the brush by impulses of one direction, or even by a steady potential, in which case it would be still more sensitive to magnetic influence.

In operating an induction coil with rapidly alternating currents, we realize with astonishment, for the first time, the great importance of the relation of capacity, self-induction and frequency as regards the general result. The effects of capacity are the most striking, for in these experiments, since the self-induction and frequency both are high, the critical capacity is very small, and need be but slightly varied to produce a very considerable change. The experimenter may bring his body in contact with the terminals of the secondary of the coil, or attach to one or both terminals insulated bodies of very small bulk, such as bulbs, and he may produce a considerable rise or fall of potential, and greatly affect the flow of the current through the primary. In the experiment before shown, in which a brush appears at a wire attached to one terminal, and the wire is vibrated when the experimenter brings his insulated body in contact with the other terminal of the coil, the sudden rise of potential was made evident.

I may show you the behavior of the coil in another manner which possesses a feature of some interest. I have here a little light fan of aluminum sheet, fastened to a needle and arranged to rotate freely in a metal piece screwed to one of the terminals of the coil. When the coil is set to work, the molecules of the air are rhythmically attracted and repelled. As the force with which they are repelled is greater than that with which they are attracted, it results that there is a repulsion exerted on the surfaces of the fan. If the fan were made simply of a metal sheet, the repulsion would be equal on the opposite sides, and would produce no effect. But if one of the opposite surfaces is screened, or if, generally speaking, the bombardment on this side is weakened in some way or other, there remains the repulsion exerted upon the other, and the fan is set in rotation. The screening is best effected by fastening upon one of the opposing sides of the fan insulated conducting coatings, or, if the fan is made in the shape of an ordinary propeller screw, by fastening on one side, and close to it, an insulated metal plate. The static screen may however, be omitted and simply a thickness of insulating, material fastened to one of the sides of the fan.

To show the behavior of the coil, the fan may be placed upon the terminal and it will readily rotate when the coil is operated by currents of very high frequency. With a steady potential, of course, and even with alternating currents of very low frequency, it would not turn, because of the very slow exchange of air and, consequently, smaller bombardment; but in the latter case it might turn if the potential were excessive. With a pin wheel, quite the opposite rule holds good; it rotates best with a steady potential, and the effort is the smaller the higher the frequency. Now, it is very easy to adjust the conditions so that the potential is normally not sufficient to turn the fan, but that by connecting the other terminal of the coil with an insulated body it rises to a much greater value, so as to rotate the fan, and it is likewise possible to stop the rotation by connecting to the terminal a body of different size, thereby diminishing the potential.

Instead of using the fan in this experiment, we may use the "electric" radiometer with similar effect. But in this case it will be found that the vanes will rotate only at high exhaustion or at ordinary pressures; they will not rotate at moderate pressures, when the air is highly conducting. This curious observation was made conjointly by Professor Crookes and myself. I attribute the result to the high conductivity of the air, the molecules of which then do not act as independent carriers of electric charges, opt act all together as a single conducting body. In such case, of course, if there is any repulsion at all of the molecules from the vanes, it must be very small. It is possible, however, that the result is in part due to the fact that the greater part of the discharge passes from the leading-in wire through the highly conducting bas, instead of passing off from the conducting vanes.

In trying the preceding experiment with the electric radiometer the potential should not exceed a certain limit, as then the electrostatic attraction between the vanes and the glass of the bulb may be so great as to stop the rotation.

A most curious feature of alternate currents of high frequencies and potentials is that they enable us to perform many experiments by the use of one wire *only*. In many respects this feature is of great interest.

In a type of alternate current motor invented by me some years ago I produced rotation by inducing, by means of a single alternating current passed through a motor circuit, in the mass or other circuits of the motor, secondary currents, which, jointly with the primary or inducing current, created a moving field of force. A simple but crude form of such a motor is obtained by winding upon an iron core a primary, and close to it a secondary coil, joining the ends of the latter and placing n freely movable metal disc within the influence of the field produced by both. The iron core is employed for obvious reasons, but it is not essential to the operation. To improve the motor, the iron core is made to encircle the armature. Again to improve, the secondary coil is made; to overlap partly the primary, so that it cannot free itself from a strong inductive action of the latter, repel its lines as it may. Once more to improve, the proper difference of phase is obtained between the primary and secondary currents by a condenser, self-induction, resistance or equivalent windings.

I had discovered, however, that rotation is produced by means of a single coil and core; my explanation of the phenomenon, and leading thought in trying the experiment, being that there must be a true time lag in the magnetization of the core. I remember the pleasure I had when, in the writings of Professor Ayrton, which came later to my hand, I found the idea of the time lag advocated. Whether there is a true time lag, or whether the retardation is due to eddy currents circulating in minute paths, must remain an open question, but the fact is that a coil wound upon an iron core and traversed by an alternating current creates a moving field of force, capable of setting an armature in rotation. It is of some interest, in conjunction with the historical Arago experiment, to mention that in lag or phase motors I have produced rotation in the opposite direction to the moving field, which means that in that experiment the magnet may not rotate, or may even rotate in the opposite direction to the moving disc. Here, then, is a motor (diagrammatically illustrated in Fig. 17), comprising a coil and iron core, and a freely movable copper disc in proximity to the latter.

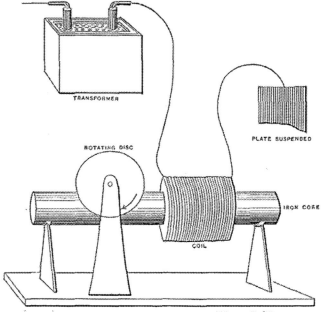

FIG. 17.—SINGLE WIRE AND "NO-WIRE" MOTOR.

To demonstrate a novel and interesting feature, I have, for a reason which I will explain, selected this type of motor. When the ends of the coil are connected to the terminals of an alternator the disc is set in rotation. But it is not this experiment, now well known, which I desire to perform. What I wish to show you is that this motor rotates with *one single* connection between it and the generator; that is to say, one terminal of the motor is connected to one terminal of the generator — in this case the secondary of a high-tension induction coil — the other terminals of motor and generator being insulated in space. To produce rotation it is generally (but not absolutely) necessary to connect the free end of the motor coil to an insulated body of some size. The experimenter's body is more than sufficient. If he touches the free terminal with an object held in the hand, a current passes through the coil and the copper disc is set in rotation. If an exhausted tube is put in series with the coil, the tube lights brilliantly, showing the passage of a strong current. Instead of the experimenter's body, a small metal sheet suspended on a cord may be used with the same result. In this case the plate acts as a condenser in series with the coil. It counteracts the self-induction of the latter and allows a strong current to pass. In such a combination, the greater the self-induction of :the coil the smaller need be the plate, and this means that a lower frequency, or eventually a lower potential, is required to operate the motor. A single coil wound upon a core has a high self-induction; for this reason principally, this type of motor was chosen to perform the experiment. Were a secondary closed coil wound upon the core, it would tend to diminish the self-induction, and then it would be necessary to employ a much higher frequency and potential. Neither would be advisable, for a higher potential would endanger the insulation of the small primary coil, and a higher frequency would result in a materially diminished torque.

It should be remarked that when such a motor with a closed secondary is used, it is not at all easy to obtain rotation with excessive frequencies, as the secondary cuts off almost completely the lines of the primary — and this, of course, the more, the higher the frequency — and allows the passage of but a minute current. In such a case, unless the secondary is closed through a condenser, it is almost essential, in order to produce rotation, to make the primary and secondary coils overlap each other more or less.

But there is an additional feature of interest about this motor, namely, it is not necessary to have even a single connection between the motor and generator, except, perhaps, through the ground; for not only is an insulated plate capable of giving off energy into space, but it is likewise capable of deriving it from an alternating electrostatic field, though in the latter case the available energy is much smaller. In this instance one of the motor terminals is connected to the insulated plate or body located within the alternating electrostatic field, and the ether terminal preferably to the ground.

It is quite possible, however, that such "no-wire" motors, as they might be called. could be operated by conduction through the rarefied air at considerable distances. Alternate currents, especially of high frequencies, pass with astonishing freedom through even slightly rarefied gases. The upper strata of the air are rarefied. To reach a number of miles out into space requires the overcoming of difficulties of a merely mechanical nature. There is no doubt that with the enormous potentials obtainable by the use of high frequencies and oil insulation luminous discharges might be passed through many miles of rarefied air, and that, by thus directing discharges energy of many hundreds or thousands of horse-power, motors or lamps might be operated at considerable distances from stationary sources. But such schemes are mentioned merely as possibilities. We shall have no need to *transmit* power at all. Ere many generations pass, cur machinery will be driven by a power obtainable at any point of the universe. This idea is not novel. Men have been led to it long *ago by* instinct or reason. It has been expressed in many ways, and in many places, in the history of old and new. We find it in the delightful myth of Antheus, who derives power from the earth; we find it among the subtle speculations of one of your splendid mathematicians, and in many hints and statements of thinkers of the present time. Throughout space there is energy. Is this energy static or kinetic? If static our hopes are in vain; if kinetic — and this we know it is, for certain — then it is a mere question of time when men will succeed in attaching their machinery to the very wheelwork of nature. Of all, living or dead, Crookes came nearest to doing it. His radiometer will turn in the light of day and in the darkness of the night; it will turn everywhere where there is heat, and heat is everywhere. But, unfortunately, this beautiful little machine, while it goes down to posterity as the most interesting, must likewise be put on record as the most inefficient machine ever invented!

The preceding experiment is only one of many equally interesting experiments which may be performed by the use of only one wire with alternate currents of high potential and frequency. We may connect an insulated line to a source of such currents, we may pass an inappreciable current over the line, and on any point of the same we are able to obtain a heavy current, capable of fusing a thick copper wire. Or we may, by the help of some artifice, decompose a solution in any electrolytic cell by connecting only one pole of the cell to the line or source of energy. Or we may, by attaching to the line, or only bringing into its vicinity, light up an incandescent lamp, an exhausted tube, or a phosphorescent bulb.

However impracticable this plan of working may appear in many cases, it certainly seems practicable, and even recommendable, in the production of light. A perfected lamp would require but little energy, and if wires were used at all we ought to be able to supply that energy without a return wire.

It is now a fact that a body may be rendered incandescent or phosphorescent by bringing it either in single contact or merely in the vicinity of a source of electric impulses of the proper character, and that in this manner a quantity of light sufficient to afford a practical illuminant may be produced. It is, therefore, to say the least, worth while to attempt to determine the best conditions and to invent the best appliances for attaining this object.

Some experiences have already been gained in this direction, and I will dwell on them briefly, in the hope that they might prove useful.

The heating of a conducting body inclosed in a bulb, and connected to a source of rapidly alternating electric impulses, is dependent on so many things of a different nature, that it would be difficult to give a generally applicable rule under which the maximum heating occurs. As regards the size of the vessel, I have lately found that at ordinary or only slightly differing atmospheric pressures, when air is a good insulator; and hence practically the same amount of energy by a certain potential and frequency is given off from the body, whether the bulb be small or large, the body is brought to a higher temperature if inclosed in a small bulb, because of the better confinement of heat in this case.

At lower pressures, when air becomes more or less conducting, or if the air be sufficiently warmed as to become conducting, the body is rendered more intensely incandescent in a large bulb, obviously because, under otherwise equal conditions of test, more energy may be given off from the body when the bulb is large.

At very high degrees of exhaustion, when the matter in the bulb becomes "radiant," a large bulb has still an advantage, but a comparatively slight one, over the small bulb. Finally, at excessively high degrees of exhaustion, which cannot be reached except

by the employment of special means, there seems to be, beyond a certain and rather small size of vessel, no perceptible difference in the heating.

These observations were the result of a number of experiments, of which one, showing the effect of the size of the bulb at a high degree of exhaustion, may be described and shown here, as it presents a feature of interest. Three spherical bulbs of 2 inches, 3 inches and 4 inches diameter were taken, and in the center of each was mounted an equal length of an ordinary incandescent lamp filament of uniform thickness. In each bulb the piece of filament was fastened to the leading-in wire of platinum, contained in a glass stem sealed in the bulb; care being taken, of course, to make everything as nearly alike as possible. On each glass stem in the inside of the bulb was slipped a highly polished tube made of aluminum sheet, which fitted the stem and was held on it by spring pressure. The function of this aluminum tube will be explained subsequently. In each bulb an equal length of filament protruded above the metal tube. It is sufficient to say now that under these conditions equal lengths of filament of the same thickness — in other words, bodies of equal bulk — were brought to incandescence. The three bulbs were sealed to a glass tube, which was connected to a Sprengel pump. When a high vacuum had been reached, the glass tube carrying the bulbs was sealed off. A current was then turned on successively on each bulb, and it was found that the filaments came to about the same brightness, and, if anything, the smallest bulb, which was placed midway between the two larger ones, may have been slightly brighter. This result was expected, for when either of the bulbs was connected to the coil the luminosity spread through the other two, hence the three bulbs constituted really one vessel. When all the three bulbs were connected in multiple arc to the coil, in the largest of them the filament glowed brightest, in the next smaller it was a little less bright, and in the smallest it only came to redness. The bulbs were then sealed off and separately tried. The brightness of the filaments was now such as would have been expected on the supposition that the energy given off was proportionate to the surface of the bulb, this surface in each case representing one of the coatings of a condenser. Accordingly, there was less difference between the largest and the middle sized than between the latter and the smallest bulb.

An interesting observation was made in this experiment. The three bulbs were suspended from a straight bare wire connected to a terminal of the coil, the largest bulb being placed at the end of the wire, at some distance from it the smallest bulb, and an equal distance from the latter the middle-sized ore. The carbons ,lowed then it both the larger bulbs about as expected, but the smallest did not get its share by far. This observation led me to exchange the position of the bulbs, and I then observed that whichever of the bulbs was in the middle it was by far less bright than it was in any other position. This mystifying result eras, of course, found to be due to the electrostatic action between the bulbs. When they were placed at a considerable distance, or when they were attached to the corners of an equilateral triangle of copper wire, they glowed about in the order determined by their surfaces.

As to the shape of the vessel, it is also of some importance, especially at high degrees of exhaustion. Of all the possible constructions, it seems that a spherical ,lobe with the refractory body mounted in its center is the best to employ. In experience it has been demonstrated that in such a globe a refractory body of a given bulk is more easily brought to incandescence than when otherwise shaped bulbs are used. There is also an advantage in giving to the incandescent body the shape of a sphere, for self-evident reasons. In any case the body should be mounted in the center, where the atoms rebounding from the glass collide. This object is best attained in the spherical bulb; but it is also attained in a cylindrical vessel with one or two straight filaments coinciding with its axis, and possibly also in parabolical or spherical bulbs with the refractory body or bodies placed in the focus or foci of the same; though the latter is not probable, as the electrified atoms should in all cases rebound normally from the surface they strike. unless the speed were excessive, in which case they *would* probably follow the general law of reflection. No matter what shape the vessel may have, if the exhaustion be low. a filament mounted in the globe is brought to the same degree of incandescence in all harts: but if the exhaustion be high and the bulb be spherical or pear-shaped, as usual, focal points form and the filament is heated to a higher degree at or near such points.

To illustrate the effect, I have here two small bulbs which are alike, only one is exhausted to a low and the other to a very high degree. When connected to the coil. the filament in the former glows uniformly throughout all its length: whereas in the latter, that portion of the filament which is in the center of the bulb ,glows far more intensely than the rest. A curious point is that the phenomenon occurs even if two filaments are mounted in a bulb, each being connected to one terminal of the coil, and. what is still more curious, if they be very near together, provided the vacuum be very high. I noted in experiments with such bulbs that the filaments would give way usually at a certain point, and in the first trials I attributed it to a defect in the carbon. But when the phenomenon occurred many times in succession I recognized its real cause.

In order to bring a refractory body unclosed in a bulb to incandescence. it is desirable, on account of economy, that all the energy sent to the bulb from the source should reach without loss the body to he heated: from there, and from nowhere else, it should be radiated. It is, of course, out of the question to reach this theoretical result, but it is possible by a proper construction of the illuminating device to approximate it more or less.

For many reasons, the refractory body is placed in the center of the bulb and it is usually supported on a glass stem containing the leading-in wire. As the potential of this wire is alternated, the rarefied gas surrounding the stem is acted upon inductively, and the glass stem is violently bombarded and heated. In this manner by far the greater portion of the energy supplied to the bulb — especially when exceedingly high frequencies are used — may be lost for the purpose contemplated. To obviate this loss, or at least to reduce it to a minimum, I usually screen the rarefied gas surrounding the stem from the inductive action of the leading-in wire by providing the stem with a tube or coating of conducting material. It seems beyond doubt that the best among metals to employ for this purpose is aluminum, on account of its many remarkable properties. Its only fault is that it is easily fusible, and, therefore, its distance from the incandescing body should be properly estimated. Usually, a thin tube, of a diameter somewhat smaller than that of the glass stem, is made of the finest aluminum sheet, and slipped on the stem. The tube is conveniently prepared by wrapping around a rod fastened in a lathe a piece of aluminum sheet of the proper size, grasping the sheet firmly with clean chamois leather or blotting paper, and spinning the rod very fast. The sheet is wound tightly around the rod, and a highly polished tube of one or three layers of the sheet is obtained. When slipped on the stem, the pressure is generally sufficient to prevent it from slipping off, but, for safety, the lower edge of the sheet may be turned inside. The upper inside corner of the sheet — that is, the one which is nearest to the refractor incandescent body — should be cut out diagonally, as it often happens that, ill consequence of the intense heat, this corner turns toward the inside and comes very near to, or in contact with, the wire, or filament, supporting the refractory body. The greater part of the energy supplied to the bulb is then used up in heating the metal tube, and the bulb is rendered useless for the purpose. The aluminum sheet should project above the glass stern more or less — one inch or so — or else, if the glass be too close to the incandescing body, it may be strongly heated and become more or less conducting, whereupon it may be ruptured, or may, by its conductivity, establish a good electrical connection between the metal tube and the leading-in wire, in which case. again, most of the energy will be lost in heating the former. Perhaps the best way is to make the top of the glass tube for about an inch, of a much smaller diameter. To still further reduce the danger arising from the heating of the glass stem, and also with the view of preventing an electrical connection between the metal tube and the electrode, I preferably wrap the stem with several layers of thin mica, which extends at least as far as the metal tube. In some bulbs I have also used an outside insulating cover.

The preceding remarks are only made to aid the experimenter in the first trials, for the difficulties which he encounters he may soon find means to overcome in his own way.

To illustrate the effect of the screen, and the advantage of using it, I have here two bulbs of the same size, with their stems, leading-in wires and incandescent lamp filaments tied to the latter, as nearly alike as possible. The stem of one bulb is provided with an aluminum tube, the stem of the other has none. Originally the two bulbs were joined by a tube which was connected to a Sprengel pump. When a high vacuum had been reached, first the connecting tube, and then the bulbs, were sealed off; they are therefore of the same degree of exhaustion. When they are separately connected to the coil giving a certain potential, the carbon filament in the bulb Provided with the aluminum screen in rendered highly incandescent, while the filament in the other bulb may, with the same potential, not even come to redness, although in reality the latter bulb takes generally more energy than the former. When they are both connected together to the terminal, the difference is even more apparent, showing the importance of the screening. The metal tube placed on the stem containing the leading-in wire performs really two distinct functions: First; it acts more or less as an electrostatic screen, thus economizing the energy supplied to the bulb; and, second, to whatever extent it may fail to act electrostatically, it acts mechanically, preventing the bombardment, and consequently intense heating and possible deterioration of the slender support of the refractory incandescent body, or of the glass stem containing the leading-in wire. I say slender support, for it is evident that in order to confine the heat more completely to the incandescing body its support should be very thin, so as to carry away the smallest possible amount of heat by conduction.

Of all the supports used I have found an ordinary incandescent lamp filament to be the best, principally because among conductors it can withstand the highest degrees of heat.

The effectiveness of the metal tube as an electrostatic screen depends largely on the degree of exhaustion.

At excessively high degrees of exhaustion — which are reached by using great care and special means in connection with the Sprengel pump — when the matter in the globe is in the ultra-radiant state, it acts most perfectly. The shadow of the upper edge of the tube is then sharply defined upon the bulb.

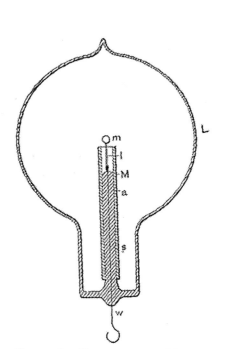 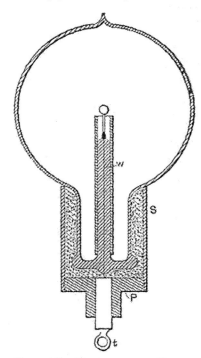

FIG. 18.—BULB WITH MICA TUBE AND ALUMINIUM SCREEN.

FIG. 19.—IMPROVED BULB WITH SOCKET AND SCREEN.

At a somewhat lower degree of exhaustion, which is about the ordinary "non-striking" vacuum, and generally as long as the matter moves predominantly in straight lines, the screen still does well. In elucidation of the preceding remark it is necessary to state that what is a "non-striking" vacuum for a coil operated, as ordinarily, by impulses, or currents, of low frequency, is not, by far, so when the coil is operated by currents of very high frequency. In such case the discharge may pass with great freedom through the rarefied gas through which a low-frequency discharge may not pass, even though the potential be much higher. At ordinary atmospheric pressures just the reverse rule holds good: the higher the frequency, the less the spark discharge is able to jump between the terminals, especially if they are knobs or spheres of some size.

Finally, at very low degrees of exhaustion, when the gas is well conducting, the metal tube not only does not act as an electrostatic screen, but even is a drawback, aiding to a considerable extent the dissipation of the energy laterally from the leading-in wire. This, of course, is to be expected. In this case, namely, the metal tube is in good electrical connection with the leading-in wire, and most of the bombardment is directed upon the tube. As long as the electrical connection is not good, the conducting tube is always of some advantage, for although it may not greatly economize energy, still it protects the support of the refractory button, and is a means for concentrating more energy upon the same.

To whatever extent the aluminum tube performs the function of a screen, its usefulness is therefore limited to very high degrees of exhaustion when it is insulated from the electrode — that is, when the gas as a whole is non-conducting, and the molecules, or atoms, act as independent carriers of electric charges.

In addition to acting as a more or less effective screen, in the true meaning of the word, the conducting tube or coating may also act, by reason of its conductivity, as sort of equalizer or dampener of the bombardment against the stem. To be explicit, I assume the action as follows: Suppose a rhythmical bombardment to occur against the conducting tube by reason of its imperfect action as a screen, it certainly must happen that some molecules, or atoms, strike the tube sooner than others. Those which came first in contact with it give un their superfluous charge, and the tube is electrified, the electrification instantly spreading over its surface. But this must diminish the energy lost in the bombardment for two reasons: first, the charge given up by the atoms spreads over a great area, and hence the electric density at any point is small, and the atoms are repelled with less energy than they would be if they would strike against a good insulator;, secondly, as the tube is electrified by the atoms which first come in contact with it, the progress of the following atoms against the tube is more or less checked by the repulsion which the electrified tube must exert upon the similarly electrified atoms. This repulsion may perhaps be sufficient to prevent a large portion of the atoms from striking the tube, but at any rate it must diminish the energy of their impact. It is clear that when the exhaustion is very low, and the rarefied gas well conducting, neither of the above effects can occur, and, on the other hand, the fewer the atoms, with the greater freedom they move; in other words, the higher the degree of exhaustion, up to a limit, the more telling will be both the effects.

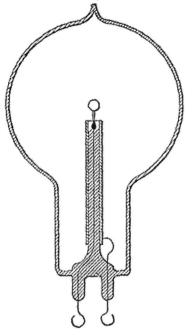

Fig. 20.—Bulb for Experiments with Conducting Tube.

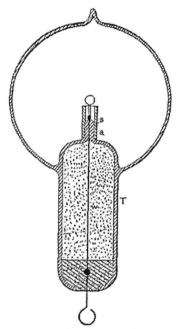

Fig. 21.—Improved Bulb with Non-Conducting Button.

What I have just said may afford an explanation of the phenomenon observed by Prof. Crookes, namely, that a discharge through a bulb is established with much greater facility when an insulator than when a conductor is present in the same. In my opinion, the conductor acts as a dampener of the motion of the atoms in the two ways pointed out; hence, to cause a visible discharge to pass through the bulb, a much higher potential is needed if a conductor, especially of much surface, be present.

For the sake of clearness of some of the remarks before made, I must now refer to Figs. 18, 19 and 20, which illustrate various arrangements with a type of bulb most generally used.

Fig. 18 is a section through a spherical bulb L, with the glass stem s, containing the leading-in wire w, which has a lamp filament l fastened to it, serving to support the refractory button m in the center. M is a sheet of thin mica wound in several layer around the stem s, and a is the aluminum tube.

Fig. 19 illustrates such a bulb in a somewhat more advanced stage of perfection. A metallic tube S is fastened by means of some cement to the neck of the tube. In the tube is screwed a plug P, of insulating material, in the center of which is fastened a metallic terminal t, for the connection to the leading-in wire w. This terminal must be well insulated from the metal tube S, therefore, if the cement used is conducting — and most generally it is sufficiently so — the space between the plug P and the neck of the bulb should be filled with some good insulating material, as mica powder.

Fig. 20 shows a bulb made for experimental purposes. In this bulb the aluminum tube is provided with an external connection, which serves to investigate the effect of the tube under various conditions. It is referred to chiefly to suggest a line of experiment followed.

Since the bombardment against the stem containing the leading-in wire is due to the inductive action of the latter upon the rarefied gas, it is of advantage to reduce this action as far as practicable by employing a very thin wire, surrounded by a very thick insulation of glass or other material, and by making the wire passing through the rarefied gas as short as practicable. To combine these features I employ a large tube T (Fig. 21), which protrudes into the bulb to some distance, and carries on the top a very short glass stem s, into which is sealed the leading-in wire w, and I protect the top of the glass stem against the heat by a small, aluminum tube a and a layer of mica underneath the same, as usual. The wire w, passing through the large tube to the outside of the bulb, should be well insulates — with a glass tube, for instance — and the space between ought to be filled out with some excellent insulator. Among many insulating powders I have tried, I have found that mica powder is the best to employ. If this precaution is not taken, the tube T, protruding into the bulb, will surely be cracked in consequence of the heating by the brushes which are apt to form in the upper part of the tube, near the exhausted globe, especially if the vacuum be excellent, and therefore the potential necessary to operate the lamp very high.

Fig. 22 illustrates a similar arrangement, with a large tube T protruding into the part of the bulb containing the refractory button m. In this case the wire leading from the outside into the bulb is omitted, the energy required being supplied through condenser coatings C C. The insulating packing P should in this construction be tightly fitting to the glass, and rather wide, or otherwise the discharge might avoid passing through the wire w, which connects the inside condenser coating to the incandescent button m.

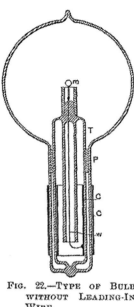

FIG. 22.—TYPE OF BULB WITHOUT LEADING-IN WIRE.

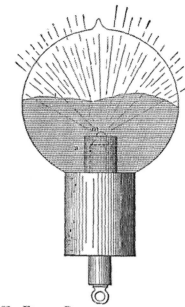

FIG. 23.—EFFECT PRODUCED BY A RUBY DROP.

The molecular bombardment against the glass stem in the bulb is a source of great trouble. As illustration I will cite a phenomenon only too frequently and unwillingly observed. A bulb, preferably a large one, may be taken, and a good conducting body, such as a piece of carbon, may be mounted in it upon a platinum wire sealed in the glass stem. The bulb may be exhausted to a fairly high degree, nearly to the point when phosphorescence begins to appear. When the bulb is connected with the coil, the piece of carbon, if small, may become highly incandescent at first, but its brightness immediately diminishes, and then the discharge may break through the glass somewhere in the middle of the stem, in the form of bright sparks, in spite of the fact that the platinum wire is in good electrical connection with the rarefied gas through the piece of carbon or metal at the top. The first sparks are singularly bright, recalling those drawn from a clear surface of mercury. But, as they heat the glass rapidly, they, of course, lose their brightness, and cease when the glass at the ruptured place becomes incandescent, or generally sufficiently hot to conduct. When observed for the first time the phenomenon must appear very curious, and shows in a striking manner how radically different alternate currents, or impulses, of high frequency behave, as compared with steady currents, or currents of low frequency. With such currents — namely, the latter — the phenomenon would of course not occur. When frequencies such as are obtained by mechanical means are used, I think that the rupture of the glass is more or less the consequence of the bombardment, which warms it up and impairs its insulating power; but with frequencies obtainable with condensers I have no doubt that the glass may give way without previous heating. Although this appears most singular at first, it is in reality what we might expect to occur. The energy supplied to the wire leading into the bulb is given off partly by direct action through the carbon button, and partly by inductive action through the glass surrounding the wire. The case is thus analogous to that in which a condenser shunted by a conductor of low resistance is connected to a source of alternating currents. As long as the frequencies are low, the conductor gets the most, and the condenser is perfectly safe; but when the frequency becomes excessive, the *role of* the conductor may become quite insignificant. In the latter case the difference of potential at the terminals of the condenser may become so great as to rupture the dielectric, notwithstanding the fact that the terminals are joined by a conductor of low resistance.

It is, of course, not necessary, when it is desired to produce the incandescence of a body inclosed in a bulb by means of these currents, that the body should be a conductor, for even a perfect non-conductor may be quite as readily heated. For this purpose it is sufficient to surround a conducting electrode with a non-conducting material, as, for instance, in the bulb described before in Fig. 21, in which a thin incandescent lamp filament is coated with a non-conductor, and supports a button of the same material on the top. At the start the bombardment goes on by inductive action through the non-conductor, until the same is sufficiently heated to become conducting, when the bombardment continues in the ordinary way.

A different arrangement used in some of the bulbs constructed is illustrated in Fig. 23. In this instance a non-conductor m is mounted in a piece of common arc light carbon so as to project some small distance above the latter. The carbon piece is connected to the leading-in wire passing through a glass stem, which is wrapped with several layers of mica. An aluminum tube a is employed as usual for screening. It is so arranged that it reaches very nearly as high as the carbon and only the non-conductor m projects a little above it. The bombardment goes at first against the upper surface of carbon, the lower parts being protected by the aluminum tube. As soon, however, as the non-conductor m is heated it is rendered good conducting, and then it becomes the center of the bombardment, being most exposed to the same.

I have also constructed during these experiments many such single-wire bulbs with or without internal electrode, in which the radiant matter was projected against, or focused upon, the body to be rendered incandescent. Fig;. 24 illustrates one of the bulbs used. It consists of a spherical globe L, provided with a long neck n, on the top, for increasing the action in some cases by the application of an external conducting coating. The globe L is blown out on the bottom into a very small bulb b, which serves to hold it firmly in a socket S of insulating material into which it is cemented. A fine lamp filament f, supported on a wire w, passes through the center of the globe L. The filament is rendered incandescent in the middle portion, where the bombardment proceeding from the lower inside surface of the globe is most intense. The lower portion of the globe, as far as the socket S reaches, is rendered conducting, either by a tinfoil coating or otherwise, and the external electrode is connected to a terminal of the coil.

The arrangement diagrammatically indicated in Fig. 24 was found to be an inferior one when it was desired to render incandescent a filament or button supported in the center of the globe, but it was convenient when the object was to excite phosphorescence.

In many experiments in which bodies of a different kind were mounted in the bulb as, for instance, indicated in Fig. 23, some observations of interest were made.

It was found, among other things, that in such cases, no matter where the bombardment began, just as soon as a high temperature was reached there was generally one of the bodies which seemed to take most of the bombardment upon itself, the other, or others, being thereby relieved. This quality appeared to depend principally on the point of fusion, and on the facility with which the body was "evaporated," or, generally speaking, disintegrated — meaning by the latter term not only the throwing off of atoms, but likewise of larger lumps. The observation made was in accordance with generally accepted notions. In a highly exhausted bulb electricity is carried off from the electrode by independent carriers, which are partly the atoms, or molecules, of the residual atmosphere, and partly the atoms, molecules, or lumps thrown off from the electrode. If the electrode is composed of bodies of different character, and if one of these is more easily disintegrated than the others, most of the electricity supplied is carried off from that body, which is then brought to a higher temperature than the others, and this the more, as upon an increase of the temperature the body is still more easily disintegrated.

It seems to me quite probable that a similar process takes place in the bulb even with a homogenous electrode, and I think it to be the principal cause of the disintegration.. There is bound to be some irregularity, even if the surface is highly polished, which, of course, is impossible with most of the refractory bodies employed as electrodes. Assume that a point of the electrode gets hotter, instantly most of the discharge passes through that point, and a minute patch is probably fused and evaporated. It is now possible that in consequence of the violent disintegration the spot attacked sinks in temperature, or that a counter force is created, as in an arc; at any rate, the local tearing off meets with the limitations incident to the experiment, whereupon the same process occurs on another place. To the eye the electrode appears uniformly brilliant, but there are upon it points constantly shifting and wandering around, of a temperature far above the mean, and this materially hastens the process of deterioration. That some such thing occurs, at least when the electrode is at a lower temperature, sufficient experimental evidence can be obtained in the following manner: Exhaust a bulb to a very high degree, so that with a fairly high potential the discharge cannot pass — that is, not *a luminous* one, for a weak invisible discharge occurs always, in all probability. Now raise slowly and carefully the potential, leaving the primary current on no more than for an instant. At a certain point, two, three, or half a dozen phosphorescent spots will appear on the globe. These places of the glass are evidently more violently bombarded than others, this being due to the unevenly distributed electric density, necessitated, of course, by sharp projections, or, generally speaking, irregularities of the electrode. But the luminous patches are constantly changing in position, which is especially well observable if one manages to produce very few, and this indicates that the configuration of the electrodes is rapidly changing.

From experiences of this kind I am led to infer that, in order to be most durable, the refractory button in the bulb should be in the form of a sphere with a highly polished surface. Such a small sphere could be manufactured from a diamond or some other crystal, but a better way would be to fuse, by the employment of extreme degrees of temperature, some oxide — as, for instance, zirconia — into a small drop, and then keep it in the bulb at a temperature somewhat below its point of fusion.

Interesting and useful results can no doubt be reached in the direction of extreme degrees of heat. How can such high temperatures be arrived at? How are the highest degrees of heat reached in nature? By the impact of stars, by high speeds and collisions. In a collision any rate of heat generation may be attained. In a chemical process we are limited. When oxygen and hydrogen combine, they fall, metaphorically speaking, from a definite height. We cannot go very far with a blast, nor by confining heat in a furnace, but in an exhausted bulb we can concentrate any amount of energy upon a minute button. Leaving practicability out of consideration, this, then, would be the means which, in my opinion, would enable us to reach the highest temperature. But a great difficulty when proceeding in this way is encountered, namely, in most cases the *body* is carried off before it can fuse and form a drop. This difficulty exists principally with an oxide such as zirconia, because it cannot be compressed in so hard a cake that it would not be carried off quickly. I endeavored repeatedly to fuse zirconia, placing it in a cup or arc light carbon as indicated in Fig. 23. It glowed with a most intense light, and the stream of the particles projected out of the carbon cup was of a vivid white; but whether it was compressed in a cake or made into a paste with carbon, it was carried off before it could be fused. The carbon cup containing the zirconia had to be mounted very low in the neck of a large bulb, as the heating of the glass by the projected particles of the oxide was so rapid that in the first trial the bulb was cracked almost in an instant when the current was turned on. The heating of the glass by the projected particles was found to be always greater when the carbon cup contained a body which was rapidly carried off — I presume because in such cases, with the same potential, higher speeds were reached, and also because, per unit of time, more matter was projected — that is, more particles would strike the glass.

The before mentioned difficulty did not exist, however, when the body mounted in the carbon cup offered great resistance to deterioration. For instance, when an oxide was first fused in an oxygen blast and then mounted in the bulb, it melted very readily into a drop.

Generally during the process of fusion magnificent light effects were noted, of which it would be difficult to give an adequate idea. Fig. 23 is intended to illustrate the effect observed with a ruby drop. At first one may see a narrow funnel of white light projected against the top of the globe, where it produces an irregularly outlined phosphorescent patch. When the point of the ruby fuses the phosphorescence becomes very powerful; but as the atoms are projected with much greater speed from the surface of the drop, soon the glass gets hot and "tired," and now only the outer edge of the patch glows. In this manner an intensely phosphorescent, sharply defined line, l, corresponding to the outline of the drop, is produced, which spreads slowly over the globe as the drop gets larger. When the mass begins to boil, small bubbles and cavities are formed, which cause dark colored spots to sweep across the globe. The bulb may be turned downward without fear of the drop falling off, as the mass possesses considerable viscosity.

I may mention here another feature of some interest, which I believe to have noted in the course of these experiments, though the observations do not amount to a certitude. It *appeared* that under the molecular impact caused by the rapidly alternating potential the body was fused and maintained in that state at a lower temperature in a highly exhausted bulb than was the case at normal pressure and application of heat in the ordinary *way* — that is, at least, judging from the quantity of the: light emitted. One of the experiments performed may be mentioned here by way of illustration. A small piece of pumice stone was stuck on a platinum wire, and first melted to it in a gas burner. The wire was next placed between two pieces of charcoal and a burner applied so as to produce an intense heat, sufficient to melt down the pumice stone into a small glass-like button. The platinum wire had to be taken of sufficient thickness to prevent its melting in the fire. While in the charcoal fire, or when held in a burner to get a better idea of the degree of heat, the button glowed with great brilliancy. The wire with the button was then mounted in a bulb, and upon exhausting the same to a high degree, the current was turned on slowly so as to prevent the cracking of the button. The button was heated to the point of fusion, and when it melted it did not, apparently, glow with the same brilliancy as before, and this would indicate a lower temperature. Leaving out of consideration the observer's possible, and even probable, error, the question is, can a body under these conditions be brought from a solid to a liquid state with evolution of *less* light?

When the potential of a body is rapidly alternated it is certain that the structure is jarred. When the potential is very high, although the vibrations may be few — say 20,000 per second — the effect upon the structure may be considerable. Suppose, for example, that a ruby is melted into a drop by a steady application of energy. When it forms a drop it will emit visible and invisible waves, which will be in a definite ratio, and to the eye the drop will appear to be of a certain brilliancy. Next, suppose we diminish to any degree we choose the energy steadily supplied, and, instead, supply energy which rises and falls according to a certain law. Now, when the drop is formed, there will be emitted from it three different kinds of vibrations — the ordinary visible, and two kinds of invisible waves: that is, the ordinary dark waves of all lengths, and, in addition, waves of a well defined character. The latter would not exist by a steady supply of the energy; still they help to jar and loosen the structure. If this really be the case, then the ruby drop will emit relatively less visible and more invisible waves than before. Thus it would seem that when a platinum wire, for instance, is fused by currents alternating with extreme rapidity, it emits at the point of fusion less light and more invisible radiation than it does when melted by a steady current, though the total energy used up in the process of fusion is the same in both cases. Or, to cite another example, a lamp filament is not capable of withstanding as long with currents of extreme frequency as it does with steady currents, assuming that it be worked at the same luminous intensity. This means that for rapidly alternating currents the filament should be shorter and thicker. The higher the frequency — that is, the greater the departure from the steady flow — the worse it would be for the filament. But if the truth of this remark were demonstrated, it would be erroneous to conclude that such a refractory button as used in these bulbs would be deteriorated quicker by currents of extremely high frequency than by steady or low frequency currents. From experience I may say that just the opposite holds good: the button withstands the bombardment better with currents of very high frequency. But this is due to the fact that a high frequency discharge passes through a rarefied gas with much greater freedom than a steady or low frequency discharge, and this will say that with the former we can work with a lower potential or with a less violent impact. As long, then, as the gas is of no consequence, a steady or low frequency current is better; but as soon as the action of the gas is desired and important, high frequencies are preferable.

In the course of these experiments a great many trials were made with all kinds of carbon buttons. Electrodes made of ordinary carbon buttons were decidedly more durable when the buttons were obtained by the application of enormous pressure. Electrodes prepared by depositing carbon in well known ways did not show up well; they blackened the globe very quickly. From many experiences I conclude that lamp filaments obtained in this manner. can be advantageously used only with low potentials and low frequency currents. Some kinds of carbon withstand so well that, in order to bring them to the point of fusion, it is necessary to employ very small buttons. In this case the observation is rendered very difficult on account of the intense heat produced. Nevertheless there can be no doubt that all kinds of carbon are fused under the molecular bombardment, but the liquid state must be one of great instability. Of all the bodies tried there were two which withstood best — diamond and Carborundum. These two showed up about equally, but the latter was preferable, for many reasons. As it is more than likely that this body is not yet generally known, I will venture to call your attention to it.

It has been recently produced by Mr. E. G. Acheson, of Monongahela City, Pa., U. S. A. It is intended to replace ordinary diamond powder for polishing precious stones, etc., and I have been informed that it accomplishes this object quite successfully. I do not know why the name "Carborundum" has been given to it, unless there is something in the process of its manufacture which justifies this selection. Through the kindness of the inventor, I obtained a short while ago some samples which I desired to test in regard to their qualities of phosphorescence and capability of withstanding high degrees of heat.

Carborundum can be obtained in two forms — in the form of "crystals" and of powder. The former appear to the naked eye dark colored, but are very brilliant; the latter is of nearly the same color as ordinary diamond powder, but very much finer. When viewed under a microscope the samples of crystals given to me did not appear to have any definite form, but rather resembled pieces of broken up egg coal of fine quality. The majority were opaque, but there were some which were transparent and colored. The crystals are a kind of carbon containing some impurities; they are extremely hard, and withstand for a long time even an oxygen blast. When the blast is directed against them they at first form a cake of some compactness, probably in consequence of the fusion of impurities they contain. The mass withstands for a very long time the blast without further fusion; but a slow carrying off, or burning, occurs, and, finally, a small quantity of a glass-like residue is left, which, I suppose, is melted alumina. When compressed strongly they conduct very well, but not as well as ordinary carbon. The powder, which is obtained from the crystals in some way, is practically non-conducting. It affords a magnificent polishing material for stones.

The time has been too short to. make a satisfactory study of the properties of this product, but enough experience has been gained in a few weeks I have experimented upon it to say that it does possess some remarkable properties in many respects. It withstands excessively high degrees of heat, it is little deteriorated by molecular bombardment, and it does not blacken the globe as ordinary carbon does. The only difficulty which I have found in its use in connection with these experiments was to find some binding material which would resist the heat and the effect of the bombardment as successfully as Carborundum itself does.

I have here a number of bulbs which I have provided with buttons of Carborundum To make such a button of Carborundum crystals I proceed in the following manner: I take an ordinary lamp filament and dip its point in tar, or some other thick substance or paint which may be readily carbonized. I next pass the point of the filament through the crystals, and then hold it vertically over a hot plate. The tar softens and forms a drop on the point of the filament, the crystals adhering to the surface of the drop. By regulating the distance from the plate the tar is slowly dried out and the button becomes solid. I then once more dip the button in tar and hold it again over a plate until the tar is evaporated, leaving only a hard mass which firmly binds the crystals. When a_ larger button is required I repeat the process several times, and I generally also cover the filament a certain distance below the button with crystals. The button being mounted in a bulb, when a good vacuum has been reached, first a weak and then a strong discharge is passed through the bulb to carbonize the tar and expel all gases, and later it is brought to a very intense incandescence.

When the powder is used I have found it best to proceed as follows: I make a thick paint of Carborundum and tar, and pass a lamp filament through the paint. Taking then most of the paint off by rubbing the filament against a piece of chamois leather, I hold it over a hot plate until the tar evaporates and the coating becomes firm. I repeat this process as many times as it is necessary to obtain a certain thickness of coating. On the point of the coated filament I form a button in the same manner.

There is no doubt that such a button — properly prepared under great pressure — of Carborundum, especially of powder of the best quality, will withstand the effect of the bombardment fully as well as anything we know. The difficulty is that the binding material gives way, and the Carborundum is slowly thrown off after some time. As it does not seen to blacken the globe in the least, it might be found useful for coating the filaments of ordinary incandescent lamps, and I think that it is even possible to produce thin threads or sticks of Carborundum which will replace the ordinary filaments in an incandescent lamp. A Carborundum coating seems to be more durable than other coatings, not only because the Carborundum can withstand high degrees of heat, but also because it seems to unite with the carbon better than any other material I have tried. A coating of zirconia or any other oxide, for instance, is far more quickly destroyed. I prepared buttons of diamond dust in the same manner as of Carborundum, and these came in durability nearest to those prepared of Carborundum, but the binding paste gave way much more quickly in the diamond buttons: this, however, I attributed to the size and irregularity of the grains of the diamond.

It was of interest to find whether Carborundum possesses the quality of phosphorescence. One is, of course, prepared to encounter two difficulties: first, as regards the rough product, the "crystals," they are good conducting, and it is a fact that conductors do not phosphorescence; second, the powder, being exceedingly fine, would not be apt to exhibit very prominently this quality, since we know that when crystals, even such as diamond or ruby, are finely powdered, they lose the property of phosphorescence to a considerable degree.

The question presents itself here, can a conductor phosphoresce? What is there in such a body as a metal, for instance, that would deprive it of the quality of phosphorescence, unless it is that property which characterizes it as a conductor? for it is a fact that most of the phosphorescent bodies lose that quality when they are sufficiently heated to become more or less conducting. Then, if a metal be in a large measure, or perhaps entirely, deprived of that property, it should be capable of phosphorescence. Therefore it is quite possible that at some extremely high frequency, when behaving practically as a non-conductor, a metal or any other conductor might exhibit the quality of phosphorescence, even though it be entirely incapable of phosphorescing under the impact of a low-frequency discharge. There is, however, another possible way how a conductor might at least *appear* to phosphoresce.

Considerable doubt still exists as to what really is phosphorescence, and as to whether the various phenomena comprised under this head are due to the same causes. Suppose that in an exhausted bulb, under the molecular impact, the surface of a pied of metal or other conductor is rendered strongly luminous, but at the same time it is found that it remains comparatively cool, would not this luminosity be called phosphorescence? Now such a result, theoretically at least, is possible, for it is a mere question of potential or speed. Assume the potential of the electrode, and consequently the speed of the projected atoms, to be sufficiently high, the surface of the metal piece against which the atoms are projected would be rendered highly incandescent, since the process of heat generation would be incomparably faster than that of radiating or conducting away from the surface of the collision. In the eye of the observer a single impact of the atoms would cause an instantaneous flash, but if the impacts were repeated with sufficient rapidity they would produce a continuous impression upon his retina. To him then the surface of the metal would appear continuously incandescent and of constant luminous intensity, while in reality the light would be either intermittent or at least chancing periodically in intensity. The metal piece would rise in temperature until equilibrium was attained — that is, until the energy continuously radiated would equal that intermittently supplied. But the supplied energy might under such conditions not be sufficient to bring the body to any more than a very moderate mean temperature, especially if the frequency of the atomic impacts be very low — just enough that the fluctuation of the intensity of the light emitted could not be detected by the eye. The body would now, owing to the manner in which .the energy is supplied, emit a strong light, and yet be at a comparatively very low mean temperature. How could the observer call the luminosity thus produced? Even if the analysis of the light would teach him something definite, still he would probably rank it under the phenomena of phosphorescence. It is conceivable that in such a way both conducting and non-conducting bodies may be maintained at a certain luminous intensity, but the energy required would very greatly vary with the nature and properties of the bodies.

These and some foregoing remarks of a speculative nature were made merely to bring out curious features of alternate currents or electric impulses. By their help we may cause a body to emit *more* light, while at a certain mean temperature, than it would emit if brought to that temperature by a steady supply; and, again, we may bring a body to the point of fusion, and cause it to emit *less* light than when fused by the application of energy in ordinary ways. It all depends on how we supply the energy, and what kind of vibrations we set up: in one case the vibrations are more, in the other less, adapted to affect our sense of vision.

Some effects, which I had not observed before, obtained with Carborundum in the first trials, I attributed to phosphorescence, but in subsequent experiments it appeared that it was devoid of that quality. The crystals possess a noteworthy feature. In a bulb provided with a single electrode in the shape of a small circular metal disc, for instance, at a certain degree of exhaustion the electrode is covered with a milky film, which is separated by a dark space from the glow filling the bulb. When the metal disc is covered with Carborundum crystals, the film is far more intense, and snow-white. This I found later to be merely an effect of the bright surface of the crystals, for when an aluminum electrode was highly polished it exhibited more or less the same phenomenon. I made a number of experiments with the samples of crystals obtained, principally because it would have been of special interest to find that they are capable of phosphorescence, on account of their being conducting. I could not produce phosphorescence distinctly, but I must remark that a decisive opinion cannot be formed until other experimenters have gone over the same ground.

The powder behaved in some experiments as though it contained alumina, but it did not exhibit with sufficient distinctness the red of the latter. Its dead color brightens considerably under the molecular impact, but I am now convinced it does not phosphoresce. Still, the tests with the powder are not conclusive, because powdered Carborundum probably does not behave like a phosphorescent sulphide, for example, which could be finely powdered without impairing the phosphorescence, but rather like powdered ruby or diamond, and therefore it would be necessary, in order to make a decisive test, to obtain it in a large lump and polish up the surface.

If the Carborundum proves useful in connection with these and similar experiments, its chief value will be found in the production of coatings, thin conductors, buttons, or other electrodes capable of withstanding extremely high degrees of heat.

The production of a small electrode capable of withstanding enormous temperatures I regard as of the greatest importance in the manufacture of light. It would enable us to obtain, by means of currents of very high frequencies, certainly 20 times, if not more, the quantity of light which is obtained in the present incandescent lamp by the same expenditure of energy. This estimate may appear to many exaggerated, but in reality I think it is far from being so. As this statement might be misunderstood I think it necessary to expose clearly the problem with which in this line of work we are confronted, and the manner in which, in my opinion, a solution will be arrived at.

Any one who begins a study of the problem will be apt to think that what is wanted in a lamp with an electrode is a very high degree of incandescence of the electrode. There he will be mistaken. The high incandescence of the button is a necessary evil, but what is really wanted is the high incandescence of the gas surrounding the button. In other words, the problem in such a lamp is to bring a mass of gas to the highest possible incandescence. The higher the incandescence, the quicker the mean vibration, the greater is the economy of the light production. But to maintain a mass of gas at a high degree of incandescence in a glass vessel, it will always be necessary to keep the incandescent mass away from the glass; that is, to confine it as much as possible to the central portion of the globe.

In one of the experiments this evening a brush was produced at the end of a wire. This brush was a flame, a source of heat and light. It did not emit much perceptible heat, nor did it glow with an intense light; but is it the less a flame because it does not scorch my hand? Is it the less a flame because it does not hurt my *eye by* its brilliancy? The problem is precisely to produce in the bulb such a flame, much smaller in size, but incomparably more powerful. Were there means at hand for producing electric impulses of a sufficiently high frequency, and for transmitting them, the bulb could be done away with, unless it were used to protect the electrode, or to economize the energy by confining the heat. But as such means are not at disposal, it becomes necessary to place the terminal in a bulb and rarefy the air in the same. This is done merely to enable the apparatus to perform the work which it is not capable of performing at ordinary air pressure. In the bulb we are able to intensify the action to any degree — so far that the brush emits a powerful light.

The intensity of the light emitted depends principally on the frequency and potential of the impulses, and on the electric density on the surface of the electrode. It is of the greatest importance to employ the smallest possible button, in order to push the density very far. Under the violent impact of the molecules of the gas surrounding it, the small electrode is of course brought to an extremely high temperature, but around it is a mass of highly incandescent gas, a flame photosphere, many hundred times the volume of the electrode. With a diamond, Carborundum or zirconia button the photosphere can be as much as one thousand times the volume of the button. Without much reflecting one would think that in pushing so far the incandescence of the electrode it would be instantly volatilized. But after a careful consideration he would find that, theoretically, it should not occur, and in this fact — which, however, is experimentally demonstrated — lies principally the future value of such a lamp.

At first, when the bombardment begins, most of the work is performed on the surface of the button, but when a highly conducting photosphere is formed the button is comparatively relieved. The higher the incandescence of the photosphere the more in approaches in conductivity to that of the electrode, and the more, therefore, the solid and the gas form one conducting body. The consequence is that the further is forced the incandescence the more work, comparatively, is performed on the gas, and the less on the electrode. The formation of a powerful photosphere is consequently the very means for protecting the electrode. This protection, of course, is a relative one, and it should not be thought that by pushing the incandescence higher the electrode is actually less deteriorated. Still, theoretically, with extreme frequencies; this result must be reached, but probably at a temperature too high for most of the refractory bodies known. Given, then, an electrode which can withstand to a very high limit the. effect of the bombardment and outward strain, it would be safe no matter how much it is forced beyond that limit. In an incandescent lamp quite different considerations apply. There the gas is not at all concerned: the whole of the work is performed on the filament; and the life of the lamp diminishes so' rapidly with the increase of the degree of incandescence that economical reasons compel us to work it at a low incandescence. But if an incandescent lamp is operated with currents of very high frequency, the action of the gas cannot be neglected, and the rules for the most economical working must be considerably modified.

In order to bring such a lamp with one or two electrodes to a great perfection, it is necessary to employ impulses of very high frequency. The high frequency secures, among others, two chief advantages, which have a most important bearing upon the economy of the light production. First, the deterioration of the electrode is reduced by reason of the fact that we employ a great many small impacts, instead of a few violent ones, which shatter quickly the structure; secondly, the formation of a large photosphere is facilitated.

In order to reduce the deterioration of the electrode to the minimum, it is desirable that the vibration be harmonic, for any suddenness hastens the process of destruction. Ar. electrode lasts much longer when kept at incandescence by currents, or impulses, obtained from a high-frequency alternator, which rise and fall more or less harmonically, than by impulses obtained from a disruptive discharge coil. In the latter case there is no doubt that most of the damage is done by the fundamental sudden discharges.

One of the elements of loss in such a lamp is the bombardment of the globe. As the potential is very high, the molecules are projected with great speed; they strike the glass, and usually excite a strong phosphorescence. The effect produced is very pretty, but for economical reasons it would be perhaps preferable to prevent, or at least reduce to the minimum, the bombardment against the globe, as in such case it is, as a rule not the object to excite phosphorescence, and as some loss of energy results from the bombardment. This loss in the bulb is principally dependent on the potential of the impulses and on the electric density on the surface of the electrode. In employing very high frequencies the loss of energy by the bombardment is greatly reduced, for, first the potential needed to perform a given amount of work is much smaller; and, secondly, by producing a highly conducting photosphere around the electrode, the same result is obtained as though the electrode were much larger, which is equivalent to a smaller electric density. But be it by the diminution of the maximum potential or of the density, the gain is effected in the same manner, namely, by avoiding violent shocks, which strain the glass much beyond its limit of elasticity. If the frequency could be brought high enough, the loss due to the imperfect elasticity of the glass would be entirely negligible. The loss due to bombardment of the globe may, however, be reduced by using two electrodes instead of one. In such case each of the electrodes may be connected to one of the terminals; or else, if it is preferable to use only one wire, one electrode may be connected to one terminal and the other to the ground or to an insulated body of some surface, as, for instance, a shade on the lamp. In the latter case, unless some judgment is used, one of the electrodes might glow more intensely than the other.

But on the whole I find it preferable when using such high frequencies to employ only one electrode and one connecting wire. I am convinced that the illuminating device of the near future will not require for its operation more than one lead, and, at any rate, it will have no leading-in wire, since the energy required can be as well transmitted through the glass. In experimental bulbs the leading-in wire is most generally used on account of convenience, as in employing condenser coatings in the manner indicated in Fig. 22, for example, there is some difficulty in fitting the parts, but these difficulties would not exist if a great many bulbs were manufactured; otherwise the energy can be conveyed through the glass as well as through a wire, and with these high frequencies the losses are very small. Such illuminating devices will necessarily involve the use of very high potentials, and this, in the eyes of practical men, might be an objectionable feature. Yet, in reality, high potentials are not objectionable — certainly not in the least as far as the safety of the devices is concerned.

There are two ways of rendering an electric appliance safe. One is to use low potentials, the other is to determine the dimensions of the apparatus so that it is safe no matter how high a potential is used. Of the two the latter seems to me the better way, for then the safety is absolute, unaffected by any possible combination of circumstances which might render even a low-potential appliance dangerous to life and. property. But the practical conditions require not only the judicious determination of the dimensions of the apparatus; they likewise necessitate the employment of energy of the proper kind. It is easy, for instance, to construct a transformer capable of giving, when operated from an ordinary alternate current machine of low tension, say 50,000 volts, which might be required to light a highly exhausted phosphorescent tube, so that, in spite of the high potential, it is perfectly safe, the shock from it producing no inconvenience. Still, such a transformer would be expensive, and in itself inefficient; and, besides, what energy was obtained from it would not be economically used for the production of light. The economy demands the employment of energy in the form of extremely rapid

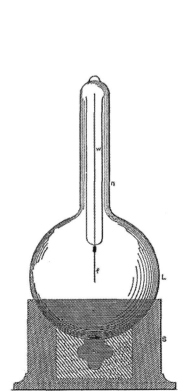

FIG. 24.—BULB WITHOUT LEADING-IN WIRE, SHOWING EFFECT OF PROJECTED MATTER.

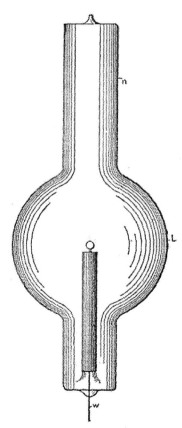

FIG. 25.—IMPROVED EXPERIMENTAL BULB.

vibrations. The problem of producing light has been likened to that of maintaining a certain high-pitch note by means of a bell. It should be said *a barely audible* note; and even these words would not express it, so wonderful is the sensitiveness of the eye. We may deliver powerful blows at long intervals, waste a good deal of energy, and still not get what we want; or we may keep up the note by delivering frequent gentle taps, and get nearer to the object sought by the expenditure of much less energy. In the production of light, as fast as the illuminating device is concerned, there can be only one rule — that is, to use as high frequencies as can be obtained; but the means for the production and conveyance of impulses of such character impose, at present at least, great limitations. Once it is decided to use very high frequencies, the return wire becomes unnecessary, and all the appliances are simplified. By the use of obvious means the same result is obtained as though the return wire were used. It is sufficient for this purpose to bring in contact with the bulb, or merely in the vicinity of the same, an insulated body of some surface. The surface need, of course, be the smaller, the higher the frequency and potential used, and necessarily, also, the higher the economy of the lamp or other device.

This plan of working has been resorted to on several occasions this evening. So, for instance, when the incandescence of a button was produced by grasping the bulb with the hand, the body of the experimenter merely served to intensify the action. The bulb used was similar to that illustrated in Fig. 19, and the coil was excited to a small potential, not sufficient to bring the button to incandescence when the bulb was hanging from the wire; and incidentally, in order to perform the experiment in a more suitable manner, the button was taken so large that a perceptible time had to elapse before, upon grasping the bulb, it could be rendered incandescent. The contact with the bulb was, of course, quite unnecessary. It is easy, by using a rather large bulb with an exceedingly small electrode, to adjust the conditions so that the latter is brought to bright incandescence by the mere approach of the experimenter within a few feet of the bulb, and that the incandescence subsides upon his receding.

In another experiment, when phosphorescence was excited, a similar bulb was used. Here again, originally, the potential was not sufficient to excite phosphorescence until the action was intensified — in this case, however, to present a different feature, by touching the socket with a metallic object held in the hand. The electrode in the bulb was a carbon button so large that it could not be brought to incandescence, and thereby spoil the effect produced by phosphorescence.

Again, in another of the early experiments, a bulb was used as illustrated in Fig. 12. In this instance, by touching the bulb with one or two fingers, one or two shadows of the stem inside were projected against the glass, the touch of the finger producing the same result as the application of an external negative electrode under ordinary circumstances.

In all these experiments the action was intensified by augmenting the capacity at the end of the lead connected to the terminal. As a rule, it is not necessary to resort to such means, and would be quite unnecessary with still higher frequencies; but when it is desired, the bulb, or tube, can be easily adapted to the purpose.

In Fig. 24, for example, an experimental bulb L is shown, which is provided with a neck n on the top for the application of an external tinfoil coating, which may be connected to a body of larger surface. Such a lamp as illustrated in Fig. 25 may also be lighted by connecting the tinfoil coating on the neck n to the terminal, and the leading-in wire w to an insulated plate. If the bulb stands in a socket upright, as shown in the cut, a shade of conducting material may be slipped in the neck n, and the action thus magnified.

A more perfected arrangement used in some of these bulbs is illustrated in Fig. 26. In this case the construction of the bulb is as shown and described before, when reference was made to Fig. 19. A zinc sheet Z, with a tubular extension T, is slipped over the metallic socket S. The bulb hangs downward from the terminal t, the zinc sheet Z, performing the double office of intensifier and reflector. The reflector is separated from the terminal t by an extension of the insulating plug P.

A similar disposition with a phosphorescent tube is illustrated in Fig. 27. The tube T is prepared from two short tubes of a different diameter, which are sealed on the ends. On the lower end is placed an outside conducting coating C, which connects to the wire w. The wire has a hoof; on the upper end for suspension, and passes through the center of the inside tube, which is filled with some good and tightly packed insulator. On the outside of the upper end of the tube T is another conducting coating C_1, upon which is slipped a metallic reflector Z, which should be separated by a thick insulation from the end of wire w.

The economical use of such a reflector or intensifier would require that all energy supplied to an air condenser should be recoverable, or, in other words, that there should not be any losses, neither in the gaseous medium nor through its action elsewhere. This is far from being so, but, fortunately, the losses may be reduced to anything desired. A few remarks are necessary on this subject, in order to make the experiences gathered in the course of these investigations perfectly clear.

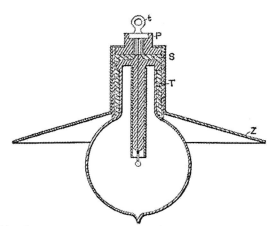

FIG. 26.—IMPROVED BULB WITH INTENSIFYING REFLECTOR.

Suppose a small helix with many well insulated turns, as in experiment Fig. 17, has one of its ends connected to one of the terminals of the induction coil, and the other to a metal plate, or, for the sake of simplicity, a sphere, insulated in space. When the coil is set to work, the potential of the sphere is alternated, and the small helix now behaves as though its free end were connected to the other terminal of the induction coil. If an iron rod be held within the small helix it is quickly brought to a hi, eh temperature, indicating the passage of a strong current through the helix: How does the insulated sphere act in this case? It can be a condenser, storing and returning the energy supplied to it, or it can be a mere sink of enemy; and the conditions of the experiment determine whether it is more one or the other. The sphere being charged to a high potential, it acts inductively upon the surrounding air, or whatever gaseous medium there might be. The molecules, or atoms, 'which are near the sphere are of course more attracted, and move through a greater distance than the farther ones. When the nearest molecules strike the sphere they are repelled, and collisions occur at all distances within the inductive action of the sphere. It is now clear that, if the potential be steady, but little loss of energy can be caused in this way, for the molecules which are nearest to the sphere, having had an additional charge imparted to them by contact, are not attracted until they have parted, if not with all, at least with most of the additional charge, which can be accomplished only after a great many collisions. From the fact that with a steady potential there is but little loss in dry air, one must come to such a conclusion. When the potential of the sphere, instead of being steady, is alternating, the conditions are entirely different. In this case a rhythmical bombardment occurs, no matter whether the molecules after coming in contact with the sphere lose the imparted charge or not; what is more, if the charge is not lost, the impacts are only the *more* violent, Still if the frequency of the impulses be very small, the loss caused by the impacts and collisions would not be serious unless the potential were excessive. But when extremely high frequencies and more or less high potentials are used, the loss may be very great. The total energy lost per unit of time is proportionate to the product of the number of impacts per second, or the frequency and the energy lost in each impact. But the energy of an impact must be 'proportionate to the square of the electric density of the sphere, since the charge imparted to the molecule is proportionate to that density. I conclude from this that the total energy lost must be proportionate to the product of the frequency and the square of the electric density; but this law needs experimental confirmation. Assuming the preceding considerations to be true, then, by rapidly alternating the potential of a body immersed in an insulating gaseous medium, any amount of energy may be dissipated into space. Most of that energy then, I believe, is not dissipated in the form of long ether waves, propagated to considerable distance, as is thought most generally, but is consumed — in the case of an insulated sphere, for example — in impact and collisional losses — that is, heat vibrations — on the surface and in the vicinity of the sphere. To reduce the dissipation it is necessary to work with a small electric density — the smaller the higher the frequency.

But since, on the assumption before made, the loss is diminished with the square of the density, and since currents of very high frequencies involve considerable waste when transmitted through conductors, it follows that, on the whole, it is better to employ one wire than two. Therefore, if motors, lamps, or devices of any kind are perfected, capable of being advantageously operated by currents of extremely high frequency. economical reasons will make it advisable to use only one wire, especially if the distance are great.

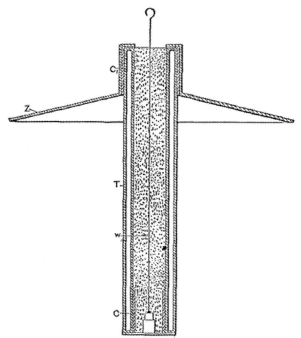

FIG. 27.—PHOSPHORESCENT TUBE WITH INTENSIFYING REFLECTOR.

When energy is absorbed in a condenser the same behaves as though its capacity were increased. Absorption always exists more or less, but generally it is small and of no consequence as long as the frequencies are not very great. In using extremely high frequencies, and, necessarily in such case, also high potentials, the absorption — or, what is here meant more particularly by this term, the loss of energy due to the presence of z gaseous medium — is an important factor to be considered, as the energy absorbed ill the air condenser may be any fraction of the supplied energy. This would seem to make it very difficult to tell from the measured or computed capacity of an air condenser its actual capacity or vibration period, especially if the condenser is of very small surface and is charged to a very high potential. As many important results are dependent upon the correctness of the estimation of the vibration period, this subject demands the most careful scrutiny of other investigators. To reduce the probable error as much as possible in experiments of the kind alluded to, it is advisable to use spheres or plates of large surface, so as to make the density exceedingly small. Otherwise, when it is practicable, an oil condenser should be used in preference. In oil or other liquid dielectrics there are seemingly no such losses as in gaseous media. It being impossible to exclude entirely the gas in condensers with solid dielectrics, such condensers should be immersed in oil, for economical reasons if nothing else; they can then be strained to the utmost and will remain cool. In Leyden jars the loss due to air is comparatively small, as the tinfoil coatings are large, close together, and the charged surfaces not directly exposed; but when the potentials are very high, the loss may be more or less considerable at, or near, the upper edge of the foil, where the air is principally acted upon. If the jar be immersed in boiled-out oil, it will be capable of performing four times the amount of work which it can for any length of time when used in the ordinary way, and the loss will lot inappreciable.

It should not be thought that the loss in heat in an air condenser is necessarily associated with the formation *of visible* streams or brushes. If a small electrode, inclosed in an unexhausted bulb, is connected to one of the terminals of the coil, streams can be seen to issue from the electrode and the air in the bulb is heated; if, instead of a small electrode, a large sphere is inclosed in the bulb, no streams are observed, still the air is heated.

Nor should it be thought that the temperature of an air condenser would give even an approximate idea of the loss in heat incurred, as in such case heat must be given off much more quickly, since there is, in addition to the ordinary radiation, a very active carrying away of heat by independent carriers going on, and since not only the apparatus, but the air at some distance from it is heated in consequence of the collisions which must occur.

Owing to this, in experiments with such a coil, a rise of temperature can be distinctly observed only when the body connected to the coil is very small. But with apparatus on a larger scale, even a body of considerable bulk would be heated, as, for instance, the body of a person; and I think that skilled physicians might make observations of utility in such experiments, which, if the apparatus were judiciously designed, would not present the slightest danger.

A question of some interest, principally to meteorologists, presents itself here. How does the earth behave? The earth is an air condenser, but is it a perfect or a very imperfect one — a mere sink of energy? There can be little doubt that to such small disturbance as might be caused in an experiment the earth behaves as an almost perfect condenser. But it might be different when its charge is set in vibration by some sudden disturbance occurring in the heavens. In such case, as before stated, probably *only* little of the energy of the vibrations set up would be lost into space in the form of long ether radiations, but most of the energy, I think, would spend itself in molecular impacts and collisions, and pass off into space in the form of short heat, and possibly light, waves. As both the frequency of the vibrations of the charge and the potential are in all probability excessive, the energy converted into heat may be considerable. Since the density must be unevenly distributed, either in consequence of the irregularity of the earth's surface, or on account of the condition of the atmosphere in various places, the effect produced would accordingly vary from place to place. Considerable variations in the temperature and pressure of the atmosphere may in this manner be caused at any point of the surface of the earth. The variations *may be* gradual or very sudden, according to the nature of the general disturbance, and may produce rain and storms, or locally modify the weather in any way.

From the remarks before made one may see what an important factor of loss the air in the neighborhood of a charged surface becomes when the electric density is great and the frequency of the impulses excessive. But the action as explained implies that the air is insulating — that is, that it is composed of independent carriers immersed in an insulating medium. This is the case only when the air is at something like ordinary or greater, or at extremely small, pressure. When the air is slightly rarefied and conducting, then true conduction losses occur also. In such case, of course, considerable energy may be dissipated into space even with a steady potential, or with impulses of low frequency, if the density is very great.

When the gas is at very low pressure, an electrode is heated more because higher speeds can be reached. If the gas around the electrode is strongly compressed, the displacements, and consequently the speeds, are very small, and the heating is insignificant. But if in such case the frequency could be sufficiently increased, the electrode would be brought to a high temperature as well as if the gas were at very low pressure; in

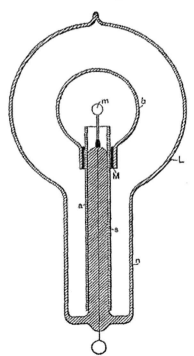

FIG. 28.—LAMP WITH AUXILIARY BULB FOR CONFINING THE ACTION TO THE CENTRE.

fact, exhausting the bulb is only necessary because we cannot produce (and possibly not convey) currents of the required frequency.

Returning to the subject of electrode lamps, it is obviously of advantage in such a lamp to confine as much as possible the heat to the electrode by preventing the circulation of the gas in the bulb. If a very small bulb be taken, it would confine the heat better than a large one, but it might not be of sufficient capacity to be operated from the coil, or, if so, the glass might get too hot. A simple way to improve in this direction is to employ a globe of the required size, but to place a small bulb, the diameter of which is properly estimated, over the refractory button contained in the globe. This arrangement is illustrated in Fig. 28.

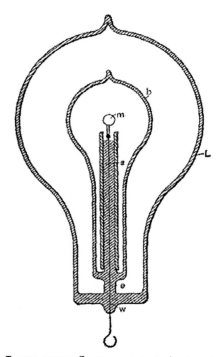

FIG. 29.—LAMP WITH INDEPENDENT AUXILIARY BULB.

The globe L has in this case a large neck n, allowing the small bulb b to slip through. Otherwise the construction is the same as shown in Fig. 18, for example. The small bulb is conveniently supported upon the stem s, carrying the refractory button m. It is separated from the aluminum tube a by several layers of mica M. in order to prevent the cracking of the neck by the rapid heating of the aluminum tube upon a sudden turning on of the current. The inside bulb should be as small as possible when it is desired to obtain light only by incandescence of the electrode. If it is desired to produce phosphorescence, the bulb should be larger, else it would be apt to get too hot, and the phosphorescence would cease. In this arrangement usually only the small bulb shows phosphorescence, as there is practically no bombardment against the outer globe. In some of these bulbs constructed as illustrated in Fig. 28 the small tube was coated with phosphorescent paint, and beautiful effects were obtained. Instead of making the inside bulb large, in order to avoid undue heating, it answers the purpose to make the electrode m larger. In this case the bombardment is weakened by reason of the smaller electric density.

Many bulbs were constructed on the plan illustrated in Fig. 29. Here a small bulb b, containing the refractory button m, upon being exhausted to a very high degree was sealed in a large globe L, which was then moderately exhausted and sealed off. The principal advantage of this construction was that it allowed of reaching extremely high vacua, and, at the same time use a large bulb. It was found, in the course of experiences with bulbs such as illustrated in Fig. 29, that it was well to make the stem s near the seal at a very thick, and the leading-in wire w thin, as it occurred sometimes that the stem at a was heated and the bulb was cracked. Often the outer globe L was exhausted only just enough to allow the discharge to pass through, and the space between the bulbs appeared crimson, producing a curious effect. In some cases, when the exhaustion in globe L was very low, and the air good conducting, it was found necessary, in order to bring the button m to high incandescence, to place, preferably on the upper part of the neck of the globe, a tinfoil coating which was connected to an insulated body, to the ground, or to the other terminal of the coil, as the highly conducting air weakened the effect somewhat, probably by being acted upon inductively from the wire w, where it entered the bulb at e. Another difficulty — which, however, is always present when the refractory button is mounted in a very small bulb — existed in the construction illustrated in Fig. 29, namely, the vacuum in the bulb b would be impaired in a comparatively short time.

The chief idea in the two last described constructions was to confine the heat to the central portion of the globe by preventing the exchange of air. An advantage is secured, but owing to the heating of the inside bulb and slow evaporation of the glass the vacuum is hard to maintain, even if the construction illustrated in Fig. 28 be chosen, in which both bulbs communicate.

But by far the better way — the ideal way — would be to reach sufficiently high frequencies. The higher the frequency the slower would be the exchange of the air, and I think that a frequency may be reached at which there would be no exchange whatever of the air molecules around the terminal. We would then produce a flame in which there would be no carrying away of material, and a queer flame it would be, for it would be rigid! With such high frequencies the inertia of the particles would come into play. As the brush, or flame, would gain rigidity in virtue of the inertia of the particles, the exchange of the latter would be prevented. This would necessarily occur, for, the number of the impulses being augmented, the potential energy of each would diminish, so that finally only atomic vibrations could be set up, and the motion of translation through measurable space would cease. Thus an ordinary gas burner connected to a source of rapidly alternating potential might have its efficiency augmented to a certain limit, and this for two reasons — because of the additional vibration imparted, and because of a slowing down of the process of carrying off. But the renewal being rendered difficult, and renewal being necessary to maintain the *burner,* a continued increase of the frequency of the impulses, assuming they could be transmitted to and impressed upon the flame, would result in the "extinction" of the latter, meaning by this term only the cessation of the chemical process.

I think, however, that in the case of an electrode immersed in a fluid insulating medium, and surrounded by independent carriers of electric charges, which can be acted upon inductively, a sufficiently high frequency of the impulses would probably result in a gravitation of the gas all around toward the electrode. For this it would be only necessary to assume that the independent bodies are irregularly shaped; they would then turn toward the electrode their side of the greatest electric density, and this would be a position in which the fluid resistance to approach would be smaller than that offered to the receding.

The general opinion, I do not doubt, is that it is out of the question to reach any such frequencies as might — assuming some of the views before expressed to be true —produce any of the results which I have pointed out as mere possibilities. This may he so, but in the course of these investigations, from the observation of many phenomena I have gained the conviction that these frequencies would be much lower than one is apt to estimate at first. In a flame we set up light vibrations by causing molecules, or atoms, to collide. But what is the ratio of the frequency of the collisions and that o; the vibrations set up? Certainly it must be incomparably smaller than that of the knocks of the bell and tile sound vibrations, or that of the discharges and the oscillations of the condenser. We may cause the molecules of the gas to collide by the use of alternate electric impulses of high frequency, and so we may imitate the process in a flame; and from experiments with frequencies which we are now able to obtain, I think that the result ,is producible with impulses which are transmissible through a conductor.

In connection with thoughts of a similar nature, it appeared to be of great interest; to demonstrate the rigidity of a vibrating gaseous column. Although with such low frequencies as, say 10,000 per second, which I was able to obtain without difficulty from a specially constructed alternator, the task looked discouraging at first, I made a series of experiments: The trials with air at ordinary pressure led to no result, but with air moderately rarefied I obtain what I think to be an unmistakable experimental evidence of the property sought for. As a result of this kind might lead able investigators to conclusions of importance I will describe one of the experiments performed.

It is well known that when a tube is slightly exhausted the discharge may be passed through it in the form of a thin luminous thread. When produced with currents of low frequency, obtained from a coil operated as usual, this thread is inert. If a magnet be approached to it, the part near the same is attracted or repelled, according to the direction of the lines of force of the magnet. It occurred to me that if such a thread would be produced with currents of very high frequency, it should be more or less rigid, and as it was visible it could be easily studied. Accordingly I prepared a tube about 1 inch in diameter and 1 meter long, with outside coating at each end. The tube was exhausted to a point at which by a little working the thread discharge could be obtained. It must be remarked here that the general aspect of the tube, and the degree of exhaustion, are quite different than when ordinary low frequency currents are used. As it was found preferable to work with one terminal, the tube prepared was suspended from the end of a wire connected to the terminal, the tinfoil coating being connected to the wire, and to the lower coating sometimes a small insulated plate was attached. When the thread was formed it extended through the upper part of the tube and lost itself in the lower end. If it possessed rigidity it resembled, not exactly an elastic cord stretched tight between two supports, but a cord suspended from a height with a small weight attached at the end. When the finger or a magnet was approached to the upper end of the luminous thread, it could be brought locally out of position by electrostatic or magnetic action; and when the disturbing object was very quickly removed, an analogous result was produced, as though a suspended cord would be displaced and quickly released near the point of suspension. In doing this the luminous thread was set in vibration, and two very sharply marked nodes, and a third indistinct one, were formed. The vibration, once set up, continued for fully eight minutes, dying gradually out. The speed of the vibration often varied perceptibly, and it could be observed that the electrostatic attraction of the glass affected the vibrating thread; but it was clear that the electrostatic action was not the cause of the vibration, for the thread was most generally stationary, and could always be set in vibration by passing the finger quickly near the upper part of the tube. With a magnet the thread could be split in two and both parts vibrated. By approaching the hand to the lower coating of the tube, or insulated plate if attached, the vibration was quickened; also, as far as I could see, by raising the potential or frequency. Thus, either increasing the frequency or passing a stronger discharge of tile same frequency corresponded to a tightening of the cord. I did not obtain any experimental evidence with condenser discharges. A luminous band excited in a bulb by repeated discharges of a Leyden jar must possess rigidity, and if deformed and suddenly released should vibrate. But probably the amount of vibrating matter is so small that in spite of the extreme sped the inertia cannot prominently assert itself. Besides, the observation in such a case is rendered extremely difficult on account of the fundamental vibration.

The demonstration of the fact — which still needs better experimental confirmation — that a vibrating gaseous column possesses rigidity, might greatly modify the views of thinkers. When with low frequencies and insignificant potentials indications of that property may be noted, how must a gaseous medium behave under the influence of enormous electrostatic stresses which may be active in the interstellar space, and which may alternate with inconceivable rapidity? The existence of such an electrostatic, rhythmically throbbing force — of a vibrating electrostatic field — would show a possible way how solids might have formed from the ultra-gaseous uterus, and how transverse and all kinds of vibrations may be transmitted through a gaseous medium. filling all space. Then, ether might be a true fluid, devoid of rigidity, and at rest, it being merely necessary as a connecting link to enable interaction. What determines the rigidity of a body? It must be the speed and the amount of moving matter. In a gas the speed may be considerable, but the density is exceedingly small; in a liquid the speed would be likely to be small, though the density may be considerable; and in both cases the inertia resistance offered to displacement is practically *nil*. But place a gaseous (or liquid) column in an intense, rapidly alternating electrostatic field, set the particles vibrating with enormous speeds, then the inertia resistance asserts itself. A body might move with more or less freedom through the vibrating mass, but as a whole it would be rigid.

There is a subject which I must mention in connection with these experiments: it is that of high vacua. This is a subject the study of which is not only interesting, but useful, for it may lead to results of great practical importance. In commercial apparatus, such as incandescent lamps, operated from ordinary systems of distribution, a much higher vacuum than obtained at present would not secure a very great advantage. In such a case the work is performed on the filament and the gas is little concerned; the improvement, therefore, would be but trifling. But when we begin to use very high frequencies and potentials, the action of the gas becomes all important, and the degree of exhaustion materially modifies the results. As long as ordinary coils, even very large ones, were used, the study of the subject was limited, because just at a point when is became most interesting it had to be interrupted on account of the "non-striking" vacuum being reached. But presently we are able to obtain from a small disruptive discharge coil potentials much higher than even the largest coil was capable of giving, and, what is more, we can make the potential alternate with great rapidity. Both of these results enable us now to pass a luminous discharge through almost any vacua obtainable, and the field of our investigations is greatly extended. Think we as we may, of all the possible directions to develop a practical

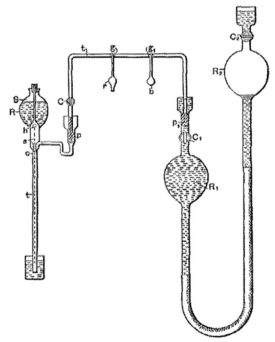

FIG. 30.—APPARATUS USED FOR OBTAINING HIGH DEGREES OF EXHAUSTION.

illuminant, the line of high vacua seems to be the most promising at present. But to reach extreme vacua the appliances must be much more improved, and ultimate perfection will not be attained until we shall have discarded the mechanical and perfected an *electrical* vacuum pump. Molecules and atoms can be thrown out of a bulb under the action of an enormous potential: *this* will be the principle of the vacuum pump of the future. For the present, we must secure the best results we can with mechanical appliances. In this respect, it might not be out of the way to say a few words about the method of, and apparatus for, producing excessively high degrees of exhaustion of which I have availed myself in the course of these investigations. It is very probable that other experimenters have used similar arrangements; but as it is possible that there may be an item of interest in their description, a few remarks, which will render this investigation. more complete, might be permitted.

The apparatus is illustrated in a drawing shown in Fig. 30. S represents a Sprengel pump, which has been specialty constructed to better suit the work required. The stopcock which is usually employed has been omitted, and instead of it a hollow stopper . s has been fitted in the neck of the reservoir R. This stopper has a small hole h, through which the mercury descends; the size of the outlet to be properly determined with respect to the section of the fall tube t, which is sealed to the reservoir instead of being connected to it in the usual manner. This arrangement overcomes the imperfections and troubles which often arise from the use of the stopcock on the reservoir and the connection of the latter with the fall tube.

The pump is connected through a U-shaped tube t to a very large reservoir R_1. Especial *care* was taken in fitting the grinding surfaces of the stoppers p and p_1, and both of these and the mercury caps above them were made exceptionally long. After the U-shaped tube was fitted and put in place, it was heated, so as to soften and take off the strain resulting from imperfect fitting. The U-shaped tube was provided with a stopcock C, and two ground connections g and g_1 one for a small bulb b, usually containing caustic potash, and the other for the receiver r, to be exhausted.

The reservoir R_1 was connected by means of a rubber tube to a slightly larger reservoir R., each of the two reservoirs being provided with a stopcock C_1 and C_2, respectively. The reservoir R_2 could be raised and lowered by a wheel and rack, and the range of its motion was so determined that when it was filled with mercury and the stopcock C_2 closed, so as to form a Torricellian vacuum in it when raised, it could be lifted so high that the mercury in reservoir R_1 would stand a little above stopcock C_1 and when this stopcock was closed and the reservoir R_2 descended, so as to form a Torricellian vacuum in reservoir R_1, it could be lowered so far as to completely empty the latter, the mercury filling the reservoir R_2 up to a little above stopcock C_2.

The capacity of the pump and of the connections was taken as small as possible relatively to the volume of reservoir, R_1, since, of course, the degree of exhaustion depended upon the ratio of these quantities.

With this apparatus I combined the usual means indicated by former experiments for the production of very high vacua. In most of the experiments it was convenient to use caustic potash. I may venture to say, in regard to its use, that much time is saved and a more perfect action of the pump insured by fusing and boiling the potash as soon as, or even before, the pump settles down. If this course is not followed the sticks, as ordinarily employed, may give moisture off at a certain very slow rate, and the pump may work for many hours without reaching a very high vacuum. The potash was heated either by a spirit lamp or by passing a discharge through it, or by passing a current through a wire contained in it. The advantage in the latter case was that the heating could be more rapidly repeated.

Generally the process of exhaustion was the following: — At the start, the stopcocks C and C_1 being open, and all other connections closed, the reservoir R_2 was raised so far that the mercury filled the reservoir R_1 and a part of the narrow connecting U-shaped tube. When the pump was set to work, the mercury would, of course, quickly rise in the tube, and reservoir R_2 was lowered, the experimenter keeping the mercury at about the same level. The reservoir R_2 was balanced by a long spring which facilitated the operation, and the friction of the parts was generally sufficient to keep it almost in any position. When the Sprengel pump had done its work, the reservoir R_2 was further lowered and the mercury descended in R_1 and filled R_2, whereupon stopcock C_2 was closed. The air adhering to the walls of R_1 and that absorbed by the mercury was carried off, and to free the mercury of all air the reservoir R_2 was for a loaf; time worked up and down: During this process some air, which would gather below stopcock C_2 was expelled from R_2 by lowering it far enough and opening the stopcock, closing the latter again before raising the reservoir. When all the air had been expelled from the mercury, and no air would gather in R_2 when it was lowered, the caustic potash was resorted to. The reservoir R_2 was now again raised until the mercury in R_1 stood above stopcock C_1. The caustic potash was fused and boiled, and the moisture partly carried off by the pump and partly re-absorbed; and this process of heating and cooling was repeated many times, and each time, upon the moisture being absorbed or carried off, the reservoir R_2 was for a long time raised and lowered. In this manner all the moisture was carried off from the mercury, and both the reservoirs were in proper condition to be used. The reservoir R_2 was then again raised to the top, and the pump was kept working for a long time. When the highest vacuum obtainable with the pump had been reached the potash bulb was usually wrapped with cotton which was sprinkled with ether so as to keep the potash at a very low temperature, then the reservoir R_2 was lowered, and upon reservoir R_1 being emptied the receiver r was quickly sealed up.

When a new bulb was put on, the mercury was always raised above stopcock C_1, which was closed, so as to always keep the mercury and both the reservoirs in fine condition, and the mercury was never withdrawn from R_1 except when the pump had reached the highest degree of exhaustion. It is necessary to observe this rule if it is desired to use the apparatus to advantage.

By means of this arrangement I was able to proceed very quickly, and when the apparatus was in perfect order it was possible to reach the phosphorescent stage in a small bulb in less than 15 minutes, which is certainly very quick work for a small laboratory arrangement requiring all in all about 100 pounds of mercury. With ordinary small bulbs the ratio of the capacity of the pump, receiver, and connections, and that of reservoir R was about 1—20, and the degrees of exhaustion reached were necessarily very high, though I am unable to make a precise and reliable statement how far the exhaustion was carried.

What impresses the investigator most in the course of these experiences is the behavior of gases when subjected to great rapidly alternating electrostatic stresses. But lie must remain in doubt as to whether the effects observed are due wholly to the molecules, or atoms, of the gas which chemical analysis discloses to us, or whether there enters into play another medium of a gaseous nature, comprising atoms, or molecules, immersed in a fluid pervading the space. Such a medium surely must exist, and I am convinced that, for instance, even if air were absent, the surface and neighborhood of a body in space would be heated by rapidly alternating the potential of the body; but no such heating of the surface or neighborhood could occur if all free atoms were removed and only a homogeneous, incompressible, and elastic fluid — such as ether is supposed to be — would remain, for then there would be no impacts, no collisions. In such a case, as far as the body itself is concerned, only frictional losses in the inside could occur.

It is a striking fact that the discharge through a gas is established with ever increasing freedom as the frequency of the impulses is augmented. It behaves in this respect quite contrarily for a metallic conductor. In the latter the impedance enters prominently into play as the frequency is increased, but the gas acts much as a series of condensers would: the facility with which the discharge passes through seems to depend on the rate of change of potential. If it act so, then in a vacuum tube even of great length, and no matter how strong the current, self-induction could not assert itself to any appreciable degree. We have, then, as far as we can now see, in the gas a conductor which is capable of transmitting electric impulses of any frequency which we may be able to produce. Could the frequency be brought high enough, then a queer system of electric distribution, which would be likely to interest bas companies, might be realized: metal pipes filled with gas — the metal being the insulator, the gas the conductor — supplying phosphorescent bulbs, or perhaps devices as yet uninvented. It is certainly possible to take a hollow core of copper, rarefy the gas in the same and by passing impulses of sufficiently high frequency through a circuit around it, bring the gas inside to a high degree of incandescence; but as to the nature of the forces there would be considerable uncertainty, for it would be doubtful whether with such impulses the copper core would act as a static screen. Such paradoxes and apparent impossibilities we encounter at every step in this line of work, and therein lies, to a great extent, the charm of the study.

I have here a short and wide tube which is exhausted to a high degree and covered with a substantial coating of bronze, the coating allowing barely the light to shine through. A metallic clasp, with a hook for suspending the tube, is fastened around the middle portion of the latter, the clasp being in contact with the bronze coating. I now want to light the gas inside by suspending the tube on a wire connected to the coil. Any one who would try the experiment for the first time, not having any previous experience, would probably take care to be quite alone when making the trial, for fear that he might become the joke of his assistants. Still, the bulb lights in spite of the metal coating, and the light can be distinctly perceived through the latter. A long tube covered with aluminum bronze lights when held in one hand — the other touching the terminal of the coil — quite powerfully. It might be objected that the coatings are not sufficiently conducting; still, even if they were highly resistant, they ought to screen the gas. They certainly screen it perfectly in a condition of rest, but not by far perfectly when the charge is surging in the coating. But the loss of energy which occurs within the tube, notwithstanding the screen, is occasioned principally energy the presence of the gas. Were we to take a large hollow metallic sphere and fill it with a perfect incompressible fluid dielectric, there would be no loss inside of the sphere, and consequently the inside might be considered as perfectly screened, though the potential be very rapidly alternating. Even were the sphere filled with oil, the loss would be incomparably smaller ,than when the fluid is replaced by a gas, for in the latter case the force produces displacements; that means impact and collisions in the inside.

No matter what the pressure of the gas may be, it becomes an important factor in the heating of a conductor when the electric density is ,great and the frequency very high. That in the heating of conductors by lightning discharges air is an element of great importance, is almost as certain as an experimental fact. I may illustrate the action of the air by the following experiment: I take a short tube which is exhausted to a moderate degree and has a platinum wire running through the middle from one end to the other. I pass a steady or low frequency current through the wire, and it is heated uniformly in all parts. The heating here is due to conduction, or frictional losses, and the has around the wire has — as far as we can see — no function to perform. But now let me pass sudden discharges, or a high frequency current, through the wire. Again the wire is heated, this time principally on the ends and least in the middle portion; and if the frequency of the impulses, or the rate of change, is high enough, the wire might :,,; well be cut in the middle as not, for practically all the heating is due to the rarefied gas. Here the gas might only act as a conductor of no impedance diverting the current from the wire as the impedance of the latter is enormously increased, and merely heating the ends of the wire by reason of their resistance to the passage of the discharge. But it is not at all necessary that the gas in the tube should be conducting; it might be at an extremely low pressure, still the ends of the wire would be heated — as, however, is ascertained by experience — only the two ends would in such case not be electrically connected through the gaseous medium. Now what with these frequencies and potentials occurs in an exhausted tube occurs in the lightning discharges at ordinary pressure. We only need remember one of the facts arrived at in the course of these investigations, namely, that to impulses of very high frequency the gas at ordinary pressure behaves much in the same manner as though it were at moderately low pressure. I think that in lightning discharges frequently wires or conducting objects are volatilized merely because air i5 present, and that, were the conductor immersed in an insulating liquid, it would be safe, for then the energy would have to spend itself somewhere else. From the behavior of gases to sudden impulses of high potential I am led to conclude that there can be no surer way of diverting a lightning discharge than by affording it a passage through a volume of gas, if such a thing can be done in a practical manner.

There are two more features upon which I think it necessary to dwell in connection with these experiments — the "radiant state" and the "non-striking vacuum."

Any one who has studied Crookes' work must have received the impression that the "radiant state" is a property of the gas inseparably connected with an extremely high degree of exhaustion. But it should be remembered that the phenomena observed in an exhausted vessel are limited to the character and capacity of the apparatus which is made use of. I think that in a bulb a molecule, or atom, does not precisely move in a straight line because it meets no obstacle, but because the velocity imparted to it is sufficient to propel it in a sensibly straight line. The mean free path is one thing, but the velocity — the energy associated with the moving body — is another, and under ordinary circumstances I believe that it is a mere question of potential or speed. A disruptive discharge coil, when the potential is pushed very far, excites phosphorescence and projects shadows, at comparatively low degrees of exhaustion. In a lightning discharge, matter moves in straight lines at ordinary pressure when the mean free path is exceedingly small, and frequently images of wires or other metallic objects have been produced by the particles thrown off in straight lines.

I have prepared a bulb to illustrate by an experiment the correctness of these assertions. .In a globe L (Fig. 31), I have mounted upon a lamp filament f a piece of lime l. The lamp filament is connected with a wire which leads into the bulb, and the general construction of the latter is as indicated in Fig. 19, before described. The bulb being suspended from a wire connected to the terminal of the coil, and the latter being set to work, the lime piece l and the projecting parts of the filament. fare bombarded. The degree of exhaustion is just such that with the potential the coil is capable of giving phosphorescence of the glass is produced, but disappears as soon as the vacuum is impaired. The lime containing moisture, and moisture being given off as soon as heating occurs, the .phosphorescence lasts only for a few moments. When the lime has been sufficiently heated, enough moisture has been given off to impair materially the vacuum of the bulb. As the bombardment goes on, one point of the lime piece is more heated. than other points, and the result is that finally practically all the discharge passes through that point which is intensely heated, and a white stream of lime particles (Fig. 31) then breaks forth from that point. This stream is composed of "radiant" matter, yet the degree of exhaustion is low. But the particles move in straight lines because the velocity imparted to them is great, and this is due to three causes — to the great electric density, the high temperature of the small point, and the fact that the particles of the lime are easily torn and thrown off — far more easily than those of carbon. With frequencies such as we are able to obtain, the particles are bodily thrown off and projected to a considerable distance, but with sufficiently high frequencies no such thing would occur: in such case only a stress would spread or a vibration would be propagated through the bulb. It would be out of the question to reach any such frequency on the assumption that the atoms move with the speed of light; but I believe that such a thing is impossible; for this an enormous potential would be required. With potentials which we are able to obtain; even with a disruptive discharge coil, the speed must be quite insignificant.

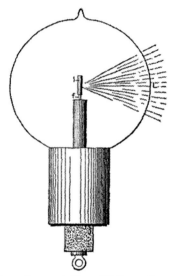

FIG. 31.—BULB SHOWING RADIANT LIME STREAM AT LOW EXHAUSTION.

As to the "non-striking vacuum," the point to be noted is that it can occur only with low frequency impulses, and it is necessitated by the impossibility of carrying off enough energy with such impulses in high vacuum since the few atoms which are around the terminal upon coming in contact with the same are repealed and kept at a distance for a comparatively long period of time, and not enough work can be performed to render the effect perceptible to the eye. If the difference of potential between the terminals is raised, the dielectric breaks down. But with very high frequency impulses there is no necessity for such breaking down, since any amount of work can be performed by continually agitating the atoms in the exhausted vessel, provided the frequency is high enough. It is easy to reach — even with frequencies obtained from an alternator as here, used — a stage at which the discharge does not pass between two electrodes in a narrow tube, each of these being connected to one of the terminals of the coil, but it is difficult to reach a point at which a luminous discharge would not occur around each electrode.

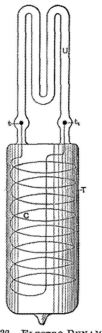

FIG. 32.—ELECTRO-DYNAMIC INDUCTION TUBE.

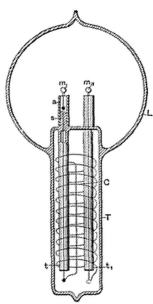

FIG. 33.—ELECTRO-DYNAMIC INDUCTION LAMP.

A thought which naturally presents itself in connection with high frequency currents, is to make use of their powerful electrodynamic inductive action to produce light effects in a sealed glass globe. The leading-in wire is one of the defects of the present incandescent lamp, and if no other improvement were made, that imperfection at least should be done away with. Following this thought, I have carried on experiments in various directions, of which some were indicated in my former paper. I may here mention one or two more lines of experiment which have been followed up.

Many bulbs were constructed as shown in Fig. 32 and Fig. 33.

In Fig. 32 a wide tube T was sealed to a smaller W-shaped tube U, of phosphorescent glass. In the tube T was placed a coil C of aluminum wire, the ends of which were provided with small spheres t and $t1$ of aluminum, and reached into the U tube. The tube T was slipped into a socket containing a primary coil through which usually the discharges of Leyden jars were directed, and the rarefied gas in the small U tube was excited to strong luminosity by the high-tension currents induced in the coil C. When Leyden jar discharges were used to induce currents in the coil C, it was found necessary to pack the tube T tightly with insulating powder, as a discharge would occur frequently between the turns of the coil, especially when the primary was thick and the air gap, through which the jars discharged, large, and no little trouble was experienced in this way.

In Fig. 33 is illustrated another form of the bulb constructed. In this case a tube T is sealed to a globe L. The tube contains a coil C, the ends of which pass through two small glass tubes t and t_1 which are sealed to the tube T. Two refractory buttons m and m_1 are mounted on lamp filaments which are fastened to the ends of the wires passing through the glass tubes t and t_1. Generally in bulbs made on this plan the globe I. communicated with the tube T. For this purpose the ends of the small tubes t and t_1 were just a trifle heated in the burner, merely to hold the wires, but not to interfere with the communication. The tube T, with the small tubes, wires through the same, and the refractory buttons in and nil, was first prepared, and then scaled to globe L, whereupon the coil C was slipped in and the connections made to its ends. The tube was then packed with insulating powder, jamming the latter as tight as possible up to very nearly the end, then it was closed and only a small hole left through which the remainder of the powder was introduced, and finally the end of the tube was closed. Usually in bulbs constructed as shown in Fig. 33 an aluminum tube a was fastened to the upper end s of each of the tubes t and tl, in order to protect that end against the heat. The buttons in and *in,* could be brought to any degree of incandescence by passing the discharges of Leyden jars around the coil C. In such bulbs with two buttons a very curious effect is produced by the formation of the shadows of each of the two buttons.

Another line of experiment, which has been assiduously followed, was to induce by electro-dynamic induction a current or luminous discharge in an exhausted tube or bulb. This matter has received such able treatment at the hands of Prof. J. J. Thomson that I could add but little to what he has trade known, even had I made it the special subject of this lecture. Still, since experiences in this line have gradually led me to the present views and results, a few words must be devoted here to this subject.

It has occurred, no doubt, to many that as a vacuum tube is made longer the electromotive force per unit length of the. tube, necessary to pass a luminous discharge through the latter, gets continually smaller; therefore, if the exhausted tube be made long enough, even with low frequencies a luminous discharge could be induced in such a tube closed upon itself. Such a tube might be placed around a hall or on a ceiling, and at once a simple appliance capable of giving considerable light would be obtained. But this would be an appliance hard to manufacture and extremely unmanageable. It would not do to snake the tube up of small lengths, because there would be with ordinary frequencies considerable loss in the coatings, and besides, if coatings were used, it would be better to supply the current directly to the tube by connecting the coatings to a transformer. But even if all objections of such nature were removed, still, with low frequencies the light conversion itself would be inefficient; as I have before stated. In using extremely high frequencies the length of the secondary — in other words, the size of the vessel — can be reduced as far as desired, and the efficiency of the light conversion is increased, provided that means are invented for efficiently obtaining such high frequencies. Thus one is led, from theoretical and practical considerations, to the use of high frequencies, and this means high electromotive forces and small currents in the primary. When he works with condenser charges — and they are the only means up to the present known for reaching these extreme frequencies — he gets to electromotive forces of several thousands of volts per turn of the primary. He cannot multiply the electro-dynamic inductive effect by taking more turns in the primary, for he arrives at the conclusion that the best way is to work with one single turn — though he must sometimes depart from this rule — and he must get along with whatever inductive effect lie can obtain with one turn. But before he has long experimented with the extreme frequencies required to set up in a small bulb an electromotive force of several thousands of volts he realizes the great importance of electrostatic effects, and these effects grow relatively to the electro-dynamic in significance as the frequency is increased.

Now, if anything is desirable in this case, it is to increase the frequency, and this would make it still worse for the electro-dynamic effects. On the other hand, it is easy to exalt the electrostatic action as far as one likes by taking more turns on the

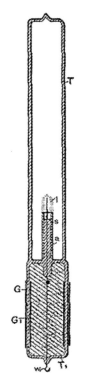

Fig. 34.—Tube with Filament Rendered Incandescent in an Electrostatic Field.

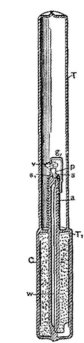

Fig. 35.—Crookes' Experiment in Electrostatic Field.

secondary, or combining self-induction and capacity to raise the potential. It should also be remembered that, in reducing the current to the smallest value and increasing the potential, the electric impulses of high frequency can be more easily transmitted through a conductor.

These and similar thoughts determined me to devote more attention to the electrostatic phenomena, and to endeavor to produce potentials as high as possible, and alternating as fast as they could be made to alternate. I then found that I could excite vacuum tubes at considerable distance from a conductor connected to a properly constructed coil, and that I could, by converting the oscillatory current of a condenser to a higher potential, establish electrostatic alternating fields which acted through the whole extent of a room, lighting up a tube no matter where it was held in space. I thought I recognized that I had made a step in advance, and I have persevered in this line; but I wish to. say that I share with all lovers of science and progress the one and only desire — to reach a result of utility to men in any direction to which thought or experiment may lead me. I think that this departure is the right one, for I cannot see, from the observation of the phenomena which manifest themselves as the frequency is increased, what there would remain to act between two, circuits conveying, for instance, impulses of several hundred millions per second, except electrostatic forces. Even with such trifling frequencies the energy would be practically all potential, and my conviction has grown strong that, to whatever kind of motion light may be due, it is produced by tremendous electrostatic stresses vibrating with extreme rapidity.

Of all these phenomena observed with currents, or electric impulses, of high frequency, the most fascinating for an audience are certainly those which are noted in an electrostatic field acting through considerable distance, and the best an unskilled lecturer can do is to begin and finish with the exhibition of these singular effects. I take a tube in the hand and move it about, and it is lighted wherever I may hold it; throughout space the invisible forces act. But I may take another tube and it might not light, the vacuum being very high. I excite it by means of a disruptive discharge coil, and now it will light in the electrostatic field. I may put it away for a few weeks or months, still it retains the faculty of being excited. What change have I produced in the tube in the act of exciting it? If a motion imparted to the atoms, it is difficult to perceive how it can persist so long without being arrested by frictional losses; and if a strain exerted in the dielectric, such as a simple electrification would produce, it is easy to see how it may persist indefinitely but very difficult to understand why such a condition should aid the excitation when we have to deal with potentials which are rapidly alternating.

Since I have exhibited these phenomena for the first time, I have obtained some other interesting effects. For instance, I have produced the incandescence of a button, filament, or wire enclosed in a tube. To get to this result it was necessary to economize the energy which is obtained from the field and direct most of it on the small body to be rendered incandescent. At the beginning the task appeared difficult, but the experiences gathered permitted me to reach the result easily. In Fig. 34 and Fig. 35 two such tubes are illustrated which are prepared for the occasion. In Fig. 34 a short tube T_1, sealed to another long tube T, is provided with a stem s, with a platinum wire sealed in the latter. A very thin lamp filament l is fastened to this wire, and connection to the outside is made through a thin copper wire w. The tube is provided with outside and inside coatings, C and C_1 respectively, and is filled as far as the coatings reach with conducting, and the space above with insulating powder. These coatings are merely used to enable me to perform two experiments with the tube — namely, to produce the effect desired either by direct connection of the body of the experimenter or of another body to the wire w, or by acting inductively through the glass. The stem s is provided with an aluminum tube a, for purposes before explained, and only a small part of the filament reaches out of this tube. By holding the tube T1 anywhere in the electrostatic field the filament is rendered incandescent.

A more interesting piece of apparatus is illustrated in Fig. 35. The construction is the same as before, only instead of the lamp filament a small platinum wire p, sealed in a stem s, and bent above it in a circle, is connected to the copper wire w which is joined to an inside coating C. A small stern s_1 is provided with a needle, on the point of which is arranged to rotate very freely a very light fan of mica v. To prevent the fan from falling out, a thin stem of glass g is bent properly and fastened to the aluminum tube. When the glass tube is held anywhere in the electrostatic field the platinum wire becomes incandescent, and the mica vanes are rotated very fast.

Intense phosphorescence may be excited in a bulb by merely connecting it to a plate within the field, and the plate need not be any larger than an ordinary lamp shade. The phosphorescence excited with these currents is incomparably more powerful than with ordinary apparatus. A small phosphorescent bulb, when attached to a wire connected to a coil, emits sufficient light to allow reading ordinary print at a distance of five to six paces. It was of interest to see how some of the phosphorescent bulbs of Professor Crookes would behave with these current;, and he has had the kindness to lend me a few for the occasion. The effects produced are magnificent, especially by the sulphide of calcium and sulphide of zinc. From the disruptive discharge coil they glow intensely merely by holding them in the hand and connecting the body to the terminal of the coil.

To whatever results investigations of this kind may lead, their chief interest ties for the present in the possibilities they offer for the production of an efficient illuminating device. In no branch of electric industry is an advance more desired than in the manufacture of light. Every thinker, when considering the barbarous methods employed, the deplorable losses incurred in our best systems of light production, must have asked himself, What is likely to be the light of the future? Is it to be an incandescent solid, as in the present lamp, or an incandescent gas, or a phosphorescent body, or something like a burner, but incomparably more efficient?

There is little chance to perfect a gas burner; not, perhaps, because human ingenuity has been bent upon that problem for centuries without a radical departure having been made — though this argument is not devoid of force — but because in a burner the higher vibrations can never be reached except by passing through all the low ones. For how is a flame produced unless by a fall of lifted weights? Such process cannot be maintained without renewal, and renewal is repeated passing from low to high vibrations. One way only seems to be open to improve a burner, and that is by trying to reach higher degrees of incandescence. Higher incandescence is equivalent to a quicker vibration; that means more light from the same material, and that, again, means more economy. In this direction some improvements have been made:, but the progress is hampered by many limitations. Discarding, then, the burner, there remain the three ways first mentioned which are essentially electrical.

Suppose the light of the immediate future to be a solid rendered incandescent by electricity. Would it not seem that it is better to employ a small button than a frail filament? From many considerations it certainly must be concluded that a button is capable of a higher economy, assuming, of course, the difficulties connected with the operation of such a lamp to be effectively overcome. But to light such a lamp we require a high potential; and to get this economically we must use high frequencies.

Such considerations apply even more to the production of light by the incandescence of a gas, or by phosphorescence. in all cases we require high frequencies and high potentials. These thoughts occurred to me a long time ago.

Incidentally we gain, by the use of very high frequencies, many advantages, such as a higher economy in the light production, the possibility of working with one lead, the possibility of doing away with the leading-in wire, etc.

The question is, how far can we go with frequencies? Ordinary conductors rapidly lose the facility of transmitting electric impulses when the frequency is greatly increased. Assume the means for the production of impulses of very great frequency brought to the utmost perfection, every one will naturally ask how to transmit them when the necessity arises. In transmitting such impulses through conductors we must remember that we have to deal with *pressure* and *flora,* in the ordinary interpretation of these terms. Let the pressure increase to an enormous value, and let the flow correspondingly diminish, then such impulses — variations merely of pressure, as it were — can no doubt be transmitted through a wire even if their frequency be many hundreds of millions per second. It would, of course, be out of question to transmit such impulses through a wire immersed in a gaseous medium, even if the wire were provided with a thick and excellent insulation for most of the energy would be lost in molecular bombardment and consequent heating. The end of the wire connected to the source would be heated, and the remote end would receive but a trifling part of the energy supplied. The prime necessity, then, if such electric impulses are to be used, is to find means to reduce as much as possible the dissipation.

The first thought is, employ the thinnest possible wire surrounded by the thickest practicable insulation. The next thought is to employ electrostatic screens. The insulatiou of the wire may be covered with a thin conducting coating and the latter concocted to the ground. But this would not do, as then all the energy would pass through the conducting coating to the ground and nothing would get to the end of the wire. If a ground connection is made it can only be made through a conductor offering an enormous impedance, or through a condenser of extremely small capacity. This, however, does not do away with other difficulties.

If the wave length of the impulses is much smaller than the length of the wire, then corresponding short waves will be sent up in the conducting coating, and it will be more or less the same as though the coating were directly connected to earth. It is therefore necessary to cut up the coating in sections much shorter than the wave length. Such an arrangement does not still afford a perfect screen, but it is ten thousand times better than none. I think it preferable to cut up the conducting coating in small sections, even if the current waves be much longer than the coating.

If a wire were provided with a perfect electrostatic screen, it would be the same as though all objects were removed from it at infinite distance. The capacity would then be reduced to the capacity of the wire itself, which would be very small. It would then be possible to send over the wire current vibrations of very high frequencies at enormous distance without affecting; greatly the character of the vibrations. A perfect screen is of course out of the question, but I believe that with a screen such as I have just described telephony could be rendered practicable across the Atlantic. According to my ideas, the gutta-percha covered wire should be provided with a third conducting coating subdivided in sections. On the top of this should be again placed a layer of gutta-percha and other insulation, and on the top of the whole the armor. But such cables will not be constructed, for ere long intelligence — transmitted without wires —will throb through the earth like a pulse through a living organism. The wonder is that, with the present state of knowledge and the experiences gained, no attempt is being made to disturb the electrostatic or magnetic condition of the earth, and transmit, if nothing else, intelligence.

It has been my chief aim in presenting these results to point out phenomena or features of novelty, and to advance ideas which I am hopeful will serve as starting points of new departures. It has been my chief desire this evening to entertain you with some novel experiments. Your applause, so frequently and generously accorded. has told me that I have succeeded.

In conclusion, let me thank you most heartily for your kindness and attention, and assure you that the honor I have had in addressing such a distinguished audience, the pleasure I have had in presenting these results to a gathering of so many able men — and among them also some of those in whose work for many years past I have found enlightenment and constant pleasure — I shall never forget.

The "Drehstrom" Patent
The Electrical World - N. Y. — Oct. 8, 1892

In the last issue of *the Electrical World* I find an article on my "Drehstrom" patent which appeared originally is *Industries,* and is, I believe, from the pen of the able editor of that journal. Some of the statements made are such as to cause an erroneous opinion to gain ground, which I deem it my duty to prevent — a disagreeable one I may say, as I do not like to express my opinion on a patent, especially if it is my own.

It may be, as the writer states, that the theory of the action of my motor advanced in my paper before the American Institute in May, 1888, is a clumsy one, but this theory was formed by me a number of years before the practical results were announced, the patents being applied for only after it was undoubtedly demonstrated that the motor could fairly compete in efficiency with the direct current motor, and that the invention was one of commercial value. These patents were taken out with the help of some of the ablest attorneys in the United States, well versed in electrical matters; the specifications were drawn up with great care, in view of the importance of the invention, and with proper regard to the state of the art at that period, and had the patents been carefully studied by others there would not hare been various features of my system reinvented, and several inventors, would have been spared at this late date a keen disappointment.

The writer apprehends that it might be difficult for a non-technical judge to decide whether a motor with two or more separate fields and armatures, coupled together mechanically, does or does not fall under my patent. I do not share his apprehension, Judges are highly educated men, and it does not require much technical knowledge to convince one that it is the same whether two belts driving a rigid arbor are close together or far apart. Nor do I think that it is necessary for the honorable judge to be a partisan of the armature reaction theory in order to recognize the identity of the two arrangements referred to by the writer of the article in question. Indeed, I would seriously doubt the sincerity of a man capable of clear conceptions were he to uphold that the arrangements are essentially different, even if the case should stand exactly as he assumes by way of illustration of " puzzles likely to arise." For where is there a difference? Take, for instance, a form of my two-phase motor. There are two sets of field magnets, one at the *neutral* parts of the other. One of the sets, therefore, might as well be removed and placed a distance sideways, but long experience shows that in output, efficiency, cost of construction and in general mechanical respects such an arrangement is inferior. The two sets are connected inductively through the armature body or the windings thereon. Part of the period one set of field magnets acts as a generator, setting up induction currents, which circulate in the field of force of the other set, which may be looked upon as a motor. Part of the period again, the second set becomes the generator and the first the motor, the action being at the same time such that the generated currents are always passed in a definite direction with respect to the field ; they are commutated as it were, and a tendency to rotate in a given direction is imparted to the armature. Now place two fields side by side and connect properly the armature windings. Are not the fields again inductively connected? Do not the currents set up by one field cause currents to circulate in the other, and is the action not exactly the same in both cases? This is a fact, no matter what theory is adhered to. The writer says that in the case of two separate structures there is really nothing which may be called rotation of the field. But is there any such thing, when the two structures are merged in one? Is it not in accordance with accepted notions to conceive the imaginary lines as surging simply in the pole projections in exactly the same manner in both the arrangements? Irrespective of the view taken, be it even the more unfair to the inventor, no one is permitted to go so far as to make him responsible, in such a case, for theories and interpretations of his invention. Theories may come and go, but the motor works, a practical result is achieved and the art is advanced through his pains and efforts. But what I desire to point out principally is that in the article above referred to the writer is only assuming a case which cannot occur. He is evidently judging the state of things from my short paper before the American Institute. This paper was written in a hurry, in fact only shortly before the meeting of the Institute, and I was unable to do full justice even to those features upon which as employee of a company owning the invention, I was permitted to dwell. Allow me to observe that my patent specification was written up more carefully than my paper and the view taken in it is a broader and truer one. While the "clumsy" theory was adapted as the beat in explanation of the action of the motor, the invention is *not* represented as dependent entirely on that theory; and in showing a three phase motor with six projections, where it was manifestly more consistent with the accepted popular ideas to assume the "lines of force" as simply surging in the projecting pole pieces, this view was distinctly and advisedly taken, as the following quotation from my foundation patent will show : "The variations in the strength and intensity of the currents transmitted through these circuits (lines and armature) and traversing the coils of the motor produce a steadily progressive shifting of the *resultant attractive force* exerted by the poles upon the armature and consequently keep the armature rapidly rotating." There is, in this instance, no question of a rotating field in the common acceptance of the term of the *resultant attractive force* there is a question simply of a diagram of force, and it is immaterial for the operation whether the fields are close together or far apart, or even whether, or not, they are inductively connected.

I do not think that in Germany, where the Patent Office is proverbially strict in upholding the rights of the inventor, an illegitimate and unfair appropriation of the invention by others will be tolerated by the courts.

The Ewing High-Frequency Alternator and Parsons Steam Engine
Electrician - London — Dec. 17, 1892

In your issue of November 18 I find a description of Prof. Ewing's high-frequency alternator, which has pleased me chiefly because it conveyed to me the knowledge that he, and with him, no doubt, other scientific men, is to investigate the properties of high-frequency currents. With apparatus such as you describe, shortly a number of experimenters, more competent than myself, will be enabled to go over the ground as yet but imperfectly explored, which will undoubtedly result in the observation of novel facts and elimination of eventual errors.

I hope it will not be interpreted as my wishing to detract anything from Prof. Ewing's merit if I state the fact that for a considerable time past I have likewise thought of combining the identical steam turbine with a high-frequency alternator. *Anch' io sono pittore.* I had a number of designs with such turbines, and would have certainly carried them out had the turbines been here easily and cheaply obtainable, and had my attention not been drawn in a different direction. As to the combination to which you give a rather complicated name, I consider it an excellent one. The advantages of using a high speed are especially great in connection with such alternators. When a belt is used to drive, one must resort to extraordinarily large diameters in order to obtain the necessary speed, and this increases the difficulties and cost of construction in an entirely unreasonable proportion. In the machine used in my recent experiments the weight of the active parts is less than 50 pounds, but there is an additional weight of over 100 pounds in the supporting frame, which a very careful constructor would have probably made much heavier. When running at its maximum speed, and with a proper capacity in the armature circuit, two and a one-half horse-power can be performed. The large diameter (30 inches), of course, has the advantage of affording better facility for radiation; but, on the other hand, it is impossible to work with a very small clearance.

I have observed with interest that Prof. Ewing has used a magnet with alternating poles. In my first trials I expected to obtain the best results with a machine of the Mordey type - that is, with one having pole projections of the same polarity. My idea was to energize the field up to the point of the maximum permeability of the iron and vary the induction around that point. But I found that with a very great number of pole projections such a machine would not give good results, although with few projections, and with an armature without iron, as used by Mordey, the results obtained were excellent. Many experiences of similar nature made in the course of my study demonstrate that the ordinary rules for the magnetic circuit do not hold good with high frequency currents. In ponderable matter magnetic permeability, and also specific inductive capacity, must undergo considerable change when the frequency is varied within wide limits. This would render very difficult the exact determination of the energy dissipated in iron cores by very rapid cycles of magnetization, and of that in conductors and condensers, by very quick reversals of current. Much valuable work remains to be done in these fields, in which it is so easy to observe novel phenomena, but so difficult to make quantitative determinations. The results of Prof. Ewing's systematical research will be awaited with great interest.

It is gratifying to note from his tests that the turbines are being rapidly improved. Though I am aware that the majority of engineers do not favor their adoption. I do not hesitate to say that I believe in their success. I think their principle uses, in no distant future, will be in connection with alternate current motors, by means of which it is easy to obtain a constant and, in any desired ratio, reduced speed. There are objections to their employment for driving direct current generators, as the commutators must be a source of some loss and trouble, on account of the very great speed; but with an alternator there is no objectionable feature whatever. No matter how much one may be opposed to the introduction of the turbine, he must have watched with surprise the development of this curious branch of the industry, in which Mr. Parsons-has been a pioneer, and everyone must wish him the success which his skill has deserved.

Nikola Tesla

On the Dissipation of the Electrical Energy of the Hertz Resonator
The Electrical Engineer — December 21, 1892

Anyone who, like myself, has had the pleasure of witnessing the beautiful demonstrations with vibrating diaphragms which Prof. Bjerknes, exhibited in person at the Paris Exposition in 1880, must have admired his ability and painstaking care to such a degree, as to have an almost implicit faith in the correctness of observations made by him. His experiments "On the Dissipation of the Electrical Energy of the Hertz Resonator," which are described in the issue of Dec. 14, of *The Electrical Engineer,* are prepared in the same ingenious and skillful manner, and the conclusions drawn from them are all the more interesting as they agree with the theories put forth by the most advanced thinkers. There can not be the slightest doubt as to the truth of these conclusions, yet the statements which follow may serve to explain in part the results arrived at in a different manner; and with this object in view I venture to call attention to a condition with which, in investigations such as those of Prof. Bjerknes, the experimenter is confronted.

The apparatus, oscillator and resonator, being immersed in air, or other discontinuous medium, there occurs—as I have pointed out in the description of my recent experiments before the English and French scientific societies—dissipation of energy by what I think might be appropriately called electric sound waves or sound-waves of electrified air. In Prof. Bjerknes's experiments principally this dissipation in the resonator need be considered, though the sound-waves—if this term be permitted—which emanate from the surfaces at the oscillator may considerably affect the observations made at some distance from the latter. Owing to this dissipation the period of vibration of an air-condenser can not be accurately determined, and I have already drawn attention to this important fact. These waves are propagated at right angles from the charged surfaces when their charges are alternated, and dissipation occurs, even if the surfaces are covered with thick and excellent insulation. Assuming that the "charge" imparted to a molecule or atom either by direct contact or inductively is proportionate to the electric density of the surface, the dissipation should be proportionate to the square of the density and to the number of waves per second. The above assumption, it should be stated, does not agree with some observations from which it appears that an atom can not take but a certain maximum charge; hence, the charge imparted may be practically independent of the density of the surface, but this is immaterial for the present consideration. This and other points will be decided when accurate quantitative determinations, which are as yet wanting, shall be trade. At present it appears certain from experiments with, high-frequency currents, that this dissipation of energy from a wire, for instance, is not very far from being proportionate to the frequency of the alternations, and increases very rapidly when the diameter of the wire is made exceedingly small. On the latter point the recently published results of Prof. Ayrton and H. Kilgour on "The Thermal Emissivity of Thin Wires in Air" throw a curious light. Exceedingly thin wires are capable of dissipating a comparatively very great amount of energy by the agitation of the surrounding air, when they are connected to a source of rapidly alternating potential. So in the experiment cited, a thin hot wire is found to be capable of emitting an extraordinarily great amount of heat, especially at elevated temperatures. In the case of a hot wire it must of course be assumed that the increased emissivity is due to the more rapid convection and not, to any, appreciable degree, to an increased radiation. Were the latter demonstrated, it would show that a wire, made hot by the application of heat in ordinary ways, behaves in some respects like one, the charge of which is rapidly alternated, the dissipation of energy per unit of surface kept at a certain temperature depending on the curvature of the surface. I do not recall any record of experiments intended to demonstrate this, yet this effect, though probably very small, should certainly be, looked for.

A number of observations showing the peculiarity, of very thin wires were made in the course of my experiments. I noted, for instance, that in the well-known Crookes instrument the mica vanes are repelled with comparatively greater force when the incandescent platinum wire is exceedingly thin. This observation enabled me to produce the spin of such vanes mounted in a vacuum tube when the latter was placed in an alternating electrostatic field. This however does not prove anything in regard to radiation, as in a highly exhausted vessel tile phenomena are principally due to molecular bombardment or convection.

When I first undertook to produce the incandescence of a wire enclosed in a bulb, by connecting it to only one of the terminals of a high tension transformer, I could not succeed for a long time. On one occasion I had mounted in a bulb a thin platinum wire, but my apparatus was not adequate to produce the incandescence. I made other bulbs, reducing the length of the wire to a small fraction; still I did not succeed. It then occurred to me that it would be desirable to have the surface of the wire as large as possible, yet the bulk small, and I provided a bulb with an exceedingly thin wire of a bulk about equal to that of the short but much thicker wire. On turning the current on the bulb the wire was instantly fused. A series of subsequent experiments showed that when the diameter of the wire was exceedingly small, considerably more energy would be dissipated per unit surface at all degrees of exhaustion than was to be expected, even on the assumption that the energy given off was in proportion to the square of the electric density. There is likewise evidence which, though not possessing the certainty of an accurate quantitative determination, is nevertheless reliable because it is the result of a great many observations, namely, that with the increase of the density the dissipation is more rapid for thin than for thick wires.

The effects noted in exhausted vessels with high-frequency currents are merely diminished in degree when the air is at ordinary pressure, but heating and dissipation occurs, as I have demonstrated, under the ordinary atmospheric conditions. Two very thin wires attached to the terminals of a high-frequency coil are capable of giving off an appreciable amount of energy. When the density is very great, the temperature of the wires may be perceptibly raised, and in such case probably the greater portion of the energy which is dissipated owing to the presence of a discontinuous medium is transformed into heat at the surface or in close proximity to the wires. Such heating could not occur in a medium possessing either of the two qualities, namely, perfect incompressibility or perfect elasticity. In fluid insulators, such as oils, though they are far from being perfectly incompressible or elastic to electric displacement, the heating is much smaller because of the continuity of the fluid.

When the electric density of the wire surfaces is small, there is no appreciable local heating, nevertheless energy is dissipated in air, by waves, which differ from ordinary sound-waves only because the air is electrified. These waves are especially conspicuous when the discharges of a powerful battery are directed through a short and thick metal bar, the number of discharges per second being very small. The experimenter may feel the impact of the air at distances of six feet or more from the bar, especially if be takes the precaution to sprinkle the face or hands with ether. These waves cannot be entirely stopped by the interposition of an insulated metal plate.

Most of the striking phenomena of mechanical displacement, sound, heat and light which have been observed, imply the presence of a medium of a gaseous structure that is one consisting of independent carriers capable of free motion.

When a glass plate is placed near a condenser the charge of which is alternated, the plate emits a sound. This sound is due to the rhythmical impact of the air against the plate. I have also found that the ringing of a condenser, first noted by Sir William Thomson, is due to the presence of the air between or near the charged surfaces.

When a disruptive discharge coil is immersed in oil contained in a tank, it is observed that the surface of the oil is agitated. This may be thought to be duo to the displacements produced in the oil by the changing stresses, but such is not the case. It is the air above the oil which is agitated and causes the motion of the latter; the oil itself would remain at rest. The displacements produced in it by changing electrostatic stresses are insignificant; to such stresses it may be said to be compressible to but a very small degree. The action of the air is shown in a curious manner for if a pointed metal bar is taken in the hand and held with the point close to the oil, a hole two inches deep is formed in the oil by the molecules of the air, which are violently projected from the point.

The preceding statements may have a general bearing upon investigations in which currents of high frequency and potential are made use of, but they also have a more direct bearing upon the experiments of Prof. Bjerknes which are here considered, namely, the "skin effect," is increased by the action of the air. Imagine a wire immersed in a medium, the conductivity of which would be some function of the frequency and potential difference but such, that the conductivity increases when either or bout of these elements are increased. In such a medium, the higher the frequency and potential difference, the greater wilt be the current which will find its way through the surrounding medium, and the smaller the part which will pass through the central portion of the wire: In the case of a wire immersed in air and traversed by a high-frequency current, the facility with which the energy is dissipated may be considered as the equivalent of the conductivity; and the analogy would be quite complete, were it not that besides the air another medium is present, the total dissipation being merely modified by the presence of the air to an extent as yet not ascertained. Nevertheless, I have sufficient evidence to draw the conclusion, that the results obtained by Prof. Bjerknes are affected by the presence of air in the following manner: 1. The dissipation of energy is more rapid when the resonator is immersed in air than it would be in a practically continuous medium, for instance, oil. 2. The dissipation owing to the presence of air renders the difference between magnetic and non-magnetic metals more striking. The first conclusion follows directly from the preceding remarks; the second follows front the two facts that the resonator receives always the same amount of energy, independent of the nature of the metal, and that the magnetism of the metal increases the impedance of the circuit. A resonator of magnetic metal behaves virtually as though its circuit were longer. There is a greater potential difference set up per unit of length; although this rosy not show itself in the deflection of the electrometer owing to the lateral dissipation. The effect of the increased impedance is strikingly illustrated in the two experiments of Prof. Bjerknes when copper is deposited upon an iron wire, and next iron upon a copper wire. Considerable thickness of copper deposit was required in the former experiment, but very little thickness of iron in the latter, as should be expected.

Taking the above views, I believe, that in the experiments of Prof. Bjerknes which lead him to undoubtedly correct conclusions, the air is a factor fully as important, if not more so, than the resistance of the metals.

On Light and Other High Frequency Phenomena
A lecture delivered before the Franklin Institute, Philadelphia, February, 1893, and before The National Electric Light Association, St. Louis, March, 1893.

Introductory — Some Thoughts on the Eye

When we look at the world around us, on Nature, we are impressed with its beauty and grandeur. Each thing we perceive, though it may be vanishingly small, is in itself a world, that is, like the whole of the universe, matter and force governed by law, — a world, the contemplation of which fills us with feelings of wonder and irresistibly urges us to ceaseless thought and inquiry. But in all this vast world, of all objects our senses reveal to us, the most marvelous, the most appealing to our imagination, appears no doubt a highly developed organism, a thinking being. If there is anything fitted to make us admire Nature's handiwork, it is certainly this inconceivable structure, which performs its innumerable motions of obedience to external influence. To understand its workings, to get a deeper insight into this Nature's masterpiece, has ever been for thinkers a fascinating aim, and after many centuries of arduous research men have arrived at a fair understanding of the functions of its organs and senses. Again, in all the perfect harmony of its parts, of the parts which constitute the material or tangible of our being, of all its organs and senses, the eye is the most wonderful. It is the most precious, the most indispensable of our perceptive or directive organs, it is the great gateway through which all knowledge enters the mind. Of all our organs, it is the one, which is in the most intimate relation with that which we call intellect. So intimate is this relation, that it is often said, the very soul shows itself in the eye.

It can be taken as a fact, which the theory of the action of the eye implies, that for each external impression, that is, for each image produced upon the retina, the ends of the visual nerves, concerned in the conveyance of the impression to the mind, must be under a peculiar stress or in a vibratory state, It now does not seem improbable that, when by the power of thought an image is evoked, a distinct reflex action, no matter how weak, is exerted upon certain ends of the visual nerves, and therefore upon the retina. Will it ever be within human power to analyze the condition of the retina when disturbed by thought or reflex action, by the help of some optical or other means of such sensitiveness that a clear idea of its state might be gained at any time? If this were possible, then the problem of reading one's thoughts with precision, like the characters of an open book, might be much easier to solve than many problems belonging to the domain of positive physical science, in the solution of which many, if not the majority: of scientific men implicitly believe. Helmholtz has shown that the fundi of the eye are themselves, luminous, and he was able to *see*, in total darkness, the movement of his arm by the light of his own eyes. This is one of the most remarkable experiments recorded in the history of science, and probably only a few men could satisfactorily repeat it, for it is very likely, that the luminosity of the eyes is associated with uncommon activity of the brain and great imaginative power. It is fluorescence of brain action, as it were.

Another fact having a bearing on this subject which has probably been noted by many, since it is stated in popular expressions, but which I cannot recollect to have found chronicled as a positive result of observation is, that at times, when a sudden idea or image presents itself to the intellect, there is a distinct and sometimes painful sensation of luminosity produced in the eye, observable even in broad daylight.

The saying then, that the soul shows itself in the eye, is deeply founded, and we feel that it expresses a great truth. It has a profound meaning even for one who, like a poet or artist, only following; his inborn instinct or love for Nature, finds delight in aimless thoughts and in the mere contemplation of natural phenomena, but a still more profound meaning for one who, in the spirit of positive scientific investigation, seeks to ascertain the causes of the effects. It is principally the natural philosopher, the physicist, for whom the eye is the subject of the most intense admiration.

Two facts about the eye must forcibly impress the mind of the physicist, notwithstanding he may think or say that it is an imperfect optical instrument, forgetting, that the very conception of that which is perfect or seems so to him, has been gained through this same instrument. First, the eye is, as far as our positive knowledge goes, the only organ which is *directly* affected by that subtle medium, which as science teaches us, must fill all space; secondly, it is the most sensitive of our organs, incomparably more sensitive to external impressions than any other.

The organ of hearing implies the impact of ponderable bodies, the organ of smell the transference of detached material particles, and the organs of taste. and of touch or force, the direct contact, or at least some interference of ponderable matter, and this is true even in those instances of animal organisms, in which some of these organs are developed to a degree of truly marvelous perfection. This being so, it seems wonderful that the organ of sight solely should be capable of being stirred by that, which all our other organs are powerless to detect, yet which plays an essential part in all natural phenomena, which transmits all energy and sustains all motion and, that most intricate of all, life, but which has properties such that even a scientifically trained mind cannot help drawing a distinction between it and all that is called matter. Considering merely this, and the fact that the eye, by its marvelous power, widens our otherwise very narrow range of perception far beyond the limits of the small world which is our own, to embrace myriads of other worlds, suns and stars in the infinite depths of the universe, would make it justifiable to assert, that it is an organ of a higher order. Its performances are beyond comprehension. Nature as far as we know never produced anything more wonderful. We can get barely a faint idea of its prodigious power by analyzing what it does and by comparing. When ether waves impinge upon the human body, they produce the sensations of warmth or cold, pleasure or pain, or perhaps other sensations of which we are not aware, and any degree or intensity of these sensations, which degrees are infinite in number, hence an infinite number of distinct sensations. But our sense of touch, or our sense of force, cannot reveal to us these differences in degree or intensity, unless they are very great. Now we can readily conceive how an organism, such as the human, in the eternal process of evolution, or more philosophically speaking, adaptation to Nature, being constrained to the use of only the sense of touch or force, for instance, might develop this sense to such a degree of sensitiveness or perfection, that it would be capable of distinguishing the minutest differences in the temperature of a body even at some distance, to a hundredth, or thousandth, or millionth part of a degree. Yet, even this apparently impossible performance would not begin to compare with that of the eye, which is capable of distinguishing and conveying to the mind in a single instant innumerable peculiarities of the body, be it in form, or color, or other respects. This power of the eye rests upon two thins, namely, the rectilinear propagation of the disturbance by which it is effected, and upon its sensitiveness. To say that the eye is sensitive is not saying anything. Compared with it, all other organs are monstrously crude. The organ of smell which guides a dog on the trail of a deer,. the organ of touch or force which guides an insect in its wanderings, the organ of hearing, which is affected by the slightest disturbances of the air, are sensitive organs, to be sure, but what are they compared with the human eye! No doubt it responds to the faintest echoes or reverberations of the medium; no doubt, it brings us tidings from other worlds, infinitely remote, but in a language we cannot as yet always understand. And why not? Because we live in a medium filled with air and other gases, vapors and a dense mass of solid particles flying about. These play an important part in many phenomena; they fritter away the energy of the vibrations before they can reach the eye; they too, arc the carriers of germs of destruction, they get into our lungs and other organs, clog up the channels and imperceptibly, yet inevitably, arrest the stream of life. Could we but do away with all ponderable matter in the, line of .sight of the telescope, it would reveal to us undreamt of marvels. Even the unaided eye, I think; would he capable of distinguishing in the pure medium, small objects at distances measured probably by hundreds or perhaps thousands of miles.

But there is something else about the eye which impresses us still more than these wonderful features which we observed, viewing it from the standpoint of a physicist, merely as an optical instrument, — something which appeals to us more than its marvelous faculty of being directly affected by the vibrations of the medium, without interference of gross matter, and snore than its inconceivable sensitiveness and discerning power. It is its significance in the processes of life. No matter what one's views on nature and life may be, he must stand amazed when, for the first time in his, thoughts, he realizes the importance of the eye in the physical processes and mental performances of the human organism. And how could it be otherwise, when he realizes, that the eye is the means through which the human race has acquired the entire knowledge it possesses, that it controls all our motions, more still, and our actions.

There is no way of acquiring knowledge except through the eye. What is the foundation of all philosophical systems of ancient and modern times, in *fact,* of all the philosophy of min? I *am I think; I think, therefore I am.* But how could I think and how would I know that I exist, if I had not the eye? For knowledge involve.; consciousness; consciousness involves ideas, conceptions; conceptions involve pictures or images, and images the sense of vision, and therefore the organ of sight. But how about blind men, will be asked? Yes, a blind man may depict in magnificent poems, forms and scenes from real life, from a world he physically does not see. A blind man may touch the keys of an instrument with. unerring precision, may model the fastest boat, may discover and invent, calculate and construct, may do still greater wonders — but all the blind men who have done such thinks have descended from those who had seeing *eyes*. Nature may reach the same result in many ways. Like a wave in the physical world, in the infinite ocean of the medium which pervades all, so in the world of organism:, in life, an impulse started proceeds onward, at times, may be, with the speed of light, at times, again, so slowly that for ages and ages it seems to stay; passing through processes of a complexity inconceivable to men, but in ;ill its forms, in all its stages, its energy. ever and ever integrally present. A single ray of light from a distant star falling upon the eye of a tyrant in by-gone times, may have altered the course. of his life, may have changed the destiny of nations, may have transformed the surface of the globe, so intricate, so inconceivably complex are the processes in Nature. In no way can we get such an overwhelming idea of the grandeur of Nature, as when we consider, that in accordance with the law of the conservation of energy, throughout the infinite, the forces are in a perfect balance, and hence the energy of a single thought may determine the motion of a Universe. It is not necessary that every individual, not even that every generation or many generations, should have the physical instrument of sight, in order to be able to form images and to think, that is, form ideas or conceptions; but sometime or other, during the process of evolution, the *eye* certainly must have existed, else thought, as we understand it, would be impossible; else conceptions, like spirit, intellect, mind, call it as you may, could not exist. It is conceivable, that in some other world, in some other beings, the eye is replaced by a different organ, equally or more perfect, but these beings cannot be men,

Now what prompts us all to voluntary motions and actions of any kind? Again the eye. If I am conscious of the motion, I must have an idea or conception, that is, an image, therefore the eye. If I am not precisely conscious of the motion, it is, because the images are vague or indistinct, being blurred by the superimposition of many. But when I perform the motion, does the impulse which prompts me to the action come from within or from without? The greatest physicists have not disdained to endeavor to answer this and similar questions and have at tunes abandoned themselves to the delights of pure and unrestrained thought. Such questions are generally considered not to belong to the realm of positive physical science, but will before long be annexed to its domain. Helmholtz has probably thought more on life than any modern scientist. Lord Kelvin expressed his belief that life's process is electrical and that there is a force inherent to the organism ,and determining its motions. just as much as I am convinced of any physical truth I am convinced that the motive impulse must come from the outside. For, consider the lowest organism we know — and there are probably many lower ones — an aggregation of a few cells only. If it is capable of voluntary motion it can perform an infinite number of motions, all definite and precise. But now a mechanism consisting of a finite number of parts and few at that, cannot perform are infinite number of definite motions, hence the impulses which govern its movements must come from the environment. So, the atom, the ulterior element of the Universe's structure, is tossed about in space eternally, a play to external influences, like a boat in a troubled sea. Were it to stop its motion *it would die:* hatter at rest, if such a thin; could exist, would be matter dead. Death of matter! Never has a sentence of deeper philosophical meaning been uttered. This is the way in which Prof. Dewar forcibly expresses it in the description of his admirable experiments, in which liquid oxygen is handled as one handles water, and air at ordinary pressure is made to condense and even to solidify by the intense cold: Experiments, which serve to illustrate, in his language, the last feeble manifestations of life, the last quiverings of matter about to die. But human eyes shall not witness such death. There is no death of matter, for throughout the infinite universe, all has to move, to vibrate, that is, to live.

I have made the preceding statements at the peril of treading upon metaphysical ground; in my desire to introduce the subject of this lecture in a manner not altogether uninteresting, I may hope, to an audience such as I have the honor to address. But now, then, returning to the subject, this divine organ of sight, this indispensable instrument for thought and all intellectual enjoyment, which lays open to us the marvels of the universe, through which we have acquired what knowledge we possess, and which prompts us to, and controls, all our physical and mental activity. By what is it affected? By light! What is light?

We have witnessed the great strides which have been made in all departments of science in recent years. So great lave been the advances that we cannot refrain from asking ourselves, Is this all true; or is it but a dream? Centuries ago men have *lived,* have thought, discovered, invented, and have believed that *they* were soaring, while they were merely proceeding at a snail's pace. So we too may be mistaken. But taking the truth of the observed events as one of the implied facts of science, we must rejoice in the, immense progress already made and still more in the anticipation of what must come, judging from the possibilities opened up by modern research. There is, however, an advance which we have been witnessing, which must be particularly gratifying to every lover of progress. It is not a discovery, or an invention, or an achievement in any particular direction. It is an advance in all directions of scientific thought and experiment I mean the generalization of the natural forces and phenomena, the looming up of a certain broad idea on the scientific horizon. It is this idea which has, however, long ago taken possession of the most advanced minds, to which I desire to call your attention, and which I intend to illustrate in a general way, in these experiments, as the first step in answering the question "What is light?" and to realize the modern meaning of this word.

It is beyond the scope of my lecture to dwell upon the subject of light in general, my object being merely to bring presently to your notice a certain class of light effects and a number of phenomena observed in pursuing the study of these effects. But to he consistent in my remarks it is necessary to state that, according to that idea, now, accepted by the majority of scientific men as a positive result of theoretical and experimental investigation, the various forms or manifestations of energy which were generally designated as "electric" or more precisely "electromagnetic" are energy manifestations of the same nature as those of radiant heat and light. Therefore the phenomena of light and heat and others besides these, may be called electrical phenomena. Thus electrical science has become the mother science of all and its study has become all important. The *day* when we shall know exactly what "electricity" is, will chronicle an event probably greater, more important than any other recorded in the history of the human race. The time will come when the comfort, the very existence, perhaps, or man will depend upon that wonderful agent. For our existence and comfort we require heat, light and mechanical power. How do we now get all these? We get them from fuel, we get them by consuming material. What will man do when the forests disappear, when the coal fields are exhausted? Only one thing according to our present knowledge will remain; that is, to transmit power at great distances. Men will go to the waterfalls, to the tides, which are the stores of an infinitesimal part of Nature's immeasurable energy. There will they harness the energy and transmit the same to their settlements, to warm their homes by, to give them light, and to keep their obedient slaves, the machines, toiling. But how will they transmit this energy if not by electricity? judge then, if the comfort, nay, the very existence, of man will not depend on electricity. I am aware that this view is not that of a practical engineer, but neither is it that of an illusionist, for it is certain, that power transmission, which at present is merely a stimulus to enterprise, will some day be a dire necessity.

It is more important for the student, who takes up the study of light phenomena, to make himself thoroughly acquainted with certain modern views, than to peruse entire books on the subject of light itself, as disconnected from these views. Were I therefore to make these demonstrations before students seeking information — and for the sake of the few of those who may be present, give me leave to so assume — it would be my principal endeavor to impress these views upon their minds in this series of experiments.

It might be sufficient for this purpose to perform a simple and well-known experiment. I might take a familiar appliance, a Leyden jar, charge it from a frictional machine, and then discharge it. In explaining to you its permanent state when charged, and its transitory condition when discharging, calling your attention to the forces which enter into play and to the various phenomena they produce, and pointing out the relation of the forces and phenomena, I might fully succeed in illustrating that modern idea. No doubt, to the thinker, this simple experiment would appeal as much as the most magnificent display. But this is to be an experimental demonstration, and one which should possess, besides instructive, also entertaining features and as such, a simple experiment, such as the one cited, would not go very far towards the attainment of the lecturer's aim. I must therefore choose another way of illustrating, more spectacular certainly, but perhaps also more instructive. Instead of the frictional machine and Leyden jar, I shall avail myself in these experiments, ,of an induction coil of peculiar properties, which was described in detail by me in a lecture before the London Institution of Electrical Engineers, in Feb., 1892. This induction coil is capable of yielding currents of enormous potential differences, alternating with extreme rapidity. With this apparatus I shall endeavor to show you three distinct classes of effects, or phenomena, and it is my desire that each experiment, while serving for the purposes of illustration, should at the. same time teach us some novel truth, or show us some novel aspect of this fascinating science. But before doing this, it seems proper and useful to dwell upon the apparatus employed, and method of obtaining the high potentials and high-frequency currents which are made use of in these experiments.

On the Apparatus and Method of Conversion

These high-frequency currents are obtained in a peculiar manner. The method employed was advanced by me about two years ago in an experimental lecture before the American Institute of Electrical Engineers. A number of ways, as practiced in the laboratory, of obtaining these currents either from continuous or low frequency alternating currents, is diagrammatically indicated in Fig. 1, which will be later described in detail. The general plan is to charge condensers, from a direct or alternate-current source, preferably of high-tension, and to discharge them disruptively while observing well-known conditions necessary to maintain the oscillations of the current. In view of the general interest taken in high-frequency currents and effects producible by them, it seems to me advisable to dwell at some length upon this method of conversion. In order to give you a clear idea of the action, I will suppose that a continuous-current generator is employed, which is often very convenient. It is desirable that the generator should possess such high tension as to be able to break through a small air space. If this is not the case, then auxiliary means have to be resorted to, some of which will be indicated subsequently. When the condensers are charged to a certain potential, the air, or insulating space, gives way and a disruptive discharge: occurs. There is then a sudden rush of current and generally a large portion of accumulated electrical energy spends itself. The condensers are thereupon quickly charged and the same process is repeated in more or less rapid succession. To produce such sudden rushes of current it is necessary to observe certain conditions. If the rate at which the condensers are discharged is the same as that at which

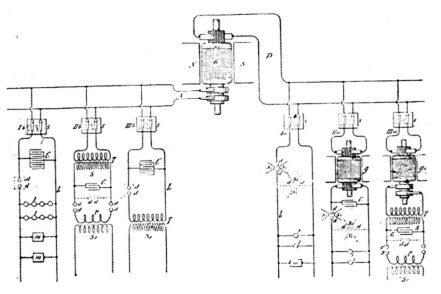

they are charged, then, clearly, in the assumed case the condensers do not come into play. If the rate of discharge be smaller than the rate of charging, then, again, the condensers cannot play an important part. But if, on the contrary, the rate of discharging is greater than that of charging, then a succession of rushes of current is obtained. It is evident that, if the rate at which the energy is dissipated by the discharge is very much greater than the rate of supply to the condensers, the sudden rushes will be comparatively few, with long-time intervals between. This always occurs when a condenser of considerable capacity is charged by means of a comparatively small machine. If the rates of supply and dissipation are not widely different, then the rushes of current will be in quicker succession, and this the more, the more nearly equal both the rates are, until limitations incident to each case and depending upon a number of causes are reached. Thus we are able to obtain from a continuous-current generator as rapid a succession of discharges as we like. Of course, the higher the tension of the generator, the smaller need be the capacity of the condensers, and for this reason, principally, it is of advantage to employ a generator of very high tension. Besides, such a generator permits the attaining of greater rates of vibration.

The rushes of current may be of the same direction under the conditions before assumed, but most generally there is an oscillation superimposed upon the fundamental vibration of the current. When the conditions are so determined that there are no oscillations, the current impulses are unidirectional and thus a means is provided of transforming a continuous current of high tension, into a direct current of lower tension, which I think may find employment in the arts.

This method of conversion is exceedingly interesting and I was much impressed by its beauty when I first conceived it. It is ideal in certain respects. It involves the employment of no mechanical devices of any kind, and it allows of obtaining currents of any desired frequency from an ordinary circuit, direct or alternating. The frequency of the fundamental discharges depending on the relative rates of supply and dissipation can be readily varied within wide limits, by simple adjustments of these quantities, and the frequency of the superimposed vibration by the determination of the capacity, self-induction and resistance of the circuit. The potential of the currents, again, may be raised as high as *any* insulation is capable of withstanding safely by combining capacity and self-induction or by induction in a secondary, which need have but comparatively few turns.

As the conditions are often such that the intermittence or oscillation does not readily establish itself, especially when a direct current source is employed, it is of advantage to associate an interrupter with the arc, as I have, some time ago, indicated the use of an air-blast or magnet, or other such device readily at hind. The magnet is employed with special advantage in the conversion of direct currents, as it is then very effective. If the primary source is an alternate current generator. it is desirable, as I have stated on another occasion, that the frequency should be low, and that tile current forming the arc be large, in order to render the magnet more effective.

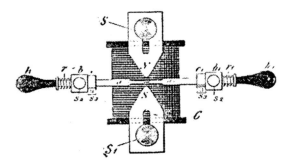

A form of such discharger with a magnet which has been found convenient, and adopted after some trials, in the conversion of direct currents particularly, is illustrated in Fig. 2. N S are the pole pieces of a very strong magnet which is excited by a coil c. The pole pieces are slotted for adjustment and can be fastened in any position by screws s sl. The discharge rods d d1, thinned down on the ends in order to allow a closer approach of the magnetic pole pieces, pass through the columns of brass b b1 and are fastened in position by screws $s_2 \, s_2$. Springs $r \, r_1$ and collars $c \, c_1$ are slipped on the rods, the latter serving to set the points of the rods at a certain suitable distance by means of screws $s_3 \, s_3$ and the former to draw the points apart. When it is desired to start the arc, one of the large rubber handles $h \, h_1$ is tapped quickly with the hand, whereby the points of the rods are brought in contact but are instantly separated by the springs $r \, r_1$. Such an arrangement has been found to be often necessary, namely in cases when the E. M. F. was not large enough to cause the discharge to break through the gap, and also when it was desirable to avoid short circuiting of the generator by the metallic contact of the rods. The rapidity of the interruptions of the current with a magnet depends on the intensity of the magnetic field and on the potential difference at the end of the arc. The interruptions are generally in such quick succession as to produce a musical sound. Years ago it was observed that when a powerful induction coil is discharged between the poles of a strong magnet, the discharge produces a loud noise not unlike a small pistol shot. It was vaguely stated that the spark was intensified by the presence of the magnetic field. It is now clear that the discharge current, flowing for some time, was interrupted a great number of times by the magnet, thus producing the sound. The phenomenon is especially marked when the field circuit of a large magnet or dynamo is broken in a powerful magnetic field.

When the current through the gap is comparatively large, it is of advantage to slip on the points of the discharge rods pieces of *very* hard carbon and let the arc play between the carbon pieces. This preserves the rods, and besides has the advantage of keeping the air space hotter, as the heat is not conducted away as quickly through the carbons, and the result is that a smaller E. M. F. in the arc gap is required to maintain a succession of discharges.

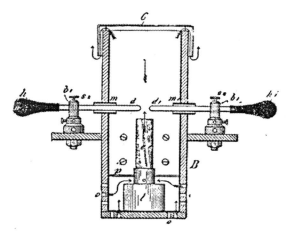

Another form of discharger, which may be employed with advantage in some cases, is illustrated in Fig. 3. In this form the discharge rods $d\ d_1$ pass through perforations in a wooden box B, which is thickly coated with mica on the inside, as indicated by the heavy lines. The perforations are provided with mica tubes $m\ m_1$ of some thickness, which are preferably not in contact with the rods $d\ d_1$. The box has a cover c which is a little larger and descends on the outside of the box. The spark gap is warmed by a small lamp l contained in the box. A plate p above the lamp allows the drought to pass only through the chimney a of the lamp, the air entering through holes oo in or near the bottom of the box and following the path indicated by the arrows. When the discharger is in operation, the door of the box is closed so that the light of the arc is not visible outside. It is desirable to exclude the light as perfectly as possible, as it interferes with some experiments. This form of discharger is simple and very effective when properly manipulated. The air being warmed to a certain temperature, has its insulating power impaired; it becomes dielectrically weak, as it were, and the consequence is that the arc can be established at much greater distance. The arc should, of course, be sufficiently insulating to allow the discharge to pass through the gap *disruptively*. The arc formed under such conditions, when long, may be made extremely sensitive, and the weal: drought through the lamp chimney a is quite sufficient to produce rapid interruptions. The adjustment is made by regulating the temperature and velocity of the drought. Instead of using the lamp, it answers the purpose to provide for a drought of warm air in other ways. A very simple way which has been practiced is to enclose the arc in a long vertical tube, with plates on the top and bottom for regulating the temperature and velocity of the air current. Some provision had to be made for deadening the sound.

The air may be rendered dielectrically weak also by rarefaction. Dischargers of this kind have likewise been used by me in connection with a magnet. A large tube is for this purpose provided with heavy electrodes of carbon or metal, between which the discharge is made to pass, the tube being placed in a powerful magnetic field The exhaustion of the tube is carried to a point at which the discharge breaks through easily, but the pressure should be more than 75 millimeters, at which the ordinary thread discharge occurs. In another form of discharger, combining the features before mentioned, the discharge was made to pass between two adjustable magnetic pole pieces, the space between them being kept at an elevated temperature.

It should be remarked here that when such, or interrupting devices of any kind, are used and the currents are passed through the primary of a disruptive discharge coil, it is not, as a rule, of advantage to produce a number of interruptions of the current per second greater than the natural frequency of vibration of the dynamo supply circuit, which is ordinarily small. It should also be pointed out here, that while the devices mentioned in connection with the disruptive discharge are advantageous under certain conditions, they may be sometimes a source of trouble, as they produce intermittences and other irregularities in the vibration which it would be very desirable to overcome.

There is, I regret to say, in this beautiful method of conversion a defect, which fortunately is not vital, .and which I have been gradually overcoming. I will best call attention to this defect and indicate a fruitful line of work, by comparing the electrical process with its mechanical analogue. The process may be illustrated in this manner. Imagine a tank; with a wide opening at the bottom, which is kept closed by spring pressure, but so that it snaps off *sudden/y* when the liquid in the tank has reached a certain height. Let the fluid be supplied to the tank by means of a pipe feeding at a certain rate. When the critical height of the liquid is reached, the spring gives way and the bottom of the tank drops out. Instantly the liquid falls through the wide opening, and the spring, reasserting itself, closes the bottom again. The tank is now filled, and after a certain time interval the same process is repeated. It is clear, that if the pipe feeds the fluid quicker than the bottom outlet is capable of letting it pass through, the bottom will remain off and the tank; will still overflow. If the rates of supply are exactly equal, then the bottom lid will remain partially open and no vibration of the same and of the liquid column will generally occur, though it might, if started by some means. But if the inlet pipe does not feed the fluid fast enough for the outlet, then there will be always vibration. Again, in such case, each time the bottom flaps up or down, the spring and the liquid column, if the pliability of the spring and the inertia of the moving parts are properly chosen, will perform independent vibrations. In this analogue the fluid may be likened to electricity or electrical energy, the tank to the condenser, the spring to the dielectric, and the pipe to the conductor through which electricity is supplied to the condenser. To make this analogy quite complete it is necessary to make the assumption, that the bottom, each time it gives way, is knocked violently against a non-elastic stop, this impact involving some loss of energy; and that, besides, some dissipation of energy results due to frictional losses. In the preceding analogue the liquid is supposed to be under a steady pressure. If the presence of the fluid be assumed to vary rhythmically, this may be taken as corresponding to the case of an alternating current. The process is then not quite as simple to consider, but the action is the same in principle.

It is desirable, in order to maintain the vibration economically, to reduce the impact and frictional losses as much as possible. As regards the latter, which in the electrical analogue correspond to the losses due to the resistance of the circuits, it is impossible to obviate them entirely, but they can be reduced to a minimum by a proper selection of the dimensions of the circuits and by the employment of thin conductors in the form of strands. But the loss of energy caused by the first breaking through of the dielectric — which in the above example corresponds to the violent knock of the bottom against the inelastic stop — would be more important to overcome. At the moment of the breaking through, the air space has a very high resistance, which is probably reduced to a very small value when the current has reached some strength, and the space is brought to a high temperature. It would materially diminish the loss of energy if the space were always kept at an extremely high temperature, but then there would be no disruptive break. By warming the space moderately by means of a lamp or otherwise, the economy as far as the arc is concerned is sensibly increased. But the magnet or other interrupting device does not diminish the loss in the arc. Likewise, a jet of air only facilitates the carrying off of the energy. Air, or a gas in general, behaves curiously in this respect. When two bodies charged to a very high potential, discharge disruptively through an air space, any amount of energy may be carried off by the air. This energy is evidently dissipated by bodily carriers, in impact and collisional losses of the molecules. The exchange of the molecules in the space occurs with inconceivable rapidity. A powerful discharge taking place between two electrodes, they may remain entirely cool, and yet the loss in the air may represent any amount of energy. It is perfectly practicable, with very great potential differences in the gap, to dissipate several horse-power in the arc of the discharge without even noticing a small increase in the temperature of the electrodes. All the frictional losses occur then practically in the air. If the exchange of the air molecules is prevented, as by enclosing the air hermetically, the bas inside of the vessel is brought quickly to a high temperature, even with a very small discharge. It is difficult to estimate how much of the energy is lost in sound waves, audible or not, in a powerful discharge. When the currents through the gap are large, the electrodes may become rapidly heated, but this is not a reliable measure of the energy wasted in the arc, as the loss through the yap itself may be comparatively small. The air or a gas in general is at ordinary pressure at least, clearly not the best medium through which a disruptive discharge should occur. Air or other gas under great pressure is of curse a much more suitable medium for the discharge gap. I have carried on long-continued experiments in this direction, unfortunately less practicable on account of the difficulties and expense in getting air under great pressure. But even if the medium in the discharge space is solid or liquid, still the same losses take place, though they are generally smaller, for just as soon as the arc is established, the solid or liquid is volatilized. Indeed, !here is no body known which would not be disintegrated by the arc, and it is an open question among scientific men, whether an arc discharge could occur at all in the air itself without the particles of the electrodes being torn off. When the current through the gap is very small and the arc very long, I believe that a relatively considerable amount of heat is taken up in the disintegration of the electrodes, which partially on this account may remain quite cold.

The ideal medium for a discharge gap should only *crack.* and the ideal electrode should be of some material which cannot be disintegrated. With small currents through the gap it is best to employ aluminum, but not when the currents are large. The disruptive break in the air, or more or less in any ordinary medium, is not of the nature of a crack, but it is rather comparable to the piercing of innumerable bullets through a mass offering great frictional remittances to the motion of the bullets, this involving considerable loss of energy. A medium which would merely crack when strained electrostatically — and this possibly might be the case with a perfect vacuum, that is, pure ether — would involve a very small loss in the gap, so small as to be entirely negligible, at least theoretically, because a crack may be produced by an infinitely small displacement. In exhausting an oblong bulb provided with two aluminum terminals, with the greatest care, I have succeeded in producing such a vacuum that the secondary discharge of a disruptive discharge coil would break disruptively through the bulb in the form of fine spark streams. The curious point was that the discharge would completely ignore the terminals and start far behind the two aluminum plates which served as electrodes. This extraordinary high vacuum could only be maintained for a very short while. To return to the ideal medium; think, for the sake of illustration, of a piece of glass or similar body clamped in a vice, and the latter tightened more and more. At a certain point a minute increase of the pressure will cause the ;glass to crack. The loss of energy involved in splitting the glass may be practically nothing, for though the force is great, the displacement need be but extremely small. Now imagine that the glass would possess the property of closing again perfectly the crack upon a minute diminution of the pressure. This is the way the dielectric in the discharge space should behave. But inasmuch as there would be always some loss in the gap, the medium, which should be continuous should exchange through the gap at a rapid rate. In the preceding example, the glass being perfectly closed, it would mean that the dielectric in the discharge space possesses a great insulating power; the glass being cracked, it would signify that the medium in the space is a good conductor. The dielectric should vary enormously in resistance by minute variations of the E. M. F. across the discharge space. This condition is attained, but in an extremely imperfect manner, by warming the air space to a. certain critical temperature, dependent on the E. M. F. across the yap, or by otherwise impairing the insulating power of the air. But as a matter of fact the air does never break down *disruptively,* if this term be rigorously interpreted, for before the sudden rush of the current occurs, there is always a weak current preceding it, which rises first gradually and then with comparative suddenness. That is the reason why the rate of change is very much greater when glass, for instance, is broken through, than when the break takes place through an air space of equivalent dielectric strength. As a medium for the discharge space, a solid, or even a liquid, would be preferable therefore. It is somewhat difficult to conceive of a solid body which would possess the property of closing instantly after it has been cracked. But a liquid, especially under great pressure, behaves practically like a solid, while it possesses the property of closing the crack. Hence it was thought that a liquid insulator might be more suitable as a dielectric than air. Following out this idea, a number of different forms of dischargers in which a variety of such insulators, sometimes under great pressure, were employed, have been experimented upon. It is thought sufficient to dwell in a few words upon one of the forms experimented upon. One of these dischargers is illustrated in Figs. 4a and 4b.

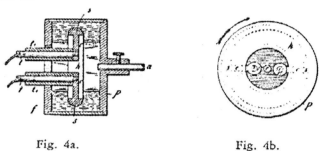

Fig. 4a.					Fig. 4b.

A hollow metal pulley P (Fig. 4a), was fastened upon an arbor a, which by suitable means was rotated at a considerable speed. On the inside of the pulley, but disconnected from the same, was supported a thin disc h (which is shown thick for the sake of clearness), of hard rubber in which there were embedded two metal segments s s with metallic extensions e e into which were screwed conducting terminals t t covered with thick tubes of hard rubber t t. The rubber disc b with its metallic segments s s, was finished in a lathe, and its entire surface highly polished so as to offer the smallest possible frictional resistance to the motion through a fluid. In the hollow of the pulley an insulating liquid such as a thin oil was poured so as to reach very nearly to the opening left in the flange f, which was screwed tightly on the front side of the pulley. The terminals t t, were connected to the opposite coatings of a battery of condensers so that the discharge occurred through the liquid. When the pulley was rotated, the liquid was forced against the rim of the pulley and considerable fluid pressure resulted. In this simple way the discharge gap was filled with a medium which behaved practically like a solid, which possessed the duality of closing instantly upon the occurrence of the break, and which moreover was circulating through the gap at a rapid rate. Very powerful effects were produced by discharges of this kind with liquid interrupters, of which a number of different forms were made. It was found that, as expected, a longer spark for a given length of wire was obtainable in this way than by using air as an interrupting device. Generally the speed, and therefore also the fluid pressure, was limited by reason of the fluid friction, in the form of discharger described, but the practically obtainable speed was more than sufficient to produce a number of breaks suitable for the circuits ordinarily used. In such instances the metal pulley P was provided with a few projections inwardly, and a definite number of breaks was then produced which could be computed from the speed of rotation of the pulley. Experiments were also carried on with liquids of different insulating power with the view of reducing the loss in the arc. When an insulating liquid is moderately warmed, the loss in the arc is diminished.

A point of some importance was noted in experiments with various discharges of this kind. It was found, for instance, that whereas the conditions maintained in these forms were favourable for the production of a great spark length, the current so obtained was not best suited to the production of light effects. Experience undoubtedly has shown, that for such purposes a harmonic rise and fall of the potential is preferable. Be it that a solid is rendered incandescent, or phosphorescent, or be it that energy is transmitted by condenser coating through the glass, it is quite certain that a harmonically rising and falling potential produces less destructive action, and that the vacuum is more permanently maintained. This would be easily explained if it were ascertained that the process going on in an exhausted vessel is of an electrolytic nature.

In the diagrammatical sketch, Fig. 1, which has been already referred to, the cases which are most likely to be met with in practice are illustrated. One has at his disposal either direct or alternating currents from a supply station. It is convenient for an experimenter in an isolated laboratory to employ a machine G, such as illustrated, capable of giving both kinds of currents. In such case it is also preferable to use a machine with multiple circuits, as in many experiments it is useful and convenient to have at one's disposal currents of different phases. In the sketch, D represents the direct and A the alternating circuit. In each of these, three branch circuits are shown, all of which are provided with double line switches s s s s s. Consider first the direct current conversion; 1a represents the simplest case. If the E. M. F. of the generator is sufficient to break through a small air space, at least when the latter is warmed or otherwise rendered poorly insulating, there is no difficulty in maintaining a vibration with fair economy by judicious adjustment of the capacity, self-induction and resistance of the circuit L. containing the devices *l l m*. The magnet N, S, can be in this case advantageously combined with the air space, The discharger d d with the magnet may be placed either way, as indicated by the full or by the dotted lines. The circuit 1a with the connections and devices is supposed to possess dimensions such as are suitable for the maintenance of a vibration. But usually the E. M. F. on the circuit or branch 1a will be something like a 100 volts or so, and in this case it is not sufficient to break through the gap. Many different means may be used to remedy this by raising the E. M. F. across the gap. The simplest is probably to insert a large self-induction coil in series with the circuit L. When the arc is established, as by the discharger illustrated in Fig. 2, the magnet blows the arc out the instant it is formed. Now the extra current of the break, being of high E. M. F., breaks through the gap, and a path of low resistance for the dynamo current being again provided, there is a sudden rush of current from the dynamo upon the weakening or subsidence of the extra current. This process is repeated in rapid succession, and in this manner I have maintained oscillation with as low as 50 volts, or even less, across the gap. But conversion on this plan is not to be recommended on account of the too heavy currents through the gap and consequent heating of the electrodes; besides, the frequencies obtained in this way are low, owing to the high self-induction necessarily associated with the circuit. It is very desirable to have the E. M. F. as high as possible, first, in order to increase the economy of the conversion, and secondly, to obtain high frequencies. The difference of potential in this electric oscillation is, of course, the equivalent of the stretching force in the mechanical vibration of the spring. To obtain very rapid vibration in a circuit of some inertia, a great stretching force or difference of potential is necessary. Incidentally, when the E. M. F. is very great, the condenser which is usually employed in connection with the circuit need but have a small capacity, and many other advantages are gained. With a view of raising the E. M. F. to a many times greater value than obtainable from ordinary distribution circuits, a rotating transformer g is used, as indicated at I 1a. Fig. 1, or else a separate high potential machine is driven by means of a motor operated from the generator G. The latter plan is in fact preferable, as changes are easier made. The connections from the high tension winding are quite similar to those in branch la with the exception that a condenser C, which should be adjustable, is connected to the high tension circuit. Usually, also, an adjustable self-induction coil in series with the circuit has been employed in these experiments. When the tension of the currents is very high, the magnet ordinarily used in connection with the discharger is of comparatively small value, as it is quite easy to adjust the dimensions of the circuit so that oscillation is maintained. The employment of a steady E. M. F. in the high frequency conversion affords some advantages over the employment of alteration E. M. F., as the adjustments are much simpler and the action can be easier controlled. But unfortunately one is limited by the obtainable potential difference. The winding also breaks down easily in consequence of the sparks which form between the sections of the armature or commutator when a vigorous oscillation takes place. Besides, these transformers are expensive to build. It has been found by experience that it is best to follow the plan illustrated at Met, In this arrangement a rotating transformer ,g, is employed to convert the low tension direct currents into low frequency alternating currents, preferably also of small tension. The tension of the currents is then raised in a stationary transformer T. The secondary s of this transformer is connected to an adjustable condenser C which discharges through the gap or discharger d d, placed in either of the ways indicated, through the primary P of a disruptive discharge coil, the high frequency current being obtained from the secondary s of this coil, as described on previous occasions. This will undoubtedly be found the cheapest and most convenient way of converting direct currents.

The three branches of the circuit A represent the usual cases met in practice when alternating currents are converted. In Fig. I b a condenser C, generally of large capacity, is connected to the circuit L containing the devices *l l, m m* The devices m m are supposed to be of high self-induction so as to bring the frequency of the circuit more or less to that of the dynamo. In this instance the discharger *d d* should best have a number of makes and breaks per second equal to twice the frequency of the dynamo. If not so, then it should have at least a number equal to a multiple or even fraction of the dynamo frequency. It should be observed, referring to I b, that the conversion to a high potential is also effected when the discharger *d d,* which is shown in the sketch, is omitted. But the effects which are produced by currents which rise instantly to high values, as in a disruptive discharge, are entirely different from those produced by dynamo currents which rise and fall harmonically. So, for instance, there might be in a given case a number of makes and breaks at $d\ d$ equal to just twice the frequency of the dynamo, or in other words, there may be the same number of fundamental oscillations as would be produced without the discharge gap, and there might even not be any quicker superimposed vibration; yet the differences of potential at the various points of the circuit, the impedance and other phenomena, dependent upon the rate of change, will bear no similarity in the two cases. Thus, when working with currents discharging disruptively, the element chiefly to be considered is not the frequency, as a student might be apt to believe, but the rate of change per unit of time. With low frequencies in a certain measure the same effects may be obtained as with high frequencies, provided the rate of change is sufficiently great. So if a low frequency current is raised to a potential of, say, 75,000 volts, and the high tension current passed through a series of high resistance lamp filaments, the importance of the rarefied gas surrounding the filament is clearly noted, as will be seen later; or, if a low frequency current of several thousand amperes is passed through a metal bar, striking phenomena of impedance are observed, just as with currents of high frequencies. But it is, of course, evident that with low frequency currents it is impossible to obtain such rates of change per unit of time as with high frequencies, hence the effects produced by the latter are much more prominent. It is deemed advisable to make the preceding remarks, inasmuch as many more recently described effects have been unwittingly identified with high frequencies. Frequency alone in reality does not mean anything, except when an undisturbed harmonic oscillation is considered.

In the branch III b a similar disposition to that in Ib is illustrated, with the difference that the currents discharging through the gap $d\ d$ are used to induce currents in the secondary s of a transformer T. In such case tale secondary should be provided with an adjustable condenser for the purpose of tuning it to the primary.

II b illustrates a plan of alternate current high frequency conversion which is most frequently used and which is found to be most convenient. This plan has been dwelt upon in detail on previous occasions and need not be described here.

Some of these results were obtained by the use of a high frequency alternator. A description of such machines will be found in my original paper before the American Institute of Electrical Engineers, and in periodicals of that period, notably in The Electrical Engineer of March 18, 1891.

I will now proceed with the experiments.

On Phenomena Produced By Electrostatic Force

The first class of effects I intend to show you are effects produced by electrostatic force. It is the force which governs the motion of the atoms, which causes them to collide and develop the life-sustaining energy of heat and light, and which causes them to aggregate in an infinite variety of ways, according to Nature's fanciful designs, and to form all these wondrous structures we perceive around us; it is, in fact, if our present views be true, the most important force for us to consider in. Nature. As the term *electrostatic* might imply a steady electric condition, it should be remarked, that in these experiments the force is not constant, but varies at a rate which may be considered moderate, about one million times a second, or thereabouts. This enables me to produce many effects which are not producible with an unvarying force.

When two conducting bodies are insulated and electrified, we say that an electrostatic force is acting between them. This force manifests itself in attractions, repulsions and stresses in the bodies and space or medium without. So great may be the strain exerted in the air, or whatever separates the two conducting bodies, that it may break down, and we observe sparks or bundles of light or streamers, as they are called. These streamers form abundantly when the force through the air is rapidly varying. I will illustrate this action of electrostatic force in a novel experiment in which I will employ the induction coil before referred to. The coil is contained in a trough filled with oil, and placed under the table. The two ends of .the secondary wire pass through the two thick columns of hard rubber which protrude to solve height above the table. It is necessary to insulate the ends or terminals of the secondary heavily with hard rubber, because even dry wood is by far ton poor an insulator for these currents of enormous potential differences. On one of the terminals of the coil, I ,have placed a large sphere of sheet brass, which is connected to a larger insulated brass plate, in order to enable me to perform the experiments under conditions, which, as you will see, are more suitable for this experiment. I now set the coil to work and approach the free terminal with a metallic object held in my hand, this simply to avoid burns As I approach the metallic object to a distance of eight or tell inches, a torrent of furious sparks breaks forth from the end of the secondary wire, which passes through the rubber column. The sparks cease when the metal in my hand touches the wire. My arm is now traversed *by* a powerful electric current, vibrating at about the rate of one million times a second. All around me the electrostatic force makes itself felt, and the air molecules and particles of dust flying about are acted upon and are hammering violently against my body. So great is this agitation of the particles, that when the lights are turned out you may see streams of feeble light appear on some parts of my body. When such a streamer breaks out on any part of the body, it produces a sensation like the pricking of a needle. Were the potentials sufficiently high and the frequency of the vibration rather low, the skin would probably be ruptured under the tremendous strain, and the blood would rush out with great force in the form of fine spray or jet so thin as to be invisible, just as oil will when placed on the positive terminal of a Holtz machine. The breaking through of the skin though it may seem impossible at first, would perhaps occur, by reason of the tissues finder the skin being incomparably better conducting. This, at least, appears plausible, judging from some observations.

I can make these streams of light visible to all, by touching with the metallic object one of the terminals as before, and approaching my free hand to the brass sphere, which is connected to the second terminal of the coil. As the hand is approached, the air between it and the sphere, or in the immediate neighborhood, is more violently agitated, and you see streams of light now break forth from my finger tips and from the whole hand (Fig. 5). Were I to approach the hand closer, powerful sparks would jump from the brass sphere to my hand, which might be injurious. The streamers offer no particular inconvenience, except that in the ends of the finger tips a burning sensation is felt. They should not be confounded with those produced by an influence machine, because in many respects they behave differently. I have attached the brass sphere and plate to one of the terminals in order to prevent the formation of visible streamers on that terminal, also in order to prevent sparks from jumping at a considerable distance. Besides, the attachment is favorable for the working of the coil.

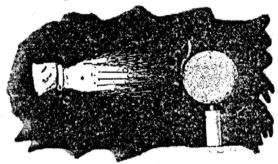

The streams of light which you have observed issuing from my hand are due to a potential of about 200,000 volts, alternating in rather irregular intervals, sometimes like a million times a second. A vibration of the same amplitude, but four times as fast, to maintain which over 3,000,000 volts would be required, would be mare than sufficient to envelop my body in a complete sheet of flame. But this flame would not burn me up; quite contrarily, the probability is ,that I would not be injured in the least. Yet a hundredth part of that energy, otherwise directed; would be amply sufficient to kill a person.

The amount of energy which may thus be passed into the body of a person depends on the frequency and potential of the currents, and by making both of these very great, a vast amount of energy may be passed into the body without causing any discomfort, except perhaps, in the arm, which is traversed by a true conduction current. The reason why no pain in the body is felt, and no injurious effect noted, is that everywhere, if a current be imagined to flow through the body, the direction of its flow would be at right angles to the surface; hence the body of the experimenter offers an enormous section to the current, and the density is very small, with the exception of the arm, perhaps, where the density may be considerable. Lout if only a small fraction of that energy would be applied in such a way that a current would traverse the body in the same manner as a low frequency current, a shock would be received which might be fatal. A direct or low frequency alternating current is fatal, I think, principally because its distribution through the body is not uniform, as it must divide itself in minute streamlets of great density, whereby some organs are vitally injured. That such a process occurs I have not the least doubt, though no evidence might apparently exist, or be found upon examination. The surest to injure and destroy life, is a continuous current, but the most painful is an alternating current of very low frequency. The expression of these views, which are the result of long continued experiment and observation, both with steady and varying currents, is elicited by the interest which is at present taken in this subject, and by the manifestly erroneous ideas which are daily propounded in journals on this subject.

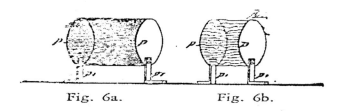

Fig. 6a.   Fig. 6b.

I may illustrate an effect of the electrostatic force by another striking experiment, but before, I must call your attention to one or two facts. I have said before, that whet; the medium between two oppositely electrified bodies is strained beyond a certain limit it gives way and, stated in popular language, the opposite electric charges unite and neutralize each other. This breaking down of the medium occurs principally when the force acting between the bodies is steady, or varies at a moderate rate. Were the variation sufficiently rapid, such a destructive break would not occur, no matter how great the force, for all the energy would be spent in radiation, convection and mechanical and chemical action. Thus the *spark* length, or greatest distance which *a spark* will jump between the electrified bodies is the smaller, the greater the variation or time rate of change. But this rule may be taken to be true only in a general way, when comparing rates which are widely different.

I will show you by an experiment the difference in the effect produced by a rapidly varying and a steady or moderately varying force. I have here two large circular brass plates *p p* (Fig. 6a and Fig. 6b), supported on movable insulating stands on the table, connected to the ends of the secondary of a coil similar to the one used before. I place the plates ten or twelve inches apart and set the coil to work. You see the whole space between the plates, nearly two cubic feet, filled with uniform light, Fig. 6a. This light is due to the streamers you have seen in the first experiment, which are now much more intense. I have already pointed out the importance of these streamers in commercial apparatus and their still greater importance in some purely scientific investigations, Often they are to weak to be visible, but they always exist, consuming energy and modifying the action of .the apparatus. When intense, as they are at present, they produce ozone in great quantity, and also, as Professor Crookes has pointed out, nitrous acid. So quick is the chemical action that if a coil, such as this one, is worked for a very long time it will make the atmosphere of a small room unbearable, for the eyes and throat are attacked. But when moderately produced, the streamers refresh the atmosphere wonderfully, like a thunder-storm, and exercises unquestionably a beneficial effect.

In this experiment the force acting between the plates changes in intensity and direction at a very rapid rate. I will now make the rate of change per unit time much smaller. This I effect by rendering the discharges through the primary of the induction coil less frequent, and also by diminishing the rapidity of the vibration in the secondary. The former result is conveniently secured by lowering the E. M. F. over the air gap in the primary circuit, the latter by approaching the two brass plates to a distance of about three or four inches. When the coil is set to work, you see no streamers or light between the plates, yet the medium between them is under a tremendous strain. I still further augment the strain by raising the E. M F. in the primary circuit, and soon you see the air give away and the hall is illuminated by a shower of brilliant and noisy sparks, Fig. 6b. These sparks could be produced also with unvarying force; they have been for many years a familiar phenomenon, though they were usually obtained from an entirely different apparatus. In describing these two phenomena so radically different in appearance, I have advisedly spoken of a "force" acting between the plates. It would be in accordance with the accepted views to say, that there was an "alternating E. M. F.," acting between the plates. This term is quite proper and applicable in all cases where there is evidence of at least a possibility of an essential interdependence of the electric state of the plates, or electric action in their neighbourhood. But if the plates were removed to an infinite distance, or if at a finite distance, there is no probability or necessity whatever for such dependence. I prefer to use the term "electrostatic force," and to say that such a force is acting around *each* plate or electrified insulated body in general. There is an inconvenience in using this expression as the term incidentally, means a steady electric condition; but a proper nomenclature will eventually settle this difficulty.

I now return to the experiment to which I have already alluded, and with which I desire to illustrate a striking effect produced by a rapidly varying electrostatic force. I attach to the end of the wire, l (Fig. 7), which is in connection with one of the terminals of the secondary of the induction coil, an exhausted bulb b. This bulb contains a thin carbon filament f, which is fastened to a platinum wire w, sealed in the glass and leading outside of the bulb, where it connects to the wire l. The bulb may be exhausted to any degree attainable with ordinary apparatus. Just a moment before, you have witnessed the breaking down of the air between the charged brass plates. You know that a plate of glass, or any other insulating material, would break down in like manner. Had I therefore a metallic coating attached to the outside of the bulb, or placed near the same, and were this coating connected to the other terminal of the coil, would be prepared to see the glass give way if the strain were sufficiently increased. Even were the coating not connected to the other terminal, but to an insulated plate, still, if you have followed recent developments, you would naturally expect a rapture of the glass.

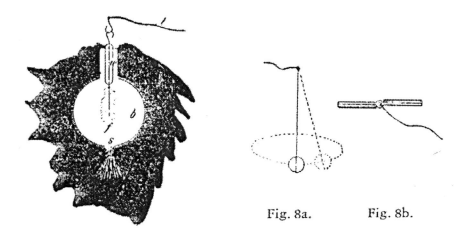

Fig. 8a. Fig. 8b.

But it will certainly surprise you to note that under the action of the varying electrostatic force, the glass gives way when all other bodies are removed from the bulb. In fact, all the surrounding bodies we perceive might be removed to an infinite distance without affecting the result in the slightest. When the coil is set to work, the glass is invariably broken through at the seal, or other narrow channel, and the vacuum is quickly impaired. Such a damaging break would not occur with a steady force, even if the same were many times greater. The break is due to the agitation of the molecules of the gas within the bulb, and outside of the same. This agitation, which is generally most violent in the narrow pointed channel near the seal, causes a heating and rupture of the glass. This rupture would, however, not occur, not even with a varying force, if the medium filling the inside of the bulb, and that surrounding it, were perfectly homogeneous. The break occurs much quicker if the top of the bulb is drawn out into a fine fibre. In bulbs used with these coils such narrow, pointed channels must therefore be avoided.

When a conducting body is immersed in air, or similar insulating medium, consisting of, or containing, small freely movable particles capable of being electrified, and when the electrification of the body is made to undergo a very rapid change — which is equivalent to saying that the electrostatic force acting around the body is varying in intensity, — the small particles are attracted and repelled, and their violent impacts against the body may cause a mechanical motion of the latter. Phenomena of this kind are noteworthy, inasmuch as they have not been observed before with apparatus such as has been commonly in use. If a very light conducting sphere be suspended oil an exceedingly fine wire, and charged to a steady potential, however high, the sphere will remain at rest. Even if the potential would be rapidly varying, provided that the small particles of matter, molecules or atoms, are evenly distributed, no motion of the sphere should result. But if one side of the conducting sphere is covered with a thick insulating *layer,* the impacts of the particles will cause the sphere to move about, generally in irregular curves, Fig. 8a. In like manner, as I have shown on a previous occasion, a fan of sheet metal, Fig. 8b, covered partially with insulating material as indicated, and placed upon the terminal of the coil so as to turn freely on it, is spun around.

All these phenomena you have witnessed and others which will be shown later, are due to the presence of a medium like air, and would not occur in a continuous medium. The action of the air may be illustrated still better by the following experiment. I take a glass tube t, Fig. 9, of about an inch in diameter, which has a platinum wire w sealed in the lower end, and to which is attached a thin lamp filament f. I connect the wire with the terminal of the coil and set the coil to work. The platinum wire is now electrified positively and negatively in rapid succession and the wire and air inside of the tube is rapidly heated by the impacts of the particles, which may be so violent as to render the filament incandescent. But if I pour oil in the tube, just as soon as the wire is covered with the oil, all action apparently ceases and there is no marked evidence of heating. The reason of this is that the oil is a practically continuous medium. The displacements in such a continuous medium are, with these frequencies, to all appearance incomparably smaller than in air, hence the work performed in such a medium is insignificant. But oil would behave very differently with frequencies many times as great, for even though the displacements be small, if the frequency were much greater, considerable work might be performed in the oil.

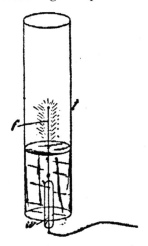
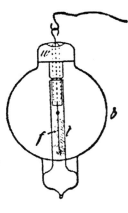

The electrostatic attractions and repulsions between bodies of measurable dimensions are, of all the manifestations of this force, the first so-called *electrical* phenomena noted. But though they have been known to us for many centuries, the precise nature of the mechanism concerned in these actions is still unknown to us, and has not been even quite satisfactorily explained. What kind of mechanism must that be? We cannot help wondering when we observe two magnets attracting and repelling each other with a force of hundreds of pounds with apparently nothing between them. We have in our commercial dynamos magnets capable of sustaining in mid-air tons of weight. But what are even these forces acting between magnets when compared with the tremendous attractions and repulsions produced by electrostatic force, to which there is apparently no limit as to intensity. In lightning discharges bodies are often charged to so high a potential that they are thrown away with inconceivable force and torn asunder or shattered into fragments. Still even such effects cannot compare with the attractions and repulsions which exist between charged molecules or atoms, and which are sufficient to project them with speeds of many kilometres a second, so that under their violent impact bodies are rendered highly incandescent and are volatilized. It is of special interest for the thinker who inquires into the nature of these forces to note that whereas the actions between individual molecules or atoms occur seemingly under any conditions, the attractions and repulsions of bodies of measurable dimensions imply a medium possessing insulating properties. So, if air; either by being rarefied or heated, is rendered more or less conducting, these actions between two electrified bodies practically cease, while the actions between the individual atoms continue to manifest themselves.

An experiment may serve as an illustration and as a means of bringing out other features of interest: Some time ago I showed that a lamp filament or wire mounted in a bulb and connected to one of the terminals of a high tension secondary coil is spinning, the top of the filament generally describing a circle. This vibration was very energetic when the air in the bulb was at ordinary pressure and became less energetic when the air in the bulb was strongly compressed. It ceased altogether when the air was exhausted so as to become comparatively good conducting. I found at that time that no vibration tool: place when the bulb was very highly exhausted. But I conjectured that the vibration which I ascribed to the electrostatic action between the walls of the bulb and the filament should take place also in a highly exhausted bulb. To test this under conditions which were more favourable, a bulb like the one in Fig. 10; was constructed. It comprised a globe b, in the neck of which was sealed a platinum wire w, carrying a thin lamp filament f. In the lower part of the globe a tube t was sealed so as to surround the filament. The exhaustion was carried as far as it was practicable with the apparatus employed.

This bulb verified my expectation, for the filament was set spinning when the current was turned on, and became incandescent. It also showed another interesting feature, bearing upon the preceding remarks, namely, when the filament had been kept incandescent some time, the narrow tube and the space inside were brought to an elevated temperature, and as the gas in the tube then became conducting, the electrostatic attraction between the glass and the filament became *very* weak or ceased, and the filament came to rest, When it came to rest it would glow far more intensely. This was probably due to its assuming the position in the centre of the tube where the molecular bombardment was most intense, and also partly to the fact that the individual impacts were more violent and that no part of the supplied energy was converted into mechanical movement. Since, in accordance with accepted views, in this experiment the incandescence must be attributed to the impacts of the particles, molecules or atoms in tire heated space, these particles must therefore, in order to explain such action, be assumed to behave as independent carriers of electric charges immersed in an insulating medium; yet there is no attractive force between the glass tube and the filament because the space in the tube is, as a whole, conducting.

It is of some interest to observe in this connection that whereas the attraction between two electrified bodies may cease owing to the impairing of the insulating power of the medium in which they are immersed, the repulsion between the bodies may still be observed. This may be explained in a plausible way. When the bodies are placed at some distance in a poorly conducting medium, such as slightly warmed or rarefied air, and are suddenly electrified, opposite electric charges being imparted to them, these charges equalize more or less by leakage through the air. But if the bodies are similarly electrified, there is less opportunity afforded for such dissipation, hence the repulsion observed in such case is greater than the attraction. Repulsive actions in a gaseous medium are however, as Prof. Crookes has shown, enhanced by molecular bombardment.

On Current Or Dynamic Electricity Phenomena

So far, I have considered principally effects produced by a varying electrostatic force in an insulating medium, such as air. When such a force is acting upon a conducting body of measurable dimensions, it causes within the same, or on its surface, displacements of the electricity, and gives rise to electric currents, and these produce another kind of phenomena, some of which I shall presently endeavor to illustrate. In presenting this second class of electrical effects, I will avail myself principally of such as are producible without any return circuit, hoping to interest you the more by presenting these phenomena in a more or less novel aspect.

It has been a long time customary, owing to .the limited experience with vibratory currents, to consider an electric current as something circulating in a closed conducting path. It was astonishing at first to realize that a current may flow through tile conducting path even if the latter be interrupted; and it was still more surprising to learn, that sometimes it may be even easier to make a current flow under such conditions than through a closed path. But that old idea is gradually disappearing, even among practical men, and will soon be entirely forgotten:

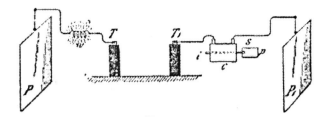

If I connect an insulated metal plate P, Fig. 11, to one of the terminals T of the induction coil by means of a wire, though this plate be very well insulated, a current passes through the wire when the coil is set to work. First I wish to give you evidence that there *is* a current passing through the connecting wire. An obvious way of demonstrating this is to insert between the terminal of the coil and the insulated plate a very thin platinum or German silver wire w and bring the latter to incandescence or fusion by the current. This requires a rather large plate of else current impulses of very high potential and frequency. Another way is to take a coil C, Fig. 11, containing many turns of thin insulated wire and to insert the same in the path of the current to the plate. When I connect one of the ends of the coil to the wire leading to another insulated plate P_1, and its other end to the terminal T_1 of the induction coil, and set the latter to work, a current passes through the inserted coil C and the existence of the current may be made manifest in various ways. For instance, I insert an iron core I within the coil. The current being one of very high frequency, will, if it be of some strength, soon bring the iron core to a noticeably higher temperature, as the hysteresis and current losses are great with such high frequencies. One might take a core of sole size, laminated or not, it would matter little; but ordinary iron .wire 1/16-th or 1/8-th of an inch thick is suitable for the purpose. While the induction coil is working, a current traverses the inserted coil and only a few moments are sufficient to bring the iron wire I to an elevated temperature sufficient to soften the sealing wax .s and cause a paper washer p fastened by it to the iron wire to fall off. Put with the apparatus such as I have here, other, much more interesting, demonstrations of this kind can be made. I have a secondary s, Fig;. 12, of coarse wire, wound upon a coil similar to the first. In the preceding experiment the current through the coil C, Fig. 11, was very small, but there, being many turns a strong heating effect was, nevertheless, produced in the iron wire. Had I passed that current through a conductor in order to show the heating of the latter, the current might have been too small to produce the effect desired. But with this coil provided with a secondary winding, I can now transform the feeble current of high tension which passes through the primary P into a strong secondary current of low tension. and this current wilt quite certainly do what I expect. In a small glass tube (t, Fig. 12), I have enclosed a coiled platinum wire, w, this merely in order to protect the wire. On each end of the glass tube is scaled a terminal of stout wire to which one of the ends of the platinum wire w, is connected. I join the terminals of the secondary coil to these terminals and insert the primary p, between the insulated plate P_1, and the terminal T_1, of the induction coil as before. The latter being set to work, instantly the platinum wire w is rendered incandescent and can be fused, even if it be very thick.

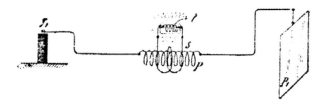

Instead of the platinum wire I now take an ordinary 50-volt 16 c p. lamp. When I set the induction coil in operation the lamp filament is brought to high incandescence. It is, however, not necessary to use the insulated plate, for the lamp (l Fig 13) is rendered incandescent even if the plate P1 be disconnected. The secondary may also he connected to the primary as indicated by the dotted line in Fig. 13, to do away more or less with the electrostatic induction or to modify the action otherwise.

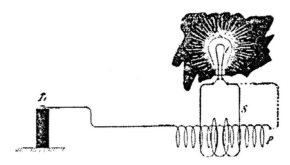

I may here call attention to a number of interesting observations "with the lamp. First, I disconnect one of the terminals of the lamp from the secondary s. When the induction coil plays, a glow is noted which fills the whole bulb. This glow is due to electrostatic induction. It increases when the bulb is grasped with the hated, and the capacity of the experimenter's body thus added to the secondary circuit. The secondary, in effect, is equivalent to a metallic coating, which would be placed near the primary. If the secondary, or its equivalent. the coating, were placed symmetrically. to the primary, the electrostatic induction would be nil under ordinary conditions, that is, when a primary return circuit is used, as both halves would neutralize each other. The secondary is in fact placed symmetrically to the primary, but the action of both halves of the latter, when only one of its ends is connected to the induction coil, is not exactly equal; hence electrostatic induction takes place, and hence the glow in the bulb. I can nearly equalize the action of both halves of the primary by connecting the other, free end of the same to the insulated plate, as in the preceding experiment. When the plate is connected, the glow disappears. With a smaller plate it would not entirely disappear and then it would contribute to the brightness of the filament when the secondary is closed, by warming the air in the bulb.

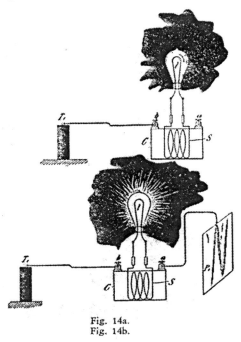

Fig. 14a.
Fig. 14b.

To demonstrate another interesting feature, I have adjusted the coils used in a certain way. I first connect both the terminals of the lamp to the secondary, one end of the primary being connected to the terminal T_1 of the induction coil and the other to tae insulated plate P, as before. When the current is turned on, the lamp glows brightly, as shown in Fig. 14b, in which C is a fine wire coil and s a coarse wire secondary wound upon it. If the insulated plate P_1 is disconnected, leaving one of the ends a of the primary insulated, the filament becomes dark or generally it diminishes in brightness (Fig. 14a). Connecting again the plate P_1 and raising the frequency of the current, I make the filament quite dark or barely red (Fig. 15b). Once more I will disconnect the plate. One will of course infer that when the plate is disconnected, the current through the primary will be weakened, that therefore the E. M. F. will fall in the secondary .s and that the brightness of the lamp will diminish. This might be the case and the result can be secured by an easy adjustment of the coils; also by varying the frequency and potential of the currents.

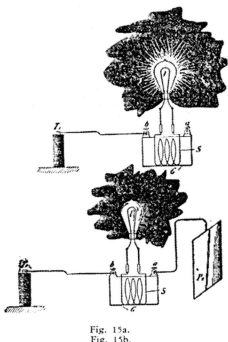

Fig. 15a.
Fig. 15b.

But it is perhaps of greater interest to note, that the lamp increases in brightness when the plate is disconnected (Fig; 15a). In this case all the energy the primary receives is now sunk info it, like the charge of a battery in an ocean cable, but most of that energy is recovered through the secondary and used to light the lamp. The current traversing the primary is strongest at the end b which is connected to the terminal T, of the induction coil, and diminishes in strength towards the remote end a. But the dynamic inductive effect exerted upon the secondary s is now greater than before, when the suspended plate was connected to the primary. These results might have been produced by a number of causes. For instance, the plate P_1 being connected, the reaction from the coil C may be such as to diminish the potential at the terminal T_1 of the induction coil, and therefore weaken the current through the primary of the coil C. Or the disconnecting of the plate may diminish the capacity effect with relation to the primary of the latter coil to such an extent that the current through it is diminished, though the potential at the terminal T_1 of the induction coil may be the same or even higher. Or the result might have been produced by the change of please of the primary and secondary currents and consequent reaction. But the chief determining factor is the relation of the self-induction and capacity of coil C and plate P_1 and the frequency of the currents. The greater brightness of the filament in Fig. 15a. is, however, in part due to the heating of the rarefied gas in the lamp by electrostatic induction; which, as before remarked, is greater when the suspended plate is disconnected.

Still another feature of some interest I tray here bring to your attention. When the insulated date is disconnected and the secondary of the coil opened, by approaching. a small object to the secondary, but very small sparks can be drawn from it, showing that the electrostatic induction is stn all in this case. But upon the secondary being closed upon itself or through the lamp, the filament ,glowing brightly, strong sparks are obtained from the secondary. The electrostatic induction is now much greater, because the closed secondary determines a greater flow of current through the primary and principally through that half of it which is connected to the induction coil. If now the bulb be grasped with the hand, the capacity of the secondary with reference to the primary is augmented by the experimenter's body and the luminosity of the filament is increased, the Incandescence now being due partly to the flow of current through the filament and partly to the molecular bombardment of the rarefied gas in the bulb.

The preceding experiments will have prepared one for the next following results of interest, obtained in the course of these investigations. Since I can pass a current through an insulated wire merely by connecting one of its ends to the source of electrical energy, since I can induce by it another current, magnetize all iron core, and, in short, perform all operations as though a return circuit were used, clearly I can also drive a motor by the aid of only one wire. On a former occasion I have described a simple form of motor comprising a single exciting coil, an iron core and disc. Fig. 16 illustrates a modified way of operating such an alternate current motor by currents induced in a transformer connected to one lead, and several other arrangements of circuits for operating a certain class of alternating motors founded on the action of currents of differing phase. In view of the present state of the art it is thought sufficient to describe these arrangements in a few words only. The diagram, Fig. 16 II., shows a primary coil P, connected with one of its ends to the line L leading from a high tension transformer terminal T_1. In inductive relation to this primary P is a secondary s of coarse wire in the circuit of which is a coil C. The currents induced in the secondary energize the iron core I, which is preferably, but not necessarily, subdivided, and set the metal disc d in rotation. Such a motor M_2 as diagrammatically shown in Fig. 16 II., has been called a "magnetic lag motor," but this expression may be objected to by those who attribute the rotation of the disc to *eddy* currents circulating in minute paths when the core I is finally subdivided. In order to operate such a motor effectively on the plan indicated, the frequencies should not he too high, not more than four or five thousand, though the rotation is produced even with ten thousand per second, or more.

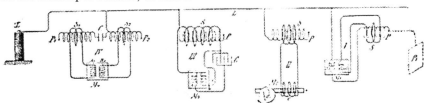

In Fig. 16 I., a motor M1 having two energizing circuits, A and B, is diagrammatically indicated. The circuit A is connected to the line L and in series with it is a primary P, which may have its free end connected to an insulated plate P_1, such connection being indicated by the dotted lines. The other motor circuit B is connected to the secondary s which is in inductive relation to the primary P. When the transformer terminal T_1 is alternately electrified, currents traverse the open line L and also circuit A and primary P. The currents through the latter induce secondary currents in the circuit S, which pass through the energizing coil B of the motor. The currents through the secondary S and those through the primary P differ in phase 90 degrees, or nearly so, and are capable of rotating an armature placed in inductive relation to the circuits A and B.

In Fig. 16 III., a similar motor M 3 with two energizing circuits A_1 and B_1 is illustrated. A primary P, connected with one of its ends to the line L has a secondary S, which is preferably wound for a tolerably high E. M. F., and to which the two energizing circuits of the motor are connected, one directly to the ends of the secondary and the other through a condenser C, by the action of which the currents traversing the circuit A_1 and B_1 are made to differ in phase.

In Fig. 16 IV., still another arrangement is shown. In this case two primaries P_1 and P_2 are connected to the line L, one through a condenser C of small capacity, and the other directly. The primaries are provided with secondaries S_1 and S_2 which are in series with the energizing circuits, A_2 and B_2 and a motor M_3 the condenser C again serving to produce the requisite difference in the phase of the currents traversing the motor circuits. As such phase motors with two or more circuits are now well known in the art, they have been here illustrated diagrammatically. No difficulty whatever is found in operating a motor in the manner indicated, or in similar ways; and although such experiments up to this day present only scientific interest, they may at a period not far distant, be carried out with practical objects in view.

It is thought useful to devote here a few remarks to the subject of operating devices of all kinds by means of only one leading wire. It is quite obvious, that when high-frequency currents are made use of, ground connections are — at least when the E. M. F. of the currents is great — better than a return wire. Such ground connections are objectionable with steady or low frequency currents on account of destructive chemical actions of the former and disturbing influences exerted by both on the neighboring circuits; but with high frequencies these actions practically do not exist. Still, even ground connections become, superfluous when the E. M. F. is very high, for soon a condition is reached, when the current may be passed more economically through open, than through closed, conductors. Remote as might seem an industrial application of such single wire transmission of energy to one not experienced in such lines of experiment, it will not seem so to anyone who for some time has carried on investigations of such nature. Indeed I cannot see why such a plan, should not be practicable. Nor should it be thought that for carrying out such a plan currents of very high frequency are expressly required, for just as soon as potentials of say 30,000 volts are used, the single wire transmission may be effected with low frequencies, and experiments have been made by me from which these inferences are trade.

When the frequencies are very high it has been found in laboratory practice quite easy to regulate the effects in the manner shown in diagram Fig. 17. Here two primaries P and P_1 are shown, each connected with one of its ends to the line L and with the other end to the condenser plates C and C, respectively. Near these are placed other condenser plates C_1 and C_1, the former being connected to the line L and the latter to an insulated larger plate P_2. On the primaries are wound secondaries S and S_1, of coarse wire, connected to the devices d and l respectively. By varying the distances of the condenser plates C and C_1, and C and C_1 the currents through the secondaries S and S_1 are varied in intensity. The curious feature is the great sensitiveness, the slightest change in the distance of the plates producing considerable variations in the intensity or strength of the currents. The sensitiveness may be rendered extreme by making the frequency such, that the primary itself, without any plate attached to its free end, satisfies, in conjunction with the closed secondary, the condition of resonance. In such condition an extremely small change in the capacity of the free terminal produces great variations. For instance, I have been able to adjust the conditions so that the mere approach of a person to the coil produces a considerable change in the brightness of the lamps attached to the secondary. Such observations and experiments possess, of course, at present, chiefly scientific interest, but they may soon become of practical importance.

Very high frequencies are of course not practicable with motors on account of the necessity of employing iron cores. But one may use sudden discharges of low frequency and thus obtain certain advantages of high-frequency currents without rendering the iron core entirely incapable of following the changes and without entailing a very great expenditure of energy in the core. I have found it quite practicable to operate with such low frequency disruptive discharges of condensers, alternating-current motors. A certain class of such motors which I advanced a few years ago, which contain closed secondary circuits, will rotate quite vigorously when the discharges are directed through the exciting coils: One reason that such a motor operates so well with these discharges is that the difference of phase between the primary and secondary currents is 90 degrees, which is generally not the case with harmonically rising and falling currents of low frequency. It might not be without interest to show an experiment with a simple motor of this kind, inasmuch as it is commonly thought that disruptive discharges are unsuitable for such purposes. The motor is illustrated in Fig. 18. It comprises a rather large iron core I with slots on the top into which are embedded thick copper washers C C. In proximity to the core is a freely-movable metal disc D. The core is provided with a primary exciting coil C_1 the ends a and b of which are connected to the terminals of the secondary S of an ordinary transformer, the primary P of the latter being connected to an alternating distribution circuit or generator G of low or moderate frequency. The terminals of the secondary S are attached to a condenser C which discharges through an air gap $d\ d$ which may be placed in series or shunt to the coil C_1. When the conditions are properly chosen the disc D rotates Keith considerable effort and the iron core I does not get very perceptibly hot. With currents from a high-frequency alternator, on the contrary, the core gets rapidly hot and the disc rotates with a much smaller effort. To perform the experiment properly it should be first ascertained that the disc D is not set in rotation when the discharge is not occurring at $d\ d$. It is preferable to use a large iron core and a condenser of large capacity so as to bring the superimposed quicker oscillation to a very low pitch or to do away with it entirely. By observing certain elementary rules I have also found it practicable to operate ordinary series or shunt direct-current motors with such disruptive discharges, and this can be done with or without a return wire.

Impedance Phenomena

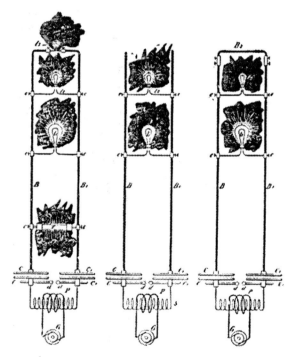

Figs. 19a, 19b and 19c.

Among the various current phenomena observed, perhaps the most interesting are those of impedance presented by conductors to currents varying at a rapid rate. In my first paper before the American Institute of Electrical Engineers, I have described a few striking observations of this kind. Thus I showed that when such currents or sudden discharges are passed through a thick metal bar there may be points on the bar only a few inches apart, which have a sufficient potential difference between them to maintain at bright incandescence an ordinary filament lamp. I have also described the curious behavior of rarefied gas surrounding a conductor, due to such sudden rushes of current. These phenomena have since been more carefully studied and one or two novel experiments of this kind are deemed of sufficient interest to be described here.

Referring to Fig. 19a, B and B_1 are very stout copper bars connected at their lower ends to plates C and C_1, respectively, of a condenser, the opposite plates of the latter being connected to the terminals of the secondary S of a high-tension transformer, the primary P of which is supplied with alternating currents from an ordinary low-frequency dynamo G or distribution circuit. The condenser discharges through an adjustable gap $d\ d$ as usual. By establishing a rapid vibration it was found quite easy to perform the following curious experiment. The bars B and B_1 were joined at the top by a low-voltage lamp l_3 a little lower was placed by means of clamps $C\ C$, a 50-volt lamp l_2; and still lower another 100-volt lamp l_1; and finally, at a certain distance below the latter lamp, an exhausted tube T. By carefully determining the positions of these devices it was found practicable to maintain them all at their proper illuminating power. Yet they were all connected in multiple arc to the two stout copper bars and required widely different pressures. This experiment requires of course some time for adjustment but is quite easily performed.

In Figs. 19b and 19c, two other experiments are illustrated which, unlike the previous experiment, do not require very careful adjustments. In Fig. 19b, two lamps, l1 and *l*2, the former a 100-volt and the latter a 50-volt are placed in certain positions as indicated, the 100-volt lamp being below the 50-volt lamp. When the arc is playing at d d and the sudden discharges are passed through the bars B B_1, the 50-volt lamp will, as a rule, burn brightly, or at least this result is easily secured, while the 100-volt lamp will burn very low or remain quite dark, Fig. 19b. Now the bars B B_1 may be joined at the top by a thick cross bar B_2 and it is quite easy to maintain the 100-volt lamp at full candle-power while the 50-volt lamp remains dark, Fig. 19c. These results, as I have pointed out previously, should not be considered to be due exactly to frequency but rather to the time rate of change which may be great, even with low frequencies. A great many other results of the same kind, equally interesting, especially to those who are *only* used to manipulate steady currents, may be obtained and they afford precious clues in investigating the nature of electric currents.

In the preceding experiments I have already had occasion to show some light phenomena and it would now be proper to study these in particular; but to make this investigation more complete I think it necessary to make first a few remarks on the subject of electrical resonance which has to be always observed in carrying out these experiments.

On Electrical Resonance

The effects of resonance are being more and more noted by engineers and are becoming of great importance in the practical operation of apparatus of all kinds with alternating currents. A few general remarks may therefore be made concerning these effects. It is clear, that if we succeed in employing the effects of resonance practically in the operation of electric devices the return wire will, as a matter of course, become unnecessary, for the electric vibration may be conveyed with one wire just as well as, and sometimes even better than, with two. The question first to answer is, then, whether pure resonance effects are producible. Theory and experiment both show that such is impossible in Nature, for as the oscillation becomes more and more vigorous, the losses in the vibrating bodies and environing media rapidly increase and necessarily check the vibration which otherwise would go on increasing forever. It is a fortunate circumstance that pure resonance is not producible, for if it were there is no telling what dangers might not lie in wait for the innocent experimenter. But to a certain degree resonance is producible, the magnitude of the effects being limited by the imperfect conductivity and imperfect elasticity of the media or, generally stated, by frictional losses. The smaller these losses, the more striking are the effects. The same is the case in mechanical vibration. A stout steel bar may be set in vibration by drops of water falling, upon it at proper intervals; and with glass, which is more perfectly elastic, the resonance effect is still more remarkable, for a goblet may be burst by singing into it a Tote of the proper pitch. The electrical resonance is the more perfectly attained, the smaller the resistance or the impedance of the conducting path and the more perfect the dielectric. In a Leyden jar discharging through a short stranded cable of thin wires these requirements are probably best fulfilled, and the resonance effects are , therefore very prominent. Such is not the case with dynamo machines, transformers and their circuits, or with commercial apparatus in general in which the presence of iron cores complicates the action or renders it impossible. In regard to Leyden jars with which resonance effects are frequently demonstrated, I would say that the effects observed are often *attributed* but are seldom *due* to true resonance, for an error is quite easily made in this respect. This may be undoubtedly demonstrated by the following experiment. Take, for instance, two large insulated metallic plates or spheres which I shall designate A and B; place them at a certain small distance apart and charge them from a frictional; or influence machine to a potential so high that just a slight increase of the difference of potential between them will cause the small air or insulating space to break down. This is easily reached by malting a few preliminary trials. If now another plate — fastened on an insulating handle and connected by a wire to one of the terminals of a high tension secondary of an induction coil, which is maintained in action by an alternator (preferably high frequency) — is approached to one of the charged bodies A or B, so a s to be nearer to either one of them, the discharge will invariably occur between them; at least it will, if the potential of the coil in connection with the plate is sufficiently high. But the explanation of this will soon be found in the fact that the approached plate acts inductively upon the bodies A and B and causes a spark to pass between them. When this spark occurs, the charges which were previously imparted to these bodies from the influence machine, must needs be lost, since the bodies are brought in electrical connection through the arc formed. Now this arc is formed whether there be resonance or not. But *even* if the spark would not be produced, still there is an alternating E. M. F. set up between the bodies when the plate is brought near one of them; therefore the approach of the plate, if it *does* not always actually, will, at any rate, *tend* to break down the air space by inductive action. Instead of the spheres or plates A and B we may take the coatings of a Leyden jar with the same result, and in place of the machine, — which is a high frequency alternator preferably, because it is more suitable for the experiment and also for the argument, — we may take another Leyden jar or battery of jars. When such jars are discharging through a circuit of low resistance the same is traversed by currents of very high frequency. The plate may now be connected to one of the coatings of the second jar, and when it is brought near to the first jar just previously charged to a high potential from an influence machine, the result is the same as before, and the first jar will discharge through a small air space upon the second being caused to discharge. But both jars and their circuits need not be tuned any closer than a basso profundo is to the note produced by a mosquito, as small sparks will be produced through the air space, or at least the latter will be considerably more strained owing to the setting up of an alternating E. M. F. by induction, which takes place when one of the jars begins to discharge. Again another error of a similar nature is quite easily made. If the circuits of the two jars are run parallel and close together, and the experiment has been performed of discharging one by the other, and now a coil of wire be added to one of the circuits whereupon the experiment does not succeed, the conclusion that this is due to the fact that the circuits are now not tuned, would be far from being safe. For the two circuits act as condenser coatings and the addition of the coil to one of them is equivalent to bridging them, at the point where the coil is placed, by a small condenser, and the effect of the latter might be to prevent the spark from jumping through the discharge space by diminishing the alternating E. M. F. acting across the same. All these remarks, and many more which might be added but for fear of wandering too far from the subject, are made with the pardonable intention of cautioning the unsuspecting student, who might gain an entirely unwarranted opinion of his skill at seeing every experiment succeed; but they are in no way thrust upon the experienced as novel observations.

In order to make reliable observations of electric resonance effects it is very desirable, if not necessary, to employ an alternator giving currents which rise and fall harmonically, as in working with make and break currents the observations are not always trustworthy, since many phenomena, which depend on the rate of change, may be produced with widely different frequencies. Even when making such observations with an alternator one is apt to be mistaken. When a circuit is connected to an alternator there are an indefinite number of values for capacity and self-induction which, in conjunction, will satisfy the condition of resonance. So there are in mechanics an infinite number of tuning forks which will respond to a note of a certain pitch, or loaded springs which have a definite period of vibration. But the resonance will be most perfectly attained in that case in which the motion is effected with the greatest freedom. Now in mechanics, considering the vibration in the common medium — that is, air — it is of comparatively little importance whether one tuning fork be somewhat larger than another, because the losses in the air are not very considerable. One may, of course, enclose a tuning fork in an exhausted vessel and by thus reducing the air resistance to a minimum obtain better resonant action. Still the difference would not be very great. But it would make a great difference if the tuning fork were immersed in mercury. In the electrical vibration it is of enormous importance to arrange the conditions so that the vibration is effected with the greatest freedom. The magnitude of the resonance effect depends, under otherwise equal conditions, on the quantity of electricity set in motion or on the strength of the current driven through the circuit. But the circuit opposes the passage of .the currents by reason of its impedance and therefore, to secure the best action it is necessary to reduce the impedance to a minimum. It is impossible to overcome it entirely, but merely in part, for the ohmic resistance cannot be overcome. But when the frequency of the impulses is very great, the flow of the current is practically determined by self-induction. Now self-induction can be overcome by combining it with capacity. If the relation between these is such, that at the frequency used they annul each other, that is, have such values as to satisfy the condition of resonance, and the greatest quantity of electricity is made to flow through the external circuit, then the best result is obtained. It is simpler and safer to join the condenser in series with the self-induction. It is clear that in such combinations there will be, for a given frequency, and considering only the fundamental vibration, values which will give the best result, with the condenser in shunt to the self-induction coil; of course more such values than with the condenser in series. But practical conditions determine the selection. In the latter case in performing the experiments one may take a small self-induction and a large capacity or a small capacity and a large self-induction, but the latter is preferable, because it is inconvenient to adjust a large capacity by small steps. By taking a coil with a very large self-induction the critical capacity is reduced to a very small value, and the capacity of the coil itself may be sufficient. It is easy, especially by observing certain artifices, to wind a coil through which the impedance will be reduced to the value of the ohmic resistance only; and for any coil there is, of course, a frequency at which the maximum current will be made to pass through the coil. The observation of the relation between self-induction, capacity and frequency is becoming important in the operation of alternate current apparatus, such as transformers or motors, because by a judicious determination of the elements the employment of an expensive condenser becomes unnecessary. Thus it is possible to pass through the coils of an alternating current motor under the normal working conditions the required current with a low E. M. F. and do away entirely with the false current, and the larger the motor, the easier such a plan becomes practicable; but it is necessary for this to employ currents of very high potential or high frequency.

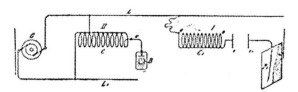

In Fig. 20 I. is shown a plan which has been followed in the study of the resonance effects by means of a high frequency alternator. C_1 is a coil of many turns, which is divided into small separate sections for the purpose of adjustment. The final adjustment was made sometimes with a few thin iron wires (though this is not always advisable) or with a closed secondary. The coil C_1 is connected with one of its ends to the line L from the alternator G and with the other end to one of the plates C of a condenser C C_1, the plate (C_1) of the latter being connected to a much larger plate P_1. In this manner both capacity and self-induction were adjusted to suit the dynamo frequency.

As regards the rise of potential through resonant action, of course, theoretically, it may amount to anything since it depends on self-induction and resistance and since these may have any value. But in practice one is limited in the selection of these values and besides these, there are other limiting causes. One may start with, say, 1,000 volts and raise the E. M. F. to 50 times that value, but one cannot start with 100,000 and raise it to ten times that value because of the losses in the media which are great, especially if the frequency is high. It should be possible to start with, for instance, two volts from a high or low frequency circuit of a dynamo and raise the E. M. F. to many hundred times that value. Thus coils of the proper dimensions might be connected each with only one of its ends to the mains from a machine of low E. M. F., and though the circuit of the machine would not be closed in the ordinary acceptance of the term, yet the machine might be burned out if a proper resonance effect would be obtained. I have not been able to produce, nor have I observed with currents from a dynamo machine, such great rises of potential. It is possible, if not probable, that with currents obtained from apparatus containing iron the disturbing influence of the latter is the cause that these theoretical possibilities cannot be realized. But if such is the case I attribute it solely to the hysteresis and Foucault current losses in the core. Generally it was necessary to transform upward, when the E. M. F. was very low, and usually an ordinary form of induction coil was employed, but sometimes the arrangement illustrated in Fig. 20 II., has been found to be convenient. In this case a coil C is made in a great many sections, a few of these being used as a primary. In this manner both primary and secondary are adjustable. One end of the coil is connected to the line L_1 from the alternator, and the other line L is connected to the intermediate point of the coil. Such a coil with adjustable primary and secondary will be found also convenient in experiments with the disruptive discharge. When true resonance is obtained the top of the wave must of course be on the free end of the coil as, for instance, at the terminal of the phosphorescence bulb B. This is easily recognized by observing the potential of a point on the wire w near to the coil.

In connection with resonance effects and the problem of transmission of energy over a single conductor which was previously considered, I would *say a* few words on a subject which constantly fills my thoughts and which concerns the welfare of all. I mean the transmission of intelligible signals or perhaps even power to any distance without the use of wires. I am becoming daily more convinced of the practicability of the scheme; and though I know full well that the great majority of scientific men will not believe that such results can be practically and immediately realized, yet I think that all consider the developments in recent years by a number of workers to have been such as to encourage thought and experiment in this direction. My conviction has grown so strong, that I no loner look upon this plan of energy or intelligence transmission as a mere theoretical possibility, but as a serious problem in electrical engineering, which must be carried out some day. The idea of transmitting intelligence without wires is the natural outcome of the most recent results of electrical investigations. Some enthusiasts have expressed their belief that telephony to any distance by induction through the air is possible. I cannot stretch my imagination se far, but I do firmly believe that it is practicable to disturb by means of powerful machines the electrostatic condition of the earth and thus transmit intelligible signals and perhaps power. In fact, what is there against the carrying out of such a scheme? We now know that electric vibration may be transmitted through a single conductor. Why then not try to avail ourselves of the earth for this purpose? We need not be frightened by the idea of distance. To the weary wanderer counting the mile-posts the earth may appear very large but to that happiest of all men, the astronomer, who ,gazes at the heavens and by their standard judges the magnitude of our globe, it appears very small. And so I thinly it must seem to the electrician, for when he considers the speed with which an electric disturbance is propagated through the earth all his ideas of distance must completely vanish.

A point of great importance would be first to know what is the capacity of the earth? and what charge does it contain if electrified? Though we have no positive evidence of a charged body existing in space without other oppositely electrified bodies being near, there is a fair probability 'that the earth is such a body, for by whatever process it was separated from other bodies — and this is the accepted view of its origin — it must have retained a charge, as occurs in all processes of mechanical separation. If it be a charged body insulated in space its capacity should be extremely small, less than one-thousandth of a farad. But the upper strata of the air are conducting, and so, perhaps, is the medium in free space beyond the atmosphere, and these may contain an opposite charge. Then the capacity might be incomparably greater. In any case it is of the greatest importance to get an idea of what quantity of electricity the earth contains. It is difficult to say whether we shall ever acquire this necessary knowledge, but there is hope that we may, and that is, by means of electrical resonance. If ever we can ascertain at what period the earth's charge, when disturbed, oscillates with respect to an oppositely electrified system or known circuit, we shall know a fact possibly of the greatest importance to the welfare of the human race. I propose to seek for the period by means of an electrical oscillator, or a source of alternating electric currents. One of the terminals of the source would be connected to earth as, for instance, to the city water mains, the other to an insulated body of large surface. It is possible that the outer conducting air strata, or free space, contain an opposite charge and that, together with the earth, they form a condenser of very large capacity. In such case the period of vibration may be very low and an alternating dynamo machine might serve for the purpose of the experiment. I would then transform the current to a potential as high as it would be found possible and connect the ends of the high tension secondary to the ground and to the insulated body. By varying the frequency *of* the currents and carefully observing the potential of the insulated body and watching for the disturbance at various neighbouring points of the earth's surface resonance might be detected. Should, as the majority of scientific men in all probability believe, the period be extremely small, then a dynamo machine would not do and a proper electrical oscillator would hate to be produced and perhaps it insight not be possible to obtain such rapid vibrations. But whether this be possible or not, and whether the earth contains a charge or not, and whatever may be its period of vibration, it certainly is possible — for of this we have daily evidence — to produce some electrical disturbance sufficiently powerful to be perceptible by suitable instruments at any point of the earth's surface.

Assume that a source of alternating currents be connected, as in Fig. 21, with one of its terminals to earth (conveniently to the water mains) and with the other to a body of large surface P. When the electric oscillation is set up there will be a movement of electricity in and out of P, and alternating currents will pass through the earth, converging to, or diverging from, the point C where the ground connection is trade. In this manner neighboring points on the earth's surface within a certain radius will be disturbed. But the disturbance will diminish with the distance, and the distance at which the effect will still be perceptible will depend on the quantity of electricity set in motion. Since the body P is insulated, in order to displace a considerable quantity, the potential of the source must be excessive, since there would be limitations as to the surface of P. The conditions might be adjusted so that the generator or source S will set up the swine electrical movement as though its circuit were closed. Thus it is certainly practicable to impress an electric vibration at least of a certain low period upon the earth by means of proper machinery. At what distance such a vibration might be made perceptible can only be conjectured. I have on another occasion considered the question how the earth might behave to electric disturbances. There is no doubt that, since in such an experiment the electrical density at the surface could be but extremely small considering the size of the earth, the air would not act as a very disturbing factor, and there would be not much energy lost through the action of the air, which would be the case if the density were great. Theoretically, then; it could not require a great amount of energy to produce a disturbance perceptible at great distance, or even all over the surface of the globe. Now, it is quite certain that at any point within a certain radius of the source S a properly adjusted self-induction and capacity device can be set in action by resonance. But not only can this be clone, but another source S_1 Fig. 21, similar to S, or any number of such sources, can be set to work in synchronism with the latter, and the vibration thus intensified and spread over a large area, or a flow of electricity produced to or from the source S_1 if the same be of opposite phase to the source S. I think that beyond doubt it is possible to operate electrical devices in a city through the ground or pipe system by resonance from an electrical oscillator located at a central point. But the practical solution of this problem would be of incomparably smaller benefit to man than the realization of the scheme of transmitting intelligence, or perhaps power, to any distance through the earth or environing medium. If this is at all possible, distance does not mean anything. Proper apparatus must first be produced by means of which the problem can be attacked and I have devoted much thought to this subject. I am firmly convinced that i: can be done and hope that we shall live to see it done.

On The Light Phenomena Produced By High-frequency Currents Of High Potential And General Remarks Relating To The Subject

Returning now to the light effects which it has been the chief object to investigate, it is thought proper to divide these effects into four classes: 1. Incandescence of a solid. 2. Phosphorescence. 3. Incandescence or phosphorescence of a rarefied gas; and 4. Luminosity produced in a gas at ordinary pressure. The first question is: How are these luminous effects produced? In order to answer this question as satisfactorily as I am able to do in the light of accepted views and with the experience acquired, and to add some interest to this demonstration, I shall dwell here upon a feature which I consider of great importance, inasmuch as it promises, besides, to throw a better light upon the nature of most of the phenomena produced by high-frequency electric currents. I have on other occasions pointed out the great importance of the presence of the rarefied gas, or atomic medium in general, around the conductor through which alternate currents of high frequency are passed, as regards the heating of the conductor by the currents. My experiments, described some time ago, have shown that, the higher the frequency and potential difference of the currents, the more important becomes the rarefied gas in which the conductor is immersed, as a factor of the heating. The potential difference, however, is, as I then pointed out, a more important element than the frequency. When both of these are sufficiently high, the heating may be almost entirely due to the presence of the rarefied gas. The experiments to follow will show the importance of the rarefied gas, or, generally, of gas at ordinary or other pressure as regards the incandescence or other luminous effects produced by currents of this kind.

 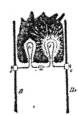 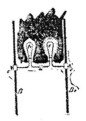

Fig. 22a. Fig. 22b. Fig. 22c.

I take two ordinary 50-volt 16 C. P. lamps which are in every respect alike, with the exception, that one has been opened at the top and the air has filled the bulb, while the other is at the ordinary degree of exhaustion of commercial lamps. When I attach the lamp which is exhausted to the terminal of the secondary of the coil, which I have already used, as in experiments illustrated in Fig. 15a for instance, and turn on the current, the filament, as you have before seen, comes to high incandescence. When I attach the second lamp, which is filled with air, instead of the former, the filament still glows, but much less brightly. This experiment illustrates only in part the truth of the statements before made. The importance of the filament's being immersed in rarefied gas is plainly noticeable but not to such a degree as might be desirable. The reason is that the secondary of this coil is wound for low tension, having only 150 turns, and the potential difference at the terminals of the lamp is therefore small. Were I to take another coil with many more turns in the secondary, the effect would be increased, since it depends partially on the potential difference, as before remarked. But since the effect likewise depends on the frequency, it may be properly stated that it depends on the time rate of the variation of the potential difference. The greater this variation, the more important becomes the gas as an element of heating. I can produce a much greater rate of variation in another way, which, besides, has the advantage of doing away with the objections, which might be made in the experiment just shown, even if both the lamps were connected in series or multiple arc to the coil, namely, that in consequence of the reactions existing between the primary and secondary coil the conclusions are rendered uncertain. This result I secure by charging, from an ordinary transformer which is fed from the alternating current supply station, a battery of condensers, and discharging the latter directly through a circuit of small self-induction, as before illustrated in Figs. 19a, 19b and 19c.

In Figs. 22a, 22b and 22c, the heavy copper bars BB_1 are connected to the opposite coatings of a battery of condensers, or generally in such way, that the high frequency or sudden discharges are made to traverse them. I connect first an ordinary 50-volt incandescent lamp to the bars by means of the clamps C C. The discharges being; passed through the lamp, the filament is rendered incandescent, though the current through it is very small, and would not be nearly sufficient to produce a visible effect under the conditions of ordinary use of the lamp. Instead of this I now attach to the bars another lamp exactly like the first, but with the seal broken off, the bulb being therefore filled with air at ordinary pressure. When the discharges are directed through the filament, as before, it does not become incandescent. But the result might still be attributed to one of the many possible reactions. I therefore connect both the lamps in multiple arc as illustrated in Fig. 22a. Passing tile discharges through both the lamps, again the filament in the exhausted lamp l glows very brightly while that in the non-exhausted lamp l_1 remains dark, as previously. But it should not be thought that the latter lamp is' taking only a small fraction of the energy supplied to both the lamps; on the contrary, it may consume a considerable portion of the energy and it may become even hotter than the one which burns brightly. In this experiment the potential difference at the terminals of the lamps varies in sign theoretically three to four million times a second. The ends of the filaments are correspondingly electrified, and the gas in the bulbs is violently agitated and a large portion of the supplied energy is thus converted into heat. In the non-exhausted bulb, there being a few million times more gas molecules than in the exhausted one, the bombardment, which is most violent at the ends of the filament, in the neck of the bulb, consumes a large portion of the energy without producing any visible effect. The reason is that, there being many molecules, the bombardment is quantitatively considerable, but the individual impacts are not very violent, as the speeds of the molecules are comparatively small owing to the small free path. In the exhausted bulb, on the contrary, the speeds are *very* great, and the individual impacts are violent and therefore better adapted to produce a visible effect. .Besides, the convection of heat is greater in the former bulb. In both the bulbs the current traversing the filaments is very small, incomparably smaller than that which they require on an ordinary low-frequency circuit. The potential difference, however, at the ends of the filaments is very great and might be possibly 20,000 volts or more, if the filaments were straight and their ends far apart. In the ordinary lamp a spark generally occurs between the ends of the filament or between the platinum wires outside, before such a difference of potential can be reached.

It might be objected that in the experiment before shown the lamps, being in multiple arc, the exhausted lamp might take a much larger current and that the effect observed might not be exactly attributable to the action of the gas in the bulbs. Such objections will lose much weight if I connect the lamps in series, with the wine result. When this is done and the discharges are directed through the filaments, it is again noted that the filament in the non-exhausted bulb l_1 remains dark, while that in the exhausted one (l) glows even snore intensely than under its normal conditions of working, Fig. 22b. According to general ideas the current through the filaments should now be the same, were it not modified by the presence of the gas around the filaments.

At this juncture I may point out another interesting feature, which illustrates the effect of the rate of change of potential of the currents. I will leave the two lamps connected in series to the bars BB_1, as in the previous experiment, Fig. 22b, but will presently reduce considerably the frequency of the currents, which was excessive in the experiment just before shown. This I may do by inserting a self-induction coil in the path of the discharges, or by augmenting the capacity of the condensers. When I now pass these low-frequency discharges through the lamps, the exhausted lamp l again is as bright as before, but it is noted also that the non-exhausted lamp l_1 glows, though not quite as intensely as the other. Reducing the current through the lamps, I may bring the filament in the latter lamp to redness, and, though the filament in the exhausted lamp 1 is bright, Fig. 22c, the degree of its incandescence is much smaller than in Fig. 22b; when the currents were of a much higher frequency.

In these experiments the gas acts in two opposite ways in determining the degree of the incandescence of the filaments, that is, by convection and bombardment. The higher the frequency and potential of the currents, the more important becomes the bombardment. The convection on the contrary should be the smaller, the higher the frequency. When the currents are steady there is practically no bombardment, and convection may therefore with such currents also considerably modify the degree of incandescence and produce results similar to those just before shown. Thus, it two lamps exactly alike, one exhausted and one not exhausted, are connected in multiple arc or series to a direct-current machine, the filament in the non-exhausted lamp will require a considerably greater current to be rendered incandescent. This result is entirely due to convection, and the effect is the more prominent the thinner the filament. Professor Ayrton and Mr. Kilgour some time ago published quantitative results concerning the thermal emissivity by radiation and convection in which the effect with thin wires was clearly shown. This effect may be strikingly illustrated by preparing a number of small, short, glass tubes, each containing through its axis the thinnest obtainable platinum wire. If these tubes be highly exhausted, a number of them may be connected in multiple arc to a direct-current machine and all of the wires may be kept at incandescence with a smaller current than that required to render incandescent a single one of the wires if the tube be not exhausted. Could the tubes be so highly exhausted that convection would be nil, then the relative amounts of heat given off by convection and radiation could be determined without the difficulties attending thermal quantitative measurements. If a source of electric impulses of high frequency and very high potential is employed, a still greater number of the tubes may be taken and the wires rendered incandescent by a current not capable of warming perceptibly a wire of the same size immersed in air at ordinary pressure, and conveying the energy to all of them.

I may here describe a result which is still more interesting, and to which I have been led by the observation of these phenomena. I noted that small differences in the density of the air produced a considerable difference in the degree of incandescence of the wires, and I thought that, since in a tube, through which a luminous discharge is passed, tile gas is generally not of uniform density, a very thin wire contained in the tube might be rendered incandescent at certain places of smaller density of the gas, while it would remain dark at the places of greater density, where the convection would be greater and the bombardment less intense. Accordingly a tube t was prepared, as illustrated in Fig. 23, which contained through the middle a very fine platinum wire w. The tube was exhausted to a moderate degree and it was found that when it was attached to the terminal of a high-frequency coil the platinum wire w, would indeed, become incandescent in patches, as illustrated in Fig. 23. Later a number of these tubes with one or more wires were prepared, each showing this result. The effect was best noted when the striated discharge occurred in the tube, but was also produced when the striae were not visible, showing that, even then, the gas in the tube was, not or uniform density. The position of the striae was generally such, that the rarefactions corresponded to the places of incandescence or greater brightness on the wire w. But in a few instances it was noted, that the bright spots on the wire were covered by the dense parts of the striated discharge as indicated by l in Fig. 23, though the effect was barely perceptible. This was explained in a plausible way by assuming that the convection was not widely different in the dense and rarefied places, and that the bombardment was greater on the dense places of the striated discharge. It is, in fact, often observed in bulbs, that under certain conditions a thin wire is brought to higher incandescence when the air is not too highly rarefied. This is the case when the potential of the coil is not high enough for the vacuum, but the result may be attributed to many different causes. In all cases this curious phenomenon of incandescence disappears when the tube, or rather the wire, acquires throughout a uniform temperature.

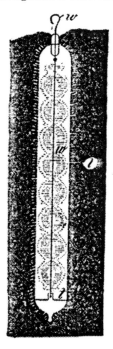
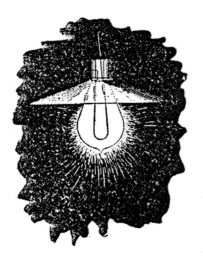

Disregarding now the modifying effect of convection there are then two distinct causes which determine the incandescence of a wire or filament with varying currents, that is, conduction current and bombardment. With steady currents we have to deal only with the former of these two causes, and the heating effect is a minimum, since the resistance is least to steady flow. When the current is a varying one the resistance is greater, and hence the heating effect is increased. Thus if the rate of change of the current is very great, the resistance may increase to such an extent that the filament is brought to incandescence with inappreciable currents, and we are able to take a short and thick block of carbon or other material and brim, it to bright incandescence with a current incomparably smaller than that required to bring to the same deuce of incandescence an ordinary thin lamp filament with a steady or low frequency current. This result is important, and illustrates how rapidly our views on these subjects are changing, and how quickly our field of knowledge is extending. In the art of incandescent lighting, to view this result in one aspect only, it has been commonly considered as an essential requirement for practical success, that the lamp filament should be thin and of high resistance. But now we know that the resistance of the filament to the steady flow does not mean anything; the filament might as well be short and thick; for if it be immersed in rarefied gas it will become incandescent by the passage of a small current. It all depends on the frequency and potential of the currents. We may conclude from this, that it would be of advantage, so far as the lamp is considered, to employ high frequencies for lighting, as they allow the use of short and thick filaments and smaller currents.

If a wire or filament be immersed in a homogeneous medium, all the heating is due to true conduction current, but if it be enclosed in an exhausted vessel the conditions are entirely different. Here the gas begins to act and the heating effect of the conduction current, as is shown in many experiments, may be *very* small compared with that of the bombardment. This is especially the case if the circuit is not closed and the potentials are of course *very* high. Suppose that a fine filament enclosed in an exhausted vessel be connected with one of its ends to the terminal of a high tension coil and with its other end to a large insulated plate. Though the circuit is not closed, the filament, as I have before shown, is brought to incandescence. If the frequency and potential be comparatively low, the filament is heated by the current passing *through it.* **If** the frequency and potential, and principally the latter, be increased, the insulated plate need be but very small, or may be done away with entirely; still: the filament will become incandescent, practically all the heating being then due to the bombardment. A practical way of combining both the effects of conduction currents and bombardment is Illustrated in Fig. 24, in which an ordinary lamp is shown provided with a very thin filament which has one of the ends of the latter connected to a shade serving the purpose of the insulated plate, and the other end to the terminal of a high tension source. It should not be thought that only rarefied gas is an important factor in the heating of a conductor by varying currents, but gas at ordinary pressure may become important, if the potential difference and frequency of the currents is excessive. On this subject I have already stated, that when a conductor is fused by a stroke of lightning, the current through it may be exceedingly small, not even sufficient to heat the conductor perceptibly, were. the latter immersed in a homogeneous medium.

From the preceding it is clear that when a conductor of high resistance is connected to the terminals of a source of high frequency currents of high potential, there may occur considerable dissipation of energy, principally at the ends of the conductor, in consequence of the action of the gas surrounding the conductor. Owing to this, the current through a section of the conductor at a point midway between its ends may be much smaller than through a section near the ends. Furthermore, the current passes principally through the outer portions of the conductor, but this effect is to be distinguished from the skin effect as ordinarily interpreted, for the latter would, or should, occur also in a continuous incompressible medium. If a great many incandescent lamps are connected in series to a source of such currents, the lamps at the ends may burn brightly, whereas those in the middle may remain entirely dark. This is due principally to bombardment, as before stated. But even if the currents be steady, provided the difference of potential is very great, the lamps at the end will burn more brightly than those in the middle. In such case there is no rhythmical bombardment, and the result is produced entirely by leakage. This leakage or dissipation into space when the tension is high, is considerable when incandescent lamps are used, and still more considerable with arcs, for the latter act like flames. Generally, of course, the dissipation is much smaller with steady, than with varying, currents.

I have contrived an experiment which illustrates in an interesting manner the effect of lateral diffusion. If a very long tube is attached to the terminal of a high frequency coil, the luminosity is greatest near the terminal and falls off gradually towards the remote end. This is more marked if the tube is narrow.

A small tube about one-half inch in diameter and twelve inches long (Fig. 25), has one of its ends drawn out into a fine fibre f nearly three feet long. The tube is placed in a brass socket T which can be screwed on the terminal T_1 of the induction coil. The discharge passing through the tube first illuminates the bottom of the same, which is of comparatively large section; but through the long glass fibre the discharge cannot pass. But gradually the rarefied gas inside becomes warmed and more conducting and the discharge spreads into the glass fibre. This spreading is so slow, that it may take half a minute or more until the discharge has worked through up to the top of the glass fibre, then presenting the appearance of a strongly luminous thin thread. By adjusting the potential at the terminal the light may be made to travel upwards at any speed. Once, however, the glass fibre is heated, the discharge breaks through its entire length instantly. The interesting point to be noted is that, the higher the frequency of the currents, or in other words, the greater relatively the lateral dissipation, at a slower rate may the light be made to propagate through the fibre. This experiment is best performed with a highly exhausted and freshly made tube. When the tube has been used for some time the experiment often fails. It is possible that the gradual and slow impairment of the vacuum is the cause. This slow propagation of the discharge through a very narrow glass tube corresponds exactly to the propagation of heat through a bar warmed at one end. The quicker the heat is carried away laterally the longer time it will take for the heat to warm the remote end. When the current of a low frequency coil is passed through the fibre from end to end, then the lateral dissipation is small and the discharge instantly breaks through almost without exception.

After these experiments and observations which have shown the importance of the discontinuity or atomic structure of the medium and which will serve to explain, in a measure at least, the nature of the four kinds of light effects producible with these currents, I may now give you an illustration of these effects. For the sake of interest I may do this in a manner which to many of you might be novel. You have seen: before that we may now convey the electric vibration to a body by means of a single wire or conductor of any hind. Since the human frame is conducting I may convey the vibration through my body.

First, as in some previous experiments, I connect my body with one of the terminals of a high-tension transformer and take in my hand an exhausted bulb which contains a small carbon button mounted upon a platinum wire leading to the outside of the bulb, and the button is rendered incandescent as soon as the transformer is set to work (Fig. 26). I may place a conducting shade on the bulb which serves to intensify the action, but it is not necessary. Nor is it required that the button should be in conducting connection with the hand through a wire leading through the glass, for sufficient energy may be transmitted through the glass itself by inductive action to render the button incandescent.

Next I take a highly exhausted bulb containing a strongly phosphorescent body, above which is mounted a small plate of aluminum on a platinum wire leading to the outside, and the currents flowing through my body excite intense phosphorescence in the bulb (Fig. 27). Next again I take in my hand a simple exhausted tube, and in the same manner the gas inside the tube is rendered highly incandescent or phosphorescent (Fig. 28). Finally, I may take in my hand a wire, bare or covered with thick insulation, it is quite immaterial; the electrical vibration is so intense as to cover the wire with a luminous film (Fig. 29).

A few words must now be devoted to each of these phenomena. In the first place, I will consider the incandescence of a button or of a solid in general, and dwell upon some facts which apply equally to all these phenomena. It was

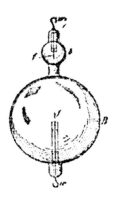
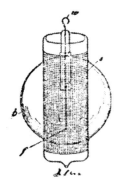

pointed out before that when a thin conductor, such as a lamp filament, for instance, is connected with one of its ends to the terminal of a transformer of high tension the filament is brought to incandescence partly by a conduction current and partly by bombardment. The shorter and thicker the filament the more important becomes the latter, and finally, reducing the filament to a mere button, all the heating must practically be attributed to the bombardment. So in the experiment before shown, the button is rendered incandescent by the rhythmical impact of freely movable small bodies in the bulb. These bodies may be the molecules of the residual gas, particles of dust or lumps torn from the electrode; whatever they are, it is certain that the heating of the button is essentially connected with the pressure of such freely movable particles, or of atomic matter in general in the bulb. The heating is the more intense the greater the number of impacts per second and the greater the energy of each impact. Yet the button would be heated also if it were connected to a source of a steady potential. In such a case electricity mould be carried away from the button by the freely movable carriers or particles flying about, and the quantity of electricity thus carried away might be sufficient to bring the button to incandescence by its passage through the latter. But the bombardment could not be of great importance in such case. For this reason it would require a comparatively very great supply of energy to the button to maintain it at incandescence with a steady potential. The higher the frequency of the electric impulses the more economically can the button be maintained at incandescence. One of the chief reasons why this is so, is, I believe, that with impulses of very high frequency there is less exchange of the freely movable carriers around the electrode and this means, that in the bulb the heated matter is better confined to the neighborhood of the button. If a double bulb, as illustrated in Fig. 30 be made, comprising a large globe B and a small *one b,* each containing as usual a filament f mounted on a platinum wire w *and* w_1 *it* is found, that if the filaments $f f$ be exactly alike, it requires less energy to keep the filament in the globe b at a certain degree of incandescence, than that in the globe B. This is due to the confinement of the movable particles around the button. In this case it is also ascertained, that the filament in the small globe b is less deteriorated when maintained a certain length of time at incandescence. This is a necessary consequence of the fact that the gas in the small bulb becomes strongly heated and therefore a very good conductor, and less work is then performed on the button, since the bombardment becomes less intense as the conductivity of the gas increases. In this construction, of course, the small bulb becomes very hot and when it reaches an elevated temperature the convection and radiation on the outside increase. On another occasion I have shown bulbs in which this drawback was largely avoided. In these instances a very small bulb, containing a refractory button, was mounted in a large globe and the space between the walls of both was highly exhausted. The outer large globe remained comparatively cool in such constructions, When the large globe was on the pump and the vacuum between the walls maintained permanent by the continuous action of the pump, the outer globe would remain quite cold, while the button in tile small bulb was kept at incandescence. But when the seal was made, and the button in the small bulb maintained incandescent some length of time, the large globe too would become warmed. Prom this I conjecture that if vacuous space (as Prof. Dewar finds) cannot convey heat, it is so merely in virtue of our rapid motion through space or, generally speaking, by the motion of the medium relatively to us, for a permanent condition could not be maintained without the medium being constantly renewed. A vacuum cannot, according to all evidence, be permanently maintained around a hot *body.*

In these constructions, before mentioned, the small bulb inside would, at least in the first stages, prevent all bombardment against the outer large globe. It occurred to me then to ascertain how a metal sieve would behave in this respect, and several bulbs, as illustrated in Fig. 31, were prepared for this purpose. In a globe b, was mounted a thin filament f (or button) upon a platinum wire w passing through a glass stem and leading to the outside of the globe. The filament f was surrounded by a metal sieve s. It was found in experiments with such bulbs that a sieve with wide meshes apparently did not in the slightest affect the bombardment against the globe b. When the vacuum was high, the shadow of the sieve was clearly projected against the globe and the latter would get hot in a short while. In some bulbs the sieve ,v was connected to a platinum wire sealed in the glass. When this wire was connected to the other terminal of the induction coil (the E. M. F. being kept low in this case), or to an insulated plate, the bombardment against the outer globe b was diminished. By taking a sieve with fine meshes the bombardment against the globe b was always diminished, but even then if the exhaustion was carried very far, and when the potential of the transformer was very high, the globe b would be bombarded and heated quickly, though no shadow of the sieve was visible, owing to the smallness of the meshes. But a glass tube or other continuous *body* mounted so as to surround the filament, did entirely cut off the bombardment and for a while the outer globe b would remain perfectly cold. Of course when the glass tube was sufficiently heated the bombardment against the outer globe could be noted at once. The experiments with these bulbs seemed to show that the speeds of the projected molecules or particles must be considerable (though quite insignificant when compared with that of light), otherwise it would be difficult to understand how *they* could traverse a fine metal sieve without being affected, unless it were found that such small particles or atoms cannot be acted upon directly at measurable distances. In regard to the speed of the projected atoms, Lord Kelvin has recently estimated it at about one kilometre a second or thereabouts in an ordinary Crookes bulb. As the potentials obtainable with a disruptive discharge coil are much higher than with ordinary coils, the speeds must, of course, be much greater when the bulbs are lighted from such a coil. Assuming the speed to be as high as five kilometres and uniform through the whole trajectory, as it should be in a very highly exhausted vessel, then if the alternate electrifications of the electrode would be of a frequency of five million, the greatest distance a particle could get away from the electrode would be one millimetre, and if it could be acted upon directly at that distance, the exchange of electrode matter or of the atoms would be very slow and there would be practically no bombardment against the bulb. This at least should be so, if the action of an electrode upon the atoms of the residual gas would be such as upon electrified bodies which we can perceive. A hot body enclosed in a n exhausted bulb produces always atomic bombardment, but a hot body has no definite rhythm, for its molecules perform vibrations of all kinds.

If a bulb containing a button or filament be exhausted as high as is possible with tile greatest care and by the use of the best artifices, it is often observed that the discharge cannot, at first, break through, but after some time, probably in consequence of some changes within the bulb, the discharge finally passes through and the button is rendered incandescent. In fact, it appears that the higher the degree of exhaustion the easier is the incandescence produced. There seem to be no other causes to which the incandescence might be attributed in such case except to the bombardment or similar action of the residual gas, or of particles of matter in general. But if the bulb be exhausted with the greatest care can these play an important part? Assume the vacuum in the bulb to be tolerably perfect, the great interest then centres in the question: Is the medium which pervades all space continuous or atomic? If atomic, then the heating of a conducting button or filament in an exhausted vessel might be due largely to ether bombardment, and then the heating of a conductor in general through which currents of high frequency or high potential are passed must be modified by the behaviour of such medium; then also the skin effect, the apparent increase of the ohmic resistance, etc., admit, partially at least, of a different explanation.

It is certainly more in accordance with many phenomena observed with high frequency currents to hold that all space is pervaded with free atoms, rather than to assume that it is devoid of these, and dark and cold, for so it must be, if filled with a continuous medium, since in such there can be neither heat nor light. Is then energy transmitted by independent carriers or by the vibration of a continuous medium? This important question is by no means as yet positively answered. But most of the effects which are here considered, especially the light effects, incandescence, or phosphorescence, involve the presence of free atoms and would be impossible without these.

In regard to the incandescence of a refractory button (or filament) in an exhausted receiver, which has been one of the subjects of this investigation, the chief experiences, which may serve as a guide in constructing such bulbs, may be summed up as follows: 1. The button should be as small as possible, spherical, of a smooth or polished surface, and of refractory material which withstands evaporation best. 2. The support of the button should be very thin and screened by an aluminum and mica sheet, as I have described on another occasion. 3. The exhaustion of the bulb should be as high as possible. 4. The frequency of the currents should be as high as practicable. 5. The currents should be of a harmonic rise and fall, without sudden interruptions. 6. The heat should be confined to the button by inclosing the same in a small bulb or otherwise. 7. The space between the walls of the small bulb and the outer globe should be highly exhausted.

Most of the considerations which apply to the incandescence of a solid just considered may likewise be applied to phosphorescence. Indeed, in an exhausted vessel the phosphorescence is, as a rule, primarily excited by the powerful beating of the electrode stream of atoms against the phosphorescent body. Even in many cases, where there is no evidence of such a bombardment, I think that phosphorescence is excited by violent impacts of atoms, which are not necessarily thrown off from the electrode but are acted upon from the same inductively through the medium or through chains of other atoms. That mechanical shocks play an important part in exciting phosphorescence in a bulb may be seen from the following experiment. If a bulb, constructed as that illustrated in Fig. 10, be taken and exhausted with the greatest care so that the discharge cannot pass, the filament *f* acts by electrostatic induction upon the tube *t* and the latter is set in vibration. If the tube *o* be rather wide, about an inch or so, the filament may be so powerfully vibrated that whenever it hits the glass tube it excites phosphorescence. But the phosphorescence ceases when the filament comes to rest. The vibration can be arrested and again started by varying the frequency of the currents. Now the filament has its own period of vibration, and if the frequency of the currents is such that there is resonance, it is easily set vibrating, though the potential of the currents be small. I have often observed that the filament in the bulb is destroyed by such mechanical resonance. The filament vibrates as a rule so rapidly that it cannot be seen and the experimenter may at first be mystified. When such an experiment as the one described is carefully performed, the potential of the currents need be extremely small, and for this reason I infer that the phosphorescence is then due to the mechanical shock of the filament against the glass, just as it is produced by striking a loaf of sugar with a knife. The mechanical shock produced by the projected atoms is easily noted when a bulb containing a button is grasped in the hand and the current turned on suddenly. I believe that a bulb could be shattered by observing the conditions of resonance.

In the experiment before cited it is, of course, open to say, that the glass tube, upon coming in contact with the filament, retains a charge of a certain sign upon the point of contact. If now the filament main touches the glass at the same point while it is oppositely charged, the charges equalize under evolution of light. But nothing of importance would be gained by such an explanation. It is unquestionable that the initial charges given to the atoms or to the glass play some part in exciting phosphorescence. So for instance, if a phosphorescent bulb be first excited by a high frequency coil by connecting it to one of the terminals of the latter and the degree of luminosity be noted and then the bulb be highly charged from a Holtz machine by attaching it preferably to the positive terminal of the machine, it is found that when the bulb is again connected to the terminal of the high frequency coil, the phosphorescence is far more intense. On another occasion I have considered the possibility of some phosphorescent phenomena in bulbs being produced by the incandescence of an infinitesimal layer on the surface of the phosphorescent body. Certainly the impact of the atoms is powerful enough to produce intense incandescence by the collisions, since they bring quickly to a high temperature a body of considerable bulk. If any such effect exists, then the best appliance for producing phosphorescence in a bulb, which we know so far, is a disruptive discharge coil giving an enormous potential with but few fundamental discharges, say 25-30 per second, just enough to produce a continuous impression upon the eye. It is a fact that such a coil excites phosphorescence under almost any condition and at all degrees of exhaustion, and I have observed effect; which appear to be due to phosphorescence even at ordinary pressures of the atmosphere, when the potentials are extremely high. But if phosphorescent light is produced by the equalization of charges of electrified atoms (whatever this may mean ultimately), then the higher the frequency of the impulses or alternate electrifications, the more economical will be the light production. It is a long known and noteworthy fact that all the phosphorescent bodies are poor conductors of electricity and heat, and that all bodies cease to emit phosphorescent light when they are brought to a certain temperature. Conductors on the contrary do not possess this quality. There are but few exceptions to the rule. Carbon is one of them. Becquerel noted that carbon phosphoresces at a certain elevated temperature preceding the dark red. This phenomenon may be easily observed in bulbs provided with a rather large carbon electrode (say, a sphere of six millimetres diameter). If the current is turned on after a few seconds, a snow white film covers the electrode, just before it bets dark red. Similar effects are noted with other conducting bodies, but many scientific men will probably not attribute them to true phosphorescence. Whether true incandescence has anything to do with .phosphorescence excited by atomic impact or mechanical shocks still remains to be decided, but it is a fact that all conditions, which tend to localize and increase the heating effect at the point of impact, are almost invariably the most favourable for the production of phosphorescence. So, if the electrode be very small, which is equivalent to saying in general, that the electric density is great; if the potential be high, and if the gas be highly rarefied, all of which things imply high speed of the projected atoms, or matter, and consequently violent impacts — the phosphorescence is very intense. If a bulb provided with a large and small electrode be attached to the terminal of an induction coil, the small electrode excites phosphorescence while the large one may not do so, because of the smaller electric density and hence smaller speed of the atoms. A bulb provided with a large electrode may be grasped with the hand while the electrode is connected to the terminal of the coil and it may not phosphoresce; but if instead of grasping the bulb with the hand, the same be touched with a pointed wire, the phosphorescence at once spreads through the bulb, because of the great density at the point of contact. With low frequencies it seems that gases of great atomic weight excite more intense phosphorescence than those of smaller weight, as for instance, hydrogen. With high frequencies the observations are not sufficiently reliable to draw a conclusion. Oxygen, as is well-known, produces exceptionally strong effects, which may be in part due to chemical action. A bulb with hydrogen residue seems to be most easily excited. Electrodes which are most easily deteriorated produce more intense phosphorescence in bulbs, but the condition is not permanent because of the impairment of the vacuum and the deposition of the electrode matter upon the phosphorescent surfaces. Some liquids, as oils, for instance, produce magnificent effects of phosphorescence (or fluorescence?), but they last only a few seconds. So if a bulb has a trace of oil on the walls and the current is turned on, the phosphorescence only persists for a few moments until the oil is carried away. Of all bodies so far tried, sulphide of zinc seems to be the most susceptible to phosphorescence. Some samples, obtained through the kindness of Prof. Henry in Paris, were employed in many of these bulbs. One of the defects of this sulphide is, that it loses its quality of emitting light when brought to a temperature which is by no means high. It can therefore, be used only for feeble intensities. An observation which might deserve notice is, that when violently bombarded from an aluminum electrode it assumes a black color, but singularly enough, it returns to the original condition when it cools down.

The most important fact arrived at in pursuing investigations in this direction is, that in all cases it is necessary, in order to excite phosphorescence with a minimum amount of energy, to observe certain conditions. Namely, there is always, no matter what the frequency of the currents, degree of exhaustion and character of the bodies in the bulb, a certain potential (assuming the bulb excited from one terminal) or potential difference (assuming the bulb to be excited with both terminals) which produces the most economical result. If the potential be increased, considerable energy may be wasted without producing any more light, and if it be diminished, then again the light production is not as economical. The exact condition under which the best result is obtained seems to depend on many things of a different nature, and it is to be yet investigated by other experimenters, but it will certainly have to be observed when such phosphorescent bulbs are operated, if the best results are to be obtained.

Coming now to the most interesting of these phenomena, the incandescence or phosphorescence of gases, at low pressures or at the ordinary pressure of the atmosphere, we must seek the explanation of these phenomena in the same primary causes, that is, in shocks or impacts of the atoms. Just as molecules or atoms beating upon a solid body excite phosphorescence in the same or render it incandescent, so when colliding among themselves they produce similar phenomena. But this is a very insufficient explanation and concerns only the crude mechanism. Light is produced by vibrations which go on at a rate almost inconceivable. If we compute, from the energy contained in the form of known radiations in a definite space the force which is necessary to set up such rapid vibrations, we find, that though the density of the ether be incomparably smaller than that of any body we know, even hydrogen, the force is something surpassing comprehension. What is this force, which in mechanical measure may amount to thousands of tons per square inch? It is electrostatic force in the light of modern views. It is impossible to conceive how a body of measurable dimensions could be charged to so high a potential that the force would be sufficient to produce these vibrations. Long before any such charge could be imparted to the body it would be shattered into atoms. The sun emits light and heat, and so does an ordinary flame or incandescent filament, but in neither of these can the force be accounted for if it be assumed that it is associated with the body as a whole. Only in one way may we account for it, namely, by identifying it with the atom. An atom is so small, that if it be charged by coming in contact with an electrified body and the charge be assumed to follow the same law as in the case of bodies of measurable dimensions, it must retain a quantity of electricity which is fully capable of accounting for these forces and tremendous rates of vibration. But the atom behaves singularly in this respect — it always takes the same "charge."

It is very likely that resonant vibration plays a most important part in all manifestations of energy in nature. Throughout space all matter is vibrating, and all rates of vibration are represented, from the lowest musical note to the highest pitch of the chemical rays, hence an atom, or complex of atoms, no matter what its period, must find a vibration with which it is in resonance. When we consider the enormous rapidity of the light vibrations, we realize the impossibility of producing such vibrations directly with any apparatus of measurable dimensions, and we are driven to the only possible means of attaining the object of setting up waves of light by electrical means and economically, that is, to affect the molecules or atoms of a gas, to cause them to collide and vibrate. We then must ask ourselves — How can free molecules or atoms be affected ?

It is a fact that they can be affected by electrostatic force, as is apparent in many of these experiments. By varying the electrostatic force we can agitate the atoms, and cause them to collide accompanied by evolution of heat and light. It is not demonstrated beyond doubt that vie can affect them otherwise. If a luminous discharge is produced in a closed exhausted tube, do the atoms arrange themselves in obedience to any other but to electrostatic force acting in straight lines from atom to atom? Only recently I investigated the mutual action between two circuits with extreme rates of vibration. When a battery of a few jars © c c c, Fig. 32), is discharged through a primary P of low resistance (the connections being as illustrated in Figs. 19a, 19b and 19c), and the frequency of vibration is many millions there are great differences of potential between points on the primary not more than a few inches apart. These differences may be 10,000 volts per inch. if not more, taking the maximum value of the h. M. F. The secondary S is therefore acted upon by electrostatic induction, which is in such extreme cases of much greater importance than the electro-dynamic. To such sudden impulses the primary as well as the secondary are poor conductors, and therefore great differences of potential may be produced by electrostatic induction between adjacent points on the secondary. Then sparks may jump between the wires and streamers become visible in the dark if the light of the discharge through the spark gap d d be carefully excluded. If now we substitute a closed vacuum tube for the metallic secondary S, the differences of potential produced in the tube by electrostatic induction from the primary are fully sufficient to excite portions of it; but as the points of certain differences of potential on the primary are not fixed, but are generally constantly changing in position, a luminous band is produced in the tube, apparently not touching the glass, as it should, if the points of maximum and minimum differences of potential were fixed on the primary. I do not exclude the possibility of such a tube being

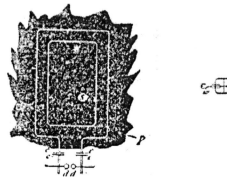
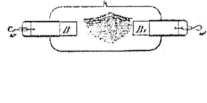

excited only by electro-dynamic induction, for very able physicists hold this view; but in my opinions, there is as yet no positive proof given that atoms of a gas in a closed tube may arrange themselves in chains under the action of an: electromotive impulse produced by electro-dynamic induction in the tube. I have been unable so far to produce striae in a tube, however long, and at whatever degree of exhaustion, that is, striae at right angles to the supposed direction of the discharge or the axis of the tube; but I have distinctly observed in a large bulb, in which a wide luminous band was produced by passing a discharge of a battery through a wire surrounding the bulb, a circle of feeble luminosity between two luminous bands, one of which was more intense than the other. Furthermore, with my present experience I do not think that such a gas discharge in a closed tube can vibrate, that is, vibrate as a whole. I am convinced that no discharge through a gas can vibrate. The atoms of a gas behave very curiously in respect to sudden electric impulses. The gas does not seem to possess any appreciable inertia to such impulses, for it is a fact, that the higher the frequency of the impulses, with the greater freedom does the discharge pass through the gas. If the gas possesses no inertia then it cannot vibrate, for some inertia is necessary for the free vibration. I conclude from this that if a lightning discharge occurs between two clouds, there can be no oscillation, such as would be expected, considering the capacity of the clouds. But if the lightning discharge strike the earth, there is always vibration — in the earth, but not in the cloud. In a gas discharge each atom vibrates at its oven rate, but there is no vibration of the conducting gaseous mass as a whole. This is an important consideration in the great problem of producing light economically, for it teaches us that to reach this result we must use impulses of very high frequency and necessarily also of high potential. It is , a fact that oxygen produces a more intense light in a tube. Is it because oxygen atoms possess some inertia and the vibration does not die out instantly? But then nitrogen should be as good, and chlorine and vapors of many other bodies much better than *oxygen,* unless the magnetic properties of the latter enter prominently into play. Or, is the process in the tube of an electrolytic nature? Many observations certainly speak for it, the most important being that matter is always carried away from the electrodes and the vacuum in a bulb cannot be permanently maintained. If such process takes place in reality, then again must we take refuge in high frequencies, for, with such, electrolytic action should be reduced to a minimum, if not rendered entirely impossible. It is an undeniable fact that with very high frequencies, provided the impulses be of harmonic nature, like those obtained from an alternator, there is less deterioration and the vacua are more permanent. With disruptive discharge coils there are sudden rises of potential and the vacua are more quickly impaired, for the electrodes are deteriorated in a very short time. It was observed in some large tubes, which were provided with heavy carbon blocks B B_1, connected to platinum wires w w_1 (as illustrated in Fig. 33), and which were employed in experiments with the disruptive discharge instead of the ordinary air gap, that the carbon particles under the action of the powerful magnetic field in which the tube was placed, were deposited in regular fine lines in the middle of the tube, as illustrated. These lines were attributed to the deflection or distortion of the discharge by the magnetic field, but why the deposit occurred principally where the field was most intense did not appear quite clear. A fact of interest, likewise noted, was that the presence of a strong magnetic field increases the deterioration of the electrodes, probably by reason of the rapid interruptions it produces, whereby there is actually a higher E.=M. F. maintained between the electrodes.

Much would remain to be said about the luminous effects produced in gases at low or ordinary pressures. With the present experiences before us we cannot say that the essential nature of these charming phenomena is sufficiently known. But investigations in this direction are being pushed with exceptional ardor. Every line of scientific pursuit has its fascinations, but electrical investigation appears to possess a peculiar attraction, for there is no experiment or observation of any kind in the domain of this wonderful science which would not forcibly appeal to us. Yet to me it seems, that of all the many marvelous thins we observe, ti vacuum tube, excited by an electric impulse from a distant source, bursting forth out of the darkness and illuminating the room with its beautiful light, is as lovely' a phenomenon as can greet our eyes. More interesting still it appears when, reducing the fundamental discharges across the gap to a very small number and waving the tube about we produce all hinds of designs in luminous lines. So by way of amusement I take a straight long tube, or a square one, or a square attached to a straight tube, and by whirling them about in the hand, I imitate the spokes of a wheel, a Gramme winding, a drum winding, an alternate current motor winding, etc (Fig. 34). Viewed from a distance the effect is weak and much of its beauty is lost, but being near or holding the tube in the hand, one cannot resist its charm.

In presenting these insignificant results I have not attempted to arrange and coordinate them, as would be proper in a strictly scientific investigation, in which every succeeding result should be a logical sequence of the preceding, so that it might be guessed in advance by the careful reader or attentive listener. I have preferred to concentrate my energies chiefly upon advancing novel facts or ideas which might serve as suggestions to others, and this may serve as an excuse for the lack of harmony. The explanations of the phenomena have been given in good faith and in the spirit of a student prepared to find that they admit of a better interpretation. There can be no great harp t in a student taking an erroneous view, but when great minds err, the world must dearly pay for their mistakes.

The Physiological and Other Effects of High Frequency Currents
The Electrical Engineer — Feb. 11, 1893

In *the Electrical Engineer*, of January 25,1893, I note an article by Mr. A. A. C. Swinton, referring to my experiments with high frequency currents. Mr: Swinton uses in these experiments the method of converting described by me in my paper before the American Institute of Electrical Engineers, in May, 1891, and published in *the Electrical Engineer* of July 8, 1891, which has since been employed by a number of experimenters; but it has somewhat surprised me to observe that he makes use of an ordinary vibrating contact-breaker, whereas he could have employed the much simpler method of converting continuous currents into alternating currents of any frequency which was shown by me two years ago. This method does not involve the employment of any moving parts, and allows the experimenter to vary the frequency et will by simple adjustments. I had thought that most electricians were at present familiar with this mode of conversion which possesses many beautiful features.

The effects observed by Mr. Swinton are not new to rue and they might have been anticipated by those who have carefully read what I !rave written on the subject. But I cannot agree with some of the views expressed by him.

First of all, in regard to the physiological effects. I have made a clear statement at the beginning of my published studies, and my continued experience :with these currents has only further strengthened me is the opinion then expressed. I stated in my paper, before mentioned, that it is an undeniable fact that currents of very high frequency are less injurious than the low frequency currents, :but I have also taken care to prevent the idea from gaining ground that these currents are absolutely harmless, as will be evident, from the following quotation: " If received directly from a machine or from a secondary of low resistance, they (high frequency currents) produce more or less powerful effects, and may, cause serious injury, especially when used in conjunction with condensed." This refers to currents of ordinary' potential differences such as are used in general commercial practice.

As regards the currents of very high potential differences, which were employed an my, experiments, I have never considered the current's strength, but the energy which the human body eras capable of receiving without injury, and I have expressed this quite clearly on more than one occasion. For instance, I stated teat "the higher, the frequency the greater the amount of electrical energy which may be passed through the body without serious discomfort." And on another occasion when a high tension coil was short-circuited though the body of the experimenter I stated that the immunity was due to the fact that less energy was available externally to the coil when the experimenter's body joined the terminals. This is practically what Mr. Swinton expresses in another way; namely, by saying that with "high frequency currents it is possible to obtain effects with exceedingly small currents," etc.

In regard to the experiments, with lamp filaments, I have, I believe, expressed myself with equal clearness. I have pointed out come phenomena of impedance which at that time (1891) were considered very striating, and I have also pointed out the great importance of the rarefied gas surrounding the filament when we have to deal with currents of such high frequency. The heating of the filament by s comparatively small current is not, as Mr. Swinton thinks, due to its impedance or increased ohmic resistance, but principally to the presence of rarefied gas in the bulb. Ample evidence of the truth of this can be obtained in vary many experiments, and to cite them would be merely lengthening this communication unduly.

Likewise, observations made when the experimenter's body was included in the path of the discharge, are, in my opinion, not impedance, but capacity, phenomena. The spark between the hands is the shorter, the larger the surface of the body, and no spark whatever would, be, obtained if the surface of the body were sufficiently large.

I would here point out that one is apt to fall into the error of supposing that the spark which is produced between two points on a conductor, not very distant from each other, is due to the impedance of the conductor. This is certainly the case when the current is of considerable strength. But when there is a vibration along a wire which is constantly maintained, and the current is inappreciable whereas the potential at the coil terminal is exceedingly high, then lateral dissipation comes into play prominently. There is then owing to this dissipation, a rapid fall of potential along the wire and high potential differences may exist between points only a short distance apart This is of course not to be confounded with those differences of potential observed between points when there are fixed wares with ventral and nodal points maintained on a conductor. The lateral dissipation, and not the skin effect, is, I think, the reason why so great an amount of energy may be passed into the body of a person without causing discomfort.

It always affords me great pleasure to note; that something which I have suggested is being employed for some instructive or practical purpose; but I may be pardoned for mentioning that other observations made by Mr. Swinton, and by other experimenters, have recently been brought forward as novel, and arrangements of apparatus which I have suggested have been used repeatedly by some who apparently are in complete ignorance of what I have done in this direction.

From Nikola Tesla - He Writes about His Experiments in Electrical Healing
The Detroit Free Press — Feb. 16, 1896

Some weeks ago this journal published an interesting article concerning electrical oscillations as observed by the eminent scientist, Nicola Tesla. So much interest was shown in the subject that Mr. Tesla was appealed to directly and in response to that appeal he sends to *The Detroit Free Press* this open letter:

Nos. 46 & 48 E. Houston Street
New York, February 10, 1896

During the past few weeks I have received so many letters concerning the same subject that it was entirely beyond my power to answer all of them individually. In view of this I hope that I shall be excused for the delay, which I must regret, in acknowledging the receipt, and also for addressing this general communication in answer to all inquiries.

The many pressing demands which have been made upon me in consequence of exaggerated statements of the journals have painfully impressed me with the fact that there are a great many sufferers, and furthermore that nothing finds a more powerful echo than a promise held out to improve the condition of the unfortunate ones.

The members of the medical fraternity are naturally more deeply interested in the task of relieving the suffering from their pain, and, as might be expected, a great many communications have been addressed to me by physicians. To these chiefly this brief statement of the actual facts is addressed.

Some journals have confounded the physiological effects of electrical oscillations with those of mechanical vibrations, this being probably due to the circumstance that a few years ago I brought to the attention of the scientific men some novel methods and apparatus for the production of electrical oscillations which, I learn, are now largely used in some modification or other in electro-therapeutic treatment and otherwise. To dispel this erroneous idea I wish to state that the effects of purely mechanical vibrations which I have more recently observed, have nothing to do with the former.

Mechanical vibrations have often been employed locally with pronounced results in the treatment of diseases, but it seems that the effects I refer to have either not been noted at all, or if so, only to a small degree, evidently because of the insufficiency of the means which have eventually been employed in the investigations.

While experimenting with a novel contrivance, constituting in its simplest form a vibrating mechanical system, in which from the nature of the construction the applied force is always in resonance with the natural period, I frequently exposed my body to continued mechanical vibrations. As the elastic force can be made as large as desired, and the applied force used be very small, great weights, half a dozen persons, for instance, may be vibrated with great rapidity by a comparatively small apparatus.

I observed that such intense mechanical vibrations produce remarkable physiological effects. They affect powerfully the condition of the stomach, undoubtedly promoting the process of digestion and relieving the feeling of distress, often experienced in consequence of the imperfect function of the organs concerned in the process. They have a strong influence upon the liver, causing it to discharge freely, similarly to an application of a catharic. They also seem to affect the glandular system, notable in the limbs; also the kidneys and bladder, and more or less influence the whole body. When applied for a longer period they produce a feeling of immense fatigue, so that a profound sleep is induced.

The excessive tiring of the body is generally accompanied by nervous relaxation, but-there seems to be besides a specific action on the nerves.

These observations, though incomplete, are, in my own limited judgment, nevertheless positive and unmistakable, and in view of this and of the importance of further investigation of the subject by competent men I prepared about a year ago a machine with suitable adjustments for varying the frequency and amplitude of the vibrations, intending to give it to some medical faculty for investigation. This machine, together with other apparatus, was unfortunately destroyed by fire a year ago, but will be reconstructed as soon as possible.

In making the above statements I wish to disconnect myself with the extraordinary opinions expressed in some journals which I have never authorized and which, though they may have been made with good intent, cannot fail to be hurtful by giving rise to visionary expectations.

Yours very truly,
N. Tesla

On Roentgen Rays
Electrical Review — March 11, 1896

One can not help looking at that little bulb of Crookes with a feeling akin to awe, when he considers all that it has done for scientific progress — first, the magnificent results obtained by its originator; next, the brilliant work of Lenard, and finally the wonderful achievements of Roentgen. Possibly it may still contain a grateful Asmodeus, who will be lot out of his narrow prison cell by a lucky student. At times it has seemed to me as though I myself heard a whispering voice, and I have searched eagerly among my dusty bulbs and bottles. I fear my imagination has deceived me, but there they are still, my dusty bulbs, and I am still listening hopefully.

After repeating Professor Roentgen's beautiful experiments, I have devoted my energies to the investigation of the nature of the radiations and to the perfecting of the means for their production. The following is a brief statement which, I hope, will be useful, of the methods employed and of the most notable results arrived at in these two directions.

In order to produce the most intense effects we have first to consider that, whatever their nature, they depend necessarily on the intensity of the cathode streams. These again being dependent on the magnitude of the potential, it follows that the highest attainable electrical pressure is desirable.

To obtain high potentials we may avail ourselves of an ordinary induction coil, or of a static machine, or of a disruptive discharge coil. I have the impression that most of the results in Europe have been arrived at through the employment of a static machine or Ruhmkorff coil. But since these appliances can produce only a comparatively small potential, we are naturally thrown on the use of the disruptive discharge coil as the most effective apparatus. With this .there is practically no limit to the spark length, and the only requirement is that the experimenter should possess a certain knowledge and skill in the adjustments of the circuits, particularly as to resonance, as I have pointed out in my earlier writings on this subject.

After constructing a disruptive coil suitable for any kind of current supply, direct or alternating the experimenter comes to the consideration as to what kind of bulb to employ. Clearly, if we put two electrodes in a bulb, or use one inside and another outside electrode, we limit the potential, for the presence not only of the anode but of any conducting object has the effect of reducing the practicable potential on the cathode. Thus, to secure the result aimed at, one is driven to the acceptance of a single electrode bulb; the other terminal being as far remote as possible.

Obviously, an inside electrode should be employed to get the highest velocity of the cathode streams, for the bulbs without inside terminals are much less efficient for this special object in consequence of the loss through the glass. A popular error seems to exist in regard to the concentration of the *rays* by concave electrodes. This, if anything, is a disadvantage. There are certain specific arrangements of the disruptive coil and circuits, condensers and static screens for the bulb, on which I have given full particulars on previous occasions.

Having selected the induction apparatus and type of bulb, tie next important consideration is the vacuum. On this subject I am able to make known a fact with which I have long been acquainted, and of which I have taken advantage in the production of vacuum jackets and all sorts of incandescent bulbs, and which I subsequently found to be of the utmost importance,. not to say essential, for the production of intense Roentgen shadows. I refer to a method of rarefaction by electrical means to any degree desirable far beyond that obtainable by mechanical appliances.

Though this result can be reached by the use of a static machine as, well as of an ordinary induction coil giving a sufficiently high potential, I have found that by far the most suitable apparatus, and one which secures the quickest action, is a disruptive coil: It is best to proceed in this way: The bulb is first exhausted by means of an ordinary vacuum pump to a rather high degree, though my experiences have shown that this is not absolutely necessary, as I have also found it possible to rarefy, beginning from low pressure. After being taken down from the pump, the bulb is attached to the terminal of the disruptive coil, preferably of high frequency of vibration, and usually the following phenomena are noted. First, there is a milky light spreading through the bulb, or possibly for a moment the glass becomes phosphorescent, if the bulb has been exhausted to a high degree. At any rate, the phosphorescence generally subsides quickly and the white light settles around .the electrode, whereupon a dark space forms at some distance from the latter. Shortly afterward the light assumes a reddish. color and the terminal grows very hot. This heating, however, is observed only with powerful apparatus. It is well to watch the bulb carefully and regulate the potential at this stage, as the electrode might be quickly consumed.

After some time the reddish light subsides, the streams becoming again white, whereupon they get weaker and weaker, wavering around the electrode until they finally disappear: Meanwhile, the phosphorescence of the glass grows more and more intense, and the spot where the stream strikes the wall becomes very hot, while the phosphorescence around the electrode ceases and the latter cools down to such an extent that the glass near it may be actually ice-cold to the touch. The gas in the bulb has then reached the required degree of rarefaction. The process may be hastened by repeated heating arid cooling and by the employment of a small electrode. It should be added that bulbs with external electrodes may be treated in the same way. It may be also of interest to state that under certain conditions, which I am investigating more closely, the pressure of the gas ,in a vessel may be augmented by electrical means.

I believe that the disintegration of the electrode, which invariably takes. place, is connected with a notable diminution of the temperature. From the point on, when the electrode gets cool, the bulb is in a very good condition for producing the Roentgen shadows. Whenever the electrode is equally, if not hotter than the glass, it is a sure indication that the vacuum is not high enough, or else that the electrode is too small. For very effective working, the inside surface of the wall, where the cathode stream strikes, should appear as if the glass were in a fluid state.

As a cooling medium I have found best to employ jets of cold air. By this means it is possible to operate successfully a bulb with a very thin wall, while the passage of the rays is not materially impeded.

I may state here that the experimenter need not be deterred from using a glass bulb, as I believe the opacity of glass as well as the transparency of aluminum are somewhat exaggerated, inasmuch as I have found that a very thin aluminum sheet throws a marked shadow, while, on the other hand, I have obtained impressions through a thick glass plate.

The above method is valuable not only as a means of obtaining the high vacua desired, but it is still more important, because the phenomena observed throw a light on the results obtained by Lenard and Roentgen.

Though the phenomenon of rarefaction under above conditions admits of different interpretations, the chief interest renters on one of them, to which I adhere — that is, on the actual expulsion of the particles through the walls of the bulb. I have lately observed that the latter commences to act properly upon the sensitive plate only from the point when the exhaustion begins to be noticeable, and the effects produced are the strongest when the process of exhaustion is most rapid, even though the phosphorescence might not appear particularly bright. Evidently, then, the two effects are closely connected, and I am getting more and more convinced that we have to deal with a stream of material particles, which strike the sensitive plate with great velocities. Taking as a basis the estimate of .Lord Kelvin on the speed of projected 'particles in a Crookes' bulb, we arrive easily by the employment of very high potentials to speeds of as much as a hundred kilometers a second. Now, again, the old question arises: Are the particles from the electrode ,or from the charged surface generally, including the case of an external electrode, projected through the glass or aluminum walls, or do they merely hit the .inner surface and cause particles from the outside of the wall to fly off, acting in a purely mechanical way, as when a row of ivory balls is struck? So far, most of the phenomena indicate that they are projected through the wall of the bulb, of whatever material it may be, and I am seeking for still more conclusive evidence in this direction.

It may not be known that even an ordinary streamer, breaking out suddenly and under great pressure from the terminal of a disruptive coil, passes through a thick glass plate as though the latter were not present. Unquestionably, with such coils pressures are practicable which will, project the particles in straight lines even under atmospheric pressure. I have obtained distinct impressions in free air, not by streamers, as some experimenters have done, using static machines or induction coils, but by actual projection, the formation of streamers being absolutely prevented by careful static screening.

A peculiar thing about the Roentgen rays is that from low frequency to the highest obtainable there seems to be no difference in the quality of the effects produced, except that they are more intense when the frequency is higher, which is very likely due to the fact that in such case the maximum pressures on the cathode are likewise higher. This is only possible on the assumption that the effects on the sensitive plate are due to projected particles, or else to vibrations far beyond any frequency which we are able to obtain by means of condenser discharges. A powerfully excited bulb is enveloped in a cloud of violet light, extending for more than a foot around it, but outside of this visible phenomenon there is no positive evidence of the existence of waves similar to those of light. On the other hand, the fact that the opacity bears some proportion to the density of the substance speaks strongly for material streams, and the same may be said of the effect discovered by Prof. J. J. Thomson. It is to be hoped that all doubts will shortly be dispelled.

A valuable evidence of the nature of the radiations and progress in the direction of obtaining strong impressions on the plate might be arrived at by perfecting plates especially sensitive to mechanical shock or impact. There are chemicals suitable for this, and the development in this direction may lead to the abandonment of the present plate. Furthermore, if we have to deal with stream. of material particles, it seems not impossible to project upon the plate a suitable substance to insure the best chemical action.

With apparatus as I have described, remarkable impressions on the plate are produced. An idea of the intensity of the effects may be gained when I mention that it is easy to obtain shadows with comparatively short exposures at distances of many feet, while at small distances and with thin objects, exposures of a few seconds are practicable. The annexed print is a shadow of a copper wire projected at a distance of 11 feet through a wooden cover over the sensitive plate. This was the first shadow taken with my improved apparatus in my laboratory. A similar impression was obtained through the body of the experimenter, a plate of glass; nearly three-sixteenths of an inch thick, a thickness of wood of fully two inches and through a distance of about four feet. I may remark, however, that when these impressions were taken, my apparatus was working under extremely unfavorable conditions, which admitted of so great improvements that I am hopeful to magnify the effects many times.

The bony structure of birds, rabbits and the like is shown within the least detail, and even the hollow of the bones is clearly visible. In a plate of a rabbit under exposure of an hour, not only every detail of the skeleton is visible, but likewise a clear outline of the abdominal cavity and the location of the lungs, the fur and many other features. Prints of even large birds show the feathers quite distinctly.

Clear shadows of the bones of human limbs are obtained by exposures ranging from a quarter of an hour to an hour, and some plates have shown such an amount of detail that it is almost impossible to believe that we have to deal with shadows only. For instance, a picture of a foot with a shoe on it was taken, and every fold of the leather, trousers, stocking, etc., is visible, while the flesh and bones stand out sharply. Through the body of the experimenter the shadows of small buttons and like objects are quickly obtained, while with an exposure of from one to one and a half hour the ribs, shoulder-bones and the bones of the upper arm appear dearly, as is shown in the annexed print. It is now demonstrated beyond any doubt that small metallic objects or bony or chalky deposits can be infallibly detected in any part of the body.

An outline of the skull is easily obtained with an exposure of 20 to 40 minutes. In one instance an exposure of 40 minutes gave dearly not only the outline, but the cavity of the eye, the chin and cheek and nasal bones, the lower jaw and connections to the upper one, the vertebral column and connections to the skull, the flesh and even the hair. By exposing the head to a powerful radiation strange effects have been noted. For instance, I find that there is a tendency to sleep and the time seems to pass away quickly. There is a general soothing effect, and I have: felt a sensation of warmth in the upper part of the head. An assistant independently confirmed the tendency to sleep and a quick lapse of time. Should these remarkable effects be verified by men with keener sense of observation, I shall still more firmly believe in the existence of material streams penetrating the skull. Thus it may be possible by these strange appliances to project a suitable chemical into any part of the body.

Roentgen advanced modestly his results, warning against too much hope. Fortunately his apprehensions were groundless, for, although we have to all appearance to deal with mere shadow projections, the possibilities of the application of his discovery are vast. I am happy to have contributed to the development of the great art he has created.

On Roentgen Rays the Latest Results
Electrical Review — March 18, 1896

To the Edition of Electrical Review

Permit me to say that I was slightly disappointed to note in your issue of March 11 the prominence you have deemed to accord to my youth and talent, while the ribs and other particulars of Fig. 1, which, with reference to the print accompanying my communication, I described as clearly visible, were kept modestly in the background: I also regretted to observe an error in one of the captions, the more so; as I must ascribe it to my own text. I namely stated on page 135, third column, seventh line: "A similar impression was obtained through the body of the experimenter, etc., through a distance of four feet." The impression here referred to was a similar one to that shown in Fig. 2, whereas the shadow in Fig. 1 was taken through a distance of 18 inches. I state this merely for the sake of correctness of my communication, but, as far as the general truth of the fact of taking such a shadow at the distance given is concerned, your caption might as well stand, for I am producing strong shadows at distances of 40 *feet*. I repeat, 40 feet and even more. Nor is ,this all. So strong are the actions on the film that provision must be made to guard the plates in my photographic a department, located on the floor above, a distance of fully :60 feet, from being spoiled by long exposure to the stray rays. Though during my investigations I have performed many experiments which seemed extraordinary, I am deeply astonished observing these unexpected manifestations, and still more so, as even now I see before me the possibility, not to say certitude, of augmenting the effects with my apparatus at least tenfold! What may we then expect? We have to deal here, evidently, with a radiation of astonishing power, and the inquiry into its nature becomes more and more interesting and important. Here is an unlooked-for result of an action which, though wonderful in itself, seemed feeble and entirely incapable of such expansion, and affords a good example of the fruitfulness of original discovery. These effects upon the sensitive plate at so great a distance I attribute to the employment of a bulb with a single terminal, which permits the use of practically any desired potential and the attainment of extraordinary speeds of the projected particles. With such a bulb it is also evident that the action upon a fluorescent screen is proportionately greater than when the usual kind of tube is employed, and I have already observed enough to feel sure that great developments are to be looked for in this direction. I consider Roentgen's discovery; of enabling us to see, by the use of a fluorescent screen, through an opaque substance, even a more beautiful one than the recording upon the plate.

Since my previous communication to you I have made considerable progress, and can presently announce one more result of importance. I have lately obtained shadows *by reflected rays only,* thus demonstrating beyond doubt that the Roentgen rays possess this property. One of the experiments may be cited here. A thick copper tube, about a foot long, was taken and one of its ends tightly closed by the plate-holder containing a sensitive plate, protected by a fiber cover as usual. Near the open end of the copper tube was placed a thick plate of glass at an angle of 45 degrees to the axis of the tube. A single-terminal bulb was then suspended above the glass plate at a distance of about eight inches, so that the bundle of rays fell upon the latter at in angle of 45 degrees, and the supposedly reflected rays passed along the axis of the copper tube. An exposure of 45 minutes gave a dear and sharp shadow of a metallic object. This shadow was produced by the reflected rays, as the direct action was absolutely excluded, it having been demonstrated that even under the severest tests with much stronger actions no impression whatever could be produced upon the film through a thickness of copper equal to that of the tube. Concluding from the intensity of the action by comparison with an equivalent effect due to the direct rays, I find that approximately two per cent of the latter were reflected from the glass plate in this experiment. I hope to be able to report shortly and more fully on this and other subjects.

In my attempts to contribute my humble share to the knowledge of the Roentgen phenomena, I am finding more and more evidence in support of the theory of moving material particles. It is not my intention, however, to advance at present any view as to the bearing of such a fact upon the present theory of light, but I merely seek to establish the fact of the existence of such material streams in so far as these isolated effects are concerned. I have already a great many indications of a bombardment occurring outside of the bulb, and I am arranging some crucial tests which, I hope, will be successful. The calculated velocities fully account for actions at distances of as much as 100 feet from the bulb, and that the projection through the glass takes place seems evident from the process of exhaustion, which I have described in my previous communication. An experiment which is illustrative in this respect, and which I intended to mention, is the following: If we attach a fairly exhausted bulb containing an electrode to the terminal of a disruptive coil, we observe small streamers breaking through the sides of the glass. Usually such a streamer will break through the seal and crack the bulb, whereupon the vacuum is impaired; but, if the seal is placed above the terminal, or if some other provision is made to prevent the streamer from passing through the glass at that point, it often occurs that the stream breaks out through the side of the bulb, producing a fine hole. Now, the extraordinary thing is that, in spite of the connection to the outer atmosphere, the air can not rush into the bulb as long as the hole is very small. The glass at the place where the rupture has occurred may grow very hot — to such a degree as to soften; but it will not collapse, but rather bulge out, showing that a pressure from the inside greater than that of the atmosphere exists. On frequent occasions I have observed that the glass bulges out and the hole, through which the streamer rushes out, becomes so large. as to be perfectly discernible to the eye. As the matter is expelled from the bulb the rarefaction increases and the streamer becomes less and less intense, whereupon the glass doses again, hermetically sealing the opening. The process of rarefaction, nevertheless, continues, streamers being still visible on the heated place until the highest degree of exhaustion is reached, whereupon they may disappear. Here, then, we have a positive evidence that matter is being expelled through the walls of the glass.

When working with highly strained bulbs I frequently experience a sudden, and sometimes even painful, shock in the eye. Such shocks may occur so often that the eye gets inflamed, and one can not be considered overcautious if he abstains from watching the bulb too closely. I see in these shocks a further evidence of larger particles being thrown off from the bulb.

Tesla's Latest Results: He Now Produces Radiographs At A Distance of More Than Forty Feet
Electrical Review - N. Y. — *March 18, 1896*

To The Editor of Electrical Review:

Permit me to say that I was slightly disappointed to note in your issue of Mar. 11 the prominence you have deemed to accord to my youth and talent, while the ribs and other particulars of which, with reference to the print accompanying my communication, I described as clearly visible, were kept modestly in the background. I also regretted to observe an error in one of the captions, the more so, as I must ascribe it to my own text. I namely stated on page 135, third column, seventh line: "A similar impression was obtained through the body of the experimenter, etc., through a distance of four feet." The general truth of the fact of taking such a shadow at the distance given is concerned, your caption might as well stand, for I am producing strong shadows at distances of 40 feet. I repeat, 40 feet and even more. Nor is this all. So strong are the actions on the film that provisions must be made to guard the plates in my photographic department, located on the floor above, a distance of fully 60 feet, from being spoiled by long exposure to the stray rays. Though during my investigations I have performed many experiments which seemed extraordinary, I am deeply astonished observing these unexpected manifestations, and still more so, as even now I see before me the possibility, not to say certitude, of augmenting the effects with my apparatus at least tenfold! What may we then expect? We have to deal here, evidently, with a radiation of astonishing power, and the inquiry into its nature becomes more and more interesting and important. Here is an unlooked-for result of an action which, though wonderful in itself, seemed feeble and entirely incapable of such expansion, and affords a good example of the fruitfulness of original discovery. These effects upon the sensitive plate at so great a distance I attribute to the employment of a bulb with a single terminal, which permits the use of practically any desired potential and the attainment of extraordinary speeds of the projected particles. With such a bulb it is also evident that the action upon a fluorescent screen is proportionately greater than when the usual kind of tube is employed, and I have already observed enough to feel sure that great developments are to be looked for in this direction. I consider Roentgen's discovery, of enabling us to see, by the use of a fluorescent screen, through an opaque substance, even a more beautiful one than the recording upon the plate.

Since my previous communication to you I have made considerable progress, and can presently announce one more result of importance. I have lately obtained shadows by reflected rays only, thus demonstrating beyond doubt that the Roentgen rays possess this property. One of the experiments may be cited here. A thick copper tube, about a foot long, was taken and one of its ends tightly closed by the plateholder containing a sensitive plate, protected by a fiber cover as usual. Near the open end of the copper tube was placed a thick plate of glass at an angle of 45 degrees to the axis of the tube. A single-terminal bulb was then suspended above the glass plate at a distance of about eight inches, so that the bundle of rays fell upon the latter at an angle of 45 degrees, and the supposedly reflected rays passed along the axis of the copper tube. An exposure of 45 minutes gave a clear and sharp shadow of a metallic object. This shadow was produced by the reflected rays, as the direct action was absolutely excluded, it having been demonstrated that even under the severest tests with much stronger actions no impression whatever could be produced upon the film through a thickness of copper equal to that of the tube. Concluding from the intensity of the action by comparison with an equivalent effect due to the direct rays, I find that approximately two per cent of the latter were reflected from the glass plate in this experiment. I hope to be able to report shortly and more fully on this and other subjects.

In my attempts to contribute my humble share to the knowledge of the Roentgen phenomena, I am finding more and more evidence in support of the theory of moving material particles. It is not my intention, however, to advance at present any view as to the bearing of such a fact upon the present theory of light, but I merely seek to establish the fact of the existence of such material streams in so far as these isolated effects are concerned. I have already a great many indications of a bombardment occurring outside of the bulb, and I am arranging some crucial tests which, I hope, will be successful. The calculated velocities fully account for actions at distances of as much as 100 feet from the bulb, and that the projection through the glass takes place seems evident from the process of exhaustion, which I have described in my previous communication. An experiment which is illustrative in this respect, and which I intended to mention, is the following; If we attach a fairly exhausted bulb containing an electrode to the terminal of a disruptive coil, we observe small streamers breaking through the side of the glass. Usually such a streamer will break through the seal and crack the bulb, whereupon the vacuum is impaired; but, if the seal is placed above the terminal, or if some other provision is made to prevent the streamer from passing through the glass at that point, it often occurs that the stream breaks out through the side of the bulb, producing a fine hole. Now, the extraordinary thing is that, in spite of the connection to the outer atmosphere, the air can not rush into the bulb as long as the hole 'is very small. The glass at the place where the rupture has occurred may grow very hot — to such a degree as to soften; but it will not collapse, but rather bulge out, showing that a pressure from the inside greater than that of the atmosphere exists. On frequent occasions I have observed that the glass bulges out and the hole, through which the streamer rushes out, becomes so large as to be perfectly discernible to the eye. As the matter is expelled from the bulb the rarefaction increases and the streamer becomes less and less intense, whereupon the glass closes again, hermetically sealing the opening. The process of rarefaction, nevertheless, continues, streamers being still visible on the heated place until the highest degree of exhaustion is reached, whereupon they may disappear. Here, then, we have a positive evidence that matter is being expelled through the walls of the glass.

When working with highly strained bulbs I frequently experience a sudden, and sometimes even painful, shock in the eye. Such shocks may occur so often that the eye gets inflamed, and one can not be considered over-cautious if he abstains from watching the bulb too closely. I see in these shocks a further evidence of larger particles being thrown off from the bulb.

Nikola Tesla.
New York, March 14.

On Reflected Roentgen Rays
Electrical Review — April 1, 1896

In previous communications in regard to the effects discovered by Roentgen, I have confined myself to giving barely a brief outline of the most noteworthy results arrived at in the course of my investigations. To state truthfully, I have ventured to express myself, the first time, after some hesitation and consequent delay, and only when I had gained the conviction that the information I had to convey was a needful one; for, in common with others, I was not quite able to free myself of a certain feeling which one must experience when he is trespassing on ground not belonging to him. The discoverer would naturally himself arrive at most of the facts in due time, and a courteous restraint in the announcement of the results on the part of his co-workers would not be amiss. How many have sinned against me by proclaiming their achievements just as I was good and ready to do it myself! But these discoveries of Roentgen, exactly of the order of the telescope and microscope, his seeing through a great thickness of an opaque substance, his recording on a sensitive plate of objects otherwise invisible, were so beautiful and fascinating, so full of promise, that all restraint was put aside, and every one abandoned himself to the pleasures of speculation and experiment. Would but every new and worthy idea find such an echo! One single year would then equal a century of progress. A delight it would be to live in such age, but a discoverer I would not wish to be.

Amongst the facts, which I have had the honor to bring to notice, is one claiming a large share of scientific interest, as well as of practical importance. I refer to the demonstration of the property of reflection, on which I have dwelt briefly. '

Having had opportunities to make many observations during my experience with vacuum bulbs and tubes, which could not be accounted for in any plausible way on any theory of vibration as far as I could judge, I began these investigations — disinclined, but expectant to find that the effects produced are due to a stream of material particles. I had many evidences of the existence of such streams. One of these I mentioned, describing the method of electrically exhausting a tube. Such exhaustion, I have found, takes place much quicker when the glass is very thin than when the walls are thick, I presume because of the easier passage of the ions. While a few minutes are sufficient when the glass is very thin, it often takes half an hour or more if the glass be thick or the electrode very large. In accordance with this idea I have, with a view of obtaining the most efficient action, selected the apparatus, and have found at each step my supposition confirmed and my conviction strengthened.

A stream of material particles, possessing a great velocity, must needs be reflected, and I was therefore quite prepared — assuming my original idea to be true — to demonstrate sooner or later this property. Considering that the reflection should be the more complete the smaller the angle of incidence, I adopted from the outset of my investigations a tube or bulb b of the form shown in Fig. 1. It was made of very thick glass, with a bottom blown as thin as possible, with the two obvious objects of restricting the radiation to the sides and facilitating the passage through the bottom. A single electrode e, in the form of a round disk of a diameter slightly less than that of the tube, was placed about an inch below the narrow neck n on the top. The leading-in conductor c was provided with a long wrapping w, so as to prevent cracking, by the formation of sparks at the point where the wire enters the bulb. It was found advantageous for a number of reasons to extend the wrapping a good distance beyond the neck, on the inside and outside as well, and to place the seal-off in the narrow neck. On other occasions I have dwelt on the employment of an

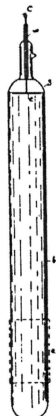

electrostatic screen in connection with such single-terminal bulbs. In the present instance the screen was preferably formed by a bronze paintings, slightly above the aluminum electrode and extending to just a little below the wrapping of the wire, so as to allow seeing constantly the end of the wrapping. Or else a small aluminum plate s, Fig. 2, was supported in the inside of the bulb above the electrode. This static screen practically doubles the effect, as it prevented all action above it. Considering, further, that the radiation sideways was restricted by the use of a very thick glass and most of it was thrown to the bottom by reflection, as I then surmised, it became evident that such a tube should prove much more efficient than one of ordinary form. Indeed, I quickly found that its power upon the sensitive plate was very nearly four times as great as that of a spherical bulb with an equivalent area of impact. This kind of tube is also very well adapted for use with two terminals by placing an external electrode e_1 as indicated by the dotted lines in Fig. 1. When the glass is taken thick the stream is sensibly parallel and concentrated. Furthermore, by making the tube as long as one desired, it was possible to employ very high potentials, otherwise impracticable with short bulbs.

The use of high potentials is of great importance, as it allows shortening considerably the time of exposure, and affecting the plate at much greater distances. I am endeavoring to determine more exactly the relation of the potential to the effect produced upon the sensitive plate. I deem it necessary to remark that the electrode should be of aluminum, as a platinum electrode, which is still persistently employed, gives inferior results and the bulb is disabled in comparatively short time. Some experimenters might find trouble in maintaining a fairly constant vacuum, owing to a peculiar process of absorption in the bulb, which has been pointed out early by Crookes, in consequence of which, by continued use, the vacuum may increase. A convenient way to prevent this I have found to be the following: The screen or aluminum plate s, Fig. 2, is placed directly upon the wrapping of the leading-in conductor c, but some distance back from the end. The right distance can be only determined by experience. If it is properly chosen, then, during the action of the bulb, the wrapping gets warmer, and a small bright spark jumps from time to time from the wire c to the aluminum plate s through the wrapping w. The passage of this spark causes gases to be formed; which slightly impair the vacuum; and in this manner, by a little skillful manipulation, the proper vacuum may be constantly maintained. Another way of getting the same result in a tube shown in Fig. 1 is to extend the wrapping so far inside that, when the bulb is' normally working, the wrapping is heated sufficiently to free gases to the required amount. It is for this purpose convenient to let the screen of bronze painting $.s$ extend just a little below the wrapping, so that the spark may be observed. There are, however, many other ways of overcoming this difficulty, which may cause some annoyance to those working with inadequate apparatus:

In order to insure the best action the experimenter should note the various stages which I have pointed out before, and through which the bulb has to pass during the process of exhaustion. He will first observe that when the Crookes phenomena show themselves most prominently there is a reddish streamer issuing from the electrode, which in the beginning covers the latter almost entirely. Up to this point the bulb practically does not affect the sensitive plate, although the glass is very hot at the point of impact. Gradually the reddish streamer disappears, and just before it ceases to be visible the bulb begins to show better action; but still the effect upon the plate is very weak. Presently a white or even bluish stream is observed, and after some time the glass on the bottom of the bulb gets a glossy appearance. The heat is still more intense and the phosphorescence through the entire bulb is extremely brilliant: One should think that such a bulb must be effective, but appearances are often deceitful, and the beautiful bulb still does not work. Even when the white. or bluish stream ceases, and the glass on the bottom is so hot as to be nearly melting, the effect on the plate is very weak. But at this stage there appears suddenly at the bottom, of the tube a star-shaped changing design, as if the electrode would throw off drops of liquid. From this moment on the power of the bulb is tenfold, and at this stage it must always be kept to give the best results:

I may remark, however, that while it may be generally stated the Crookes vacuum is not high enough for the production of the Roentgen phenomena, this is not literally true. Nor are the Crookes phenomena produced at a particular degree of exhaustion, but manifest themselves even with poor vacua, provided the potential is high enough. This is likewise true of the Roentgen effects. Naturally, to verify this, provision must be made not to overheat the bulb when the potential is raised. This is easily done by reducing the number of impulses or their duration, when raising the potential: For such experiments, it will be found of advantage to use in connection with the ordinary induction coil a rotating commutator, instead of a vibrating brake. By changing the speed of the commutator, and also regulating the duration of contact, one is enabled to adjust the conditions to suit the degree of vacuum and potential employed.

In my experiments on reflection, presently considered, I have used the apparatus shown in Fig: 2. It consists of a T-shaped box throughout; of a square cross-section. The walls are mine of lead over one-eighth of an inch thick, which, under the conditions of the experiments, was found to be entirely impervious, even by long exposures to the rays. On the top end was supported firmly the bulb b, inclosed in a glass tube t of thick Bohemian glass; which reached some distance into the lead box. The lower end of the box was tightly closed by a plate-holder P_1, containing the sensitive film p_1, protected as usual.. Finally the side end was closed by a similar plate-holder P, with the sensitive protected film p. To obtain sharp images the objects o and o_1, exactly alike, were placed in the center of the fiber cover, protecting the sensitive plates. In the central portion of the box, provision was made for inserting a plate r of material; the reflective power of which was to be tested; and the dimensions of the box were such that the reflected ray and the direct one had to go through the same distance, the reflecting plate being at an angle of 45 degrees to the incident as well as reflected ray. Care was taken to exclude all possibility of action upon the plate p, except by reflected rays, and the reflecting plate r was made to fit tight all around in the lead box, so that no rays could reach the film p_1, except by passing through the plate to be tested. In my earliest experiments on reflection I observed only the effects of reflected rays, but in this instance, on the suggestion of Prof, Wm. A. Anthony, I provided the above means for simultaneously examining the action of the direct rays, which eventually passed through the reflecting plate. In this manner it was possible to compare the amount of .the transmitted and reflected radiation. The glass tube t surrounding the bulb b served to render the stream parallel and more intense. By taking impressions at various distances I found that through a considerable distance there was but little spreading of the bundle of rays or stream of particles.

To reduce the error which is caused unavoidably by too long exposures and very small distances, I reduced the exposure to an hour, and the total distance through which the rays had to pass before reaching the sensitive plates was 20 inches, the distance from the bottom of the bulb to the reflecting plate being 13 inches.

It is needless to remark that all the precautions in regard to the sensitive plates — constancy of potential, uniform working of the bulbs, and maintenance of the same conditions in general during these tests have been taken, as far as it was practicable. The plates to be tested were made of uniform size, so as to fit the space provided in the lead box. Of the conductors the following were tested: Brass, toolsteel, zinc, aluminum copper, lead, silver, tin; and nickel, and of the insulators, lead-glass, ebonite, and mica. The summary of the observations is given in the following table:

Reflecting body	Impression by transmitted rays	Impression by reflected rays
Brass.	Strong.	Fairly strong.
Toolsteel.	Barely perceptible.	Very feeble.
Zinc.	None.	Very strong.
Aluminum.	Very strong.	None.
Copper.	None.	Fairly strong, but much less than zinc.
Lead.	None.	Very strong, but a little weaker than zinc.
Silver.	Strong, a thin plate being used.	weaker than copper.
Tin.	None.	Very strong; about like lead.
Nickel:	None.	About like copper.
Lead-glass.	Very strong.	Feeble.
Mica.	Very strong.	Very strong; about like lead.
Ebonite.	Strong.	About like copper.

By comparing, as in previous experiments, the intensity of the impression by reflected rays with an equivalent impression due to a direct exposure of the same bulb and at the same distance — that is, by calculating from the times of exposure under assumption that the action upon the plate was proportionate to the time — the following approximate results were obtained:

Reflecting body	Impression, by direct action	Impression by reflected rays
Brass	100	2
Toolsteel	100	0,5
Zinc	100	3
Aluminum	100	0
Copper	100	2
Lead	100	2,5
Silver	100	1,75
Tin	100	2,5
Nickel	100	2
Lead-glass	100	1
Mica	100	2,5
Ebonite	100	2

While these figures can be but rough approximations, there is, nevertheless, a fair probability that they are correct, in so far as the relative values of the impressions by reflected rays for the various bodies are concerned. Arranging the metals according to these values, and leaving for the moment the alloys or impure bodies out of question, we arrive at the following order: Zinc, lead, tin, copper, silver. The tin appears to reflect fully as well as lead, but, allowing for an error in the observation, we may assume that it reflects less, and in this case we find that this order is precisely the contact series of metals in air. If this proves true we shall be confronted with the most extraordinary fact. Why is zinc, for instance, the best reflector among the metals tested and why, at the same time, is it one of the foremost in the contact series? I have not as yet tried magnesium. The truth is that I was somewhat excited over these results. Magnesium should be even a better reflector than zinc, and sodium still better than magnesium. How can this singular relationship be explained? The only possible explanation seems to me at present that the bulb throws out streams of matter in some primary condition, and that the reflection of these streams is dependent upon some fundamental and electrical property of the metals. This would seem to lead to the inference that these streams must be of uniform electrification; that is, that they must be anodic or cathodic in character, but not both. Since the announcement, I believe in France for the first time, that the streams are anodic, I have investigated the subject and find that I can not agree with this contention. On the contrary, I find that anodic and cathodic streams both affect the plate, and, furthermore, I have been led to the conviction that the phosphorescence of the glass has nothing whatever to do with the photographic impressions. An obvious proof is that such impressions are produced with aluminum vessels when there is no phosphorescence, and, as regards the anodic or cathodic character, the simple fact that we can produce impressions by a luminous discharge excited by induction of a closed vessel, when there is neither anode nor cathode, would seem to dispose effectually of the assumption that the streams are issuing solely from one of the electrodes. It may, perhaps, be useful to point out here a simple fact in relation to the induction coils, which may lead an experimenter into an error. When a vacuum tube is attached to the terminals of an induction coil, both of the terminals are acted upon alike as long as the tube is not very highly exhausted. At a high degree of exhaustion both the electrodes act practically independently, and since they behave as bodies possessing considerable capacity, the consequence is that the coil is unbalanced. If the cathode, for instance, is very large, the pressure on the anode may rise considerably, and if the latter is made smaller, as is frequently the case, the electric density may be many times that on the cathode. It results from this that the anode gets very hot, while the cathode may be cool. Quite the opposite occurs if both of them are made exactly alike. But assuming the above conditions to exist, the hotter anode emits a more intense stream than the cool cathode, since the velocity of the particles is dependent on the electrical density, and likewise on the temperature.

From the previous tests air interesting observation can also be made in regard to the opacity. Far instance, a brass plate one-sixteenth inch thick proved fairly transparent while plates of zinc and copper of the same thickness showed themselves to be entirely opaque.

Since I have investigated reflection and arrived to results in this direction, I have been able to produce stronger effects by employing proper reflectors. By surrounding a bulb with a very thick glass tube the effect may be augmented very considerably. The employment of a zinc reflector in one instance showed an increase of about 40 per cent in the impression produced. I attach great practical value to the employment of proper reflectors, because by means of them we can employ any quantity of bulbs, and so produce any intensity of radiation required.

One disappointment in the course of these investigations has been the entire failure of my efforts to demonstrate refraction. I have employed lenses of all kinds and tried a great many experiments, but could not obtain any positive result.

On Roentgen Radiations
Electrical Review — April 8, 1896

Having observed the unexpected behavior of the various metals in regard to the reflection of these radiations, (see Electrical Review of April 1, 1896) I have endeavored to settle several still doubtful points. As, for the present, it appeared. chiefly desirable to establish the exact order of the metals, or conductors, in regard to their powers of reflection, leaving for further investigation the determination of the magnitude of the effects, I modified slightly the apparatus and procedure described in my communication just referred to. The reflecting plates were not made each of one metal, as before, but of two metals, the reflective power of which was to be compared. This was done by fastening upon a plate of lead the two metal plates to be investigated, so that the reflecting surface was divided in two halves by the joining line. Furthermore, to prevent any spreading and mingling of the rays reflected from both halves, I divided the lead box into two compartments by a thick lead plate through the middle. Care was taken that the density of the rays falling upon the reflecting surfaces was as uniform as possible, arid with this object in view the glass tube surrounding the bulb was lifted up so as to just expose the half-spherical bottom of the latter. The bulb was placed as exactly as it was practicable in the center, so that both halves of the reflecting plate were equally exposed to the radiations.

Having failed to obtain, in former experiments, a record for iron owing to an oversight, I tried to ascertain its position in the series by comparing it with copper, using a plate made up of iron and copper. The experiments showed that iron reflected about as much as copper, but which metal reflected better was impossible to determine with safety by this method. Next I endeavored to find whether tin or lead was a better reflector, by the same method. Three experiments were performed, and in each case the metals behaved nearly alike, but tin appeared just a trifle better. Finally I investigated the properties of magnesium as compared with zinc. In fact, the experiments showed that magnesium reflected a little better.

I am not yet satisfied, in view of the importance of this relation of the metals with the means employed, and will try to devise an apparatus which will do away with all the defects of the present. The time of the exposure I have found practicable to reduce to a few minutes by the help of a fluorescent paper.

In my previous communications I have barely hinted at the practical importance of the use of suitable reflectors. One would be apt to conclude that, since under the conditions of the previously described experiments, zinc, for instance, reflected only three per cent of the incident rays, the gain secured by the employment of such zinc reflector would be small. This, of course, would be an erroneous conclusion. First of all it should be remembered that in the instances mentioned before the angle of incidence was 45 degrees, and that for larger angles a much greater portion of the .rays would be reflected. The exact law of reflection is still to be determined. Now, let us suppose the shadow of an object is, taken at a distance, D. In order to get a sharp shadow we must take this distance not less than two feet, and I am finding it more and more necessary to adopt a still greater distance. If, for the sake of simplicity, we assume a spherical bulb and electrode, the radiation will be uniform on all sides, and any element of a surface of a sphere of radius D, drawn around the electrode, will receive an equal quantity of rays. The total surface of this sphere will be 4 pi D^2. The object, the shadow of which is to be taken, may have a small area a, which gets only an insignificant part of the total rays emitted, this part being given by the proportion $\frac{a}{4\pi D^2}$. In reality we can not assume less than $\frac{a}{D^2\pi}$ a/D^2pi as effective ratio, but even then, if D is very large and the object, that is, the area a, small, this ratio may be still so small that evidently, by the use of a proper reflector, we can, easily concentrate upon the area a an amount of rays several times exceeding that which would fall upon it without the use of a reflector, in spite of the fact that we are able to reflect only a few per cent of the total incident rays.

As an evidence of the effectiveness of such a reflector, the annexed print of the shoulder and ribs of a man is shown. A funnel-shaped zinc reflector, two feet high, with an opening of five inches on the bottom and 23 inches at the top, was used in the experiment. A tube, similar in every respect to those previously described, was suspended in the funnel, so that just the static screen of the tube was above the former. The exact distance from the electrode to the sensitive plate was four and one-half feet. The distance from the end of the tube to the plate was three and one-half feet. The exposure lasted 40 minutes. The plate showed very strongly and clearly every bone, and shoulder and ribs, but I can not tell how clearly they will appear in the print. I selected the same object as in my first report in your columns on this investigation, so as to give a better idea of the progress made. The advance will be best appreciated by stating that the distance in this case was much more than double, while the time of exposure was less than one-half. The chief importance of a reflector consists, however, in this, that it allows the use of many bulbs without sacrifice of precision and clearness, and also the concentration of a great quantity of radiation upon a very small area.

Since the use of phosphorescent or fluorescent bodies in connection with the sensitive film has been suggested by Professors Henry and Salvioni, I have found it an easy matter to shorten the time of exposure to a few minutes, or even seconds. So far, it seems that the tungstate of calcium, recently introduced by Edison, and manufactured by Messrs. Aylsworth & Jackson, is the most sensitive body. I obtained a sample of it and used it in a series of tests. It fluoresces decidedly better than barium-platino-cyanide, but, owing to the size of the crystals and necessarily uneven distribution on the paper, it does not leave a dean impression. For use in connection with the sensitive films, it should be ground very fine, and some way should be adopted of distributing it uniformly. The paper also must adhere firmly to the film all over the plate, so as to get fairly sharp outlines. The fluorescence of this body seems to depend on a peculiar radiation, because I tested several bulbs, which otherwise worked excellently, without producing a very good result, and I almost gained a false impression.. One or two of the bulbs, however, effected it very powerfully. An impression of the hand was taken at a distance of about six feet from the bulb with an exposure of less khan one minute, and even then it was found that the plate was overexposed. I then took an impression of the chest of a man at a distance of 12 feet from the end of the tube, exposing five minutes. The developed plate showed the ribs dearly, but the outlines were not sharp. Next, I employed a tube with a zinc reflector, as before described, taking an impression of the chest of an assistant at a distance of four feet from the bulb. The latter was strained a little too much in this experiment and exploded, in consequence of the great internal pressure against the bombarded spot. This . accident will frequently occur when the bulb is strained too, high, the preceding outward sign being an increased activity and vapor like appearance of the gas in the bulb and rapid heating of the latter. The process causing the abnormal increase in the internal pressure against the glass wall seems to be due to some action opposite to that noted by Crookes and Spottiswoode, and is very rapid, and for this reason the experimenter should watch carefully for these ominous signs and instantly reduce the potential. Owing to the untimely end of the bulb in this last described experiment, the exposure lasted only one minute. Nevertheless, a very strong impression of the skeleton of the chest, showing the right and left ribs and other details, was obtained. The outlines, however, were again much less sharp than when the ordinary process without the phosphorescent intensifier was followed, although care was taken to press the fluorescent paper firmly against the film. From the foregoing it is evident that, when using the above means for shortening the time of exposure, the thickness of the object is not of very much consequence.

I obtained a still better idea of the quality of tungstate of calcium by observing the effect upon a fluorescent screen made of this chemical. Such a screen, together with a paper box, has been termed with the fanciful name "fluoroscope." It is really Salvioni's Cryptoscope with the lens omitted, which is a great disadvantage. To appreciate the performances of such a screen, it is necessary to work at night, when the eye has for a long time been used to the darkness, and made capable of noting the faint effects on the screen. In one instance the performance of this screen was particularly noteworthy. It was illuminated at a distance of 20 feet, and even at a distance of 40 feet I could still observe a faint shadow passing across the field of vision, when moving the hand in front of the instrument. Looking at a distance of about three feet from the bulb through the body of an assistant, I could distinguish easily the spinal column in the upper part of the body, which was more transparent. In the lower part of the body the column and the rest were practically not perceptible. The ribs were only very faintly seen. The bones of the neck were plainly noticeable, and I could see through the body of the assistant very easily a square plate of copper, as it was moved up and down in front of the bulb. When looking through the head I could observe only the outline of the skull and the chin-bone, yet the field of vision was still bright. Everything else appeared indistinct. This shows that improving of fluorescence will not aid us very much in the examination of the internal parts of the body. The solution rather will come through the production of very powerful radiations, capable of producing very strong. shadows. I believe I have indicated the right way to secure this result. Although it must be admitted that the performance of such a screen is remarkable with the appliances I have used, I have, nevertheless, convinced myself of its still limited value for the purpose of examination. We can distinguish the bones in the limbs, but not nearly as clearly as a photographic impression shows it. Eventually, however, with the help of strong radiations and good reflectors, such fluorescent screens may become valuable instruments for investigation.

A few weeks ago, when I observed a small screen of barium-platino-cyanide flare up at a great distance from the bulb, I told some friends that it might be possible to observe by the aid of such a screen objects passing through a street. This possibility seems to me much nearer at present than it appeared then. Forty feet is a fair width for a street, and a screen lights up faintly at that distance from a single bulb. I mention this odd idea only as an illustration of how these scientific developments may even affect our morals and customs. Perhaps we shall shortly get so used to this state of things that nobody will feel the slightest embarrassment while he is conscious that his skeleton and other particulars are being scrutinized by indelicate observers.

Fluorescent screens afford some help in getting an idea of the condition of the bulb when working. I hoped to find some evidence of refraction by means of such a screen, placing a lens between it and the bulb, and varying the focal distance. To my disappointment, although the shadow of the lens was observable at a distance of 20 feet, I could see no trace of refraction. The use of. the screen for the purpose of noting the effects of reflection and diffraction proved likewise futile.

Roentgen Ray Investigations
Electrical Review — April 22, 1896

Further investigations concerning the behavior of the various metals in regard to reflection of these radiations have given additional support to the opinion which I have before expressed; namely, that Volta's electric contact series in air is identical with that which is obtained when arranging the metals according to their powers of reflection, the most electro-positive metal being the best reflector. Confining myself to the metals easily experimented upon, this series is magnesium, lead, tin, iron, copper, silver, gold and platinum. The last named metal should be found to be the poorest, and sodium one of the best, reflectors. This relation is rendered still more interesting and suggestive when we consider that this series is approximately the same which is obtained when arranging the metals according to their energies of combination with oxygen, as calculated from their chemical equivalents.

Should the above relation be confirmed by other physicists, we shall be justified to draw the following conclusions: *First,* the highly exhausted bulb emits material streams which, impinging on a metallic surface, are reflected; *second,* these streams are formed of matter in some primary or elementary condition; *third,* these material streams are probably the same agent which is the cause of the electro-motive tension between metals in close proximity or actual contact, and they may possibly, to some extent, determine the energy of combination of the metals with oxygen; *fourth,* every metal or conductor is more or less a source of such streams; *fifth,* these streams or radiations must be produced by some radiations which exist in the medium; and *sixth,* streams resembling the cathodic must be emitted by the sun and probably also by other sources of radiant energy, such as an arc light or Bunsen burner.

The first of these conclusions, assuming the above-cited fact to be correct, is evident and incontrovertible. No theory of vibration of any kind would account for this singular relation between the powers of reflection and electric properties of the metals, Streams of projected matter coming in actual contact with the reflecting' metal surface afford the only plausible explanation.

The second conclusion is likewise obvious, since no difference whatever is observed by employing various qualities of glass for the bulb, electrodes of different metals and any kind of residual gases. Evidently, whatever the matter constituting the streams may be, it must undergo a change in the process of expulsion, or, generally speaking; projection — since the views in this regard still differ — in such a way as to lose entirely the characteristics which it possessed when forming the electrode, or wall of the bulb, or the gaseous contents of the latter.

The existence of the above relation between the reflecting and contact series forces us likewise to the third conclusion, because a mere coincidence of that kind is, to say the least, extremely improbable. Besides, the fact may be cited that there is always a difference of potential set up between two metal plates at some distance and in the path of the rays issuing from an exhausted bulb.

Now, since there exists an electric pressure of difference of potential between two metals in dose proximity or contact, we must, when considering all the foregoing, come to the fourth conclusion, namely, that the metals emit similar streams, and I therefore anticipate that, if a sensitive film be placed between two plates, say, of magnesium and copper, a true Roentgen shadow picture would be obtained after a very long exposure in the dark. Or, in general, such picture could be secured whenever the plate is placed near a metallic or conducting body, leaving for the present the insulators out of consideration. Sodium, one of the first of the electric contact series, but not yet experimented upon, should give out more of such streams than even magnesium.

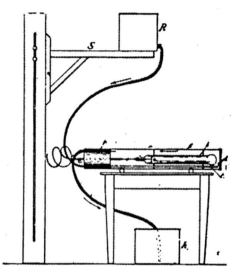

Obviously, such streams could not be forever emitted, unless there is a continuous supply of radiation from the medium in some other form; or possibly the streams which the bodies themselves emit are merely reflected streams coming from other sources. But since all investigation has strengthened the opinion advanced by Roentgen that for the production of these radiations some impact is aired, the former of the two possibilities is the more probable one, and we must assume that the radiations existing in the medium and giving rise to those here considered partake something of the nature of cathodic streams.

But if such streams exist all around us in the ambient medium, the question arises, whence do they come? The only answer is: From the sun. I infer, therefore, that the sun and other sources of radiant energy must, in a less degree, emit radiations or streams of matter similar to those thrown off by an electrode in a highly exhausted inclosure. This seems to be, at this moment, still a point of controversy. According to my present convictions a Roentgen shadow picture should, with very long exposures, be obtained from all sources of radiant energy, provided the radiations are permitted first to impinge upon a metal or other body.

The preceding considerations tend to show that the lumps of matter composing a cathodic stream in the bulb are broken up into incomparably smaller particles by impact against the wall of the latter, and, owing to this, are enabled to pass into the air. All evidence which I have so far obtained points rather to this than to the throwing off of particles of the wall itself under the violent impact of the cathodic stream. According to my convictions, then, the difference between Lenard and Roentgen rays, if there be any, lies solely in this, that the particles composing the latter are incomparably smaller and possess a higher velocity. To these two qualifications I chiefly attribute the non-deflectibility by a magnet which I believe will be disproved in the end. Both kinds of rays, however, affect the sensitive plate and fluorescent screen, only the rays discovered by Roentgen are much more effective. We know now that these rays are produced under certain exceptional conditions in a bulb, the vacuum being extremely high, and that the *range* of greatest activity is rather small.

I have endeavored to find whether the reflected rays possess certain distinctive features, and I have taken pictures of various objects with this purpose in view, but no marked difference was noted in any case. I therefore conclude that the matter composing the Roentgen rays does not suffer further degradation by impact against bodies. One of the most important tasks for the experimenter remains still to determine what becomes of the energy of these rays. In a number of experiments with rays reflected from and transmitted through a conducting of insulating plate, I found that only a small part of the rays could be accounted for. For instance, through a zinc plate, one-sixteenth of an inch thick, under an incident angle of 45 degrees, about two and one-half per cent were reflected and about three per cent transmitted through the plate, hence over 94 per cent of the total radiation remain to be accounted for. All the tests which I have been able to make have confirmed Roentgen's statement that these rays are incapable of raising the temperature of a body. To trace this lost energy and account for it in a plausible way will be equivalent to making a new discovery..,

Since it is now demonstrated that all bodies reflect more or less, the diffusion through the air is easily accounted for. Observing the tendency to scatter through the air, I have been led to increase the efficiency of reflectors by providing not one; but separated successive layers for reflection, by making the reflector of thin sheets of metal; mica or other substances. The efficiency of mica. as a reflector I attribute chiefly to the fact that it is composed of many superimposed layers which reflect individually. These many successive reflections are, in my opinion, also the cause of the scattering through the air.

In my communication to you of April 1, I have for the first time stated that these rays are composed of matter in a "primary" or elementary condition or state. I have chosen this mode of expression in order to avoid the use of the word "ether," which is usually understood in the sense. of the Maxwellian interpretation, which would not be in accord with my present convictions in regard to the nature of the radiations.

An observation which might be of some interest is the following: A few years ago I described on one occasion a phenomenon observed in highly exhausted bulbs. It is a brush or stream issuing from a single electrode under certain conditions, which rotates very rapidly in consequence of the action of the earth's magnetism. Now I have recently observed this same phenomenon in several bulbs which were capable of impressing the sensitive film and fluorescent screen very. strongly. As the brush is rapidly twirling around I have conjectured that perhaps also the Lenard and Roentgen streams axe rotating under the action of the earth's magnetizing and I am endeavoring to obtain an evidence of such motion by studying the action of a bulb in various positions with respect to the magnetic axis of the earth.

In so far as the vibrational character of the rays is concerned, I still hold that the vibration is merely that which is conditioned by the apparatus employed. With the ordinary induction coil we have almost exclusively to deal with a very low vibration impressed by the commutating device or brake. With the disruptive coil we usually have a very strong superimposed vibration in addition to the fundamental one, and it is easy to trace sometimes as much as the fourth octave of the fundamental vibration. But I can not reconcile myself with the idea of vibrations approximating or even exceeding those of light, and think that all these effects could be as well produced with a steady electrical pressure as from a battery, with the exclusion of all vibration which may, occur, even in such instance, as has been pointed out by De La Rive. In my experiments I have tried to ascertain whether a greater difference between the shadows of the bones and flesh could be obtained by employing currents of extremely high frequency, but I have been unable to discover any such effect which would be dependent on the frequency of the currents, although the latter were varied between as wide limits as :was possible. But it is a rule that the more intense the action the .sharper the shadows obtained, provided that the distance is not too small. It is furthermore of the greatest importance for the clearness of the shadows that the rays should be passed through some tubular reflector, which renders them sensibly parallel.

In order then to bring out as much detail as possible on a sensitive plate, we have to proceed in precisely the same way as if we had to deal with flying bullets hitting against a wall composed of parts of different density with the problem before us of producing as large as possible a difference in the trajectories of the bullets which pass through the various parts of the wall. Manifestly, this difference will be the greater the greater the velocity of the bullets; hence, in order 'to bring out detail, very strong radiations are required. Proceeding on this theory I have employed exceptionally thick films and developed very slowly, and in this way clearer pictures have been obtained. The importance of slow development has been first pointed out by Professor Wright, of Yale. Of course, .if Professor Henry's suggestion of the use of a fluorescent body in contact with the sensitive film is made use of, the process is reduced to an ordinary quick photographic procedure, and the above consideration does not apply.

It being desirable to produce as powerful a radiation as possible, I have continued to devote my attention' to this problem and have been quite successful. First of all, there existed limitations in the vacuum tube which did not permit the applying of as high a potential as I desired; namely, when a certain high degree of exhaustion was reached a spark would form behind the electrode, which would prevent straining the tube much higher. This inconvenience I have overcome entirely by making the wire leading to the electrode very long and passing it through a narrow channel, so that the heat from the electrode could not cause the formation of such sparks. Another limitation was imposed by streamers which would break out at the end of the tube when the potential was excessive. This latter inconvenience I have overcome either by the use of a cold blast of air along the tube, as I have mentioned before, or else by immersion of the tube in oil. The oil, as it is now well known, is a means of rendering impossible the formation of streamers by the exclusion of all air. The use of the oil in connection with the production of these radiations has been early advocated in this country by Professor Trowbridge. Originally I employed a wooden box made thoroughly tight with wax and filled with oil or other liquid, in which the tube was immersed. Observing certain specific actions, I modified and improved the apparatus, and in my later investigations I have employed an arrangement as shown in the annexed cut. A bulb b, of the kind described before, with a leading-in wire and neck much longer than here shown, was, inserted into a large and thick glass tube t. The tube was closed in front by a diaphragm d of pergament, and by a rubber plug P in the back. The plug was provided with two holes, into the lower one of which a glass tube t_1, reaching to very nearly the end of the bulb, was inserted. Oil of some kind was made to flow through rubber tubes $r\ r$ from a large reservoir R, placed on an adjustable support S, to the lower reservoir R_1, the path of the oil being clearly observable from the drawing. By adjusting the difference of the level between the two reservoirs it was easy to maintain a permanent condition of working. The outer glass tube t served in part as a reflector, while at the same time it permitted the observation of the bulb b during the action. The plug P, in which the conductor c was tightly sealed, was so arranged that it could be shifted in and out of the tube t, so as to vary the thickness of the oil traversed by the rays.

I have obtained some results with this apparatus which clearly show the advantage of such disposition. For instance, at a distance of 45 feet from the end of the bulb my assistants and myself could observe clearly the fingers of the hand through a screen of tungstate of calcium, the rays traversing about two and one half inches of oil and the diaphragm d. It is practicable with such apparatus to make photographs of small objects at a distance of 40 feet, with only a few minutes exposure, by the help of Professor Henry's method. But, even without the use of a fluorescent powder, short exposures are practicable, so that I think the use of the above method is not essential for quick procedure. I rather believe that in the practical development of this principle, if it shall be necessary, Professor Salvioni's suggestion of a fluorescent emulsion, combined with a film, will have to be adopted. This is bound to give better results than an independent fluorescent screen, and will very much simplify the process. I may say, however, that, since my last communication, considerable improvement has been made in the screens. The manufacturers of Edison's tungstate of calcium are now furnishing screens which give fairly clean pictures. The powder is fine and it is more uniformly distributed. I consider, also, that the employment of a softer and thicker paper than before is of advantage. It is just to remark that the tungstate of calcium has also proved to be an excellent fluorescent in the bulb. I tested its qualities for such use immediately and find it so "far unexcelled. Whether it will be so for a long time remains to be seen. News reaches us that several fluorescent bodies, better than the cyanides, have been discovered abroad.

Another improvement with a view of increasing the sharpness of the shadows has been proposed to me by Mr. E. R. Hewitt. He assumed that the absence of sharpness of the outlines in the shadows on the screen was due to the spread of the fluorescence frown crystal to crystal. He proposed to avoid this by using a thin aluminum plate with many parallel .grooves. Acting on this suggestion, I made some experiments with wire gauze and, furthermore, with screens made of a mixture of a fluorescent with a non-fluorescent powder. T found that the general brightness of the screen was diminished, but that with a strong radiation the shadows appeared sharper. This idea might be found capable of useful application.

By the use of the above apparatus I have been enabled to examine much better than before the body by means of the fluorescent screen. Presently the vertebral column can be seen quite clearly, even in the lower part of the body. I have also clearly noted the outlines of the hip bones. Looking in the region of the heart I have been able to locate in unmistakably. The background appeared much brighter, and this difference in the intensity of the shadow and surrounding has surprised me. The ribs I could now see on a number of occasions quite distinctly, as well as the shoulder bones. Of course, there is no difficulty whatever in observing the bones of all limbs. I noted certain peculiar effects which I attribute to the oil. For instance, the rays passed through plates of metal over one-eighth of an inch thick, and in one instance I could see quite clearly the bones of my hand through sheets of copper, iron and brass of a thickness of nearly' one-quarter of an inch. Through glass the rays seemed to pass with such freedom that, looking through the screen in a direction at right angles to the axis of the tube, the action was most intense, although the rays had to pass through a great thickness of glass and oil. A glass slab nearly one-half of an inch thick, held in front of the screen, hardly dimmed the fluorescence. When holding the. screen in front of the tube at a distance of about three feet, the head of an assistant, thrust between the screen and the tube, cast but a feeble shadow. It appeared some times as if the bones and the flesh were equally transparent to the radiations passing through, the oil. When very close to the bulb, the screen was illuminated through the body of an assistant so strongly that, when a hand was moved-in, front, I could clearly note the motion of the hand. through the body. In one instance I could even distinguish the bones of the arm.

Having observed the extraordinary transparence of the bones in some instances, I at first surmised that the rays might be vibrations of high pitch, and that the oil had in. some way absorbed a part of them. This view, however, became untenable when I found that at a certain distance from the bulb I obtained a sharp shadow of the bones. This latter observation led me to apply usefully the screen in taking impressions on the plate. Namely, in such ,case it is of advantage to first determine by means of the screen the proper distance at which the object is to be placed before taking the impression. It will be found that often the image is much clearer at a greater distance. In order, to avoid any error when observing with the screen, I have surrounded the box with thick metal plates, so as to prevent the fluorescence, in consequence of the radiations, reaching the screen from the sides. I believe that such an arrangement is absolutely necessary if one wishes to make correct observations.

During my study of the behavior of oils and other liquid insulators, which I am still continuing, it has occurred to me to investigate the important effect discovered by Prof. J. J. Thomson. He announced some time ago that all bodies traversed by Roentgen radiations become conductors of electricity. I applied a sensitive resonance test to the investigation of this phenomenon in a manner pointed out in my earlier writings on high frequency currents. A secondary, preferably not in very close inductive relation to the primary circuit, was connected to the latter and to the ground, and the vibration through the primary, was so adjusted that true resonance took place. As the secondary had a considerable number of turns, very small bodies attached to the free terminal produced considerable variations of potential on the latter. Placing a tube in a box of wood filled with oil and attaching it to the terminal, I adjusted the vibration through the primary so that resonance took place without the bulb radiating Roentgen rays to an appreciable extent. I then changed the conditions so that the bulb became very active in the production of the rays. The oil should have now, according to Prof. J. J. Thomson's statement, become a conductor and a very marked change in the vibration should have occurred. This was found not to be the case, so that we must see in the phenomenon discovered by J. J. Thomson only a further evidence that we have to deal here with streams of matter which, traversing the bodies, carry away electrical charges. But the bodies do not become conductors in the common acceptance of the term. The method I have followed is so delicate that a mistake is almost an impossibility.

Electrical Progress. On Apparatus for Cathography
Mining and Scientific Press — May 9, 1896

In order to produce the most intense effects, we have first to consider that, whatever their nature, they depend necessarily on the intensity of the cathode streams. Then, again, being dependent on the magnitude of the potential, it follows that the highest attainable electrical pressure is desirable.

To obtain high potentials we may avail ourselves of an ordinary induction coil, or of a static machine, or preferably of a disruptive discharge coil. If we put two electrodes in a bulb, or use one inside and another outside electrode, we limit the potential, for the presence not only of the anode, but o! any conducting object, has the effect of reducing the practicable potential on the 'cathode. Thus, to secure the result aimed at, one is driven to the acceptance of a single electrode bulb, the other terminal being as far remote as possible.

Having selected the induction apparatus and type of bulb, the next important consideration is the vacuum. On this - subject I am able to make known a fact with which I have long been acquainted, and of which I have taken advantage in the production of vacuum jackets and incandescent bulbs, and which I subsequently found to be of the utmost importance, not to say essential, for the production' of intense Roentgen shadows. I refer to a method of rarefaction by electrical means to any degree desirable, far beyond that obtainable by mechanical appliances.

Though this result can be reached b the use of a static machine, as well as of an ordinary induction coil giving a sufficiently high potential, I have found that by far the most suitable apparatus, and one which secures the quickest action, is a disruptive coil. It is best to proceed is this way: The bulb is first exhausted by means of an ordinary vacuum pump to a rather high degree, though my experiences have shown that this is not absolutely necessary, as I have also found it possible to rarefy, beginning from low pressure: After being taken down from the pump, the bulb is attached to the terminal of the disruptive coil, preferably of high frequency of vibration; and usually the following phenomena are noted: First, there is a milky light spreading through the bulb, or possibly for a moment the glass becomes phosphorescent, if the bulb has been exhausted to a high degree. At any rate, the phosphorescence generally subsides quickly and the white light settles around the electrode, whereupon a dark space forms at some distance from the latter, Shortly afterwards the light assumes a reddish color and the terminal grows very hot. This heating, however, is observed only with powerful apparatus. It is well to watch the bulb carefully and regulate the potential at this stage, as the electrode might be quickly consumed.

After some time the reddish light subsides, the streams becoming again white, whereupon they get weaker and weaker, wavering around the electrode until they finally disappear. Meanwhile, the phosphorescence of the glass grows more and more intense, and the spot where the stream strikes the wall becomes very hot, while the phosphorescence around the electrode ceases and the latter cools down to such an extent that the glass near it may be actually ice-cold to the touch. The gas in the bulb has then. reached the required degree of rarefaction. The process may be hastened by repeated heating and cooling and by the employment of a small electrode.

I may state here that the experimenter need not be deterred from using a glass bulb, as I believe the opacity of glass, as well as the transparency of aluminum, are somewhat exaggerate, inasmuch as I have found that" a very thin aluminum sheet throws a marked shadow, while, on the other hand, I have obtained impressions through a thick glass plate.

The above method is not only valuable as a means of obtaining the high vacua desired, but it is still more important, because the phenomena observed throw a light on the results obtained by Lenard and Roentgen.

Though the phenomenon of rarefaction under above conditions admits of different interpretations, the chief interest centers on one of them,. to which I adhere — that is, on the actual expulsion of the particles through the walls of the bulb. I have lately observes that the latter commences to act properly upon the sensitive plate only from the point when the exhaustion begin, to be noticeable, and the effects. produced are the strongest when the process of exhaustion is most rapid, even though the phosphorescence might not appear particularly bright. Evidently, then, the two effects are closely connected, and I am getting more and more convinced that we have to deal with a stream of material particles, which strike the sensitive plate with great velocities. Taking as a basis the estimate of Lord Kelvin on the speed of projected particles in a Crookes' bulb, we arrive easily by the employment of very high potentials to speeds of as much as a hundred kilometers a second.

It may not be known that even an ordinary streamer, breaking out suddenly and under great pressure from the terminal of a disruptive coil, passes through a thick glass plate as though the latter were not present. Unquestionably, with such coils pressures are practicable which will project the particles in straight lines even. under atmospheric pressure. I have obtained distinct impressions in free air, not by streamers, as some experimenters have done, using static machines or induction coils, but by actual projection, the formation of streamers being, absolutely prevented by careful static screening.

A valuable evidence of the nature of the radiations and progress in the direction of obtaining strong impressions on the plate might be arrived at by perfecting plates especially sensitive to mechanical shock or impact. There are chemicals suitable for this, and the development in this direction may lead to the abandonment of the present plate. Furthermore, if we have to deal with streams of material particles, it seems not impossible to project upon the plate a suitable substance to insure the best chemical action.

By exposing the head to a powerful radiation strange effects have been noted. For instance, I find that there is a tendency to sleep and the time seems to pass away quickly. There is a general soothing effect, and I have felt a sensation of warmth is the upper part of the head. Au assistant independently confirmed the tendency to sleep and a quick lapse of time. Should these remarkable effects be verified by men with keener sense of observation, I shall still more firmly believe in the existence of material streams penetrating the skull. Thus it may be possible to protect a suitable chemical into any part of the body.

Since my above-mentioned discoveries I have made considerable progress, and can presently announce one more result of importance. I have lately obtained shadows by redected rays only, thus demonstrating beyond doubt that the Roentgen rays possess this property. One of the experiments may be cited here. A thick copper tube, about a foot long, was taken and one of its ends tightly closed by the plate-bolder containing a sensitive plate, protected by a fiber cover as usual: Near the open end of the copper tube was placed a thick plate glass at an angle of 45° to the axis of the tube. A single terminal bulb was then suspended above the glass plate at a distance of about eight inches, so that the bundle of rays fell upon the latter at an angle of 45°, and supposedly reelected rays passed along the axis of the copper tube. An exposure of forty-five minutes gave a clear and sharp shadow of a metallic object. This shadow was produced by the reflected rays, as the direct action was absolutely excluded, it having been demonstrated that, even under the severest tests with much stronger actions, no impression whatever could be produced upon 'the film through a thickness of copper equal to that of the tube. Concluding from the intensity of the action by comparison with an equivalent effect due to the direct rays, I find that approximately two per cent of the latter were reflected from the glass plate in this experiment.

An Interesting Feature of X-Ray Radiations
Electrical Review — July 8, 1896

The following observations, made with bulbs emitting Roentgen radiations, may be of value in throwing additional light upon the nature of these radiations, as well as illustrating better properties already known. In the main these observations agree with the views which have forced themselves upon my, mind from the outset, namely, :that the rays consist of streams of minute material particles projected with great velocity. In numerous experiments I have found that the matter which, by impact within the bulb, causes the formation of the rays may come from tidier of the electrodes. Inasmuch as the latter are by continued use disintegrated to a marked degree, it seems more plausible to assume that the projected matter consists of parts of the electrodes themselves rather than of the residual gas. However, .other observations, upon which I can not dwell at present, lead to this conclusion. The lumps of projected matter are by impact further disintegrated into particles so minute as to be able to pass through the walls of the bulb: or else they tear off such particles from the walls, or generally bodies, against which they are projected. At any rate, an impact and consequent shattering seems absolutely necessary for the production of Roentgen rays. The vibration, if there be any, is only that which is impressed by the apparatus, and the vibrations can only be longitudinal.

The principal source of the rays is invariably the place of first impact within the bulb, whether it be the anode, as in some forms of tube, or an inclosed insulated body, or the glass wall. When the matter thrown off from an electrode, after striking against an obstacle, is thrown against another body, as the wall of the bulb, for instance, the place of second impact is a very feeble source of the rays.

These and other facts will be better appreciated by referring to the annexed, figure, in which a form of tube is shown used in a number of my experiments. The general form is that described on previous occasions. A single electrode e, consisting of a massive aluminum plate, is mounted on a conductor t, provided with a glass wrapping w as usual, and sealed in one of the ends of a straight tube b, about five centimeters in diameter and 30 centimeters long. The other end of the tube is blown out into a thin bulb of a slightly larger diameter, and near this end is supported on a glass stem a a funnel f of thin platinum sheet, In such bulbs I have used a number of different metals for impact with a view of increasing the intensity of the rays and also for the purpose of reflecting and concentrating them. Since, however, in a later contribution, Professor Roentgen has pointed out that platinum gives the most intense rays, I have used chiefly this metal, finding a marked increase in the effect upon the screen or sensitive plate. The particular object of the presently described construction was to ascertain whether the rays generated at the inner surface of the platinum funnel f would be brought to a focus outside of the bulb, and further, whether they would proceed in straight lines from that point. For this purpose the apex of the platinum cone was arranged to be about two centimeters outside of the bulb at o.

When the bulb was properly exhausted and set in action, the glass wall below the funnel f became strongly phosphorescent, but not uniformly, as there was a narrow ring r r on the periphery brighter than the rest, this ring being evidently due to the rays reflected from the platinum sheet.

Placing a fluorescent screen in contact or quite close to the glass wall below the funnel, the portion of the screen in the immediate neighborhood of the phosphorescent patch was brightly illuminated, the outlines being entirely indistinct. Receding now with the screen from the bulb, the strongly illuminated spot became smaller and the outlines sharper, until, when the point o was reached, the luminous part had dwindled down to a small point. Moving the screen a few millimeters beyond o caused a small dark spot to appear, which widened into a circle and became larger and larger in the same measure as the distance from the bulb was increased (see S), until, at a sufficiently large distance, the dark circle covered the entire screen. This experiment illustrated in a beautiful way the propagation in straight lines, which Roentgen originally proved by pinhole photographs. But, besides this, an important point was noted; namely, that the fluorescent glass wall emitted practically no rays, whereas, had the platinum not been present, it would have been, under similar conditions, an efficient source of the rays, for the glass, even by weak excitation of the bulb, was strongly heated. I can only explain the absence of the radiation from the glass by assuming that the matter proceeding from the surface of the platinum sheet was already in a finely divided state when it reached the glass wall. A remarkable fact is, also, that, at least by a weak excitation of the bulb, the edges of the dark circle were very sharp, which speaks strongly against diffusion. By exciting the bulb very strongly, the background became brighter and the shadow S fainter, though it continued to be plainly visible even then.

From the preceding it is evident .that, by a suitable construction of the bulb, the rays emanating from the latter may be concentrated upon any small area at some distance, and a practical advantage may be taken of this fact when producing impressions upon a plate or examining bodies by means of a fluorescent screen.

Roentgen Rays Or Streams
Electrical Review — August 12, 1896

In the original report of his epochal discoveries, Roentgen expressed his conviction that the .phenomena he observed were due to certain novel disturbances in the ether. This opinion deserves to be considered the more as it was probably formed in the first enthusiasm over the revelations, when the mind of the discoverer was capable of a much deeper insight into the nature of things.

It was known since long ago that certain dark radiations, capable of penetrating opaque bodies, existed, and when the rectilinear propagation, the action on a fluorescent screen and on a sensitive film was noted, an obvious and unavoidable inference was that the new radiations were transverse vibrations, similar to those known as light. On the other hand, it was difficult to resist certain arguments in favor of the less popular theory of material particles, especially as, since the researches of Lenard, it has become very probable that material streams, resembling the cathodic, existed in free air, Furthermore, I myself have brought to notice the fact that similar material streams — which were subsequently, upon Roentgen's announcement, found capable of producing impressions on a sensitive film — were obtainable in free air, even without the employment of a vacuum bulb, simply by the use of very high potentials, suitable for imparting to the molecules of the air or other particles a sufficiently high; velocity. In reality, such puffs or jets of particles are formed in the vicinity of every highly charged conductor, the potential of which is rapidly varying, and I have shown that, unless they are prevented, they are fatal to every condenser or high-potential transformer, no matter how thick the insulation. They also render practically valueless any estimate of the period of vibration of an electro-magnetic system by the usual made of calculation or measurement in a static condition in all cases in which the potential is very high and the frequency excessive.

It is significant that, with these and other facts before him, Roentgen inclined to the conviction that the rays he discovered were longitudinal waves of ether.

After a long and careful investigation, with apparatus excellently suited for the purpose, capable of producing impressions. at great distances, and after examining the results pointed out by other experimenters, I have come to the conclusion which I have already intimated in my former contributions to your esteemed journal, and which I now find courage to pronounce without hesitation, that the original hypothesis of Roentgen will be confirmed in two particulars; first, in regard to the longitudinal character of the disturbances; second, in regard to the medium concerned in their propagation. The present expression of my views is made solely for the purpose of preserving a faithful record of what, to my mind, appears to be the true interpretation of these new and important manifestations of energy.

Recent observations of some dark radiations from novel sources by Becquerel and others, and certain deductions of Helmholtz, seemingly. applicable to the explanation of the peculiarities of the Roentgen rays, have given additional weight to the arguments on behalf of the theory of transverse vibrations, and accordingly this interpretation of the phenomena is held in favor. But this view is still of a purely speculative character, being, as it is at present, unsupported by any conclusive experiment. Contrarily, there is considerable experimental evidence that some matter is projected with great velocity from the bulbs, this matter being in all probability the only cause of the actions discovered by Roentgen.

There is but little doubt at present that a cathodic stream within a bulb is composed of small particles of matter, thrown off with great velocity from the electrode. The velocity probably attained is estimable, and fully accountable for the mechanical and heating effects produced by the impact against the wall or obstacle within the bulb. It is, furthermore, an accepted view that the projected lumps of matter act as inelastic bodies, similarly to ever so many small lead bullets. It can be easily shown that the velocity of the stream *may* be as much as 100 kilometers a second, or even more, at least in bulbs with a single electrode, in which the practicable vacua and potentials are much higher than in the ordinary bulbs with two electrodes. But, now, matter moving with such great velocity must surely penetrate great thicknesses of the obstruction in its path, if the laws of mechanical impact are at all applicable to a cathodic stream. I have presently so much familiarized myself with this view that, if I had no experimental evidence, I would not question the fact that some matter is projected through the thin wall of a vacuum tube. The exit from the latter is, however,, the more likely to occur, as the lumps of matter must be shattered into still much smaller particles by the impact, From my experiments on reflection of the Roentgen rays, before reported, which, with powerful radiations, may be shown to exist under all angles of incidence, it appears that the lumps or molecules are indeed shattered into fragments or constituents so small as to make them lose entirely some physical properties possessed before the impact. Thus, the material composing the electrode, the wall of the bulb or obstruction of any kind placed within the latter, are of absolutely no consequence, except in so far as the intensity of the radiations is concerned. It also appears, as I have pointed out, that no further disintegration of the lumps is attendant upon a second impact. The matter composing the cathodic stream is, to all evidence, reduced to matter of some primary form, heretofore not known, as such velocities and such violent impacts have probably never been studied or even attained before these extraordinary manifestations were observed. Is it not possible that the very ether vortexes which, according to Lord Kelvin's ideal theory, compose the lumps, are dissolved, and that in the Roentgen phenomena we may witness a transformation of ordinary matter into ether? It is in this sense that,. I think, Roentgen's first hypothesis will be confirmed. In such case there can be, of course, no question of waves other than the longitudinal assumed by him, only, in my opinion, the frequency must be very small — that of the electromagnetic vibrating system — generally not more than a few millions a second. If such process of transformation does take place, it will be difficult, if not impossible, to determine the amount of energy represented in the radiations, and the statement that this amount is very small should be received with some caution.

I As to the rays exhaustively studied by Lenard, which have proved to be the nucleus of these great realizations, I hold them to be true cathodic streams, projected through the wall of the tube. Their deflectibility by a magnet shows to my mind simply that they differ but little from those within the bulb. The lumps of matter are probably large and the velocity small as compared with .that of the Roentgen rays. They should, however be capable in a minor degree of all the actions of the latter. These actions I consider to, be purely :mechanical and obtainable by other means. So, for instance, I think that if a gun loaded with mercury were fired through a thin board, the projected mercury vapor would cast a shadow of an object upon a film made especially sensitive to mechanical impact, or upon a screen of material capable of being rendered fluorescent by such impact.

The following observations made by myself and others speak more or less for the existence of the streams of matter.

I — Phenomena of Exhaustion

On this subject I have expressed myself on another occasion. It is only necessary to once more point out that the effect observed by me should not be confounded with that noted by Spottiswoode and Crookes. I explain the latter phenomenon as follows: The first fluorescence appearing when the current is turned on, is due to some organic matter almost always introduced in the bulb in the process of manufacture. A minute layer of such matter on the wall produces invariably this first fluorescence, and the latter never takes place when the bulb has been exhausted under application of a high degree of heat or when the organic matter is otherwise destroyed. Upon the disappearance of the first fluorescence .the rarefaction increases slowly, this being . a necessary result of particles being projected from the electrode and fastening themselves upon the wall. These particles absorb a large portion of the residual gas. The latter can. be again freed by the application of heat to the bulb or otherwise. So much of the effects observed by these investigators. In the instance observed by myself, there must be actual expulsion of matter, and for this speak following facts: (a) the exhaustion is quicker when the glass is thin; (b) when the potential is higher; © when the discharges are more sudden; (d) when there is no obstruction within the bulb; (e) the exhaustion takes place quickest with an aluminum or platinum electrode, the former metal giving particles moving with greatest velocity, the latter particles of greatest weight; (f) the glass wall, when softened by the heat, does not collapse, but bulges outwardly; (g) the exhaustion takes place, in some cases, even if a small perceptible hole is pierced through the glass; (h) all causes tending to impart a greater velocity to the particles hasten the process of exhaustion.

II — Relation Between Opacity and Density

The important fact pointed out early by Roentgen and confirmed by subsequent research, namely, that a body is the more opaque to the rays the denser it is, can not be explained as satisfactorily under any other assumption as that of the rays being streams of matter, in which case such simple relation between opacity and density would necessarily exist. This relation is the more important in its bearing upon the nature of the rays, as it does not at all exist in light-giving vibrations, and should consequently not be found to so marked a degree and under *all conditions with vibrations,* presumably similar to and approximating in frequency the light vibrations.

III — Definition of Shadows on Screen or Plate

When taking impressions or observing shadows while varying the intensity of the radiations, but maintaining all other conditions as nearly as possible alike, it is found that the employment of more intense radiations secures little, if any, advantage, as regards the definition of the details. At first it was thought that all' there was needed was to produce very powerful rays. But the experience was disappointing, for, while I succeeded in producing rays capable of impressing a plate at distances of certainly not less than 30 meters, I obtained but slightly better results. There was one advantage in using such intense rays, and this was that the plate could be further removed from the source, and consequently a better shadow was obtained. But otherwise nothing to speak of was gained. The screen in the dark box would be at times rendered so bright as to allow reading at some distance plainly, but the shadow was not more distinct for all that. In fact, often a very strong radiation gave a poorer impression than a weak one. Now, a fact which I have repeatedly observed and to which I attach great importance in this connection, is the following: When taking impressions at a small distance with a tube giving very intense rays, no shadow, unless a scarcely perceptible one, is obtained. Thus, for instance, the flesh and bones of the hand appear equally transparent. Increasing presently gradually the distance, it is found that the bones cast a shadow, while the flesh leaves no impression. The distance still increased, the shadow of the flesh appears, while that of the bones grows deeper, and in this neighborhood a place can be found at which the definition of the shadow is clearest. If the distance is still further continually increased, the detail is lost, and finally only a vague shadow is perceptible, showing apparently the outlines of the hand.

This often-noted fact disagrees entirely with any theory of transverse vibrations, but can be easily explained on the assumption of material streams. When the hand is near and the velocity of the stream of particles very great, both bone and flesh are easily penetrated, and the effect due to the difference in the retardation of the particles passing through the heterogeneous parts can not be observed. The screen can fluoresce only up to a certain limited intensity, and the film can be affected only to a certain small degree. When the distance is increased, or, what is equivalent, when the intensity of the radiation is reduced, the more. resisting bones begin to throw the shadow first. Upon a further increase of the distance the flesh begins likewise to stop enough of the particles to leave a trace on the screen. But in all cases, at a certain distance, manifestly that which under the conditions of the experiment gives the greatest difference in the trajectories of the particles within *the range perceptible on the screen or film,* the clearest shadow is secured.

IV — the Rays Are All of One Kind

The preceding explains the apparent existence of rays of different kind; that is, of different rates of vibration, as it is asserted. In my opinion, the velocity and possibly the size of the particles both are different, and this fully accounts for the discordant results obtained in regard to the transparency of various bodies to these *rays*. I found, for example, in many cases that aluminum was less transparent than glass, and in some instances brass appeared to be very transparent as compared with other metallic bodies. Such observations showed that it was necessary, in making the comparison, to take rigorously equal thicknesses of the bodies and place them as closely together as possible. They also showed the fallacy of comparing results obtained with different bulbs.

V — Action on the Films

Many experiments with films of different thicknesses show that decidedly more detail is obtainable with a thick film than with a thin one. This appears to me to be a further evidence in support of the above views, as the result can be easily explained when considering the preceding remarks.

VI — the Behavior of Various Bodies in Reflecting the Rays

on which I have previously dwelt, will, if verified by other experimenters, leave no room for a doubt that the radiations are streams of some matter, or possibly of ether, as before observed.

VII — The Entire Absence of Refraction

and other features possessed by the light waves has, since Roentgen's announcement, not yet been satisfactorily explained. A trace at least of such an effect would be found if the rays were transverse vibrations.

VIII — The Discharge of Conductors

by the rays shows, in so far as l have been able to follow the researches of others, that the electrical charge is taken off by the bodily carriers. It is also found that the opacity plays an important part, and the observations are mostly in accord with the above views.

IX — The Source of the Rays

is, I find, always the place of the first impact of the cathodic stream, a second impact producing little or no rays. This fact would be difficult to account for unless streams of matter are assumed to exist.

X — Shadows in Space Outside of the Bulb

An almost crucial test of the existence of material streams is afforded by the formation of shadows in space at a distance from the bulb, to which

I have called attention quite recently. I will presently refer to my preceding communication on this subject, and will only point out that such shadows could not be formed under the conditions described, except by streams of matter.

XI — All Bodies Are Transparent to Very Strong Rays

Experiments establish this fact beyond any doubt. With very intense radiations, I obtain, easily, impressions through what may be considered a great thickness of any metal. It is impossible to explain this on any theory of transverse vibrations. We can show how one or other body might allow the rays to pass through, but such explanations are not applicable to *all* bodies without exception. On the contrary, assuming material streams; such a result is unavoidable.

A great many other observations and facts might be added to the above, as further evidence in support of the above views. I have noted certain peculiarities of bodies obstructing a cathodic stream within the bulb. I have observed that the same rays are produced at all degrees of exhaustion and using bodies of vastly different physical properties, and have found a number of features in regard to the pressure, the vacuum, the residual gas, the material of the electrode, etc., all of which observations are more or less in accord with what I have stated before. I hope, however, that there is enough in the present lines to enlist the attention of others.

Mr. Tesla on Thermo Electricity
The Electrical Engineer - N. Y. — *December 23, 1896*

In a letter to the editor of the Buffalo Enquirer, Mr. Nikola Tesla replies as follows in regard to an inquiry on the subject of the future of electricity:

"The transmission of power has interested me not only as a technical problem, but far more in its bearing upon the welfare of mankind. In this sense I have expressed myself in a lecture, delivered some time ago.

"Since electrical transmission of energy is a process much more economical than any other we know of, it necessarily must play an important part in the future, no matter how the primary energy is derived from the sun. Of all the ways the utilization of a waterfall seems to be the simplest and least wasteful. Even if we could, by combining carbon in a battery, convert the work of the chemical combination into electrical energy with very high economy, such mode of obtaining power would, in my opinion, be no more than a mere makeshift, bound to be replaced sooner or later by a more perfect method, which implies no consumption of any material whatever."

On Electricity
The Address on the Occasion of the Commemoration of the Introduction of Niagara Falls Power in Buffalo At the Ellicott Club, January 12, 1897

I have scarcely had courage enough to address an audience on a few unavoidable occasions, and the experience of this evening, even as disconnected from the cause of our meeting, is quite novel to me. Although in those few instances, of which I have retained agreeable memory, my words have met with a generous reception, I never deceived myself, and knew quite well that my success was not due to any excellency in the rhetorical or demonstrative art. Nevertheless, my sense of duty to respond to the request with which I was honored a few days ago was strong enough to overcome my very grave apprehensions in regard to my ability of doing justice to the topic assigned to me. It is true, at times — even now, as I speak — my mind feels full of the subject, but I know that, as soon as I shall attempt expression, the fugitive conceptions will vanish, and I shall experience certain well known sensations of abandonment, chill and silence. I can see already your disappointed countenances and can read in them the painful regret of the mistake in your choice.

These remarks, gentlemen; are not made with the selfish desire of winning your kindness and indulgence on my shortcomings, but with the honest intention of offering you an apology for your disappointment. Nor are they made — as you might be disposed to think — in that playful spirit. which, to the enjoyment of the listeners, is often displayed by belated speakers. On the contrary, I am deeply earnest in my wish that I were capable of having the fire of eloquence kindled in me, that I might dwell in adequate terms on this fascinating science of electricity, on the marvelous development which electrical annals have recorded and which, as one of the speakers justly remarked, stamp this age as the Electrical Age, and particularly on the great event we are commemorating this day. Unfortunately, this my desire must remain unfulfilled, but I am hopeful that in my formless and incomplete statements, among the few ideas and facts I shall mention there may be something of interest and usefulness,, something befitting this unique occasion.

Gentlemen, there are a number of features clearly discernible in, and characteristic of, human intellectual progress in more recent times — features which afford grew comfort to the minds of all those who have really at heart the advancement and welfare of mankind.

First of all, the inquiry, by the aid of the microscope and electrical instruments of precision, into the nature of our organs and senses, and particularly of those through which we commune directly with the outside world and through which knowledge is conveyed to our minds, has revealed their exact construction and mode of action, which is in conformity with simple and well established physical principles and laws. Hence the observations we make and the facts we ascertain by their help are *real* facts and observations, and our knowledge is *true* knowledge. To illustrate: Our knowledge of form, for instance, is dependent upon the positive fact that light propagates in straight lines, and, owing to this, the image formed by a lens is exactly similar to the object seen. Indeed, my thoughts in such fields and directions have led me to the conclusion that most all human knowledge is based on this simple truth, since practically every idea or conception — and therefore all knowledge — presupposes visual impressions. But if light would not propagate in accordance with the law mentioned, but in conformity with any other law which we might presently conceive, whereby not only the image might not bear any likeness to the object seen, but even the images of the same object at different times or distances might not resemble each other, then. our knowledge of form would be very defective, for then we might see, for example, a three-cornered figure as a six or twelve-cornered one. With the clear understanding of the mechanism and mode of action of our organs, we remove all doubts as to *the reality* and *truth* of the impressions received from the outside, and thus we bar out — forever, we may hope — that unhealthy speculation and skepticism into which formerly even strong minds were apt to fall.

Let me tell you of another comforting feature. The progress in a measured time is nowadays more rapid and greater than it ever was before. This is quite in accordance with the fundamental law of motion, which commands acceleration and increase of momentum or. accumulation of energy under the action of a continuously acting force and tendency, and is the more true as every advance weakens the elements tending 'to produce friction and retardation. For, after all, what *is* progress, or — more correctly — development, or evolution, if not a movement, infinitely complex and often unscrutinizable, it is true, but nevertheless exactly determined in quantity as well as in quality of motion by the physical. conditions and laws governing? This feature of more recent development is best shown in the rapid merging together of the various arts and sciences by the obliteration of the hard and fast lines of separation, of borders, some of which only a few years ago seemed unsurpassable, and which, like veritable Chinese walls, surrounded every department of inquiry and barred progress. A sense of connectedness of the various apparently widely different forced and phenomena we observe is taking possession of our minds, a sense of deeper understanding of nature as a whole, which, though not yet quite clear and defined, is keen enough to inspire us with the confidence of vast realizations in the near future.

But these ,features chiefly interest the scientific man, the thinker and reasoner. There is another feature which affords us still more satisfaction and enjoyment, and which is of still more universal: interest, chiefly because of its. bearing upon the welfare of mankind. Gentlemen, there is an influence which is getting strong and stronger day by day, which shows itself more and more in all departments of human activity,. an influence most fruitful and beneficial — the influence of the artist. It was a happy day for the mass of humanity when the artist felt the desire of becoming a physician, an electrician, an engineer or mechanician or — whatnot — a mathematician or a financier; for it was he who wrought all these wonders and grandeur we are witnessing. It was he who abolished that small, pedantic, narrow-grooved school teaching which made of an aspiring student a galley-slave, and he who allowed freedom in the choice of subject of study according to one's pleasure and inclination, and so facilitated development.

Some, who delight in the exercise of the powers of criticism, call this an asymmetrical development, a degeneration or departure from the normal,. or even a degradation of the race. But they are mistaken. This is a welcome state of things, a blessing, a wise subdivision of labors, the establishment of conditions most favorable to progress. Let one concentrate all his energies in one single great effort; let him perceive a single truth, even though he be consumed by the sacred fire, then millions of less gifted men can easily follow. Therefore it is not as much quantity as quality of work which determines the magnitude of the progress.

It was the artist, too, who awakened that broad philanthropic spirit which, even in old ages, shone in the teachings of noble reformers and philosophers, that spirit which makes men in all departments and positions work not as much for any material benefit or compensation — though reason may command this also — but chiefly for the sake of success, for the pleasure there is in achieving it and for the good they might be able to do thereby to their fellow-men. Through his influence types of men are now pressing forward, impelled by a deep love for their study, men who are doing wonders in their respective branches, whose chief aim and enjoyment is the acquisition and spread of knowledge, men who look far above earthly things, whose banner is Excelsior! Gentlemen, let us honor the artist, let us thank him, let us drink his health!

Now, in all these enjoyable and elevating features which characterize modern intellectual development, electricity, the expansion of the science of electricity, hay been a most potent factor. Electrical science has revealed to us the true nature of light, has provided us with innumerable appliances and instruments of precision, and has thereby vastly added to the exactness of our knowledge. Electrical science has disclosed to us the more intimate relation existing between widely different forces and phenomena and has thus led us to a more complete comprehension of Nature and its many manifestations to our senses. Electrical science, too, by its fascination, by its promises of immense realizations, of wonderful possibilities chiefly in humanitarian respects, has attracted the attention and enlisted the energies of the artist; for where is there a field in which his God-given powers would be of greater benefit to his fellow-men than this unexplored, almost virgin, region, where, like in a silent forest, a thousand voices respond to every call?

With these comforting features, with these cheering prospects, we need not look with any feeling of incertitude or apprehension into the future. There are pessimistic men, who, with anxious faces, continuously whisper in your ear that the nations are secretly arming — arming to the teeth; that they are going to pounce upon each other at a given signal and destroy themselves; that they are all trying to outdo that victorious, great, wonderful German army, against which there is no resistance, for every German has the discipline in his very blood — every German is a soldier. But these men are in error. Look only at our recent experience with the British in that Venezuela difficulty. Two other nations might have crashed together, but not the Anglo-Saxons; they are too far ahead. The men who tell you this are ignoring forces which are continually at work, silently but resistlessly — forces which say Peace!

There is the genuine artist, who inspires us with higher and nobler sentiments, and. makes us abhor strife and carnage. There is the engineer, who bridges gulfs and chasms, and facilitates contact and equalization of the heterogeneous masses of humanity. There is the mechanic, who comes with his beautiful time and energy-saving appliances, who perfects his flying machine, not to drop a. bag of dynamite on a city or vessel, but to facilitate transport and travel. There, again, is the chemist, who opens new resources and makes existence more pleasant and secure; and there is the electrician, who sends his messages of peace to all parts of the globe. The time will not be long in coming when those men who are turning their ingenuity to inventing quick-firing guns, torpedoes and other implements of destruction — all the while assuring you that it is for the love and good of humanity — will find no takers for their odious tools, and will realize that, had they used their inventive talent to other directions, they might have reaped a far better reward than the sestertia received. And then, and none too soon, the cry will be echoed everywhere. Brethren, stop these high-handed methods of the strong, these remnants of barbarism so inimical to progress! Give that valiant warrior opportunities for displaying a more commendable courage than that he shows when, intoxicated with victory, he rushes to the destruction of his fellow-men. Let Lim toil day and night with a small chance of achieving and yet be unflinching; let him challenge the dangers of exploring the heights of the air and the depths of the sea; let him brave the dread of the plague, the heat of the tropic desert and the ice of the polar region. Turn your energies to warding off the common enemies and dangers, the perils that are all around you, that threaten you in the air you breathe, in the water you drink, in the food you consume. Is it not strange, is it not sham, that we, beings. in the highest state of development in this our world, beings with such immense powers of thought and action, we, the. masters of the globe, should be absolutely at the merry of our unseen foes, that we should not know whether a swallow of food or drink brings joy and life or pain and destruction to us! In this most modern and sensible warfare, in which the bacteriologist leads, the services electricity will render will prove invaluable. The economical production of high-frequency currents, which is now an accomplished fact, enables us to generate easily and in large quantities ozone for the disinfection of the water and the air, while certain novel radiations recently discovered give hope of finding effective remedies against ills of microbic origin, which have heretofore withstood all efforts of the physician. But let me turn to a more pleasant theme.

   I have referred to the merging together of the various sciences or departments of research, and to a certain perception of intimate connection between the manifold and apparently different forces and phenomena. Already we know, chiefly through the efforts of a bold pioneer, that light, radiant heat, electrical and magnetic actions are closely related, not to say identical. The chemist professes that the effects of combination and separation of bodies he observes are due to electrical forces, and the physician and physiologist will tell you that even life's progress is electrical. Thus electrical science has gained a universal meaning, and with right this age can claim the name "Age of Electricity."

   I wish much to tell you on this occasion — I may say I actually burn for desire of telling you — what electricity really ii, but I have very strong reasons, which my co-workers will best appreciate, to follow a precedent established by a great and venerable philosopher, and I shall not dwell on this purely scientific aspect of electricity.

There is another reason for the claim which I have before stated which is even more potent than the former, and that is the immense development in all electrical branches in more recent years and its influence upon other departments of science and industry. To illustrate this influence I only need to refer to the steam or gas engine. For more than half a century the steam engine has served the innumerable wants of man. The work it was called to perform was of such variety and the conditions in each case were so different that, of necessity, a great many types of engines have resulted. In the vast majority of cases the problem put before the engineer was no:, as it should have been, the broad one of converting the greatest possible amount of heat energy into mechanical power, but it was rather the specific problem of obtaining the mechanical power in such form as to be best suitable for general use. As the reciprocating motion of the piston was not convenient for practical purposes, except in very few instances, the piston was connected to a crank, and thus rotating motion was obtained, which was more suitable and preferable, though it involved numerous disadvantages incident to the crude and wasteful means employed. But until quite recently there were at the disposal of the engineer, for the transformation and transmission of the motion of the piston, no better ,means than rigid mechanical connections. The past few years have brought forcibly to the attention of the builder the electric motor, with its ideal features. Here was a mode of transmitting mechanical motion simpler by far, and also much more economical. Had this mode been perfected earlier, there can be no doubt that, of the many different types of engine, the majority would not exist, :for just as soon as an engine was coupled with an electric generator a type was produced capable of almost universal use. From this moment on there was no necessity to endeavor to perfect engines of special designs capable of doing special kinds of work. The engineer's task became now to concentrate all his efforts upon one type, to perfect one kind of engine — the best, the universal, the engine of the immediate future; namely, the one which is best suitable for the generation of electricity. The first efforts in this direction gave a strong impetus to the development of the reciprocating high. speed engine, and also to the turbine, which latter was a type of engine of very limited practical usefulness, but became, to a certain extent, valuable in connection with the electric generator and orator. Still, even the former engine, though improved in many particulars, is not radically changed, and even now has the same objectionable features and limitations. To do away with these as much as possible, a new type of engine is being perfected in which more favorable conditions for economy are maintained, which expands the working fluid with utmost rapidity and loses little heat on the walls, an engine stripped of all usual regulating mechanism — packings, oilers and other appendages — and forming part of an electric generator; and in this type, I may say, I have implicit faith.

The gas or explosive engine has been likewise profoundly affected by the commercial introduction of electric light and power, particularly in quite recent years. The engineer is turning his energies more and more in this direction, being attracted by the prospect of obtaining a higher thermodynamic efficiency. Much larger engines are now being built, the construction is constantly improved, and a novel type of engine, best suitable for the generation of electricity, is being rapidly evolved.

There are many other lines of manufacture and industry in which the influence of electrical development has been even more powerfully felt. So, for instance, the manufacture of a great variety of articles of metal, and especially of chemical products. The welding of metals by electricity, though involving a wasteful process, has, nevertheless, been accepted as a legitimate art, while the manufacture of metal sheet, seamless tubes and the like affords promise of much improvement. We are coming gradually; but surely, to the fusion of bodies and reduction of all kinds of ores — even of iron ores — by the use of electricity, and in each of these departments great realizations are probable. Again, the economical conversion of ordinary currents of supply into high-frequency currents opens, up new possibilities, such as the combination of the atmospheric nitrogen and the production of its compounds; for instance, ammonia and nitric acid, and their salts, by novel processes.

The high-frequency currents also bring us to the realization of a more economical system of lighting; namely by means of phosphorescent bulbs or tubes, and enable us to produce with these appliances light of practically any candle-power. Following other developments in purely electrical lines, we have all rejoiced in observing .the ;rapid strides made, which, in quite recent years, have been . beyond our most sanguine expectations. To enumerate the many advances recorded is a subject for the reviewer, but I can not pass without mentioning the beautiful discoveries of Lenard and Roentgen, particularly the latter, which have found such a powerful response throughout the scientific world that they have made us forget, for a time, the great achievement of Linde in Germany, who has effected the liquefaction of air on an industrial scale by a process of continuous cooling: the discovery of argon by Lord Rayleigh and Professor Ramsay, and. the splendid pioneer work of Professor Dewar in the field of low temperature research. The fact that the United States have contributed a very liberal share to this prodigious progress must afford to all of us great satisfaction. While. honoring the workers in other countries and all those who, by profession or inclination, are devoting themselves to strictly scientific pursuits, we have particular reasons to mention with gratitude the names of those who have so much contributed to this marvelous development of electrical industry in this country. Bell, who, by his admirable invention enabling us to transmit speech to great distances, has profoundly affected our commercial and social relations, and even our very mode of life; Edison, who, had he not done anything else beyond his early work in incandescent lighting, would have proved himself one of the greatest benefactors of the age; Westinghouse, the founder of the commercial alternating system; Brush, the great. pioneer of arc lighting; Thomson, who gave us the first practical welding machine, and who, with keen sense, contributed very materially to the development of a number of scientific and industrial branches; Weston, who once led the world in dynamo design, and now leads in the construction of electric instruments; Sprague, who, with rare energy, mastered the problem and insured the success of practical electrical railroading; Acheson, Hall, Willson and others, who are creating new and revolutionizing industries here under our very eyes at Niagara. Nor is the work of these gifted men nearly finished a t this hour. Much more is still to come, for, fortunately, most of, them are still full of enthusiasm and vigor. All of these men and many more are untiringly at work investigating new regions and opening up unsuspected and promising fields. Weekly, if not daily, we learn through the journals of a new advance into some unexplored region, where at every step success beckons friendly, and leads the toiler on to hard and harder tasks.

But among all these many departments of research, these many branches of industry, new and old, which are being rapidly expanded, there is one dominating all others in importance — one which is of the greatest significance for the comfort and welfare, not to say for the existence, of mankind, and that is the electrical transmission of power. And in this most important of all fields, gentlemen; long afterwards, when time will have placed the events in their proper perspective, and assigned men to their deserved places, the great event we are commemorating to-day will stand out as designating a new and glorious epoch in the history of humanity — an epoch grander than that marked by the advent of the steam engine. We have many a monument of past ages; we have the palaces and pyramids, the temples of the Greek and the cathedrals of Christendom. In them is exemplified the power of men; the greatness of nations, the love of art and religious devotion. But that monument at Niagara has something of its own, more in accord with our present thoughts and tendencies. It is a monument worthy of our scientific age, a true monument of enlightenment and of peace. It signifies the subjugation of natural forces to the service of. man, the discontinuance of barbarous methods, the relieving of millions from want and suffering. No matter what we attempt to do, no matter to, what fields we turn our efforts, we are dependent on power. Our economists may propose more economical systems of administration and utilization of resources, our legislators may make wiser laws and treaties, it matters little; that kind of help can be only temporary. If we want to reduce poverty and misery, we want to give to every deserving individual what is needed for a safe existence of an intelligent being, we want to provide 'More machinery, more power. Power is our mainstay, the primary source of our many-sided energies. With sufficient power at our disposal we can satisfy most of our wants and offer a guaranty for safe and comfortable existence to all, except perhaps to those who are the greatest criminals of all — the voluntarily idle.

The development and wealth of a city, the success of a nation, the progress of the whole human race, is regulated by the power available. Think of the victorious march of the British, the ,like of which history has never recorded. Apart from the qualities of the race, which have been of great moment, they own the conquest of the world to — coal. For with coal they produce their iron; coal furnishes them light and heat; coal drives the wheels of their immense manufacturing establishments, and coal propels their conquering fleets. But the stores are being more and more exhausted, the labor is getting dearer and dearer, and the. demand is continuously increasing. It must be clear to every one that soon some new source of power supply must be opened. up, or that at least the present methods must be materially improved. A great deal is expected from a more economical utilization of the stored energy of the carbon in a battery; but while the attainment of such a result would be hailed as a great achievement, it would not be as much of an advance towards the ultimate and permanent method of obtaining power as some engineers seem to believe. By reasons both of economy and convenience we are driven to the general adoption of a system of energy supply from central stations, and for such purposes the beauties of the mechanical generation of electricity can not be exaggerated. The advantages of this universally accepted method are certainly so great that the probability of replacing the engine dynamos by batteries is, in my opinion, a remote one, the more so as the high-pressure steam engine and gas engine give promise of a considerably more economical thermodynamic conversion. Even if we had this day such an economical coal battery, its introduction in central stations would by no means be assured, as its use would entail many inconveniences and drawbacks. Very likely the carbon could not be burned in its natural form as in a boiler, but would have to be specially prepared to secure uniformity in the current generation. There would be a great many cells needed to make up the electro-motive force usually required. The process of cleaning and renewal, the handling of nasty fluids and gases and the great space necessary for so many batteries would make it difficult, if not commercially unprofitable, to operate such a ‚plant in a city or densely populated district. Again, if the station be erected in the outskirts, the conversion by rotating transformers or otherwise would be a serious and unavoidable drawback. Furthermore, the regulating appliances and other accessories which would have to be provided would probably make the plant fully as much, if not more, complicated than the present. We might, of course, place the batteries at or near the coal mine, and from there transmit the energy to distant points in the form of high-tension alternating currents obtained from rotating transformers, but even in this most favorable case the process would be a barbarous one, certainly more so than the present, as it would still involve the consumption of material, while at the same time it would restrict the engineer and mechanic in the exercise of their beautiful art. As to the energy supply in small isolated places as dwellings, I have placed my confidence in the development of a light storage battery, involving the use of chemicals manufactured by cheap water power, such as some carbide or oxygen-hydrogen cell.

But we shall not satisfy ourselves simply with improving steam and explosive engines or, inventing new batteries; we have ‚something much better to work for, a greater task to fulfill. We have to evolve means for obtaining energy from stores which are forever inexhaustible, to perfect methods which do not imply consumption and waste of any material whatever. Upon this great possibility, which I have long ago recognized, upon this great problem, the practical solution of which means so much for humanity, I have myself concentrated my efforts since a number of years, and a few happy ideas which came to me have inspired me to attempt the most difficult, and given me strength and courage in adversity. Nearly six years ago my confidence had become strong enough to prompt me to an expression of hope in the ultimate solution of this all dominating problem. I have made progress since, and have passed the stage of mere conviction such as is derived from a diligent study of known facts, conclusions and calculations. I now feel sure that the realization of that idea is not far off. But precisely for this reason I feel impelled to point out here an important fact, which I hope will be remembered. Having examined for a long time the possibilities of the development I refer to, namely, that of the operation. of engines on any point of the earth by the energy of the medium, I find that even under the theoretically best conditions such a method of obtaining power can not equal in economy, simplicity and many other features the present method, involving a conversion of the mechanical energy of running water into electrical energy and the transmission of the latter in the form of currents of very high tension to great distances. Provided, therefore, that we can avail ourselves of currents of sufficiently high tension, a waterfall affords us the most advantageous means of getting power from the sun sufficient for all our wants, and this recognition has impressed me strongly with the future importance of the water power, not so much because of its commercial value, though it may be very great, but chiefly because of its bearing upon our safety and welfare. I am glad to say that also in this latter direction my efforts have not been unsuccessful, for I have devised means which will allow us the use in power transmission of electro-motive forces much higher than those practicable with ordinary apparatus. In fact, progress in this field has given me fresh hope that I shall see the fulfillment of one of my fondest dreams; namely, the transmission of power from station to station without the employment of any connecting wire. Still, whatever method of transmission be ultimately adopted, nearness to the source of power will remain an important advantage.

Gentlemen, some of the ideas I have expressed may appear to many of you hardly realizable; nevertheless, they are the result. of long-continued thought and work. You would judge them more justly if you would have devoted your life to them, as I have done. With ideas it is like with dizzy heights you climb: At first they cause you discomfort and you are anxious to get down, distrustful of your own powers; but soon the remoteness of the turmoil of life and the inspiring influence of the altitude calm your blood; your step gets firm and sure and you begin to look — for dizzier heights. I have attempted to speak to you on "Electricity," its development and influence, but I fear that I have done it much like a boy who tries to draw a likeness with a few straight lines. But I have endeavored to bring out one feature, to speak to you in one strain which I felt sure would find response in the hearts of all of you, the only one worthy of this occasion — the humanitarian. In the great enterprise at Niagara we see not only a bold engineering and commercial feat, but far more, a giant stride in the right direction as indicated both by exact science and philanthropy. Its success is a signal for the utilization of water powers all over the world, and its influence upon industrial development is incalculable. We must all rejoice in the great achievement and congratulate the intrepid pioneers who have joined their efforts and means to bring it about. It is a, pleasure to learn of the friendly attitude of the citizens of Buffalo and of the encouragement given to the enterprise by the Canadian authorities. We shall hope that other cities, like Rochester on this side and Hamilton and Toronto in Canada, will soon follow Buffalo's lead. This fortunate city herself is to be congratulated. With resources now unequaled, with commercial facilities and advantages such as few cities in the world possess, and with the enthusiasm and progressive spirit of its citizens, it is sure to become one of the greatest industrial centers of the globe.

On the Hurtful Actions of Lenard and Roentgen Tubes
Electrical Review — May 5, 1897

The rapidly extending use of the Lenard and Roentgen tubes or Crookes bulbs as implements of the physician, or as instruments of research in laboratories, makes it desirable, particularly in view of the possibility of certain hurtful actions on the human tissues, to investigate the nature of these influences, to ascertain the conditions under which they are liable to occur and — what is most important for the practitioner — to render all injury impossible by the observance of certain rules and the employment of unfailing remedies.

As I have stated in a previous communication (see Electrical Review of December 2, 1896) no experimenter need be deterred from using freely the Roentgen rays for fear of a poisonous or deleterious action, and it is entirely wrong to give room to expressions of a kind such as may tend to impede the progress and create a prejudice against an already highly beneficial and still more promising discovery; but it can not be denied that it is equally uncommendable to ignore dangers now when we know that, under certain circumstances, they actually exist. I consider it the more necessary to be aware of these dangers, as I foresee the coming into general use of novel apparatus, capable of developing rays of incomparably greater power. In scientific laboratories the instruments are usually in the hands of persons skilled in their manipulation and capable of approximately estimating the magnitude of the effects, and the omission of necessary precautions is, in the present state of our knowledge, not so much to be apprehended; but the physicians, who are keenly appreciating the immense benefits derived from the proper application of the new principle, and th= numerous amateurs who are fascinated by the beauty of the novel manifestations, who are all passionately bent upon experimentation in the newly opened up fields, but many of whom are naturally not armed with the special knowledge of the electrician — all of these are much in need of reliable information from experts, and for these chiefly the following lines are written. However, in view of the still incomplete knowledge of these *rays,* I wish the statements which follow to be considered as devoid of authoritativeness, other than that which is based on the conscientiousness of my study and the faith in the precision of my senses and observations.

Ever since Professor Roentgen's discovery was made known I have carried on investigations in the directions indicated by him, and with perfected apparatus, producing rays of much greater intensity than it was possible to obtain with the usual appliances. Commonly, my bulbs were capable of showing the shadow of a hand on a phosphorescent screen at distances of 40 or 50 feet, or even more, and to the actions of these bulbs myself and several of my assistants were exposed for hours at a time, and although the exposures took place every day, not the faintest hurtful action was noted — as long as certain precautions were taken. On the contrary, be, it a coincidence, or an effect of the rays, or the result of some secondary cause present in the operation of the bulbs — as, for example, the generation of ozone — my own health, and that of two persons who were daily under the influence of the rays, more or less, has materially improved, and, whatever be the reason, it is a fact that a troublesome cough with which Z was constantly afflicted has entirely disappeared, a similar improvement being observed on another person.

In getting the photographic impressions or studying the rays with a phosphorescent screen, I employed a plate of thin aluminum sheet or a gauze of aluminum wires, which was interposed between the bulb and the person, and connected to the ground directly or through a condenser. I adopted this precaution because it was known to me, a long time before, that a certain irritation of the skin is caused by very strong streamers, which, mostly at small distance, are formed on the body of a person through . the electrostatic influence of a terminal of alternating high potential. I found that the occurrence of these streamers and their hurtful consequence was completely prevented by the employment of a conducting object, as a sheet of wire gauze placed and connected as described. It was observed, however, that the injurious effects mentioned did not seem to diminish gradually with the distance from the terminal, but ceased abruptly, and I could give no . other explanation for the irritation of the skin which would be as plausible as that which I have expressed; namely, that the effect was due to ozone, which was abundantly produced. The latter peculiarity mentioned was also in agreement with this view, since the generation of ozone ceases abruptly at a definite distance from the terminal, making it evident that a certain intensity of action is absolutely required, as in a process of electrolytic decomposition.

In carrying further my investigations, I gradually modified the apparatus in several ways, and immediately I had opportunities to observe hurtful influences following the exposures. Inquiring now what changes I had introduced, I found that I had made three departures from the plan originally followed; First, the aluminum screen was not used; second, a bulb was employed which, instead of aluminum, contained platinum, either as electrode or impact plate; and third, the distances at which the exposures took place were smaller than usual.

It did not require a long time to ascertain that the interposed aluminum sheet was a very effective remedy against injury, for a hand could be exposed for a long time behind it without the skin being reddened, which otherwise invariably and very quickly occurred. This fact impressed me with the conviction that, whatever the nature of the hurtful influences, it was in a large measure dependent either on an electrostatic action, or electrification, or secondary effects resulting therefrom, such as are attendant to the formation of streamers. This view afforded an explanation why an observer could watch a bulb for any length of time, as long as he was holding the hand in front of the body, as in examining with a fluorescent screen, with perfect immunity to all parts of hi& body, with the exception of the hand. It likewise explained why burns were produced in some instances on the opposite side of the body, adjacent to the photographic plate, whereas portions on the directly exposed part of the body, which were much nearer to the bulb, and consequently subjected to by far stronger rays, remained unaffected. It also made it easy to understand why the patient experienced a prickling sensation on the exposed part of the body whenever an injurious action took place. Finally, this view agreed with the numerous observations that the hurtful actions occurred when air was present, clothing, however thick, affording no protection, while they practically ceased when a layer of a fluid, quite easily penetrated by the rays, but excluding all contact of the air with the skin, was used as a preventive.

Following, now, the second line of investigation, I compared bulbs containing aluminum only with those in which platinum was used besides, ordinarily as impact body, and soon there were enough evidences on hand to dispel all doubt as to the latter metal being by far the more injurious. In support of this statement, one of the experiences may be cited which, at the same time, may illustrate the necessity — of taking proper precautions when operating bulbs of very high power. In order to carry out comparative tests, two tubes were constructed of an improved Lenard pattern, in size and most' other respects' alike. Both contained. a concave cathode or reflector of nearly two inches in diameter, and both were provided with an aluminum cap or window. In one of the tubes the cathodic focus was made to coincide with the center of the cap, in the other the cathodic stream was concentrated upon a platinum wire supported on a glass stem axially with the tube a little in front of the window, and in each case the metal of the latter was thinned down in the central portion to such an extent as to be barely able to withstand the inward air pressure. In studying the action of the tubes, I exposed one hand to that containing aluminum only, and' the other to the tube with the platinum wire. On turning on the former tube, I was surprised to observe that the aluminum window emitted a clear note, corresponding to the rhythmical impact of the cathodic stream. Placing the hand quite near the window, I felt distinctly that something warm was striking it. The sensation was unmistakable, and, quite apart. from the warmth felt, differed very much from that prickling feeling produced by streamers or minute sparks. Next I examined the tube with the platinum wire. No sound was emitted by the aluminum window, all the energy of the impact being seemingly spent on the platinum wire, which became incandescent, or else the matter composing the cathodic stream was so far disintegrated that the thin metal sheet offered no material obstruction to its passage. If big lumps are hurled against a wire netting with large meshes, there is considerable pressure exerted against the netting; if, on the contrary — for illustration — the lumps are very small as compared with the meshes, the pressure might not be manifest. But, although the window did not vibrate, I felt, nevertheless, again, and distinctly, that something was impinging against the hand, and the .sensation of warmth was stronger than in the previous case. In the action on the screen there was apparently no difference between the two tubes, both rendering it very bright, and the definition of the shadows was the same, as far as it was possible to judge. I had looked through the screen at the second tube a few times, only when something detracted my attention, and it was not until about 20 minutes later, when I observed that the hand exposed to it was much reddened and swollen. Thinking that it was due to some accidental injury, I turned again to the examination of the platinum tube; thrusting the same hand close to the window, and now I felt instantly a sensation of pain, which became more pronounced when the hand was placed repeatedly near the aluminum window. A peculiar feature was that the pain appeared to be seated, not at the surface, but deep in the tissues of the hand, or rather in the bones. Although the aggregate exposure was certainly not more than half a minute, I had to suffer severe pain for a few days afterward, and some time later I observed that all the *hair was* destroyed and that the nails on the injured hand had grown anew.

The bulb containing no platinum was now experimented with, more care being taken, but soon its comparative harmlessness was manifest, for, while it reddened the skin; the injury was not nearly as severe as with the other tube. The valuable experiences thus gained were: The evidence of something hot striking the exposed member; the pain *instantly* felt; the injury produced *immediately* after the exposure, and the increased violence due, in all probability, to the presence of the platinum.

Some time afterward I observed other remarkable actions at very small distances from powerful Lenard tubes. For instance, the hand being held near the window only for a few seconds, the skin seems to become tight, or else the muscles are stiffened, for some resistance is experienced in closing the fist, but upon opening and dosing it repeatedly the sensation disappears, apparently no ill effect remaining. I have, furthermore, observed a decided influence on the nasal discharge organs similar to the effects of a cold just contracted. But the most interesting observation in this respect is the following: When such a powerful bulb is watched for some time, the head of the observer being brought very close, he soon after that experiences a sensation so peculiar that no one will fail to notice it when once his attention is called to it, it being almost as positive as touch. If one imagines himself looking at something like a cartridge, for instance, in close and dangerous proximity, and just about to explode, he will get a good idea of the sensation produced, only, in the case of the cartridge, one can not render himself an account where the feeling exactly resides, for it seems to extend all over the body, this indicating that it comes from a general awareness of danger resulting from previous and manifold experiences, and not from the anticipation of an unpleasant impression directly upon one of the organs, as the eye or the ear; but, in the case of the Lenard bulb, one can at once, and with precision, locate the sensation; it is in the head. Now, this observation might not be of any value except, perhaps, in view of the peculiarity and acuteness of the feeling, were it not that exactly the same sensation is .produced when working for some time with a noisy spark gap, or, in general, when exposing the ear to sharp noises or explosions. Since it seems impossible to imagine how the latter could cause such a sensation in any other way except by directly impressing the organs of hearing, I conclude, that a Roentgen or Lenard tube, working in perfect silence as it may, nevertheless produces violent explosions or reports and concussions, which; though they are inaudible, take some material effect upon the bony structure of the head. Their inaudibility may be sufficiently explained by the well founded assumption that not the air, but some finer medium, is concerned in their propagation.

But it was in following up the third line of inquiry into the nature of these hurtful actions, namely, in studying the influence of distance, that the most important fact was unearthed. To illustrate it popularly, I will say that the Roentgen tube acts exactly like a source of intense heat. If one places the hand near to a red-hot stove, he may be instantly injured. If he keeps the hand at

a. certain small distance, he may be able to withstand the rays for a few minutes or more, and may still be injured by prolonged exposure; but if he recedes only a little farther, where the heat is slightly less, he may withstand the heat in comfort and any length of time without receiving any injury, the radiations at that distance being too weak to seriously interfere with the life process of the skin. This is absolutely the way such a bulb acts. Beyond a certain distance no hurtful effect whatever is produced on the skin, no matter *how long* the exposure. The character of the burns is also such as might be expected from a source of high heat. I have maintained, in all deference to the opinions of others, that those who have likened the effect`s on the skin arid tissues to sunburns have misinterpreted them. There is no similarity in this respect, except in so far 'as the reddening and peeling of the skin is concerned; which may result from innumerable causes. The burns, when slight, rather resemble those people often receive when working .close to a strong fire. But when the injury is severe, it is in all appearances like that received from contact with fire or from a red-hot iron. There may be no period of incubation at all, as is evident from the foregoing, remarks; the rays taking effect immediately, not to say instantly. In a severe case the skin gets deeply colored and blackened in places, and ugly; ill-foreboding blisters form; thick layers come off, exposing the raw flesh, which; for a time; discharges freely, Burning pain, feverishness and such symptoms are of course but natural accompaniments. One single injury of this kind, in the abdominal region, to a dear and zealous assistant — the only accident that ever happened to any one but myself in all my laboratory experience — I had the misfortune to witness. It occurred before all these and other experiences were gained; following directly an exposure of five. minutes at. the fairly safe distance of 11 inches to a very highly-charged platinum tube, the protecting aluminum screen having been unfortunately omitted, and it was such as to fill me with the gravest apprehensions. Fortunately, frequent warm baths, free application of vaseline, cleaning and general bodily care soon repaired the ravages of the destructive agent, and I breathed again freely. Had I known more than I did of these injurious actions, such unfortunate exposure would not have been made; had I known less than I did, it might have been made at a smaller distance, and a serious, perhaps .irremediable, injury might have resulted. ,

I am using the first opportunity to comply with the bitter duty of recording the accident. I hope that others will do likewise, so that the most complete knowledge of these dangerous. actions may soon be acquired. My apprehensions led me to consider, with keener interest than I would have felt otherwise, what the probabilities were in such. a case of the internal tissues being seriously injured. I came to the very comforting conclusion that, no matter what the rays are ultimately recognized to be, practically all their destructive energy must spend itself on the surface of the body, the internal tissues being, in all probability, safe, unless the bulb would be placed in very close proximity to the skin, or else, that rays of far greater intensity than now producible were generated. There are many reasons why this should be so, some of which will appear dear from my foregoing statements referring to the nature of the hurtful agencies, but I may be able to cite new facts in support of this view. A significant feature of the case reported may be mentioned. It was observed that on three places, which were covered by thick bone buttons, the skin was entirely unaffected, while it was entirely destroyed under each of the small holes in the buttons. Now, it was impossible for the rays, as investigation showed, to reach these points of the skin in straight lines drawn from the bulb, and this would seem to indicate that not all the injury was due to the *rays* or radiations under consideration, which unmistakably propagate in straight lines, but that, at least in part, concomitant causes were responsible. A further experimental demonstration of this fact may be obtained in the following manner: The experimenter may excite a bulb to a suitable and rather small degree, so as to illuminate the fluorescent screen to a certain intensity at a distance of, say, seven inches. He may expose his hand at that distance, and the skin will be reddened after a certain duration of exposure. He may now force the bulb up to a much higher power, until, at a distance of 14 inches, the screen is illuminated even stronger than it was before at half that distance: The rays are now evidently stronger at the greater distance, and yet he may expose the hand a very long time, and it is safe to assert that he will not be injured. Of course, it is possible to bring forth arguments which might deprive the above demonstration of force. So, it might. be stated, that the actions on the screen or photographic plate do not give us an idea as to the density and other quantitative features of the rays, these actions being entirely of a qualitative character. Suppose the rays are formed by streams of material particles, as I believe, it is thinkable that it might be of no particular consequence, in so far as the visible impression on the screen or film is concerned, whether a trillion of particles per square millimeter strike the sensitive layer or only a million, for example; but with the actions on the skin it is different; theca must surely and very materially depend on the quantity of the streams.

As soon as the before-mentioned fact was recognized, namely, that beyond a certain distance even the most powerful tubes are incapable of producing injurious action; no matter how long the exposure may last, it became important to ascertain the safe distance. Going over all my previous experiences, I found that, very frequently, I have had tubes which at a distance of 12 feet, .for illustration, gave a strong impression of the chest of a person with an exposure of a few minutes, and many times persons have been subjected to the rays from these tubes at a distance of from 18 to 24 inches, the time of exposure varying from 10 to 45 minutes, and never the faintest trace of an injurious action was observed. With such tubes I have even made long exposures at distances of 14 inches, always, of course, through a thin sheet or wire gauze of aluminum connected to the ground, and, in each case, observing the precaution that the metal would not give any spark when the person was touching it with the hand, as it might sometimes be when the electrical vibration is of extremely high frequence, in, which we a ground connection, through a condenser of proper capacity, should be resorted to. In all these instances bulbs containing only aluminum were used; and I therefore still lack sufficient data to form an enact idea of what distance would have been safe with a platinum tube. Froth the case previously cited, we see that a grave injury, resulted at a distance of 11 inches, but I believe that, had the protecting screen been used, the injury, if any, would have-been very slight. Taking all my experiences together, I am convinced that no serious injury can result if the distance is greater than 16 inches and the impression is taken in the manner I have described.

Having been successful in a number of lines of inquiry pertaining to this new department of science, I am able at present to form a broader view of the actions of the bulbs, which; I hope, will soon assume a quite definite shape. For the present, the following brief statement may be sufficient. According to the evidences I am obtaining, the bulb, when in action, is emitting a stream of small material particles. There are some experiments which seem to indicate that these particles start from the outer wall of the bulb; there are others which seem to prove that there is an actual penetration of the wall, and, in the case of a thin aluminum window, I have now not the least doubt that some of the finely disintegrated cathodic matter is actually forced through. These streams may simply be projected to a great distance, the velocity gradually diminishing without the formation of any waves, or they may give rise to concussions and longitudinal waves. This, for the present consideration, is entirely immaterial, but, assuming the existence of such streams of particles, and disregarding such actions as might be due to the properties, chemical or physical, of the projected matter, we have to consider the following specific actions:

First. There is the thermal effect. The temperature of the electrode or impact body does not in any way give us an idea of the degree of heat of the particles, but, if we consider the probable velocities only, they correspond to temperatures which may be as high as 100,000 degrees centigrade. It may be sufficient that the particles are simply at a high temperature to produce an injurious action, and, in fact, many evidences point in this direction. But against this is the experimental fact that we can not demonstrate such a transference of heat, and no satisfactory explanation is found yet, although, in carrying my investigations in this direction, I have arrived at some results.

Second, there is the purely electrical effect. We have absolute experimental evidence that the particles or rays, to express myself generally, convey an immense amount of electricity, and I have even found a way of how to estimate and measure that amount. Now it is likewise possible that the mere fact of these particles being highly electrified is sufficient to cause the destruction of the tissue. Certainly, on contact with the skin, the electrical charges will be given off, and may give rise to strong and destructive local currents in minute paths of the tissue. Experimental results are in accord with this view, and, in pushing my inquiry in this direction, I have been still more successful than in the first. Yet, while as I have suggested before, this view explains best the action on a sensitive layer, experiment shows that, when the supposed particles traverse a grounded plate, they are not deprived entirely of their electrification, which is not satisfactorily explained.

The *third* effect to be considered is the electro-chemical. The charged particles give rise to an abundant generation of ozone and other gases, and these we know, by experiment, destroy even such a thing as rubber, and are, therefore, the most likely agent in the destruction of the skin, and the evidences are strongest in this direction, since a small layer of a fluid, preventing the contact of gaseous matter with the skin, seems to stop all action.

The *last* effect to be considered is the purely mechanical. It is thinkable that material particles, moving with great speed, may, merely by a mechanical impact and unavoidable heating at such speeds, be sufficient to deteriorate the tissues, and in such a case deeper layers might also be injured, whereas it is very probable that no such. thing would occur if any, of the former explanations would be found to hold.

Summing up my experimental experiences. and the conclusions derived from them, it would seem advisable,. first, to abandon, the use of bulbs containing platinum; second, to substitute for them a properly constructed Lenard tube, containing pure aluminum only, a tube of this kind having, besides, the advantage that it can be constructed with great mechanical precision, and therefore is capable of producing .much sharper impressions; third, to use a protecting screen of aluminum sheet, as suggested, or, instead of this, a wet cloth or a layer of a fluid; fourth, to make the exposures at distances of, at least, 14 inches, and preferably expose longer at a larger distance.

On the Source of Roentgen Rays and the Practical Construction and Safe Operation of Lenard Tubes
Electrical Review — August 11, 1897

I have for some time felt that a few indications in regard to the practical construction of Lenard tubes of improved designs, a great number of which I have recently exhibited before the New York Academy of Sciences (April 6, 1897), would be useful and timely, particularly as by their proper construction and use much of the danger attending the experimentation with the rays may be avoided. The simple precautions which I have suggested in my previous communications are seemingly disregarded, and cases of injury to patients are being almost daily reported, and in view of this only, were it for no other reason, the following lines, referring to this subject, would have been written before had not again pressing and unavoidable duties prevented me from doing so. A short and, I may say, most unwelcome interruption of the work which has been claiming my attention makes this now possible. However, as these opportunities are scarce, I will utilize the present to dwell in a few words on some other matters in connection with this subject, and particularly on a result of importance which I have reached some time ago by the aid of such a Lenard tube, and which, if I am correctly informed, I can. only in part consider as my own, since it seems that practically it has been expressed in other words by Professor Roentgen in a recent communication to the Academy of Sciences of Berlin. The result alluded to has reference to the much disputed question of the source of the Roentgen rays. As will be remembered, in the first announcement of his discovery, Roentgen was of the opinion that the rays which affected the sensitive layer emanated from the fluorescent spot on the glass wall of the bulb; other scientific men next made the cathode responsible; still others the anode, while some thought that the rays were emitted solely from fluorescent powders of surfaces, and speculations, mostly unfounded, increased to such an extent that, despairingly, one , would exclaim with the poet:

"*O glucklich wer noch hoffen kann,*
Aus diesem Meer des Irrtums aufzutauchen!"

My own experiments led me to recognize that, regardless of the location, the chief source of, these rays was the place of the *first* impact of .the projected stream of particles within the bulb. This was merely a broad statement, of which that of Professor Roentgen was a special case, as in his first experiments the fluorescent spot on the glass wall was, incidentally, the place of the first impact of the cathodic stream. Investigations carried on up to the present day have only confirmed the correctness of the above opinion, and the place of the first collision of the stream of particles — be it an anode or independent impact body, the glass wall or an aluminum window — is still found to be: the principal source of the rays. But, as will be seen presently, it is not the only source.

Since recording the above fact my efforts were directed to finding answers to the following questions: First, is it necessary that the impact body should be within the tube? Second, is it required that the obstacle in the path of the cathodic stream should be a solid or liquid? And, third, to what extent is the velocity of the stream necessary for the generation of and influence upon the character of the rays emitted?

In order to ascertain whether a body located outside of the tube and in the path or in the direction of the stream of particles was capable of producing the same peculiar phenomena as an object located inside; it appeared necessary to first show that there is an actual penetration of the particles through the wall, or otherwise that' the actions of the supposed streams; of whatever nature they might, be, were sufficiently pronounced in the outer region close to the wall of the bulb as to produce some of the effects which are peculiar to a cathodic stream. It was not difficult to obtain with a properly prepared Lenard tube, having an exceedingly thin window, many and at first surprising evidences of this character. Some of these have already, been pointed out; and it is thought sufficient to cite here one more which I have since observed. In the hollow aluminum cap A of a tube as shown in diagram Fig. 1, which will be described in detail, I placed a half-dollar silver piece, supporting it at a small distance from and parallel to the window or bottom of the cap by strips of mica in such a' manner . that it was not touching the metal of the tube, an air space being left all around it: Upon exciting the bulb for about 30 to 45 seconds by the secondary discharge of a powerful coil of a novel type now well known, it was found that the silver piece was rendered so hot as to actually scorch the hand; yet the aluminum window, which offered a very insignificant obstacle to the cathodic stream, was only moderately warmed. Thus it was shown that the silver alloy, owing to its density and thickness, took up most of the energy of the impact, being acted upon by the particles almost identically as if it had been inside of the bulb, and, what is more, indications were obtained, by observing the shadows, that it behaved like a second source of the rays, inasmuch as the outlines of the shadows, instead of being sharp and clear as when the half-dollar piece was removed, were dimmed. It was immaterial for the chief object of the inquiry to decide by more exact methods whether the cathodic particles actually penetrated the window, or whether a new and separate stream was projected from the outer side of the window. In my mind there exists not the least doubt that the former was the case, as in this respect I have been able to obtain numerous additional proofs, upon which I may dwell in the near future.

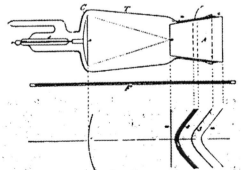

Fig. 1. — Illustrating an Experiment Revealing the Real Source of the Roentgen Rays.

I next endeavored to ascertain whether it was necessary that the obstacle outside was, as in this case, a solid body, or a liquid, or broadly, a body of measurable dimensions, and it was in investigating in this direction that I came upon the important result to which I referred in the introductory statements of this communication. I namely observed rather accidentally, although I was following up a systematic inquiry, what is illustrated in .diagram Fig. 1. The diagram shows a Lenard tube of improved design, consisting of a tube T of thick glass tapering towards the open end, or neck n, into which is fitted an aluminum cap A, and a spherical cathode e, supported on a glass stem s, and platinum wire w sealed in the opposite end of the tube as usual. The aluminum cap A, as will be observed, is not in actual contact. with the ground-glass wall, being held at a small distance from the latter by a narrow and continuous ring of tinfoil r. The outer space between the glass and the cap A is filled with cement c, in a manner which I shall later describe. F is a Roentgen screen such as is ordinarily used in making the observations.

Now, in looking upon the screen in the direction from F to T, the dark lines indicated on the lower part of the diagram were seen on the illuminated background. The curved line e and the straight line W were, of course, at once recognized as the outlines of the cathode a and the bottom of the cap A respectively, although, in consequence of a confusing optical illusion, they appeared mush closer together than they actually were. For instance, if the distance between a and o was five inches, these lines would appear on the screen about two inches apart, as nearly as I could judge by the eye. This illusion may be easily explained and is quite unimportant, except that it might be of some moment to physicians to keep this fact in mind when making examinations with the screen as, owing to the above effect, which is sometimes exaggerated to a degree hard to believe, a completely erroneous idea of the distance of the various parts of the object under 'examination might be gained, to the detriment of the surgical operation. But while the lines a and W were easily accounted for, the curved lines t, g, a were at first puzzling. Soon, however, it was ascertained that the faint line a was the shadow of the edge of the aluminum cap, the much darker line g that of the rim of the glass tube T, and t the shadow of the tinfoil ring r. These shadows on the screen F clearly showed that the agency which affected the fluorescent material was proceeding from the space outside of the bulb towards the' aluminum cap, and chiefly from the region through which the primary disturbances or streams emitted from the tube through the window were passing, which observation could not be explained in a more plausible manner than by assuming that the air and dust particles outside, in the path of the projected streams, afforded an obstacle to their passage and gave rise to impacts and collisions spreading through the air in all directions, thus producing continuously new sources of the rays. It is this fact which; in his recent communication before mentioned, Roentgen has brought out. So, at least, I have interpreted his reported statement that the rays emanate from the irradiated air. It now remains to be shown whether the air, from which carefully all foreign particles are removed, is capable of behaving as an impact body and source of the rays, in order to decide whether the generation of the latter is dependent on the presence in the air of impact particles of measurable dimensions. I have reasons to think so.

With the knowledge of this fact we are now able to form a more general idea of the process of generation of the radiations which have been discovered by Lenard and Roentgen. It may be comprised in the statement that the streams of minute material particles projected from an electrode with great velocity in encountering obstacles wherever they may be, within the bulb, in the air or other medium or in the sensitive layers themselves, give rise to rays or radiations possessing many of the properties of those known as light. If this physical process of generation of these rays is undoubtedly demonstrated as true, it will have most important consequences, as it will induce physicists to again critically examine many phenomena which are presently attributed to transverse ether waves, which may lead to a radical modification of existing views and theories in regard to these phenomena, if not as to their essence so, at least, as to the mode of their production. '

My effort to arrive at an answer to the third of the above questions led me to the establishment, by actual photographs, of the dose relationship which exists between the Lenard and Roentgen rays. The photographs bearing on this point were exhibited of a meeting of the New York Academy of Sciences — before referred to — April 6, 1897, but, unfortunately, owing to the shortness of my address, arid concentration of thought on other matters, I omitted what was most important; namely, to describe the manner in which these, photographs were obtained, an oversight which I was able to only partially repair the day following. I did, however, on that occasion illustrate and describe experiments in which was shown the deflectibility of the Roentgen rays by a magnet, which establishes a still closer relationship, if not identity, of the rays named after these two discoverers. But the description of these experiments in detail, as well as of other investigations and results in harmony with and restricted to the subject I brought before that scientific body, will appear in a longer communication which I am slowly preparing.

To bring out clearly the significance of the photographs in question, I would recall that, in some of my previous contributions to scientific societies, I have endeavored to dispel a popular opinion before existing that the phenomena known as those of Crookes were dependent on and indicative of high vacua. With this object in view, I showed that phosphorescence and most of the phenomena in Crookes bulbs were producible at greater pressures of the gases in the bulbs by the use of much higher or more sudden electro-motive impulses. Having this well demonstrated fact before me, I prepared a tube in the manner described by Lenard in his first classical communication on this subject. The tube was exhausted to a moderate degree, either by chance or of necessity, and it was found that, when operated by an ordinary high-tension coil of a low rate of change in the current, no rays of any of the two kinds could be detected, even when the tube was so highly strained .as to become very hot in a few moments. Now, I expected that, if the suddenness of the impulses through the bulb were sufficiently increased, rays would be emitted. To test this I employed a coil of a type which I have repeatedly described, in which the primary is operated by the discharges of a condenser. With such an instrument any desired suddenness of the impulses may be secured, there being practically no limit in this respect, as the energy accumulated in the condenser is the most violently explosive agent we know, and any potential or electrical pressure is obtainable: Indeed, I found that in increasing the suddenness of the electro-motive, impulses through the tube — without, however, increasing, but rather diminishing the total energy conveyed to it — phosphorescence was observed and rays began to appear, first the feebler Lenard rays and later, by pushing the suddenness far enough, Roentgen rays of great intensity, which enabled me to obtain photographs showing the finest texture of the bones. Still, the same tube, when again operated with the ordinary coil of a low rate of change in the primary current, emitted practically no rays, even when, as before stated, much more energy, as judged from the heating, was passed through it. This experience, together with the fact that I have succeeded in producing. by the use of immense electrical pressures, obtainable with certain apparatus designed for this express purpose, some impressions in free air, have led me to the conclusion that in lightning discharges Lenard and Roentgen *rays* must be generated at ordinary, atmospheric pressure.

At this juncture I realize, by a perusal of the preceding lines, that my scientific interest has dominated the practical, and that the following remarks must be devoted to the primary object of this communication — that is, to giving some data for the construction to those engaged in the manufacture of the tubes and, perhaps, a few useful hints to practicing physicians who are dependent on such information. The foregoing was, nevertheless, not lost for this object, inasmuch as it has shown how much the result obtained depends on the proper construction of the instruments, for, with ordinary implements, most of the above observations could not have been made.

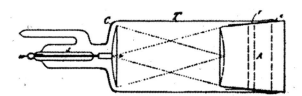

Fig. 2. — Improved Lenard Tube.

I have already described the form of tube illustrated in Fig. 1, and in Fig. 2 another still further improved design is shown. In this case the aluminum cap A, instead of having a straight bottom as before, is shaped spherically, the renter of the sphere coinciding with that of the electrode *e*, which itself, as in Fig. 1, has it's focus in the center of the window of cap A, as indicated by the dotted lines. The aluminum cap A has a tinfoil ring r, as that in Fig. 1, or else the metal of the cap is spun out on that place so as to afford a bearing of small surface between the metal and the glass. This is an important practical detail as, by making the bearing surface small, the pressure per unit of area is increased and a more perfect joint made. The ring r should be first spun out and then ground to fit the neck of the bulb. If a tinfoil ring is used instead, it may be cut out of one of the ordinary tinfoil caps obtainable in the market, care being taken that the ring is very smooth.

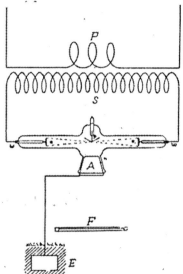

Fig. 3. — Illustrating Arrangement with Improved Double-Focus Tube for Reducing the Injurious Actions.

In Fig. 3 I have shown a modified design of tube which, as the two types before described, was comprised in the collection I exhibited. This, as will be observed, is a double-focus tube, with impact plates of iridium alloy and an aluminum cap A opposite the same. The tube is not shown because of any originality in design, but simply to illustrate a practical feature. It will be noted that the aluminum caps in the tubes described are fitted inside of the necks and not outside, as is frequently done. Long experience has demonstrated that it is practically impossible to maintain a high vacuum in a tube with an outside cap. The only way I have been able to do this in a fair measure is by cooling the cap by a jet of air, for instance, and observing the following precautions: The air jet is first turned on slightly and upon this the tube is excited. The current through the latter, and also the air pressure, are then gradually increased and brought to the normal working condition. Upon completing the experiment the air pressure and current through the tube are both gradually reduced and both so manipulated that no great differences in temperature result between the glass and aluminum cap. If those precautions are not observed the vacuum will be immediately impaired in consequence of the uneven expansion of the glass and metal.

With tubes, as these presently described, it is quite unnecessary to observe this precaution if proper care is taken in their preparation. In inserting the cap the latter is cooled down as low as it is deemed advisable without endangering the glass, and it is then, gently pushed in the neck of the tube, taking care that it sets straight.

The two most important operations in the manufacture of such a tube are, however, the thinning down of the aluminum window and the sealing in of the cap. The metal of the latter may be one thirty-second or even one-sixteenth of an inch thick, and in such case the central portion may be thinned down by a countersink tool about one-fourth of an inch in diameter as far as it is possible without tearing the sheet. The further thinning down may then be done by hand with a scraping tool; and, finally, the metal should be gently beaten down so as to surely close the pores which might permit a slow leak. Instead of proceeding in this way I have employed a cap with a hole in the center, which I have closed with a sheet of pure aluminum a few thousandths of an inch thick, riveted to the cap by means of a washer of thick metal, but the results were not quite as satisfactory.

In sealing the cap I have adopted the following procedure: The tube is fastened on the pump in the proper position and exhausted until a permanent condition is reached. The degree of exhaustion is a measure of perfection of the joint. The leak is usually considerable, but this is not so serious a defect as might be thought. Heat is now gradually applied to the tube by means of a gas stove until a temperature up to about the boiling point of sealing wax is reached. The space between the cap and the glass is then filled with sealing wax of good quality; and, when the latter begins to boil, the temperature is reduced to allow its settling in the cavity. The heat is then again, increased, and this process of heating and cooling is repeated several times until the entire cavity, upon reduction of the temperature, is found to be filled uniformly with the wax, all bubbles having disappeared. A little more wax is then put on the top and the exhaustion carried on for an hour or so, according to the capacity of the pump, by application of moderate heat much below the melting point of the wax.

A tube prepared in this manner will maintain the vacuum very well, and will last. indefinitely. If not used for a few months, it may gradually lose the high vacuum, but it can be quickly worked up. However, if after long use it becomes necessary to clean the tube, this is easily done by gently warming it and taking off the cap: The cleaning may be done first with acid, then with highly diluted alkali, next with distilled water, and finally with pure rectified alcohol.

These tubes, when properly prepared, give impressions much sharper and reveal much more detail. than those of ordinary make. It is important for the clearness of the impressions that the electrode should be properly shaped, and that the focus should be exactly in the center of the cap or slightly inside. In fitting in the cap, the distance from the electrode should be measured as exactly as possible. It should also be remarked that the thinner the window, the sharper are the impressions, but it is not advisable to make it too thin, as it is apt to melt in a point on turning on the current.

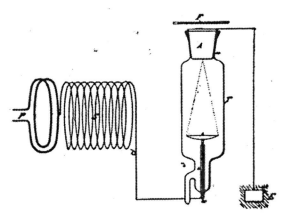

Fig. 4. — Illustrating Arrangement with a Lenard Tube for Safe Working at Close Range.

The above advantages are not the only ones which these tubes offer. They are also better adapted for purposes of examination by surgeons, particularly if used in the peculiar manner illustrated in diagrams Fig. 3 and Fig. 4, which are self-explanatory. It will be seen that in each of these the cap is connected to the ground. This decidedly diminishes the injurious action and enables also to take impressions with very short exposures of a few seconds only at dose range, inasmuch as, during the operation of the bulb, one can easily touch the cap without any inconvenience, owing to the ground connection. The arrangement shown in Fig. 4 is particularly advantageous with a form of single terminal, which coil I have described on other occasions and which is diagrammatically illustrated, P being the primary and S the secondary. In this instance the high-potential terminal is. connected to the electrode, while the cap is grounded. The tube may be placed in the position indicated in the drawing, under the operating table and quite close or even in contact with the body of the patient, if the impression requires only a few seconds as, for instance, in examining parts of the members. I have taken many impressions with such tubes and have observed no injurious action, but I would advise not to expose for longer than two or three minutes at very short distances. In this respect the experimenter should bear in mind what I have stated in previous communications. At all events it is certain that, in proceeding in the manner described, additional safety is obtained and the process of taking impressions much quickened. To cool the cap, a jet of air may be used, as before stated, or else a small quantity of water may be ,poured in the cap each time when an impression is taken. The water only slightly impairs the action of the tube, while it maintains the window at a safe temperature. I may add that the tubes are improved by providing back of the electrode a metallic coating C, shown in Fig. 3 and Fig. 4.

Tesla's Latest Advances in Vacuum-tube Lighting. Application of Tubes of High Illuminating Power to Photography and Other Purposes
Electrical Review - N. Y. — *Jan. 5, 1898*

To the Editor of *Electrical Review*:

A few years ago I began a series of experiments with a view of ascertaining the applicability of the light emitted by phosphorescent vacuum tubes to ordinary photography. The results soon showed that, even with a tube giving no more light than the equivalent of one half of a candle, objects could be easily photographed with exposures of a few minutes, and the time could be reduced at will by pushing the tube to a high candlepower. Photographs of persons were likewise obtained at that time and, if I am not mistaken, these were the first likenesses produced with this kind of illumination. However, a number of facts, not pertaining to the subject presently considered, were observed in the course of the experiments which, had they been immediately published, might have materially hastened important scientific developments which have taken place since. To dwell on these and other experimental results obtained at that time, more extensively at the first opportunity, is one of my good resolutions for the coming year. A calamity unfortunately, interrupted my labors for a short period, but as soon as I was able I took up again the thread of the investigation, which was not only interesting in connection with the principal object in view, but was also useful in many other respects. So, for instance, in making observations as to the efficiency or any peculiarity of the vacuum tubes, the photographic plate was found to be an excellent means of comparison, note taken of the distance and time of exposure, character of the phosphorescent body, degree of rarefaction and other such particulars of the moment.

A rather curious feature in the photographs obtained with tubes of moderate illuminating power, as a few candles, was that the lights and shadows came out remarkably strong, as when very short exposures are made by flashlight, but the outlines were not sharp and practically no detail was visible. By producing tubes of much greater candlepower, a notable improvement in this respect was effected, and this advance prompted me to further efforts in this direction, which finally resulted in the production of a tube of an illuminating power of equal to that of hundreds, and even thousands, of ordinary vacuum tubes. What is more, I believe that I am far from having attained the limit in the amount of light producible, and believe that this method of illumination will be eventually employed for lighthouse purposes. This probably will be considered the oddest and most unlooked-for development of the vacuum tube.

Simultaneously with this progress a corresponding improvement was made in the efficiency of the light produced. A few words on this point might not be amiss, considering that a popular and erroneous opinion still exists in regard to the power consumed by vacuum tubes lighted by ordinary means. So deeply rooted is this opinion which, I will frankly confess, I myself shared for a long time, that, shortly after my own first efforts, Sir David Solomons and Messrs. Pike & Harris undertook to introduce in England such tubes on a large scale in competition with the incandescent system of lighting. The enterprise, which was commented on in the technical periodicals, was commendable enough, but it was not difficult to foretell its fate; for although the high-frequency currents obtained from the alternator yielded better economical results than interrupted currents, and although they were obtained in a convenient and fairly economical manner, still the efficiency of the whole system was necessarily too small for competition with incandescent lamps. The reason for the great power consumption, which may often be as much as 10 times that taking place in incandescent lamps for an equivalent amount of light, are not far to seek. A vacuum tube, particularly if it be very large, offers an immense radiating surface, and is capable of giving off a great amount of energy without rising perceptibly in temperature. What still increases the dissipation of energy is the high temperature of the rarefied gas. Generally it is supposed that the particles are not brought to a high temperature, but a calculation from the amount of matter contained in the tube, leads to results which would seem to indicate that, of all the means at disposal for bringing a small amount of matter to a high temperature, the vacuum tube is the most effective. This observation may lead to valuable uses of such tubes in astronomical researches, and a line of experiment to this end was suggested to me recently by Dr. Geo. E. Hale, of the Yerkes Observatory. As compared with these disadvantages the incandescent lamp, crude and inefficient as it undoubtedly is, possesses vastly superior features. These difficulties have been recognized by me early, and my efforts during the past few years have been directed towards overcoming these defects and have finally resulted in material advances, so that I find it possible to obtain from a tube of a volume not much greater than that of a bulb of an incandescent lamp, about the same amount of light produced by the latter, without the tube becoming overheated, which is sure to take place under ordinary conditions. Both of these improvements, the increase of candle-power as well as degree of efficiency, have been achieved by gradual perfection of the means of producing economically harmonical electrical vibrations of extreme rapidity. The fundamental principle involved is now well known, and it only remains to describe the features of the system in detail, a duty with which I expect to be able to comply soon, this being another one of my good resolutions.

The purpose of the present communication is chiefly to give an idea in how far the object here aimed at was obtained. The photographs shown were taken by a tube having a radiating surface of about two hundred square inches. The frequency of the oscillations, which were obtained from an Edison direct-current supply circuit, I estimated to be about two million a second. The illuminating power of the tube approximated about one thousand candles, and the exposures ranged from two to five seconds, the distance of the object being four to five feet from the tube. It might be asked why, with so high an illuminating power, the exposures should not be instantaneous. I would not undertake to satisfactorily answer this question, which was put to me recently by a scientific man, whose visit to my laboratory I still vividly recollect. Likenesses can, of course, be obtained with instantaneous exposures, but it has been found preferable to expose longer and at a greater distance from the tube. The results so far obtained would make it appear that this kind of light will be of great value in photography, not only because the artist will be able to exactly adjust the conditions in every experiment so as to secure the best result, which is impossible with ordinary light. He will thus be made entirely

independent of daylight, and will be able to carry on his work at any hour, night or day. It might also be of value to the painter, though its use for such purposes I still consider problematical.
Nikola Tesla
New York, Jan. 3

Tesla on Animal Training by Electricity
New York Journal — Feb. 6, 1898

To the Editor of the *Journal*:

It seems to me that there are interesting possibilities in the training of animals by electricity. Of course,. it's rather out of my province, but the idea of the electrical subjugator appears feasible when one knows the power of electricity and the instinctive fear that brutes have of the unknown. And the electrical method seems more humane than those I believe are in use - the whip, red hot irons, and drugs, which are likely to do permanent injury, while the physical effects of an electric shock are soon gone, only the moral ones remaining.

The subjugator referred to will do the work, but I think an apparatus could be designed that would be less dangerous to the man. I do not desire to be understood as giving the matter deep thought, but believe that if instead of the armored backpad, the trainer used a wand, with two prongs at one end, better results would follow. This wand would be connected with the supply cables and could be applied to any part of the animal's body at will. Its operation would be precisely the same as the subjugator here illustrated, the two prongs supplying the positive and negative poles of contact found in the flattened wires. With this wand an animal could be simply shocked, stunned or killed, as required.

To cure animals of jumping at men in cages, a screen of stout but flexible wire could be stretched between the trainer and his subject, the wires to be alternately positive and negative, and connected through the regulator with the dynamo. After a couple of springs which would hurl him half insensible back into his corner, the taste for unexpected jumps would leave the brute.

[*The following article appeared with the above.-Ed*]
Prague, Jan. 22.

Science has come to aid the lion tamer in subduing the wild beast. The red hot iron will, in future, be cast aside as unnecessary and out of date. Live wires, surcharged with electricity that baffle the lion's fiercest assaults, and burn and maim him badly have taken the place of the lash and scorching iron. A lion tamer of Austria, Louis Koemmenich, has been the first to call in the assistance of the lightning to subdue wild beasts.

Koemmenich has invented what he calls the electrical subjugator. This is a shield of electric wires that fasten on the back of the lion tamer and are connected with a dynamo by a wire coil of sufficient length to allow Koemmenich to move around the cage.

In his hand he will carry a charged metal ball on an insulated handle, to be used as the red hot iron was in former days.

The dynamo is operated by an assistant outside of the cage.

Should a lion show a disposition to leap on Koemmenich, he invites attack by deliberately turning his back to the lion and apparently encouraging the onslaught. When the beast springs his paws come in contact with the electric shield, and he receives a shock of 1,500 volts from the dynamo.

The operator can, if necessary, increase the voltage so as to shock the animal to death.

Thus far the device has worked like magic. One dose of lightning is sufficient for the average lion. Whips and even hot irons they have dared, but no animal has yet troubled Koemmenich after receiving into its body 1,500 volts from the electric subjugator. Whenever Koemmenich enters the cage after an encounter with a lion that has run against the electrical subjugator, he will cower away into a corner of the cage, and never need any further punishment.

High Frequency Oscillators for Electro-Therapeutic and Other Purposes

Read at the eighth annual meeting of The American Electro-Therapeutic Association, Buffalo, N. Y., Sept. 13 to 15, 1898

Some theoretical possibilities offered by currents of very high frequency and observations which I casually made while pursuing experiments with alternating currents, as well as the stimulating influence of the work of Hertz and of views boldly put forth by Oliver Lodge determined me some time during 1889 to enter a systematic investigation of high frequency phenomena, and the results soon reached were such as to justify further efforts towards providing the laboratory with efficient means for carrying on the research in this particular field, which has proved itself so fruitful since. As a consequence alternators of special design were constructed and various arrangements for converting ordinary into high frequency currents perfected, both of which were duly described and are now — I assume — familiar.

One of the early observed and remarkable features of the high frequency currents, and one which was chiefly of interest to the physician, was their apparent harmlessness which made it possible to pass relatively great amounts of electrical energy through the body of a person without causing pain or serious discomfort. This peculiarity which, together with other mostly unlooked-for properties of these currents I had the honor to bring to the attention of scientific men first in an article in a technical journal in February, 1891, and in subsequent contributions to scientific societies, made it at once evident, that these currents would lend themselves particularly to electro-therapeutic uses.

With regard to the electrical actions in general, and by analogy it was reasonable to infer that the physiological effects, however complex, might be resolved in three classes. First the statical, that is, such as are chiefly dependent on the magnitude of electrical potential; second, the dynamical, that is, those principally dependent on the quality of electrical movement or current's strength through the body, and third, effects of a distinct nature due to electrical waves or oscillations, that is, impulses in which the electrical energy is alternately passing in more or less rapid succession through the static and dynamic forms.

Most generally in practice these different actions are coexistent, but by a suitable selection of apparatus and observance of conditions the experimenter may make one or other of these effects predominate. Thus he may pass through the body, or any part of the same, currents of comparatively large volume under a small electrical pressure, or lie may subject the body to a high electrical pressure while the current is negligibly small, or he may put the patient under the influence of electrical waves transmitted, if desired, at considerable distance through space.

While it remained for the physician to investigate the specific actions on the organism and indicate proper methods of treatment, the various ways of applying these currents to the body of a patient suggested themselves readily to the electrician.

As one cannot be too clear in describing a subject, a diagrammatic illustration of the several modes of connecting the circuits which 1 will enumerate, though obvious for the majority, is deemed of advantage.

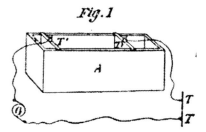

Fig. 1

The first and simplest method of applying the currents was to connect the body of the patient to two points of the generator, be it a dynamo or induction coil. Fig. 1 is intended to illustrate this case. The alternator G may be one giving from five to ten thousand complete vibrations per second, this number being still within the limit of practicability. The electromotive force — as measured by a hot wire instrument — may be from fifty to one hundred volts. To enable strong currents to be passed through the tissues, the terminals T T, which serve to establish contact with the patient's body should, of course, be of large area, and covered with cloth saturated with a solution of electrolyte harmless to the skin, or else the contacts are made by immersion. The regulation of the currents is best effected by means of an insulating trough A provided with two metal terminals T' T' of considerable surface, one of which, at least, should be movable. The trough is filled with water and an electrolytic solution is added to the same, until a degree of conductivity is obtained suitable for the experiments.

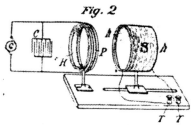

Fig. 2

When it is desired to use small currents of high tension, a secondary coil is resorted to, as illustrated in Fig. 2. I have found it from the outset convenient to make a departure from the ordinary ways of winding the coils with a considerable number of small turns. For many reasons the physician will find it better to provide a large hoop H of not less than, say, three feet in diameter and preferably more, and to wind upon it a few turns of stout cable P. The secondary coil S is easily prepared by taking two wooden hoops h h and joining them with stiff cardboard. One single layer of ordinary magnet wire, and not too thin at that, will be generally sufficient, the number of turns necessary for the particular use for which the coil is intended being easily ascertained by a few trials. Two plates of large surface, forming an adjustable condenser, may be used for the purpose of synchronizing the secondary with the primary circuit, but this is generally not necessary. In this manner a cheap coil is obtained, and one which cannot be easily injured. Additional advantages, however, will be found in the perfect regulation which is effected merely by altering the distance between the primary and secondary, for which adjustment provision should be made, and, furthermore, in the occurrence of harmonics which are more pronounced in such large coils of thick wire, situated at sonic distance from the primary.

The preceding arrangements may also be used with alternating or interrupted currents of low frequency, but certain peculiar properties of high frequency currents make it possible to apply the latter in ways entirely impracticable with the former.

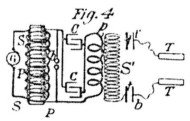

One of the prominent characteristics of high frequency or, to be snore general, of rapidly varying currents, is that they pass with difficulty through stout conductors of high self-induction. So great is the obstruction which self-induction offers to their passage that it was found practicable, as shown in the early experiments to which reference has been made, to maintain differences of potential of many thousands of volts between two points — not more than a few inches apart — of a thick copper bar of inappreciable resistance. This observation naturally suggested the disposition illustrated in Fig. 3. The source of high frequency impulses is in this instance a familiar type of transformer which may be supplied from a generator G of ordinary direct or alternating currents. The transformer comprises a primary P, a secondary S, two condensers C C which are joined in series, a loop or coil of very thick wire L and a circuit interrupting device or break b. The; currents are derived from the loop h by two contacts c c', one or both of which are capable of displacement along the wire L. By varying the distance between these contacts, any difference of potential, from a few volts to many thousands, is readily obtained on the terminals or handles T T. This mode of using the currents is entirely safe and particularly convenient, but it requires a very uniform working of the break b employed for charging and discharging the condenser.

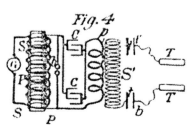

Another equally remarkable feature of high frequency impulses was found in the facility with which they are transmitted through condensers, moderate electromotive forces and very small capacities being required to enable currents of considerable volume to pass. This observation made it practicable to resort to a plan such as indicated in Fig. 4. Here the connections are similar to those shown in the preceding case, except that the condensers C C are joined in parallel. This lowers the frequency of the currents, but has the advantage of allowing the working with a much smaller difference of potential on the terminals of the secondary S. Since the latter is the chief item of expense of such apparatus and since its price rapidly increases with the number of turns .required, the experimenter will find it generally cheaper to make a sacrifice in the frequency, which, however, will be high enough for most purposes,. However, he only needs to reduce proportionately the number of turns or the length of primary p to obtain the same *frequency* as before, but the economy of transformation will be somewhat reduced in so doing and the break b will require more attention. The secondary S' of the high frequency coil has two metal plates t t of considerable surface connected to its terminals, and the current for use is derived from two similar plates t' t' in proximity to the former. Both the tension and volume of the currents taken from terminals T T may be easily regulated and in a continuous manner by simply varying the distance between the two pairs of plates t t and t' t' respectively.

A facility is also afforded in this disposition for raising or lowering the potential of one of the terminals T, irrespective of the changes produced on the other terminal, this making it possible to cause a stronger action on one or other part of the patient's body.

The physician may find it for some or other reason convenient to modify the arrangements in Figs. 2, 3 and It by connecting one terminal of the high frequency source to the ground. The *effects* will be in most respects the same, but certain peculiarities will be noted in each case. When a ground connection is made it may be of some consequence which of the terminals of the secondary is connected to the ground, as in high frequency discharges the impulses of one direction are generally preponderating.

Among the various noteworthy features of these currents there is one which lends itself especially to many valuable uses. It is the facility which they afford for conveying large amounts of electrical energy to a body entirely insulated in space. The practicability of this method of energy transmission, which is already receiving useful applications and promises to become of great importance in the near future, has helped to dispel the old notion assuming the necessity of a return circuit for the conveyance of electrical *energy* in any considerable amount. With novel appliances we are enabled to pass through a wire, entirely insulated on one end, currents strong enough to fuse it, or to convey through the wire any amount of energy to an insulated body. This mode of applying high frequency currents in medical treatment appears to me to offer the greatest possibilities at the hands of the physician. The effects produced in this manner possess features entirely distinct from those observed when the currents are applied in any of the before mentioned or similar ways.

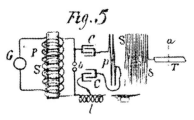

Fig. 5

The circuit connections as usually made are illustrated schematically in Fig. 5, which, with reference to the diagrams before shown, is self-explanatory. The condensers C C, connected in series, are preferably charged by a step-up transformer, but a high frequency alternator, static machine, or a direct current generator, if it be of sufficiently high tension to enable the use of small condensers, may be used with more or less success. The primary p, through which the high frequency discharges of the condensers are passed, consists of very few turns of cable of as low resistance as possible, and the secondary s, preferably at some distance from the primary to facilitate free oscillation, has one of its ends — that is the one which is nearer to the primary — connected to the ground, while the other end leads to on insulated terminal T, with which the body of the patient is connected. It is of importance in this case to establish synchronism between the oscillations in the primary and secondary circuits p and s respectively. This will be as a rule best effected by varying the self-induction of the circuit including the primary loop or coil p, for which purpose an adjustable self-induction a is provided; but in cases when the electromotive force of the generator is exceptionally high, as when a static machine is used and a condenser consisting of merely two plates offers sufficient capacity, it will be simpler to attain the same object by varying the distance of the plates.

The primary and secondary oscillations being in close synchronism, the points of highest potential will be on a part of terminal T, and the consumption of energy will occur chiefly there. The attachment of the patient's body to the terminal will in most cases very materially affect the period of oscillation in the secondary, making it longer, and a readjustment of the primary circuit will have to be made in each case to suit the capacity of the body connected with terminal T. Synchronism should always be preserved, and the intensity of the action varied' by moving the secondary coil to or from the primary, as may be desired. I know of no method which would make it possible to subject the human body to such excessive electrical pressures as are practicable with this, or of one which would enable the conveying to and giving off from the *body* without serious injury amounts of electrical energy approximating even in a remote degree those which are entirely practicable when this manner of applying the energy is resorted to. This is evidently due to the fact that action is chiefly superficial, the largest possible section being offered to the transfer of the current, or, to say more correctly, of the energy. With a very rapidly and smoothly working break I would not think it impossible to convey to the body of a person and to give off into the space energy at the rate of several horse power with impunity, while a small part of this amount applied in other ways could not fail to produce injury.

When a person is subjected to the action of such a coil, the proper adjustments being carefully observed, luminous streams are seen in the dark issuing from all parts of the body. These streams are short and of delicate texture when the number of breaks is very great and the action of the device b (Fig. 5) free of any irregularities, but when the number of breaks is small or the action of the device imperfect, long and noisy streams appear which cause some discomfort. The physiological effects produced with apparatus of this kind may be graduated from a hardly perceptible action when the secondary is at a great distance from the primary, to a most violent one when both coils are placed at a small distance. In the latter case only a few seconds are sufficient to cause a feeling of warmth all over the body, and soon after the person perspires freely. I have repeatedly, in demonstrations to friends, exposed myself longer to the action of the oscillations, and each time, after the lapse of an hour or so, an immense fatigue, of which it is difficult to give an idea, would take hold of me. It was greater than I experienced on some occasions after the most straining and prolonged bodily exertion. I could scarcely make a step and could keep the *eyes* open only with the greatest difficulty. I slept soundly afterward, and the after-effect was certainly beneficial, but the medicine was manifestly too strong to be used frequently.

One should be cautious in performing such experiments for more than one reason. At or near the surface of the skin, where the most intense action takes place, various chemical products are formed, the chief being ozone and nitrogen compounds. The former is itself very destructive, this feature being illustrated by the fact that the rubber insulation. of a wire is destroyed so quickly as to make the use of such insulation entirely impracticable. The compounds of nitrogen, when moisture is present, consist largely of nitric acid which might, by excessive application, prove burtul to tire skirl. So far. I have not noted injuries which could be traced directly to this cause, though on several occasions burns were produced in all respects similar to those 'which were later observed and attributed to the Ronttgen rays. This view is seemingly being; abandoned, having not been substantiated by experimental facts, and so also is the notion that these rays are transverse vibrations. But while investigation is being turned in what appears to be the right direction, scientific men are still at sea. This state of things impedes the progress of the physicist in these new regions and makes the already hard task of the physician still more difficult and uncertain.

One or two observations made while pursuing experiments with the apparatus described might be found as deserving mention here. As before stated, when the oscillations in the primary and secondary circuits are in synchronism, the points of highest potential are on some portion of the terminal T. The synchronism being perfect and the length of the secondary coil just equal to one-quarter of the wave length, these points will be exactly on the free end of terminal T, that is, the one situated farthest from the end of the wire attached to the terminal. If this be so and if now the period of the oscillations in the primary be shortened, the points of highest potential will recede towards the secondary coil, since the wave-length is reduced and since the attachment of one end of the secondary coil to the ground determines the position of the nodal points, that is, the points of least potential. Thus, by varying the period of vibration of the primary circuit in any manner, the points of highest potential may be shifted accordingly along the terminal T, which has been shown, designedly, long to illustrate this feature. The same phenomenon is, of course, produced if the body of a patient constitutes the terminal, and an assistant may by the motion of a handle cause the points of highest potential to shift along the body with any speed he may desire. When the action of the coil is vigorous, the region of highest potential is easily and unpleasantly located by the discomfort or pain experienced, and it is most curious to feel how the pain wanders up and down, or eventually across the body, from hand to hand, if the connection to the coil is accordingly made — in obedience to the movement of the handle controlling the oscillations. Though I have not observed any specific action in experiments of this kind, I have always felt that this effect might be capable of valuable use in electrotherapy.

Another observation which promises to lead to much more useful results is the following: As before remarked, by adopting the method described, the body of a person may be subjected without danger to electrical pressures vastly in excess of any producible by ordinary apparatus, for they may amount to several million volts, as has been shown in actual practice. Now, when a conducting body is electrified to so high a degree, small particles, which may be adhering firmly to its surface, are torn off with violence and thrown to distances which can be only conjectured. I find that not only firmly adhering matter, as paint, for instance, is thrown off, but even the particles of the toughest metals are torn off. Such actions have been thought to be restricted to a vacuous inclosure, but with a powerful coil they occur also in the ordinary atmosphere. The facts mentioned would make it reasonable to expect that this extraordinary effect which, in other ways, I have already usefully applied, will likewise prove to be of value in electro-therapy. The continuous improvement of the instruments and the study of the phenomenon may shortly lead to the establishment of a novel mode of hygienic treatment which would permit an instantaneous cleaning of the skin of a person, simply by connecting the same to, or possibly, by merely placing the person in the vicinity of a source of intense electrical oscillations, this having the effect of throwing off, in a twinkle of the eye, dust or particles of any extraneous matter adhering to the body. Such a result brought about in a practicable manner would, without doubt, be of incalculable value in hygiene and would be an efficient and time-saving substitute for a water bath, and particularly appreciated by those whose contentment consists in undertaking more than they can accomplish.

High frequency impulses produce powerful inductive actions and in virtue of this feature they lend themselves in other ways to the uses of the electro-therapeutics. These inductive effects are either electrostatic or electrodynamic. The former diminish much more rapidly with the distance — with the square of the same — the latter are reduced simply in proportion to the distance. On the other hand, the former grow with the square of intensity of the source, while the latter increase in a simple proportion with the intensity. Both of these effects may be utilized for establishing a field of strong action extending through considerable space, as through a large hall, and such an arrangement might be suitable for use in hospitals or institutions of this kind, where it is desirable to treat a number of patients at the same time.

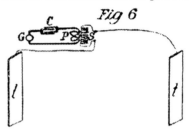

Fig: 6 illustrates the manner, as I have shown it originally, in which such a field of electrostatic action is established. In this diagram G is a generator of currents of very high frequency, C a condenser for counteracting the self-induction of the circuit which includes the primary P of an induction coil, the secondary S of which has two plates t t of large surface connected to its terminals. Well known adjustments being observed, a very strong action occurs chiefly in the space between the plates, and the body of a person is subjected to rapid variations *of* potential and surgings *of* current, which produce, even at a great distance, marked physiological effects. In my first experiments I used two metal plates as shown, but later I found it preferable to replace them by two large hollow spheres of brass covered with wax of a thickness of about two inches. The cables leading to the terminals of the secondary coil were similarly covered, so that any of them could be approached without danger of the insulation breaking down. In this manner the unpleasant shocks, to which the experimenter was exposed when using the plates, were prevented.

In Fig. 7 a plan for similarly utilizing the dynamic inductive effects of high frequency currents is illustrated. As the frequencies obtainable from an alternator are not as high as is desired, conversion by means of condensers is resorted to. The diagram will be understood at a glance from the foregoing description. It only need be stated that the primary p, through which the condensers are made to discharge, is formed by a thick stranded cable of low self-induction and resistance, and passes all around the hall. Any number of secondary coils s s s, each consisting generally of a single layer of rather thick wire, may be provided. I have found it practicable to use as many as one hundred, each being adjusted for a definite period and responding to a particular vibration passed through the primary. Such a plant I have had in use in my laboratory since *1892,* and many times it has contributed to the pleasure of my visitors and also proved itself of practical utility. On a latter occasion I had the pleasure of entertaining some of the members with experiments of this kind, and this opportunity I cannot let pass without expressing my thanks for the interest which was awakened in me by their visit, as well as for the generous acknowledgment of the courtesy by the Association. Since that time my apparatus has been very materially improved, and now I am able to create a field of such intense induction in the laboratory that a coil three feet in diameter, by careful adjustment, will deliver energy at the rate of one-quarter of a horse power, no matter where it is placed within the area inclosed by the primary loops. Long sparks, streamers and all other phenomena obtainable with induction coils are easily producible anywhere within the space, and such coils, though not connected to anything, may be utilized exactly as ordinary coils, and what is still more remarkable, they are more effective. For the past few years I have often been urged to show experiments in public, but, though I was desirous to comply with such requests, pressing work has so far made it impossible. These advances have been the result of slow but steady improvement in the details of the apparatus which I hope to be able to describe connectedly in the near future.

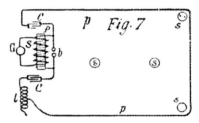

However remarkable the electrodynamic inductive effects, which I have mentioned, may appear, they *may* be still considerably intensified by concentrating the action upon a very small space. It is evident that since, as before stated, electromotive forces of many thousand volts are maintained between two points of a conducting bar or loop only a few inches long, electromotive forces of approximately the same magnitude will be set up in conductors situated near by. Indeed, I found that it was practicable in this manner to pass a discharge through a highly exhausted bulb, although the electromotive force required amounted to as much as ten or twenty thousand volts, and for a long time I followed up experiments in this direction with the object of producing light in a novel and more economical way. But the tests left no doubt that there was great energy consumption attendant to this mode of illumination, at least with the apparatus I had then at command, and, finding another method which promised a higher economy of transformation, my efforts turned in this new direction. Shortly afterward (some time in June, 1891), Prof. J. J. Thomson described experiments which were evidently the outcome of long investigation, and in which he supplied much novel and interesting information, and this made me return with renewed zeal to my own experiments. Soon my efforts were centered upon producing in a small space the most intense inductive action, and by gradual improvement in the apparatus I obtained results of a surprising character. For instance, when the end of a heavy bar of iron was thrust within a loop powerfully energized, a few moments were sufficient to raise the bar to a high temperature. Even heavy lumps of other metals were heated as rapidly as though they were placed in a furnace. When a continuous hand formed of a sheet of tin was thrust into the loop, the metal was fused instantly, the action being comparable to an explosion, and no wonder, for the frictional losses accumulated in it at the rate of possibly ten horse power. Masses of poorly conducting material behaved similarly, and when a highly exhausted bulb was pushed into the loop, the glass was heated in a few seconds nearly to the point of melting.

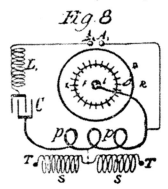

When I first observed these astonishing actions, I eras interested to study their effects upon living tissues. As may be assumed, I proceeded with all the necessary caution, and well I might, for I had the evidence that in a turn of only a few inches in diameter an electromotive force of more than ten thousand volts was produced, and such high pressure would be more than sufficient to generate destructive currents in the tissue. This appeared all the more certain as bodies of comparatively poor conductivity were rapidly heated and even partially destroyed. One may imagine my astonishment when I found that I could thrust my hand or any other part of the *body* within the loop and hold it there with impunity. More than on one occasion, impelled by a desire to make some novel and useful observation, I have willingly or unconsciously performed an experiment connected with some risk, this being scarcely avoidable in laboratory experience, but have always believed, and do so now, that I have never undertaken anything in which, according to my own estimation, the chances of being injured were so great as when I placed my head within the space in which such terribly destructive forces were at work. Yet I have done so, and repeatedly, and have felt nothing. But I am firmly convinced that there is great danger attending such experiment, and some one going just a step farther than I have gone may be instantly destroyed. For, condition may exist similar to those observable with a vacuum bulb. It may be placed in the field of the loop, however intensely energized, and so long as no path for the current is formed, it will remain cool and consume practically no energy. But the moment the first feeble current passes, most of the *energy* of the oscillations rashes to the place of consumption. If by any action whatever, a conducting path were farmed within the living tissue or bones of the head, it would result in the instant destruction of these and death of the foolhardy experimenter. Such a method of killing, if it were rendered practicable, would be absolutely painless. Now, why is it that in a space in which such violent turmoil is going on living tissue remains uninjured? One might say the currents cannot pass because of the great self-induction offered by the large conducting, mass. But this it cannot be, because a mass of metal offers a still higher self-induction and is heated just the same. One might argue the tissues offer too great a resistance. But this main cannot be the reason, for all evidence shows that the tissues conduct well enough, and besides, bodies of approximately the same resistance are raised to a high temperature. One might attribute the apparent harmlessness of the oscillations to the high specific heat of the tissue, but even a rough quantitative estimate from experiments with other bodies shows that this view is untenable. The only plausible explanation I have so far found is that the tissues are condensers. This only can account for the absence of injurious action. But it is remarkable that, as soon as a heterogeneous circuit is constituted, as by taking in the hands a bar of metal and forming a closed loop in this manner, the passage of the currents through the arms is felt, and other physiological effects are distinctly noted. The strongest action is, of course, secured when the exciting loop makes only one turn, unless the connections take up a considerable portion of the total length of the circuit, in which case the experimenter should settle upon the least number of turns by carefully estimating what he loses by increasing the number of turns, and what he gains by utilizing thus a greater proportion of the total length of the circuit. It should be borne in mind that, when the exciting coil has a considerable number of turns and is of some length, the effects of electrostatic induction may preponderate, as there may exist a very great difference of potential — a hundred thousand volts or more — between the first and last turn. However, these latter effects are always present even when a single turn is employed.

When a person is placed within such a loop, any pieces of metal, though of small bulk, are perceptibly warmed. Without doubt they would be also heated — particularly if they were of iron — when embedded in living tissue, and this suggests the possibility of surgical treatment by this method. It might be possible to sterilize wounds, or to locate, or even to extract metallic objects, or to perform other operations of this kind within the sphere of the surgeon's duties in this novel manner.

Most of the results enumerated, and many others still more remarkable, are made possible *only* by utilizing the discharges of a condenser. It is probable that but a very few — even among those who are working in these identical fields — fully appreciate what a wonderful instrument such a condenser is in reality. Let me convey an idea to this effect. One may take a condenser, small enough to go in one's vest pocket, and by skilfully using it he may create an electrical pressure vastly in excess — a hundred times greater if necessary — than any producible by the largest static machine ever constructed Or, he may take the same condenser and, using it in a. different way, he may obtain from it currents against which those of the most powerful welding machine are utterly insignificant. Those who are imbued with popular notions as to the pressure of static machines and currents obtainable with a commercial transformer, will be astonished at this statement — yet the truth of it is easy to see. Such results are obtainable, anti easily, because the condenser can discharge the stored energy in an inconceivably short time. Nothing like this property is known in physical science. A compressed spring, or a storage battery, or any other form of device capable of storing energy, cannot do this; if they could, things undreamt of at present might be accomplished by their means. The nearest approach to a charged condenser is a high explosive, as dynamite. But even the most violent explosion of such a compound. bears no comparison with the discharge or explosion of a condenser. For, while the pressures which are produced in the detonation of a chemical compound are measured in tens of tons per square inch, those which may be caused by condenser discharges may amount to thousands of tons per square inch, and if a chemical could be made which would explode as quickly as a condenser can be discharged under conditions which are realizable — an ounce of it would quite certainly be sufficient to render useless the largest battleship.

That important realizations would follow from the use of an instrument possessing such ideal properties I have been convinced since long ago, but I also recognized early that great difficulties would have to be overcome before it could replace less perfect implements now used in the arts for the manifold transformations of electrical energy. These difficulties were many. The condensers themselves, as usually manufactured, were inefficient, the conductors wasteful, the best insulation inadequate, and the conditions for the most efficient conversion were hard to adjust and to maintain. One difficulty, however, which was more serious than the others, and to which I called attention when I first described this system of energy transformation, was found in the devices necessarily used for controlling the charges and discharges of the condenser. They were wanting in efficiency and reliability and threatened to prove a decided drawback, greatly restricting the use of the system and depriving it of many valuable features. For a number of years I have tried to master this difficulty. During this time a great number of such devices were experimented upon. Many of them promised well at first, only to prove inadequate in the end. Reluctantly, I came back upon an idea on which I had worked long before. It was to replace the ordinary brushes and commutator segments by fluid contacts. I had encountered difficulties then, but the intervening years in the laboratory were not spent in vain, and I made headway. First it was necessary to provide for a circulation of the fluid, but forcing it through by a pump, proved itself impractical. Then the happy idea presented itself to make the pumping device an integral part of the circuit interrupter, unclosing both in a receptacle to prevent oxidation. Next some simple ways of maintaining the circulation, as by rotating a body of mercury, presented themselves. Then I learned how to reduce the wear and losses which still existed. I fear that these statements, indicating how much effort was spent in these seemingly insignificant details will not convey a high idea of my ability, but I confess that my patience was taxed to the utmost. Finally, though, I had the satisfaction of producing devices which are simple and reliable in their operation, which require practically no attention and which are capable of effecting a transformation of considerable amounts of energy with fair economy. It is not the best that can be done, by any means, but it is satisfactory, and I feel that the hardest task is clone.

The physician will now be able to obtain an instrument suitable to fulfil many requirements. He will be able to use it in electro-therapeutic treatment in most of the ways enumerated. He will have the facility of providing himself with coils such as he may desire to have for any particular purpose, which will give him any current or any pressure he may wish to obtain. Such coils will consist of but a few turns of wire, and tile expense of preparing them will be quite insignificant. The instrument will also enable him to generate Rontgen *rays of* much greater power than obtainable with ordinary apparatus. A tube must still be furnished by the manufacturers which will not deteriorate and which will allow to concentrate larger amounts of energy upon the electrodes. When this is clone, nothing will stand in the way of an expensive and efficient application of this beautiful discovery which must ultimately prove itself of the highest value, not only at the hands of the surgeon, but also of the electrotherapist and, what is most important, of the bacteriologist.

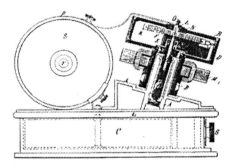

To give a general idea of an instrument in which many of the latter improvements ate embodied, I would refer to Fig. 9, which illustrates the chief parts of the same in side elevation and partially in vertical cross-section. The arrangement of the parts is the same as in the form of instrument exhibited on former occasions, only the exciting coil with the vibrating interrupter is replaced by one of the improved circuit breaker: to which reference has been made.

This device comprises a casting A with a protruding sleeve B, which in a bushing supports a freely rotatable shaft a. The latter carries an armature within a stationary field magnet M and on the top, a hollow iron pulley D, which contains the break proper. Within the shaft a, and concentrically with the same, is placed a smaller shaft b, likewise freely movable on ball-bearings and supporting a weight F. This weight being on one side and the shafts a and b inclined to the vertical, the weight remains stationary as the pulley is rotated. Fastened to the weight F is a device R in the form of a scoop with very thin walls, narrow on the end nearer to the pulley and wider on the other end. A small quantity of mercury being placed in the pulley and the latter rotated against the narrow end of the scoop, a portion of the fluid is taken up and thrown in a thin and wide stream towards the centre of the pulley. The top of the latter is hermetically closed by an iron washer, as shown, this washer supporting on a steel rod L a disk F of the same metal provided with a number of thin contact blades K. The rod L is insulated by washers N from the pulley, and for the convenience of filling in the mercury a small screw o is provided. The bolt L forming one terminal of the circuit breaker is connected by a copper strip to the primary p. The other end of the primary coil leads to one of the terminals of the condenser C, contained in a compartment of a box A, another compartment of the same being reserved for switch S and terminals of the instrument. The other terminal of the condenser is connected to the casting A and through it to pulley D. When the pulley is rotated, the contact blades K are brought rapidly in and out of contact with the stream of mercury, thus closing and opening the circuit in quick succession. With such a device it is easy to obtain ten thousand makes and breaks per second and even more. The secondary a is made of two separate coils and so arranged that it can be slipped out, and a metal strip in its middle connects it to the primary coil. This is done to prevent the secondary from breaking down when one of the terminals is overloaded, as it often happens in working Rontgen bulbs. This form of coil will withstand a very much greater difference of potential than coils as ordinarily constructed.

The motor has both field and armature built of plates, so that it can be used on alternating as well as direct current supply circuits, and the shafts are as nearly as possible vertical, so as to require the least care in oiling. Thus, the only thing which' really requires some attention is the commutator of the motor, but where alternating currents are always available, this source of possible trouble is easily done away with.

The circuit connections of the instrument have been already shown and the mode of operation explained in periodicals. The usual manner of connecting is illustrated in Fig. 3, in which $A_2 A_2$ are the terminals of the supply circuit, L, a self-induction coil for raising the pressure, which is connected in series with condenser C and primary P P. The remaining letters designate the parts correspondingly marked in Fig. 9 and will be understood with reference to the latter.

My Submarine Destroyer
New York Journal — Nov. 13, 1898

Yesterday Nikola Tesla gave to the Sunday Journal exclusively the news of his latest invention — a submarine torpedo boat. He has perfected his device after observing the defects of the torpedo boats in the recent war, and noting the fatalities of submarine boats invented up to date. His submarine boat will carry no lives to risk, but can be directed at a distance of miles from on shore or from the deck of a war ship. The power to do this will be the electric vibrations of the air used in wireless telegraphy. By this means a whole flotilla of submarine destroyers can be turned against a hostile fleet, and perhaps destroy it, without the enemy knowing how they were attacked. This seems almost incredible until the great magician of electricity explains his wonderful invention, point by point, in the following statement.

"I am now prepared to announce through the Journal my invention of a submarine torpedo boat that I am confident will be the greatest weapon of the navy from this time on.

"The almost utter uselessness of the present kind of torpedo boat has been conclusively demonstrated in the recent war. Neither the courage and skill of the Americans nor the desperate extremities of the Spaniards were able to bring the torpedo boats into successful action. These frail craft, of which so much was expected, simply made an easy target for land batteries and rapid-fire guns of opposing war ships.

"The submarine boats, on the other hand, which have up to this time been built to carry torpedoes have proved death traps for men and were consequently ineffective. The submarine boat, or, more properly speaking, the submarine destroyer, which I have invented is as compact as the torpedo itself. In fact, it is simply an enlarged torpedo shell, thirty-six and a half feet long, loaded with other torpedoes to discharge. Like a torpedo, also, it has its own propelling device. But here the likeness stops. The ordinary torpedo, once launched, plunges head 'on blindly and no known power can turn it one way or another. It hits or misses, according to the trueness with which it is aimed at its launching.

"But my submarine boat, loaded with its torpedoes, can start out from a protected bay or be dropped over a ship's side, make its devious way below the surface, through dangerous channels of mine beds, into protected harbors and attack a fleet at anchor, or go out to sea and circle about, watching for its prey, then dart upon it at a favorable moment, rush up to within a hundred feet if need be, discharge its deadly weapon and return to the hand that sent it. Yet through all these wonderful evolutions it will be under the absolute and instant control of a distant human hand on a far-off headland, or on a war ship whose hull is below the horizon and invisible to the enemy.

"I am aware that this sounds almost incredible and I have refrained from making this invention public till I had worked out every practical detail of it. In my laboratory I now have such a model, and my plans and description at the Patent Office at Washington show the full specifications of it.

"As to the mechanism which is to be stored in this submarine shell: The first and most essential thing is a motor, with storage battery to drive the propeller. Then there are smaller motors and batteries to operate the steering gear, on the same principle that an ordinary vessel is now steered by steam or electricity. Besides these there are still other storage batteries and motors to feed electric signal lights. But in order that the weight of the machinery shall not be too great to destroy the buoyancy or make the boat go too deep in the water compressed air motors will also be used to perform certain functions, such as to fill and empty the water tanks which raise the boat to the surface or sink it to any required depth. Pneumatic air or motors will also fire the torpedoes and pump out the water that may leak in at any time.

"This submarine destroyer will be equipped with six 14-foot Whitehead torpedoes. These will be arranged vertically in two rows in the bow. As one torpedo falls into position and is discharged by pneumatic force, another torpedo, by the force of gravity, falls into the position of the first one, the others above being held up by automatic arms. They can be fired as rapidly as a self-cocking revolver is emptied or at intervals of minutes or hours. The discharge takes place through a single tube, projecting straight ahead in the bow. The small amount of water which leaks through each time is caught by drain pipes and a compressed air pump instantly expels it. As each torpedo is expelled a buoyancy regulator will open the sea cocks and let enough water in the ballast tanks to make the buoyancy uniform and keep the boat at the same distance beneath the surface.

"This submarine destroyer will carry a charge of torpedoes greater than that of the largest destroyers now in use. Those vessels of five hundred tons each which cost the Government $500,000, carry but three or four torpedoes, while this simple submarine destroyer, which can be built for 448,000 to $50,000 or less, will carry six torpedoes. It will have, also, the incalculable advantage of being absolutely invisible to an enemy, and have no human lives to risk or steam boilers to blow up and destroy itself.

"All that is necessary to make this submarine boat subject to perfect control at any distance is to properly wire it, just like a modern house is wired so that a button here rings a bell, a lever there turns on the lights, a hidden wire somewhere else sets off a burglar alarm and a thermal device give a fire alarm.

"The only difference in the case of the submarine boat is in the delicacy of the instruments employed. To the propelling device, the steering gear, the signal apparatus and the mechanism for firing the torpedoes are attached little instruments which are attuned to a certain electro-magnetic synchronism.

"Then there is a similar set of synchronistic instruments all connected to the little switchboard, and placed either on shore or on an ordinary war ship. By moving the lever on the switchboard I can give the proper impulse to the submarine boat to go ahead, to reverse, throw the helm to port or starboard, rise, sink, discharge her torpedoes or return.

"It might be thought that some great power would be necessary to be projected across miles of distance and operate on the far-off boat. The power is all stored in the submarine boat itself - in its storage batteries and compressed air. All that is needed to affect the synchronistic instruments is a set of high alternating currents, which can be produced by my oscillator attached to any ordinary dynamo situated on shore or on a war ship.

"How such an apparently complicated mechanism can be operated and controlled at a distance of miles is no mystery. It is as simple as the messenger call to be found in almost any office. This is a little metal box with a lever on the outside. By moving the crank to a certain point it gives vibrating sounds and springs back, into position, and its momentary buzzing calls a messenger. But move this same crank a third further around the dial and it buzzes still longer, and pretty soon a policeman appears, summoned by its mysterious call. Again, move the crank this time to the farthest limit of the circle and scarcely has its more prolonged hum of recoil sounded when the city fire apparatus dashes up to your place at its call.

"Now, my device for controlling the motion of a distant submarine boat is exactly similar. Only I need no connecting wires between my switchboard and the distant submarine boat, for I make use of the now well-known principle of wireless telegraphy. As I move this little lever to points which I have marked on a circular dial I cause a different number of vibrations each time. In this case two waves go forth at each half turn of the lever and affect different parts of the distant destroyer's machinery.

"How such submarine destroyers should actually be used in war I leave for naval tacticians to determine. But it seems to me that they could best be operated by taking a number on board a large fast auxiliary cruiser like the St. Louis or St. Paul, launch them, several at a time, like life boats, and direct their movements from a switch board placed in the forward fighting top.

"In order that the director of the submarine destroyer may know its exact position at every movement, two masts, at bow and stern, will project up just above the water, too minute to be seen or hit by an enemy's guns by day, and by night they will carry hooded lights.

"The lookout placed in the fighting top could detect a hostile ship off on the horizon while the auxiliary cruiser's big hull is still invisible to the enemy. Starting these little destroyers out under direction of a man with a telescope, they could attack and destroy a whole armada - destroy it utterly - in an hour, and the enemy never have a sight of their antagonists or know what power destroyed them. A big auxiliary cruiser, used to carry these submarine destroyers, could also carry a cargo of torpedoes sufficient to conduct a long campaign and go half way around the world.

"She could carry the gun cotton and other explosives needed to load the torpedoes in safe magazines below the water line, and do away with much of the danger of transporting loaded torpedoes. When necessary for use the war heads could be loaded, fitted to the torpedoes, and the submarine destroyers fully equipped.

"A high, projecting headland overlooking a harbor and the sea would also be a good point on which to establish a station and have the destroyers laid up at docks below ready to start.

"That is the whole story of my latest invention. It is simple enough, you say. Of course it is, because I have worked all my life to make each one of the details so simple that it will work as easily as the electric ticker in a stock broker's office.

Letter to Editor
Electrical Engineer - N. Y. — *Nov. 24, 1898*

New York, Nov. 18, 1898
46 & 48 East Houston St.

Editor of The Electrical Engineer, 120 Liberty St., New York City:

Sir - By publishing in your columns of Nov. 17 my recent contribution to the Electro-Therapeutic Society you have finally succeeded - after many vain attempts made during a number of years - in causing me a serious injury. It has cost me great pains to write that paper, and I have expected to see it appear among other dignified contributions of its kind, and I confess, the wound is deep. But you will have no opportunity for inflicting a similar one, as I propose to take better care of my papers in the future. In what manner you have secured this one in advance of other electrical periodicals who had an equal right to the same, rests with the secretary of the society to explain.

Your editorial comment would not concern me in the least, were it not my duty to take note of it. On more than one occasion you have offended me, but in my qualities both as Christian and philosopher I have always forgiven you and only pitied you for your errors. This time, though, your offence is graver than the previous ones, for you have dared to cast a shadow on my honor.

No doubt you must have in your possession, from the illustrious men whom you quote, tangible proofs in support of your statement reflecting on my honesty. Being a bearer of great honors from a number of American universities, it is my duty, in view of the slur thus cast upon them, to exact from you that in your next issue you produce these, together with this letter, which in justice to myself, I am forwarding to other electrical journals. In the absence of such proofs, I require that, together with the preceding, you publish instead a complete and humble apology for your insulting remark which reflects on me as well as on those who honor me.

On this condition I will again forgive you; but I would advise you to limit yourself in your future attacks to statements for which you are not liable to be punished by law.

N. Tesla

Tesla Describes His Efforts in Various Fields of Work
From *The Sun, New York, November 21, 1898*

To the Editor of the Sun

Sir: Had it not been for other urgent duties, I would before this have acknowledged your highly appreciative editorial of November 13. Such earnest comments and the frequent evidences of the highest appreciation of my labors by men who are the recognized leaders of this day in scientific speculation, discovery and invention are a powerful stimulus, and I am thankful for them. There is nothing that gives me so much strength and courage as the feeling that those who are competent to judge have faith in me.

Permit me on this occasion to make a few statements which will define my position in the various fields of investigation you have touched upon.

I can not but gratefully acknowledge my indebtedness to earlier workers, as Dr. Hertz and Dr. Lodge, in my efforts to produce a practical and economical lighting system on the lines which I first disclosed in a lecture at Columbia College in 1891. There exists a popular error in regard to this light, inasmuch as it is believed that it can be obtained without generation of heat. The enthusiasm of Dr. Lodge is probably responsible for this error, which I have pointed out early by showing the impossibility of reaching a high vibration without going through the lower or fundamental tones. On purely theoretical grounds such a result is thinkable, but it would imply a device for starting the vibrations of unattainable qualities, inasmuch as it would have to be entirely devoid of inertia and other properties of matter. Though I have conceptions in this regard, I dismiss for the present this proposition as being impossible. We can not produce light without heat, but we can surely produce a more efficient light than that obtained in the incandescent lamp, which, though a beautiful invention, is sadly lacking in the feature of efficiency. As the first step toward this realization, I have found it necessary to invent some method for transforming economically the ordinary currents as furnished from the lighting circuits into electrical vibrations of great rapidity. This was a difficult problem, and it was only recently that I was able to announce its practical and thoroughly satisfactory solution. But this was not the only requirement in a system of this kind. It was necessary also to increase the intensity of the light, which at first was very feeble. In this direction, too, I met with complete success, so that at present I am producing a thoroughly serviceable and economical light of any desired intensity. I do not mean to say that this system will revolutionize those in use at present, which have resulted from the cooperation of many able men. I am only sure that it will have its fields of usefulness.

As to the idea of rendering the energy of the sun available for industrial purposes, it fascinated me early but I must admit it was only long after I discovered the rotating magnetic field that it took a firm hold upon my mind. In assailing the problem I found two possible ways of solving it. Either power was to be developed on the spot by converting the energy of the sun's radiations or the energy of vast reservoirs was to be transmitted economically to any distance. Though there were other possible sources of economical power, only the two solutions mentioned offer the ideal feature of power being obtained without any consumption of material. After long thought I finally arrived at two solutions, but on the first of these, namely, that referring to the development of power in any locality from the sun's radiations, I can not dwell at present. The system of power transmission without wires, in the form in which I have described it recently, originated in this manner. Starting from two facts that the earth was a conductor insulated in space, and that a body can not be charged without causing an equivalent displacement of electricity in the earth, I undertook to construct a machine suited for creating as large a displacement as possible of the earth's electricity.

This machine was simply to charge and discharge in rapid succession a body insulated in space, thus altering periodically the amount of electricity in the earth, and consequently the pressure all over its surface. It was nothing but what in mechanics is a pump, forcing water from a large reservoir into a small one and back again. Primarily I .contemplated only the sending of messages to great distances in this manner, and I described the scheme in detail, pointing out on that occasion the importance of ascertaining certain electrical conditions of the earth. The attractive feature of this plan was that the intensity of the signals should diminish very little with the distance, and, in fact, should not diminish at all, if it were not for certain losses occurring, chiefly in the atmosphere. As all my previous ideas, this one, too, received the treatment of Marsyas, but it forms, nevertheless, the basis of what is now known as "wireless telegraphy." This statement will bear rigorous examination, but it is not made with the intent of detracting from the merit of others. On the contrary, it is with great pleasure that I acknowledge the early work of Dr. Lodge, the brilliant experiments of Marconi, and of a later experimenter in this line, Dr. Slaby, of Berlin. Now, this idea I extended to a system of power transmission, and I submitted it to Helmholtz on the occasion of his visit to this country. He unhesitatingly said that power could certainly be transmitted in this manner, but he doubted that I could ever produce an apparatus capable of creating the high pressures of a number of million volts, which were required to attack the problem with any chance of success, and that I could overcome the difficulties of insulation. Impossible as this problem seemed at first, I was fortunate to master it in a comparatively short time, and it was in perfecting this apparatus that I came to a turning point in the development of this idea. I, namely, at once observed that the air, which is a perfect insulator for currents produced by ordinary apparatus, was easily traversed by currents furnished by my improved machine, giving a tension of something like 2,500,000 volts. A further investigation in this direction led to another valuable fact; namely, that the conductivity of the air for these currents increased very rapidly with its degree of rarefaction, and at once the transmission of energy through the upper strata of air, which, without such results as I have obtained, would be nothing more than a dream, became easily realizable. This appears all the more certain, as I found it quite practicable to transmit, under conditions such as exist in heights well explored, electrical energy in large amounts. I have thus overcome all the chief obstacles which originally stood in the way, and the success of my system now rests merely on engineering skill.

Referring to my latest invention, I wish to bring out a point which has been overlooked. I arrived, as has been stated, at the idea through entirely abstract speculations on the human organism, which I conceived to be a self-propelling machine, the motions of which are governed by impressions received through the eye. Endeavoring to construct a mechanical model resembling in its essential, material features the human body, I was led to combine a controlling device, or organ sensitive to certain waves, with a body provided with propelling and directing mechanism, and the rest naturally followed. Originally the idea interested me only from the scientific point of view, but

soon I saw that I had made a departure which sooner or later must produce a profound change in things and conditions presently existing. I hope this change will be for the good only, for, if it were otherwise, I wish that I had never made the invention. The future may or may not bear out my present convictions, but I can not refrain from saying that it is difficult for me to see at present how, with such a principle brought to great perfection, as it undoubtedly will be in the course of time, guns can maintain themselves as weapons. We shall be able, by availing ourselves of this advance, to send a projectile at much greater distance, it will not be limited in any way by weight or amount of explosive charge, we shall be able to submerge it at command, to arrest it in its flight, and call it back, and to send it out again and explode it at will, and, more than this, it will never make a miss, since all chance in this regard, if hitting the object of attack were at all required, is eliminated. But the chief feature of such a weapon is still to be told; namely, it may be made to respond only to a certain note or tune, it may be endowed with selective power. Directly such an arm is produced, it becomes almost impossible to meet it with a corresponding development. It is this feature, perhaps, more than in its power of destruction, that its tendency to arrest the development of arms and to stop warfare will reside. With renewed thanks, I remain,
Very truly, yours,
N. Tesla
New York,
November 19

On Current Interrupters
Electrical Review — March 15, 1899

To stimulate the ardor of the zealous experimenters, who believe in the revolutionary character of this discovery, it might be well to suggest one or two such simple devices for interrupting the current. For instance, a very primitive contrivance of this kind comprises a poker — yes, an ordinary poker, connected by means of a flexible cable to one of the mains of the generator, and a bathtub filled with conducting fluid which is connected in any suitable manner, through the primary of an induction coil, to the other pole of the generator. When the experimenter desires to take a Roentgen picture, he brings the end of the poker to white heat, and, thrusting the same into the bathtub, he will at once witness an astonishing phenomenon; the seething and boiling liquid making and breaking the current in rapid succession, and the powerful rays generated will at once convince him of the great practical value of this discovery. I 'night further suggest that the poker may be conveniently heated by means of a welding machine.'

Another device, entirely automatic, and, probably suitable for use in suburban districts, comprises two insulated metal plates, supported in any convenient manner, in close proximity to each other. These plates are connected through the primary of an induction coil with the determinals of a generator, and are bridged by two movable contacts joined by a flexible cable. The two contacts are both attached to the legs of a good-sized chicken standing astride on the plates. Heat being applied to the latter, muscular contractions are produced in the legs of the chicken, which thus makes and breaks the current through the induction coil. Any number, of such chickens may be provided and the contacts connected in series or multiple arc, as may be desired, thus, increasing the frequency of the impulses. In this manner fierce sparks, suitable for most purposes, may be obtained, and vacuum tubes may be operated, and these contrivances will be found a notable improvement on certain circuit-breakers of old, with which two enterprising editors undertook some years ago to revolutionize the systems of electric lighting. The enterprising editors, are wiser now. They are to be congratulated, and their readers, scientific societies and the profession, all ought to be congratulated, and — "all is well that ends well." The observant experimenter will not fail to note that the fierce sparks frighten the chickens, which are thus put into more violent spasms and muscular contractions, this again increasing the fierceness of the sparks, which, in return, causes a greater fright of the chickens and increased speed of interruptions; it is, in fact, as Kipling says:

"Interdependence absolute, foreseen, ordained, decreed,
To work, ye'll note, as any tilt an' every rate o'speed."

But to return, in all earnestness, to the "electrolytic interrupter" described, this is a device with which I am perfectly familiar, having carried on extensive experiments with the same two or three years ago. It was one of many devices which I invented in my efforts to produce an economical contrivance of this kind. The name is really not appropriate, inasmuch as any fluid, either conducting or made so in any manner, as by being rendered acid or alkaline, or by being heated, may be used. I have even found it possible, under certain conditions, to operate with mercury. The device is extremely simple, but the great waste of energy attendant upon its operation and certain other defects make it entirely unsuitable for any valuable, practical purpose, and as far as those instances are concerned, in which a small amount of energy is needed, much better results are obtained by a properly designed mechanical circuit-breaker. The experimenters are very likely deceived by finding that an induction coil gives longer sparks when this device is inserted in place of the ordinary break, but this is due merely to the fact that the break is not properly designed. Of the total energy supplied from the mains, scarcely one-fourth is obtainable of that amount, which a well constructed mechanical break furnishes in the secondary, and although I have designed many improved forms, I have found it impossible to increase materially the economy. Two improvements, however, which I found at that time necessary to introduce, I may mention for the benefit of 'those who are using the device. As will be readily noted, the small terminal is surrounded by a gaseous bubble, in which the makes and breaks are formed, generally in an irregular manner, by the liquid being driven towards the terminal at some point. The force which drives the liquid is evidently the pressure of the fluid column, and by increasing the fluid pressure in any manner the liquid is forced with greater speed towards the terminal and thus the frequency is increased. Another necessary improvement was to make a provision for preventing the acid or alkali from being carried off into the atmosphere, which always happens more or less, even if the liquid column be of some height. During my early experiments with the device, I became so interested in it that I neglected this precaution, and . I noted that the acid had attacked all 'the apparatus in my laboratory. The experimenter will conveniently carry out both of these improvements by taking a long glass tube of, say, six to eight feet in length, and arranging the interrupting device close to the bottom of the tube, with an outlet for eventually replacing the 'liquid. The high column will prevent the fumes from vitiating the atmosphere of the room, and the increased pressure will add materially to the effectiveness of the performance. If the liquid column be,. say, nine times as high, the force driving the fluid towards the contact is nine times as great, and this force is capable, under the same conditions, of driving the fluid three times as fast, hence the frequency is increased in that ratio and, .in fact, in a somewhat greater ratio, as the gaseous bubble, being compressed, is rendered smaller, and therefore the liquid is made to travel through a smaller distance. The electrode, of course, should be very small :to insure the regularity of operation, and it is not necessary to use platinum. The pressure may, however, be increased in other ways, and I have obtained some results of interest, in experiments of this kind.

As before stated, the device is very wasteful, and, while it, may be used in some instances, I consider it of little or no practical value. It will please me to be convinced of the contrary; but I do not think that I am erring. My chief reasons for this statement are that there are many other ways in which by far better results are obtainable with are equally simple, if not more so. I may mention one here, based on a different principle which is incomparably more effective, more efficient and also simpler on the whole. It comprises a fine stream of conducting fluid which is made to issue, with any desired speed, from an orifice connected with one pole of a generator, through the primary of the induction coil, against the other terminal of the generator placed at a small distance. This device gives discharges of a remarkable suddenness, and the frequency may be brought within reasonable limits, almost to anything desired. I have used this device for a long time in connection with ordinary coils and in a form of my own coil with results greatly superior in every respect to those obtainable with the form of device discussed.

The Problem of Increasing Human Energy, With Special References to the Harnessing of the Sun's Energy
The Century Illustrated Monthly Magazine — June 1900

The Onward Movement of Man — the Energy of the Movement — the Three Ways of Increasing Human Energy

Of all the endless variety of phenomena which nature presents to our senses, there is none that fills our minds with greater wonder than that inconceivably complex movement which, in its entirety, we designate as human life. Its mysterious origin is veiled in the forever impenetrable mist of, the past, its character is rendered incomprehensible by its infinite intricacy, and its destination is hidden in the unfathomable depths of the future. Whence does it come? What is it? Whither does it: tend? are the great questions which the sages of all times have endeavored to answer.

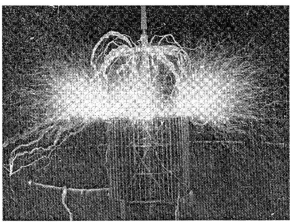

Fig: 1. This result is produced by the discharge of an electrical oscillator giving twelve million volts. The electrical pressure, alternating one hundred thousand times per second, excites the normally inert nitrogen, causing it to combine with the oxygen. The flame-like discharge shown in the photograph, measures sixty-five feet across.

Modern science says: The sun is the past, the earth is the present, the moon is the future. From an incandescent mass we have originated, and into a frozen mass we shall turn. Merciless is the law of nature, and rapidly and irresistibly we are drawn to our doom. Lord Kelvin, in his profound meditations, allows us only a, short spin of life, something like six million years, after which time the sun's bright light will have ceased to shine, and its life-giving heat will have ebbed away, and our own earth will be a lump of ice, hurrying on through the eternal night. But do not let us despair. There will still be left on it a glimmering spark of. life, and there .will be a chance to kindle a new fire on some distant star. This wonderful possibility seems, indeed, to exist, judging from Professor Dewar's beautiful experiments with, liquid air, which show that germs of organic life are not destroyed by cold, no matter how intense; consequently they may be transmitted through the interstellar space. Meanwhile the cheering lights of science and art, ever increasing in intensity, illuminate our path, and the marvels they disclose, and the enjoyments they offer, make us .Measurably forgetful of the gloomy future.

Though we may never be able to comprehend human life, we know certainly that it is a movement, of whatever nature it be. The existence of a movement unavoidably implies a body which is being moved and a force which is moving it. Hence, wherever there is life, there is a mass moved by a force. All mass possesses inertia, all force tends to persist. Owing to this universal property and condition, a body, be it at rest or in motion, tends to remain in the same state, and a force, manifesting itself anywhere and through whatever cause, produces an equivalent opposing force, and as an absolute necessity of this it follows that every movement in nature must be rhythmical. Long age this simple truth was clearly pointed out by Herbert Spencer, who arrived at it through a somewhat different process of reasoning. It is borne out in everything we perceive — in the movement of a planet, in the surging and ebbing of the tide, in the reverberations of the air, the swinging of a pendulum, the oscillations of an electric current; and in the infinitely varied phenomena of organic life. Does not the whole of human life attest it? Birth, growth, old age, and death of an individual, family, race, or nation, what is it all but a rhythm? All life-manifestation, then, even in its most intricate form, as exemplified in man, however involved and inscrutable, is *only* a movement, to which the same general laws of movement which govern throughout the physical universe must be applicable.

When we speak of man, we have a conception of humanity as a whole, and before applying scientific methods to the investigation of his movement, we must accept this as a physical fact. But can any one doubt to-day that all the millions of individuals and all innumerable types and characters constitute an entirety, a unit? Though free to think and act, we are held together, like the stars in the firmament; with ties inseparable. These ties we cannot see, but we can feel them. I cut myself in the finger, and it pains me: this finger is a part of me. I see a friend hurt, and it hurts me, too: my friend and I are one. And now I see stricken down an enemy, a lump of matter which, of all the lumps of matter in the universe, I care least for, and still it grieves me. Does this not prove that each of us is only a part of a whole?

For ages this idea has been proclaimed in the consummately wise teachings of religion, probably not alone as a means of insuring peace and harmony among men; but as a deeply founded truth. The Buddhist expresses it in one way, the Christian in another, but both say the same: We are all one. Metaphysical proofs are, however, not the only ones which we are able to bring forth in support of this idea. Science, too, recognizes this connectedness of separate individuals, though not quite in the same sense as it admits that the suns, planets, and moons of a constellation are one body, and there can be no doubt that it will be experimentally confirmed in times to come, when our means and methods for investigating physical and other states and phenomena shall have been brought to great perfection. Still more: this one human being lives on and on, The individual is ephemeral, races and nations come and pass away, but man remains. Therein lies the profound difference between the. individual and the whole. Therein, too, is to be found the partial explanation of many of those marvelous phenomena of heredity which are the result of countless centuries of feeble but persistent influence.

Conceive, then, man as a mass urged on by a force. Though this movement is not of a translatory character, implying change of place, yet the general laws of mechanical movement are applicable to it, and the energy associated with this mass can be measured, in accordance with well-known principles, by half the product of the mass with the square of a certain velocity. So, for instance, a cannon-ball which is at rest possesses a certain amount of energy in the form of heat, which we measure in a similar way. We imagine the ball to consist of innumerable minute particles, called atoms or molecules, which vibrate, or whirl around one another. We determine their masses and velocities, and from them the energy of each of these minute systems, and adding them all together, we get an idea of the total heat-energy contained in the ball, which is only seemingly at rest. In this purely theoretical estimate this energy may then be calculated by multiplying half of the total mass — that is, half of the sum of all the small masses — with the square of a velocity which is determined from the velocities of the separate particles. In like manner we may conceive of human energy being measured by half the human mass multiplied with the square of a velocity which we are not yet able to compute. But our deficiency in this knowledge will not vitiate the truth of the deductions I shall draw, which rest on the firm basis that the same laws of mass and force govern throughout nature.

Man, however, is not an ordinary mass, consisting of spinning atoms and molecules, and containing merely heat-energy. He is a mass possessed of certain higher qualities by reason of the creative principle of life with which he is endowed. His mass, as .the water in an ocean wave, is being continuously exchanged, new taking the place, of the old. Not only this, but he grows, propagates, and dies, thus altering his mass independently, both in bulk and density. What is ,most wonderful of all, he is capable of increasing or diminishing his velocity of movement by the mysterious power he possesses of appropriating more or less energy from other substance, and turning it into motive energy. But in any given moment we may ignore these slow changes and assume that human energy is measured by half the product of man's mass with the square of a certain hypothetical velocity. However we may compute this velocity, and whatever we may take as the standard of its measure, we must, in harmony with this conception, come to the conclusion that the great problem of science is, and always will be, to increase the energy thus defined. Many years ago, stimulated by the perusal of that deeply interesting work, Draper's "History of the Intellectual Development of Europe," depicting so vividly human movement, I recognized that to solve this eternal problem must ever be the chief task of the man of science. Some results of my own efforts to this end I shall endeavor briefly to describe here.

Diagram A,

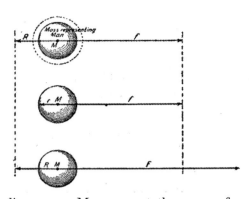

Let, then, in diagram a, M represent the mass of man. This mass is impelled in one direction by a force f, which is resisted by another partly frictional and partly negative force R, acting in a direction exactly opposite, and retarding the movement of the mass. Such an antagonistic force is present in every movement, and must be taken into consideration. The difference between these two forces is the effective force which imparts a velocity V to the mass M in the direction of the arrow on the line representing the force f. In accordance with the preceding, the human energy will then be given by the product $\frac{1}{2}MV^2 = \frac{1}{2}M \times V^2$, in which M is the total mass of man in the ordinary interpretation of the term "mass," and V is a certain hypothetical velocity, which, in the present state of science, we are unable exactly to define and determine. To increase the human energy is, therefore, equivalent to increasing this product, and there are, as will readily be seen, only three ways possible to attain this result, which are illustrated in the diagram below. The first way, shown in the top figure, is to increase the mass (as indicated by the dotted circle), leaving the two opposing forces the same. The second way is to reduce the retarding force R to a smaller value r, leaving the mass and the impelling force the same, as diagrammatically shown in the middle figure. The third way, which is illustrated in the last figure, is to increase the impelling force f to a higher value F, while the mass and the retarding force R remain unaltered. Evidently fixed limits exist as regards increase of mass and reduction of retarding force, but the impelling force can be increased indefinitely. Each of these three possible solutions presents a different aspect of the main problem of increasing human energy, which is thus divided into three distinct problems, to 6e successively considered.

The First Problem: How to Increase the Human Mass — the Burning of Atmospheric Nitrogen

Viewed generally, there are obviously two ways of increasing the mass of mankind: first, by aiding and maintaining those forces and conditions which tend to increase it; and, second, by opposing and reducing those which tend to diminish it. The mass will be increased by careful attention to health, by substantial food, by moderation, by regularity of habits, by the promotion of marriage, by conscientious attention to the children, and, generally stated, by the observance of all the many precepts and laws of religion and hygiene. But in adding new mass to the old, three cases again present themselves. Either the mass added is of the same velocity as the old, or it is of a smaller or of a higher velocity. To gain an idea of the relative importance of these cases, imagine a train composed of, say, one hundred locomotives running on a track, and suppose that, to increase the energy of the moving mass, four more locomotives are added to the train. If these four move at the same velocity at which the train is going, the total energy, will be increased four per cent.; if they are moving at only one half of that velocity, the increase will amount to only one per cent.; if they are moving at twice that velocity, the increase of energy will be sixteen per cent. This simple illustration shows that it is of the greatest importance to add mass of a higher velocity. Stated more to the point, if, for example, the children be of the same degree of enlightenment as the parents, — that is, mass of the "same velocity," — the energy will simply increase proportionately to the number added. If they are less intelligent, or advanced, or mass of "smaller velocity," there will be a very slight gain in the energy; but if they are further advanced, or mass of "higher velocity," then the new generation will add very considerably to the sum total of human energy. Any addition of mass of "smaller velocity," beyond that indispensable amount required by the law expressed in the proverb, "*Mens Sava in corpora sono,*" should be strenuously opposed. For instance, the mere development of muscle, as aimed at in some of our colleges, I consider equivalent to adding mass of "smaller velocity," and I would not commend it, although my views were different when I was a student myself. Moderate exercise, insuring the right balance between mind and body, and the highest efficiency of performance, is, of course, a prime requirement. The above example shows that the most important result to be attained is the education, or the increase of the "velocity," of the mass newly added.

Conversely, it scarcely need be stated that everything that is against the teachings of religion and the laws of hygiene is tending to decrease the mass. Whisky, wine, tea, coffee, tobacco, and other such stimulants are responsible for the shortening of the lives of many, and ought to be used with moderation. But I do not think that rigorous measures of suppression of habits followed through many generations are commendable. It is wiser to preach moderation than abstinence. We have become accustomed to these stimulants, and if such reforms are to be effected, they must be slow and gradual. Those who are devoting their energies to such ends could make themselves far more useful by turning their efforts in other directions; as, for instance, toward providing pure water.

For every person who perishes from the effects of a stimulant; at least a thousand die from the consequences of drinking impure water. This precious fluid, which daily infuses new life into us, is likewise the chief vehicle through which disease and death enter our bodies. The germs of destruction it conveys are enemies all the more terrible as they perform their fatal work unperceived. They seal our doom while we live and enjoy. The majority of people are so ignorant or careless in drinking water, and the consequences of this are so disastrous, that a philanthropist can scarcely use his efforts better than by endeavoring to enlighten those who are thus injuring themselves. By systematic purification and sterilization of the drinking-water the human mass would be very considerably increased. It should be made a rigid rule — which might be enforced by law — to boil or to sterilize otherwise the drinking-water in every household and public place. The mere filtering does not afford sufficient security against infection. All ice for internal uses should be artificially prepared from water thoroughly sterilized. The importance of eliminating germs of disease from the city water is generally recognized, but little is being done to improve the existing conditions, as no satisfactory method of sterilizing great quantities of water has as yet been brought forward. By improved electrical appliances we are now enabled to produce ozone cheaply and in large amounts, and this ideal disinfectant seems to offer a happy solution of the important question.

Gambling, business rush, and excitement, particularly on the exchanges, are causes of much mass-reduction, all the more so because the individuals concerned represent units of higher value. Incapacity of observing the first symptoms of an illness, and careless neglect of the same, are important factors of mortality. In noting carefully every new sign of approaching danger, and making conscientiously every possible effort to avert it, we are not only following wise laws of hygiene in the interest of our well-being and the success of our labors, but we are also complying with a higher moral duty. Every one should consider his body as a priceless gift from one whom he loves above all, as a marvelous work of art, of undescribable beauty and mastery beyond human conception, and so delicate and frail that a word, a breath, a look, nay, a thought, may injure it. Uncleanliness, which breeds disease and death, is not only a self-destructive but a highly immoral habit. In keeping our bodies free from infection, healthful, and pure, we are expressing our reverence for the high principle with which they are endowed. He who follows the precepts of hygiene in this spirit is proving himself, so far, truly religious. Laxity of morals is a terrible evil, which poisons both mind and body, and which is responsible for a great reduction of the human mass in some countries. Many of the present customs and tendencies are productive of similar hurtful results. For example, the society life, modern education and pursuits of women, tending to draw them away from their household duties and make men out of them, must needs detract from the elevating ideal they represent, diminish the artistic creative power, and cause sterility and a general weakening of the race. A thousand other evils might be mentioned, but all put together, in their bearing upon the problem under discussion, they would not equal a single one, the want of food, brought on by poverty, destitution, and famine. Millions of individuals die yearly for want of food, thus keeping down the mass. Even in our enlightened communities, and notwithstanding the many charitable efforts, this is still, in all probability, the chief evil. I do not mean here absolute want of food, but want of healthful nutriment.

How to provide good and plentiful food is, therefore, a most important question of the day. On general principles the raising of cattle as a means of providing food is objectionable, because, in the sense interpreted above, it must undoubtedly tend to the addition of mass of a "smaller velocity." It is certainly preferable to raise vegetables, and I think, therefore, that vegetarianism is a commendable departure from the established barbarous habit. That we can subsist on plant food and perform. our work even to advantage is not a theory, but a well-demonstrated fact. Many races living almost exclusively on vegetables are of superior physique and strength. There is no doubt that some plant food, such as oatmeal, is more economical than meat, and superior to it in regard to both mechanical and mental performance. Such food,.moreover, taxes our digestive organs decidedly less, and, in making us more contented and sociable, produces an amount of good difficult to estimate. In view of these facts every effort should be made to stop the wanton and cruel slaughter of animals, which must be destructive to our morals. To free ourselves from animal instincts and appetites, which keep us down, we should begin at the very root from which they spring: we should effect a radical reform in the character of the food.

There seems to be no *philosophical* necessity for food. We can conceive of organized beings living without nourishment, and. deriving all the energy they need for the performance of their life-functions from the ambient medium. In a crystal we have the clear evidence of the existence of a formative life-principle, and though we cannot understand the life of a crystal, it is none the less a living being. There may be, besides crystals, other such individualized, material systems of beings, perhaps of gaseous constitution, or composed of substance still more tenuous. In view of this possibility, — nay, probability, — we cannot apodictically deny the existence of organized beings on a planet merely because the conditions on the same are unsuitable for the existence of life as we conceive it. We cannot even, with positive assurance, assert that some of them might not be present here, in this our world, in the very midst of us, for their constitution and life-manifestation may be such that we are unable to perceive them.

The production of artificial food as a means for causing an increase of the human mass naturally suggests itself, but a direct attempt of this kind to provide nourishment does not appear to me rational, at least not for the present. Whether we could thrive on such food is very doubtful. We are the result of ages of continuous adaptation, and we cannot radically change without unforeseen and, in all probability, disastrous consequences. So uncertain an experiment should not be tried. By far the best way, it seems to me, to meet the ravages of the evil, would be to find ways of increasing the productivity of the soil. With this object the preservation of forests is of an importance which cannot be overestimated, and in this connection, also, the utilization of water-power for purposes of electrical transmission, dispensing in many ways with the necessity of burning wood, and tending thereby to forest preservation, is to be strongly advocated. But there are limits in the improvement to be effected in this and similar ways.

To increase materially the productivity of the soil, it must be more effectively fertilized by artificial means. The question of food-production resolves itself, then; into the question how best to fertilize the soil. What it is that made the soil is still a mystery. To explain its origin is probably equivalent to explaining the origin of life itself. The rocks, disintegrated by moisture and heat and wind and weather, were in themselves not capable of maintaining life. Some unexplained condition arose, and some new principle came into effect, and the first layer capable of sustaining low organisms, like mosses, was formed. These, by their life and death, added more of the life-sustaining quality to the soil, and higher organisms could then subsist, and so on and on, until at last highly developed plant and animal life could flourish. But though the theories are, even now, not in agreement as to how fertilization is effected, it is a fact, only too well ascertained, that the soil cannot indefinitely sustain life, and some way must be found to supply it with the substances which have been abstracted from it by the plants. The chief and most valuable among these substances are compounds of nitrogen, and the cheap production of these is, therefore, the key for the solution of the all-important food problem. Our atmosphere contains an inexhaustible amount of nitrogen, and could we but oxidize it and produce these compounds, an incalculable benefit for mankind would follow.

Long ago this idea took a powerful hold on the imagination of scientific men; but an efficient means for accomplishing this result could not be devised. The problem was rendered extremely difficult by the extraordinary inertness of the nitrogen, which refuses to combine even with oxygen. But here electricity comes to our aid: the dormant affinities of the element are awakened by an electric current of the proper quality. As a lump of coal which has been in contact with oxygen for centuries without burning will combine with it when once ignited, so nitrogen, excited by electricity, will burn. I did not succeed, however, in producing electrical discharges exciting very effectively the atmospheric nitrogen until a comparatively recent date, although I .showed, in May, 1891, in a scientific lecture, a novel form of discharge or electrical flame named "St. Elmo's hotfire," which, besides being capable of generating ozone in abundance, also possessed, as I pointed out on that occasion, distinctly the quality of exciting chemical affinities. This discharge or flame was then only three or four inches long, its chemical action was likewise very feeble, and consequently the process of oxidation of the nitrogen was wasteful. How to intensify this action was the question. Evidently electric currents of a peculiar kind had to be produced in order to render the process of nitrogen combustion more efficient.

The first advance was made in ascertaining that the chemical activity of the discharge was very considerably increased by using currents of extremely high frequency or rate of vibration. This was an important improvement, but practical considerations soon .set a definite limit to the progress in this direction. Next, the effects of the electrical pressure of the current impulses, of their wave-form and other characteristic features, were investigated. Then the influence of the atmospheric pressure and temperature and of the presence of water and other bodies was studied, and thus the best conditions for causing the most intense chemical action of the discharge and securing the highest efficiency of the process were gradually ascertained. Naturally, the improvements were not quick in coming; still, little by little, I advanced. The flame grew larger and larger, and its oxidizing action more and more intense. From an insignificant brush-discharge a few inches long it developed into a marvelous electrical phenomenon, a roaring blaze, devouring the nitrogen of the atmosphere and measuring sixty or seventy feet across. Thus slowly, almost imperceptibly, possibility became accomplishment. All is not yet done, by any means, but to what a degree my efforts have been rewarded an idea may be gained from an inspection of Fig. 1, which, with its title, is self-explanatory. The flame-like discharge visible is produced by the intense electrical oscillations which pass through the coil shown, and violently agitate the electrified molecules of the air. By this means a strong affinity is created between the two normally indifferent constituents of the atmosphere, and they combine readily, even if no further provision is made for intensifying the chemical action of the discharge. In the manufacture of nitrogen compounds by this method, of course, every possible means bearing upon the intensity of this action and the efficiency of the process will be taken advantage of, and, besides, special arrangements will be provided for the fixation of the compounds formed, as they are generally unstable, the nitrogen becoming again inert after a little lapse of time. Steam is a simple and effective means for fixing permanently the compounds. The result illustrated makes it practicable to oxidize the atmospheric nitrogen in unlimited quantities, merely by the use of cheap mechanical power and simple electrical apparatus. In this manner many compounds of nitrogen may be manufactured all over the world, at a small cost, and in any desired amount, and by means of these compounds the soil can be fertilized and its productiveness indefinitely increased. An abundance of cheap and healthful food, not artificial, but such as we are accustomed to, may thus be obtained. This new and inexhaustible source of food-supply will be of incalculable benefit to mankind, for it will enormously contribute to the increase of the human mass, and thus add immensely to human energy. Soon, I hope, the world will see the beginning of an industry which, in time to come, will, I believe, be in importance next to that of iron.

The Second Problem: How to Reduce the Force Retarding the Human Mass — the Art of Telautomatics

As before stated, the force which retards the onward movement of man is partly frictional and partly negative. To illustrate this distinction I may name, for example, ignorance, stupidity, and imbecility as some of the purely frictional forces, or remittances devoid of any directive tendency. On the other hand, visionariness, insanity, self-destructive tendency, religious fanaticism, and the like, are all forces of a negative character, acting in definite directions. To reduce or entirely to overcome these dissimilar retarding forces, radically different methods must be employed. One knows, for instance, what a fanatic may do, and one can take preventive measures, can enlighten, convince, and possibly direct him, turn his vice into virtue; but one does not know, and never can know, what a brute or an imbecile may do, and one must deal with him as with a mass, inert, without mind, let loose by the mad elements. A negative force always implies some quality, not infrequently a high one, though badly directed, which it is possible to turn to good advantage; but a directionless, frictional force involves unavoidable loss. Evidently, then, the first and general answer to the above question is: turn all negative force in the right direction and reduce all frictional force.

There can be no doubt that, of all the frictional remittances, the one that most retards human movement is ignorance. Not without reason said that man of wisdom, Buddha: "Ignorance is the greatest evil in the world." The friction which results from ignorance, and which is greatly increased owing to the numerous languages and nationalities, can be reduced only by the spread of knowledge and the unification of the heterogeneous elements of humanity. No effort could be better spent. But however ignorance may have retarded the onward movement of man in times past, it is certain that, nowadays, negative forces have become of greater importance. Among these there is one of far greater moment than any other. It is called organized warfare. When we consider the millions of individuals, often the ablest in mind and body, the flower of humanity, who are compelled to a life of inactivity and unproductiveness, the immense sums of money daily required for the maintenance of armies and war apparatus, representing ever so much of human energy, all the effort uselessly spent in the production of arms and implements of destruction, the loss of life and the fostering of a barbarous spirit, we are appalled at the inestimable loss to mankind which the existence of these deplorable conditions must involve. What can we do to combat best this great evil?

Law and order absolutely require the maintenance of organized force. No community can exist and prosper without rigid discipline. Every country must be able to defend itself, should the necessity arise. The conditions of to-day are not the result of yesterday, and a radical change cannot be effected to-morrow. If the nations would at once disarm, it is more than likely that a state of things worse than war itself would follow. Universal peace is a beautiful dream; but not at once realizable. We have seen recently that even the noble effort of the man invested with the greatest worldly power has been virtually without effect. And no wonder, for the establishment of universal peace is, for the time being, a physical impossibility. War is a negative force, and cannot be turned in a positive direction without passing through the intermediate phases. It is the problem of making a wheel, rotating one way, turn in the opposite direction without slowing it down, stopping it, and speeding it up again the other way.

It has been argued that the perfection of guns of great destructive power will stop warfare. So I myself thought for a long time, but now I believe this to be a profound mistake. Such developments will greatly modify, but not arrest it. On the contrary, I think that every new arm that is invented, every new departure that is made in this direction, merely invites new talent and skill, engages new effort, offers a new incentive, and so only gives a fresh impetus to further development. Think of the discovery of gunpowder. Can we conceive of any more radical departure than was effected by this innovation? Let us imagine ourselves living in that period: would we not have thought then that warfare was at an end, when the armor' of the knight became an object of ridicule, when bodily strength and skill, meaning so much before, became of comparatively little value? Yet gunpowder did not stop warfare; quite the opposite — it acted as a most powerful incentive. Nor do I believe that warfare can ever be arrested by any scientific or ideal development, so long as similar conditions to those now prevailing exist, because war has itself become a science, and because war involves some of the most sacred sentiments of which man is capable. In fact, it is doubtful' whether men who would not be ready to fight for a, high principle would be good for anything at all. It is not the mind which makes man, nor is it the body; it is mind and body. Our virtues and our failings are inseparable, like force and matter. When they separate, man is no more.

Another argument, which carries considerable force, is frequently made, namely, that war must soon become impossible because the means of defense are outstripping the means of attack. This is only in accordance with a fundamental law which may be expressed by the statement that it is easier to destroy than to build. This law defines human capacities and human conditions. Were these such that it would be easier to build than to destroy, man would go on unresisted, creating and accumulating without limit. Such conditions are not of this earth. A being which could do this would not be a man; it might be a god. Defense will always have the advantage over attack, but this alone, it seems to me, can never stop war. By the use of new principles of defense we can render harbors impregnable against attack, but we cannot by such means prevent two war-ships meeting in battle on the high sea. And then, if we follow this idea to its ultimate development, we are led to the conclusion that it would be better for mankind if attack and defense were just oppositely related: for if every country, even the smallest, could surround itself with a wall absolutely impenetrable, and could defy the rest of the world, a state of things would surely be brought on which would be extremely unfavorable to human progress. It is by abolishing all the barriers which separate nations and countries that civilization is best furthered.

Again, it is contended by some that the advent of the flying-machine must bring on universal peace. This, too, I believe to be an entirely erroneous view. The flying-machine is certainly coming, and very soon, but the conditions will remain the same as before. In fact, I see no reason why a ruling power, like Great Britain, might not govern the air as well as the sea. Without wishing to put myself on record as a prophet, I do not hesitate to say that the next years will see the establishment of an "air-power," and its center may not be far from New York. But, for all that, men will fight on merrily.

The ideal development of the war principle would ultimately lead to the transformation of the whole energy of war into purely potential, explosive energy, like that of an electrical condenser. In this form .the war-energy could be maintained without effort; it would need to be much smaller in amount, while incomparably more effective.

As regards the security of a country against foreign invasion, it is interesting to note that it depends only on the relative, and not on the absolute, number of the individuals or magnitude of the forces, and that, if every country should reduce the war-force in the same ratio, the security would remain unaltered. An international agreement with the object of reducing .to a minimum the war-force which, in view of the present still imperfect education of the masses, is absolutely indispensable, would, therefore, seem to be the first rational step to take .toward diminishing the force retarding human movement.

Fortunately, the existing conditions cannot continue indefinitely, for a new element is beginning to assert itself. A change for the better is imminent, and I shall now endeavor to show what, according to my ideas, will be the first advance toward the establishment of peaceful relations between nations, and by what means it will eventually be accomplished.

Let us go back to the early beginning, when the law of the stronger was the only law. The light of reason was not yet kindled, and the weak was entirely at the merry of the strong. The weak individual then began to learn how to defend himself. He made use of a club, stone; spear. sling, or bow and arrow, and in the course of time, instead of physical strength, intelligence became the chief deciding factor in the battle. The wild character was gradually softened by the awakening of noble sentiments, and so, imperceptibly, after ages of continued progress, we have come from the brutal fight of the unreasoning animal to what we call the "civilized warfare" of to-day, in which the combatants shake hands, talk in a friendly way, and smoke cigars in the entrances, ready to engage again in deadly conflict at a signal. Let pessimists say what they like, here is an absolute evidence of great and gratifying advance.

But now, what is the next phase in this evolution? Not peace as yet, by any means. The next change which should naturally follow from modern developments should be the continuous diminution of the number of individuals engaged in battle. The apparatus will be one of specifically great power, but only a few individuals will be required to operate it. This evolution will bring more and more into prominence a machine or mechanism with the fewest individuals as an element of warfare, and the absolutely unavoidable consequence of this will be the abandonment of large, clumsy, slowly moving, and unmanageable units. Greatest possible speed and maximum rate of energy-delivery by the war apparatus will be the main object. The loss of life will become smaller and smaller, and finally, the number of the individuals continuously diminishing, merely machines will meet in a contest without bloodshed, the nations being simply interested, ambitious spectators. When this happy condition is realized, peace will be assured. But, no matter to what degree of perfection rapid-fire guns, high-power cannon, explosive projectiles, torpedo-boats, or other implements of war may be brought, no matter how destructive they *may* be made, that condition can never be reached through *any* such development. All such implements require men for their operation; men are indispensable parts of the machinery. Their object is to kill and to destroy. Their power resides in their capacity for doing evil: So long as men meet in battle, there will be bloodshed. Bloodshed will ever keep up barbarous passion. To break this fierce spirit, a radical departure must be made, an entirely new principle must be introduced, something that never existed before in warfare — a principle which will forcibly, unavoidably, turn the battle into a mere spectacle, a play, a contest without loss of blood. To bring on this result men must be dispensed with: machine must fight machine. But how accomplish that which seems impossible? The answer is simple enough: produce a machine capable of acting as though it were part of a human being — no mere mechanical contrivance, comprising levers, screws, wheels, clutches, and nothing more, but a machine embodying a higher principle, which will enable it to perform its duties as though it had intelligence, experience, reason, judgment, a mind! This conclusion is the result of my thoughts and observations which have extended through virtually my~ whole life, and I shall now briefly describe how I came to accomplish that which at first seemed an unrealizable dream.

A long time ago, when I was a boy, I was afflicted with a singular trouble, which seems to have been due to an extraordinary excitability of the retina. It was the appearance of images which, by .their persistence, marred the vision of real objects and interfered with thought. When a word was said to me, the image of the object which it designated would appear vividly before my eyes, and many times it was impossible for me to tell whether the object I saw was real or not. This caused me great discomfort and anxiety, and I tried hard to free myself of the spell. But for a long time I tried in vain, and it was not, as I still clearly recollect, until I was about twelve years old that I succeeded for the first time, by an effort of the will, in banishing an image which presented itself. My happiness will never be as complete as it was then, but, unfortunately (as I thought at that time), the old trouble returned, and with it my anxiety. Here it was that the observations to which I refer began. I noted, namely, that whenever the image of an object appeared before my eyes I had seen something which reminded me of it. In the first instances I thought this to be purely accidental, but soon I convinced myself that it was not so. A visual impression, consciously or unconsciously received, invariably preceded the appearance of the image. Gradually the desire arose in me to find out, every time, what caused the images to appear, and the satisfaction of this desire soon became a necessity. The next observation I made was that, just as these images followed as a result of something I had seen, so also the .thoughts which I conceived were suggested in like manner. Again, I experienced the same desire to locate the image which caused the thought, and this search for the original visual impression soon grew to be a second nature. My mind became automatic, as it were, and in the course of years of continued, almost unconscious performance, I acquired the ability of locating every time and, as a rule, instantly the visual impression which started the thought. Nor is this all. It was not long before I was aware that also all my movements were prompted in the same way, and so, searching, observing, and verifying continuously, year after year, I have, by every thought and every act of mine, demonstrated, and do so daily, to my absolute satisfaction, that I am an automaton endowed with power of movement, which merely responds to external stimuli beating upon my sense organs, and thinks and acts and moves accordingly. I remember only one or two cases in all my life in which I was unable to locate the first impression which prompted a movement or a thought, or even a dream.

With these experiences it was only natural that, long ago, I conceived the idea of constructing an automaton which would mechanically represent me, and which would respond; as I do myself, but, of course, in a much more primitive manner, to external influences. Such an automaton evidently had to have motive power, organs for locomotion, directive organs, and one or more sensitive organs so adapted as to be excited by external stimuli. This machine would, I reasoned, perform its movements in the manner of a living being, for it would have all the chief mechanical characteristics or elements of the same. There was still the capacity for growth, propagation, and, above all, the mind which would be wanting to make the model complete. But growth was not necessary in this case, since a machine could be manufactured full-grown, so to speak. As to the capacity for propagation, it could likewise be left out of consideration, for in the mechanical model it merely signified a process of manufacture. Whether the automaton be of flesh and bone, or of wood and steel, it mattered little, provided it could perform all the duties required of it like an intelligent being. To do so, it had to have an element corresponding to the mind, which would effect the control of all its movements and operations, and cause it to act, in any unforeseen case that might present itself, with knowledge, reason, judgment, and experience. But this element I could easily embody in it by conveying to it my own intelligence, my own understanding. So this invention was evolved, and so a new art came into exist; nee, for which the name "telautomatics" has been suggested, which means the art of controlling the movements and operations of distant automatons.

Fig 2: A machine having all the bodily or translatory movements and the operations of the interior mechanism controlled from a distance without wires. The crewless boat shown in the photograph contains its own motive power, propelling and steering machinery, and numerous other accessories, all of which are controlled by transmitting from a distance, without wires, electrical oscillations to a circuit carried by the boat and adjusted to respond only to these oscillations.

This principle evidently was applicable to any kind of machine that moves on land or in the water or in the air. In applying it practically for the first time, I selected a boat (see Fig. 2). A storage battery placed within it furnished the motive power. The propeller, driven by a motor, represented the locomotive organs. The rudder, controlled by another motor likewise driven by the battery, took the place of the directive organs. As to the sensitive organ, obviously the first thought was to utilize a device responsive to rays of light, like a selenium cell, to represent the human eye. But upon closer inquiry I found that, owing to experimental and other difficulties, no thoroughly satisfactory control of the automaton could be effected by light, radiant heat, Hertzian radiations, or by rays in general, that is, disturbances which pass in straight lines through space. One of the reasons was that any obstacle coming between the operator and the distant automaton would place it beyond his control. Another reason was that the sensitive device representing the eye would have to be in a definite position with respect to the distant controlling apparatus, and this necessity would impose great limitations in the control. Still another and very important reason was that, in using rays, it would be difficult, if not impossible, to give to the automaton individual features or characteristics distinguishing it from other machines of this kind. Evidently the automaton should respond only to an individual call, as a person responds to a name. Such considerations led me to conclude that the sensitive device of the machine should correspond to the ear rather than to the eye of a human being, for in this case its actions could be controlled irrespective of intervening obstacles, regardless of its position relative to the distant controlling apparatus, and, last, but not least, it would remain deaf and unresponsive, like a faithful servant, to all calls but that of its master. These requirements made it imperative to use, in the control of the automaton, instead of light — or other rays, waves or disturbances which propagate in all directions through space, like sound, or which follow a path of least resistance, however curved. I attained the result aimed at by means of an electric circuit placed within the boat, and adjusted, or "tuned," exactly to electrical vibrations of the proper kind transmitted to it from a distant "electrical oscillator." This circuit, in responding, however feebly, to the transmitted .vibrations, affected magnets and other contrivances, through the medium of which were controlled the movements of the propeller and rudder, and also the operations of numerous other appliances.

By the simple means described the knowledge, experience, judgment — the mind, so to speak — of the distant operator were embodied in that machine, which was thus enabled to move and to perform all its operations with reason and intelligence. It behaved just like a blindfolded person obeying directions received through the ear.

The automatons so far constructed had "borrowed minds," so to speak, as each merely formed part of the distant operator who conveyed to it his intelligent orders; but this art is only in the beginning. I purpose to show that, however impossible it may now seem, an automaton maybe contrived which will have its "own mind," and by this I mean that it will be able, independent of any operator, left entirely to itself, to perform, in response to external influences affecting its sensitive organs,, a great variety of acts and operations as if it had intelligence. It will be able to follow a course laid out or to obey orders given far in advance; it will be capable of distinguishing between what it ought and what it ought not to do, and of making experiences or, otherwise stated, of recording impressions which will definitely affect its subsequent actions. In fact, I have already conceived such a plan.

Although I evolved this invention many years ago and explained it to my visitors very frequently in my laboratory demonstrations, it was not until much later, long after I had perfected it, that it became known, when, naturally enough, it gave rise to much discussion and to sensational reports. But the true significance of this new art was not grasped by the majority, nor was the great force of the underlying principle recognized. As nearly as I could judge from the numerous comments which then appeared, the results I had obtained were considered as entirely impossible. Even the few who were disposed to admit the practicability of the invention saw in it merely an automobile torpedo, which was to be used for the purpose of blowing up battleships, with doubtful success. The general impression was that I contemplated simply the steering of such a vessel by means of Hertzian or other rays. There are torpedoes steered electrically by wires, and there are means of communicating without wires, and the above was, of course, an obvious inference. Had I accomplished nothing more than this, I should have made a small advance indeed. But the art I have evolved does not contemplate merely ,the change of direction of a moving vessel; it affords a means of absolutely controlling, in every respect, all the innumerable translatory movements; as well as the operations of all the internal organs, no matter how many, of an individualized automaton. Criticisms to the effect that the control of the automaton could be interfered with were made by people who do not even dream of the wonderful results which can be accomplished by the use of electrical vibrations. The world moves slowly, and new truths are difficult to see. Certainly, by the use of this: principle, an arm for attack as well as defense may be provided, of a destructiveness all the greater as the principle is applicable to submarine and aerial vessels: There is virtually no restriction as to the amount of explosive it can carry, or as to the distance at which it can strike, and failure is almost impossible. But the force of this new principle does not wholly reside in its destructiveness. Its advent introduces into warfare an element which never existed before — a fighting-machine without men as a means of attack and defense. The continuous development in this direction must ultimately make war a mere contest of machines without men and without loss of life — a condition which would have been impossible without this new departure, and which, in my opinion, must be reached as preliminary to permanent peace. The future will either bear out or disprove these views. My ideas on this subject have been put forth with deep conviction, but in a humble spirit.

The establishment of permanent peaceful relations between nations would most effectively reduce the force retarding the human mass, and would be the best solution of this great human problem. But will the dream of universal peace ever be realized? Let us hope that it will. When all darkness shall be dissipated by the light of science, when all nations shall be merged into one, and patriotism shall be identical with religion, when there shall be one language, one country, one end, then the dream will have become reality.

The Third Problem: How to Increase the Force Accelerating the Human Mass — the Harnessing of the Sun's Energy

Of the three possible solutions of the main problem of increasing human energy, this is by far the most important to consider, not only because of its intrinsic significance, but also because of its intimate bearing on all the many elements and conditions which determine the movement of humanity. In order to proceed systematically, it would be necessary for me to dwell on all those considerations which have guided me from the outset in my efforts to arrive at a solution, and which have led me, step by step, to the results I shall now describe. As a preliminary study of the problem an analytical investigation, such as I have made, of the chief forces which determine the onward movement, would be of advantage, particularly in conveying an idea of that hypothetical "velocity" which, as explained in the beginning, is a measure of human energy; but to deal with this specifically here, as I would desire, would lead me far beyond the scope of the present subject. Suffice it to state that the resultant of all these forces is always in the direction of reason, which, therefore, determines, at any time, the direction of human movement. This is to say that every effort which is scientifically applied, rational, useful; or practical, must be in the direction in which the mass is moving. The practical, rational man, the observer, the man of business he who reasons, calculates, or determines in advance, carefully applies his effort so that when coming into effect it will be in the direction of the movement, making it thus most efficient, and in this knowledge and ability lies the secret of his success. Every new fact discovered, every new experience or new element added to our knowledge and entering into the domain of reason, affects the same and, therefore, changes the direction of the movement, which, however, must always take place along ,the resultant of all those efforts which, at that time, we designate as reasonable, that is, self-preserving, useful, profitable, or practical. These efforts concern our daily life, our necessities and comforts, our work and business, and it is these which drive man onward.

But looking at all this busy world about us, on all this complex mass as it daily throbs and moves, what is it but an immense clockwork driven by a spring? In the morning, when we rise, we cannot fail to note that all the objects about us are manufactured by machinery: the water we use is lifted by steam-power; the trains bring out breakfast from distant localities; the elevators in our dwelling and in our office building, the cars that carry us there, are all driven by power; in all our daily errands, and in our very life-pursuit, we depend upon it; all the objects we see tell us of it; and when we return to our machine-made dwelling at night, lest we should forget it, all the material comforts of our home, our cheering stove and lamp, remind us how much we depend on power. And when there is an accidental stoppage of the machinery, when the city is snow-bound, or the life-sustaining movement otherwise temporarily arrested, we are affrighted to realize how impossible it would be for us to live the life we live without motive power. Motive power means work. To increase the force accelerating human movement means, therefore, to perform more work.

So we find that the three possible solutions of the great problem of increasing human energy are answered by the three words: *food, peace, work.* Many a year I have thought and pondered, lost myself in speculations and theories, considering man as a mass moved by a force, viewing his inexplicable movement in the light of a mechanical one, and applying the simple principles of mechanics to the analysis of the same until I arrived at these solutions, only to realize that they were taught to me in my early childhood. These three words sound the key-notes of the Christian religion. Their scientific meaning and purpose are now clear .to me: food to increase the massy peace to diminish the retarding force, and work to increase the force accelerating human movement. These are the only three solutions which are possible of that great problem, and all of them have one object, one end, namely, to increase human energy. When we recognize this, we cannot help wondering how profoundly wise and scientific and how immensely practical the Christian religion is, and in what a marked contrast it stands in this respect to other religions. It is unmistakably the result of practical experiment and scientific observation which have extended through ages, while other religions seem to be the outcome of merely abstract reasoning. Work, untiring effort, useful and accumulative, with periods of rest and recuperation aiming at higher efficiency, is its chief and ever-recurring command. Thus we are inspired both by Christianity and Science to do our utmost toward increasing the performance of mankind. This most important of human problems I shall now specifically consider.

The Source of Human Energy — the Three Ways of Drawing Energy from the Sun

First let us ask: Whence comes all the motive power? What is the spring that drives all? We see the ocean rise and fall, the rivers flow, the wind, rain, hail, and snow beat on our windows, ,the trains and steamers come and go; we hear the rattling noise of carriages, the voices from the street; we feel, smell, and taste; and we think of all this. And all this movement from the surging of the mighty ocean to that subtle movement concerned in our ,thought, has but one common cause. All this energy emanates from one single center, one single source — the sun. The sun is the spring that drives all. The sun maintains all human life and supplies all human energy. Another answer we have now found to the above great question: To increase the force accelerating human movement means to turn to the uses of man more of the son's energy. We honor and revere those great men of bygone times whose names are linked with immortal achievements, who have proved themselves benefactors of humanity — the religious reformer with his wise maxims of life, the philosopher with his deep truths, the mathematician with his formulae, the physicist with his laws, the discoverer with his principles and secrets wrested from nature, the artist with his forms of the beautiful; but who honors him, the greatest of all, — who can tell the name of him, — who first turned to use the sun's energy to save the effort of a weak fellow-creature? That was man's first act of scientific philanthropy, and its consequences have been incalculable.

From the very beginning three ways of drawing energy from the sun were open to man. The savage, when he warmed. his frozen limbs at a fire kindled in some way, availed himself of the energy of the sun stored in the burning material. When he carried a bundle of branches to his cave and burned them there, he made use of the sun's stored energy transported from one to another locality. When he set sail to his canoe, he utilized the energy of the sun supplied to the atmosphere or ambient medium. There can be no doubt that the first is the oldest way. A fire, found accidentally, taught the savage to appreciate its beneficial heat. He then very likely conceived the idea of the carrying the glowing embers to his abode. Finally he learned to use the force of a swift ' current of water or air. It is characteristic of modern development that progress has been effected in the same order. The utilization of the energy stored in wood or coal, or, generally speaking, fuel, led to the steam-engine. Next a great stride in advance was made in energy-transportation by the use of electricity, which permitted the transfer of energy from one locality to another without transporting the material. But as to the utilization of the energy of the ambient medium, no radical step forward has as yet been made known.

The ultimate results of development in these three directions are: first, the burning of coal by a cold process in a battery; second, the efficient utilization of .the energy of the ambient medium; and, third, the transmission without wires of electrical energy to any distance. In whatever way these results may be arrived at, their practical application will necessarily involve an extensive use of iron, and this invaluable metal will undoubtedly be an essential element in the further development along these three lines. If we succeed in burning coal by a cold process and thus obtaining electrical energy in an efficient and inexpensive manner, we shall require in many practical uses of this energy electric motors — that is, iron. If we are successful in deriving energy from the ambient medium, we shall need, both in the obtainment and utilization of the energy, machinery — again, iron. If we realize the transmission of electrical energy without wires on an industrial scale, we shall be compelled to use extensively electric generators — once more, iron. Whatever we may do, iron will probably be the chief means of accomplishment in the near future, possibly more so than in the past. How long its reign will last is difficult to tell, for even now aluminum is looming up as a threatening competitor. But for the time being, next to providing new resources of energy, it is of the greatest importance to make improvements in the manufacture and utilization of iron. Great advances are possible in these latter directions, which, if brought about, would enormously increase the useful performance of mankind.

Great Possibilities Offered by Iron for Increasing Human Performance — Enormous Waste in Iron Manufacture

Iron is by far the most important factor in modern progress. It contributes more than any other industrial product to the force accelerating human movement. So general is the use of this metal, and so intimately is it connected with all that concerns our life, that it has become as indispensable to us as the very air we breathe. Its name is synonymous with usefulness. But, however great the influence of iron may be on the present human development, it does not add to the force urging man onward nearly as much as it might. First of all, its manufacture as now carried on is connected with an appalling waste of fuel — that is, waste of energy. Then, again, only a part of all the iron produced is applied for useful purposes. A good part of it goes to create frictional remittances, while still another large part is the means of developing negative forces greatly retarding human movement. Thus the negative force of war is almost wholly represented in iron. It is impossible to estimate with any degree of accuracy the magnitude of this greatest of all retarding forces, but it is certainly very considerable. If the present positive impelling force due to all useful applications of iron be represented by ten, for instance, I should not think it exaggeration to estimate the negative force of war, with due consideration of all its retarding influences and results, at, say, six. On the basis of this estimate the effective impelling force of iron in the positive direction would be measured by the difference of these two numbers, which is four. But if, through the establishment of universal peace, the manufacture of war machinery should cease, and all struggle for supremacy between nations should be turned into healthful, ever active and productive commercial competition, then the positive impelling force due to iron would be measured by the sum of those two numbers, which is sixteen — that is, this force would have four times its present value. This example is, of course, merely intended to give an idea of the immense increase in the useful performance of mankind which would result from a radical reform of the iron industries supplying the implements of warfare.

A similar inestimable advantage in the saving of energy available to man would be secured by obviating the great waste of coal which is inseparably connected with the present methods of manufacturing iron. In some countries, as in Great Britain, the hurtful effects of this squandering of fuel are beginning to be felt. The price of coal is constantly rising, and the poor are made to suffer more and more. Though we are still far from the dreaded "exhaustion of the coal-fields," philanthropy commands us to invent novel methods of manufacturing iron, which will not involve such barbarous waste of this valuable material from which we derive at present most of our energy. It is our duty to coming generations to leave this store of energy intact for them, or at least not to touch it until we shall have perfected processes for burning coal more efficiently. Those who are to come after us will need fuel more than we do. We should be able to manufacture the iron we require by using the sun's energy, without wasting any coal at all. As an effort to this end the idea of smelting iron ores by electric currents obtained from the energy of falling water has naturally suggested itself to many. I have myself spent much time in endeavoring to evolve such a practical process,. which would enable iron to be manufactured at small cost. After a prolonged investigation of the subject, finding that it was unprofitable to use the currents generated directly for smelting the ore, I devised a method which is far more economical.

Economical Production of Iron by a New Process

The industrial project, as I worked it out six years ago, contemplated the employment of the electric currents derived from the energy of a waterfall, not directly for smelting the ore, but for decomposing water, as, a preliminary step. To lessen the cost of the plant, I proposed to generate the currents in exceptionally cheap and simple dynamos, which I designed for this sole purpose. The hydrogen liberated in the electrolytic decomposition was to be burned or recombined with oxygen, not with that from which it was separated, but with that of the atmosphere. Thus very nearly the total electrical energy used up in the decomposition of the water would be recovered in the form of heat resulting from the recombination of the hydrogen. This heat was to be applied to the smelting of ,the ore. The oxygen gained as a by-product in the decomposition of the water I intended to use for certain other industrial purposes, which would probably yield good financial returns, inasmuch as this is the cheapest way of obtaining this gas in large quantities. In any event, it could be employed to burn all kinds of refuse, cheap hydrocarbon, or coal of the most inferior quality which could not be burned in air or be otherwise utilized to advantage, and thus again a considerable amount of heat would be made available for the smelting of the ore. To increase the economy of .the process I contemplated, furthermore, using an arrangement such that the hot metal and the products of combustion, coming out of the furnace, would give up their heat upon the cold ore going into the furnace, so that comparatively little of the heat-energy would be lost in the smelting. I calculated that probably forty thousand pounds of iron could be produced per horse-power per annum by this method. Liberal allowances were made for those losses which are unavoidable, the above quantity being about half of that theoretically obtainable. Relying on this estimate and on practical data with reference to a certain kind of sand ore existing in abundance in the region of the Great Lakes, including cost of transportation and labor, I found that in some localities iron could be manufactured in this manner cheaper than by any of the adopted methods. This result would be attained all the more surely if the oxygen obtained from the water, instead of being used for smelting the ore, as assumed, should be more profitably employed. Any new demand for this gas would secure a higher revenue from the plant, thus cheapening the iron. This project was advanced merely in the interest of industry. Some day, I hope, a beautiful industrial butterfly will come out of the dusty and shriveled chrysalis.

The production of iron from sand ore by a process of magnetic separation is highly commendable in principle, since it involves no waste of coal; but the usefulness of this method is largely reduced by the necessity of melting the iron afterward. As to the crushing of iron ore, I would consider it rational only if done by water-power, or by energy otherwise obtained without consumption of fuel. An electrolytic cold process, which would make it possible to extract iron cheaply, and also to mold it into the required forms without any fuel consumption, would, in my opinion, be a very great advance in iron manufacture. In common with some other metals, iron has so far resisted electrolytic treatment, but there can be no doubt that such a cold process will ultimately replace in metallurgy the present crude method of casting, and thus obviate the enormous; waste of fuel necessitated by the repeated heating of metal in the foundries.

Up to a few decades ago the usefulness of iron was based almost wholly on its remarkable mechanical properties, but since the advent of the commercial dynamo and electric motor its value to mankind has been greatly increased by its unique magnetic qualities. As regards the latter, iron has been greatly improved of late. The signal progress began about thirteen years ago, when I discovered that in using soft Bessemer steel instead of wrought iron, as then customary, in an alternating motor, the performance of the machine was doubled. I brought this fact .to the attention of Mr. Albert Schmid, to whose untiring efforts and ability is largely due the supremacy of American electrical machinery, and who was then superintendent of an industrial corporation engaged in this field. Following my suggestion, he constructed transformers of steel, and they showed the same marked improvement. The investigation was then systematically continued under Mr. Schmid's guidance, the impurities being gradually eliminated from the "steel" (which was only such in name, for in reality it was pure soft iron), and soon a product resulted which admitted of little further improvement.

The Coming Age of Aluminum — Doom of the Copper Industry — the Great Civilizing Potency of the New Metal

With the advances made in iron of late years we have arrived virtually at the limits of improvement. We cannot hope to increase very materially its tensile strength. elasticity, hardness, or malleability, nor can we expect to make it much better as regards its magnetic qualities. More recently a notable gain was secured by the mixture of a small percentage of nickel with the iron, but there is not much room for further advance in this direction. New discoveries may be expected, but they cannot greatly add to the valuable properties of the metal, though they may considerably reduce the cost of manufacture. The immediate future of iron is assured by its cheapness and its unrivaled mechanical and magnetic qualities. These are such that no other product can compete with it now. But there can be no doubt that, at a time not very distant, iron, in many of its now uncontested domains, will have .to pass the scepter to another: the coming age will be the age of aluminum. It is only seventy years since this wonderful metal was discovered by Woehler, and the aluminum industry, scarcely forty years old, commands already the attention of the entire world. Such rapid growth has not been recorded in the history of civilization before. Not long ago aluminum was sold at the fanciful price of thirty or forty dollars per pound; to-day it can be had in any desired amount for as many cents. What is more, the time is not far off when this price, too, will be considered fanciful, for great improvements are possible in the methods of its manufacture. Most of the metal is now produced in the electric furnace by a process combining fusion and electrolysis, which offers a number of advantageous features, but involves naturally a great waste of the electrical energy of the current. My estimates show that the price of aluminum could be considerably reduced by adopting in its manufacture a method similar to that proposed by me for the production of iron. A pound of aluminum requires for fusion only about seventy per cent. of the heat needed for melting a pound of iron, and inasmuch as its weight is only about one third of that of the latter, a volume of aluminum four times that of iron could be obtained from a given amount of heat-energy. But a cold electrolytic process of manufacture is the ideal solution. and on this I have placed my hope.

The absolutely unavoidable consequence of the advance of the aluminum industry will be the annihilation of the copper industry. They cannot exist and prosper together, and the latter is doomed beyond any hope of recovery. Even now it is cheaper to convey an electric current through aluminum wires than through copper wires; aluminum castings cost less, and in many domestic and other uses copper hasp no chance of successfully competing. A further material reduction of the price of aluminum cannot but be fatal to copper. But the progress of the former will not go on unchecked, for, as it ever happens in such cases, the larger industry will absorb the smaller one: the giant copper interests will control the pygmy aluminum interests, and the slow-pacing copper will reduce the lively gait of aluminum. This will only delay, not avoid, the impending catastrophe.

Aluminum, however, will not stop at downing copper. Before many years have passed it will be engaged in a fierce struggle with iron, and in the latter it will find an adversary not easy to conquer. The issue of the contest will largely depend on whether iron shall be indispensable in electric machinery. This the future alone can decide. The magnetism as exhibited in iron is an isolated phenomenon in nature. What it is that makes this metal behave so radically different from all other materials in this respect has not yet been ascertained, though many theories have been suggested. As regards magnetism, the molecules of the various bodies behave like hollow beams partly filled with a heavy fluid and balanced in the middle in the manner of a see-saw. Evidently some disturbing influence exists in nature which causes each molecule, like such a beam, to tilt either one or the other way. If the molecules are tilted one way, the body is magnetic; if they are tilted the other way, the body is non-magnetic; but both positions are stable, as they would be in the case of the hollow beam, owing to the rushing of the fluid to the lower end. Now, the wonderful thing is that the molecules of all known bodies went one way, while those of iron went the other way. This metal, it would seem, has an origin entirely different from that of the rest of the globe. It is highly improbable that we shall discover some other and cheaper material which will equal or surpass iron in magnetic qualities.

Unless we should make a radical departure in the character of the electric currents employed, iron will be indispensable. Yet the advantages it offers are only apparent. So long as we use feeble magnetic forces it is by far superior to any other material; but if we find ways of producing great magnetic forces, then better results will be obtainable without it. In fact, I have already produced electric transformers in which no iron is employed, and which are capable of performing ten times as much work per pound of weight as those with iron. This result is attained by using electric currents of a very high rate of vibration, produced in novel ways, instead of the ordinary currents now employed in the industries. I have also succeeded in operating electric motors without iron by such rapidly vibrating currents, but the results, so far, have been inferior to those obtained with ordinary motors constructed of iron, although theoretically the former should be capable of performing incomparably more work per unit of weight than the latter. But the seemingly insuperable difficulties which are now in the way may be overcome in the end, and then iron will be done away with, and all electric machinery will be manufactured of aluminum, in all probability, at prices ridiculously low. This would be a severe, if not a fatal, blow to iron. In many other branches of industry, as ship-building, or wherever lightness of structure is required, the progress of the new metal will be much quicker. For such uses it is eminently suitable, and is sure to supersede iron sooner or later. It is highly probable that in the course of time we shall be able to give it many of those qualities which make iron so valuable.

While it is impossible to tell when this industrial revolution will be consummated, there can be no doubt that the future belongs to aluminum, and that in times to come it will be the chief means of increasing human performance. It has in this respect capacities greater by far than those of any other metal. I should estimate its civilizing potency at fully one hundred times that of iron. This estimate, though it may astonish, is not at all exaggerated. First of all, we must remember that there is; thirty times as much aluminum as iron in bulk, available for the uses of man. This in itself offers great possibilities. Then, again, the new metal is much more easily workable, which adds to its value. In many of its properties it partakes of the character of a precious metal, which gives it additional worth. Its electric conductivity, which, for a given weight, is greater than that of any other metal, would be alone sufficient to make it one of the most important factors in future human progress. Its extreme lightness makes it far more easy to transport the objects manufactured. By virtue of this property it will revolutionize naval construction, and in facilitating transport and travel it will add enormously to the useful performance of mankind. But its greatest civilizing potency will be, I believe, in aerial travel, which is sure to be brought about by means of it. Telegraphic instruments will slowly enlighten the barbarian. Electric motors and lamps will do it more quickly, but quicker than anything else the flying-machine will do it. By rendering travel ideally easy it will be the best means for unifying the heterogeneous elements of humanity. As the first step toward this realization we should produce a lighter storage-battery or get more energy from coal.

Efforts Toward Obtaining More Energy from Coal — the Electric Transmission — the Gas-engine — the Cold-coal Battery

I remember that at one time I considered the production of electricity by burning coal in a battery as the greatest achievement toward advancing civilization, and I am surprised to find how much the continuous study of these subjects has modified my views. It now seems to me that to burn coal, however efficiently, in a battery would be a mere makeshift, a phase in the evolution toward something much more perfect. After all, in generating electricity in this manner, we should be destroying material, and this would be a barbarous process. We ought to be able to obtain the energy we need without consumption of material. But I am far from underrating the value of such an efficient method of burning fuel. At the present time most motive power comes from coal, and, either directly or by its products, it adds vastly to human energy. Unfortunately, in all the processes now adopted, the larger portion of the energy of the coal is uselessly dissipated. The best steam-engines utilize only a small part of the total energy. Even in gas-engines, in which, particularly of late, better results are obtainable, there is still a barbarous waste going on. In our electric-lighting systems we scarcely utilize one third of one per cent., and in lighting by gas a much smaller fraction of the total energy of the coal. Considering the various uses of coal throughout the world, we certainly do not utilize more than two per cent, of its energy theoretically available. The man who should stop this senseless waste would be a great benefactor of humanity, though the solution he would offer could not be a permanent one, since it would ultimately lead to the exhaustion of the store of material. Efforts toward obtaining more energy from coal are now being made chiefly in two directions — by generating electricity and by producing gas for motive-power purposes. In both of these lines notable success has already been achieved.

The advent of the alternating-current system of electric power-transmission marks an epoch in the economy of energy available to man from coal. Evidently all electrical energy obtained from a waterfall, saving so much fuel, is a net gain to mankind, which is all the more effective as it is secured with little expenditure of human effort, and as this most perfect of all known methods of deriving energy from the sun contributes in many ways to the advancement of civilization. But electricity enables us also to get from coal much more energy than was practicable in the old ways. Instead of transporting the coal to distant places of consumption, we burn it near the mine, develop electricity in the dynamos, and transmit the current to remote localities, thus effecting a considerable saving. Instead of driving the machinery in a factory in the old wasteful way by belts and shafting, we generate electricity by steam-power and operate electric motors. In this manner it is not uncommon to obtain two or three times as much effective motive power from the fuel, besides securing many other important advantages. It is in this field as much as in the transmission of energy to great distances that the alternating system, with its ideally simple machinery, is bringing about an industrial revolution. But in many lines this progress has not yet been felt. For example, steamers and trains are still being propelled by ,the direct application of steam-power to shafts or axles. A much greater percentage of the heat-energy of the fuel could be transformed in motive energy by using, in place of the adopted marine engines and locomotives, dynamos driven by specially designed high-pressure steam- or gas-engines and by utilizing the electricity generated for the propulsion. A gain of fifty to one hundred per cent. in the effective energy derived from the coal could be secured in this manner. It is difficult to understand why a fact so plain and obvious is not receiving more attention from engineers. In ocean steamers such an improvement would be particularly desirable, as it would do away with noise and increase materially the speed and the carrying capacity of the liners.

Still more energy is now being obtained from coal by the latest improved gas-engine, the economy of which is, on the average, probably twice that of the best steam-engine. The introduction of the gas-engine is very much facilitated by the importance of the gas industry: With the increasing use of the electric light more and more of, the gas is utilized for heating and motive-power purposes. In many instances gas is manufactured close to the coal-mine and conveyed to distant places of consumption, a considerable saving both in the cost of transportation and in utilization of the energy of the fuel being thus effected. In the present state of the mechanical and electrical arts the most rational way of deriving energy from coal is evidently to manufacture gas close to the coal store, and to utilize it, either on the spot or elsewhere, to generate electricity for industrial uses in dynamos driven by gas-engines. The commercial success of such a plant is largely dependent upon the production of gas-engines of great nominal horse-power, which, judging from the keen activity in this field, will soon be forthcoming. Instead of consuming coal directly, as usual, gas should be manufactured from it and burned to economize energy.

But all such improvements cannot be more than passing phases in the evolution toward something far more perfect, for ultimately we must succeed in obtaining electricity from coal in a more direct way, involving no great loss of its heat-energy. Whether coal can be oxidized by a cold process is still a question. Its combination with oxygen always evolves heat, and whether the energy of the combination of the carbon with another element can be turned directly into electrical energy has riot yet been determined. Under certain conditions nitric acid will burn the carbon, generating an electric current, but the solution does not remain cold. Other means of oxidizing coal have been proposed, but they have offered no promise of leading to an efficient process. My own lack of success has been complete, though perhaps not quite so complete as that of some who have "perfected" the cold-coal battery. This problem is essentially one for the chemist to solve. It is not for the physicist; who determines all his results in advance, so that, when the experiment is tried, it cannot fail. Chemistry, though a positive science, does not yet admit of a solution by such positive methods as those which are available in the treatment of many physical problems. The result, if possible, will be arrived at through patient trying rather than through deduction or calculation. The time will soon come, however, when the chemist will be able to follow a course clearly mapped out beforehand, and when the process of his arriving at a desired result will be purely constructive. The cold-coal battery would give a great impetus to electrical development; it would lead very shortly to a practical flying-machine, and would enormously enhance the introduction of the automobile. But these and *many* other problems will be better solved, and in a more scientific manner, by a light-storage battery.

Energy from the Medium — the Windmill and the Solar Engine — Motive Power from Terrestrial Heat — Electricity from Natural Sources

Besides fuel, there is abundant material from which we might eventually derive power. An immense amount of energy is locked up in limestone, for instance, and machines can be driven by liberating the carbonic acid through sulphuric acid or otherwise. I once constructed such an engine, and it operated satisfactorily.

But, whatever our resources of primary energy may be in the future, we must, to be rational, obtain it without consumption of any material. Long ago I came to this conclusion, and to arrive at this result only two ways, as before indicated, appeared possible — either to turn to use the energy of the sun stored in the ambient medium, or to transmit, through the medium, the sun's energy to distant places from some locality where it was obtainable without consumption of material. At that time 'I at once rejected the latter method as entirely impracticable, and turned . to examine the possibilities of the former.

It is difficult to believe, but it is, nevertheless, a fact, that since time immemorial man has had at his disposal a fairly good machine which has enabled him to utilize the energy of the ambient medium. This machine is the windmill. Contrary to popular belief, the power obtainable from wind is very considerable. Many a deluded inventor has spent years of his life in endeavoring to "harness the tides," and some have even proposed to compress air by tide- or wave-power for supplying energy, never understanding the signs of the old windmill on the hill, as it sorrowfully waved its arms about and bade them stop. The fact is that a wave- or tide-motor would have, as a rule, but a small chance of competing commercially with the windmill, which is by far the better machine, allowing a much greater amount of energy to be obtained in a simpler way. Wind-power has been, in old times, of inestimable value to man, if for nothing else but for enabling him to cross the seas, and it is even now a very important factor in travel and transportation. But there are great limitations in this ideally simple method of utilizing the sun's energy. The machines are large for a given output, and the power is intermittent, thus necessitating the storage of energy and increasing the cost of the plant.

A far better way, however, to obtain power would be to avail ourselves of the sun's rays, which beat the earth incessantly and supply energy at a maximum rate of over four million horse-power per square mile. Although the average energy received per square mile in any locality during the year is only a small fraction of that amount, yet an inexhaustible source of power would be opened up by the discovery of some efficient method of utilizing the energy of the rays. The only rational way known to me at the time when I began the study of this subject was to employ some kind of heat- or thermodynamic engine, driven by a volatile fluid evaporated in a boiler by the heat of the rays. But closer investigation of this method, and calculation, showed that, notwithstanding the apparently vast amount of energy received from the sun's rays, only a small fraction of that energy could be actually utilized in this manner. Furthermore, the energy supplied through the sun's radiations is periodical, and the same limitations as in the use of the windmill I found to exist here also. After a long study of this made of obtaining motive power from the sun, taking into account the necessarily large bulk of the boiler, the low efficiency of the heat-engine, the additional cost of storing the energy, and other drawbacks, I came to the conclusion that the "solar engine," a few instances excepted, could not be industrially exploited with success.

Another way of getting motive power from the medium without consuming any material would be to utilize the heat contained in the earth, the water, or the air for driving an engine. It is a well-known fact that the interior portions of the globe are very hot, the temperature rising, as observations show, with the approach to the center at the rate of approximately 1^0 C. for every hundred feet of depth. The difficulties of sinking shafts and placing boilers at depths of, say, twelve thousand feet, corresponding to an increase in temperature of about 120^0 C., are not insuperable, and we could certainly avail ourselves in this way of the internal heat of the globe. In fact, it would not be necessary to go to any depth at all in order to derive energy from the stored terrestrial heat. The superficial layers of the earth and the air strata close to the same are at a temperature sufficiently high to evaporate some extremely volatile substances, which we might use in our boilers instead of water. There is no doubt that a vessel might be propelled on the ocean by an engine driven by such a volatile fluid, no other energy being used but the heat abstracted from the water. But the amount of power which could be obtained in this manner would be, without further provision, very small.

Electricity produced by natural causes is another source of energy which might be 'rendered available. Lightning discharges involve great amounts of electrical energy, which we could utilize by transforming and storing it. Some years ago I made known a method of electrical transformation which renders the first part of this task easy, but the storing of the energy of lightning discharges will be difficult to accomplish. It is well known, furthermore, that electric currents circulate constantly through the earth, and that there exists between the earth and any air stratum a difference of electrical pressure, which varies in proportion to .the height.

In recent experiments I have discovered two novel facts of importance in this connection. One of these facts is that an electric current is generated in a wire extending from the ground to a great height by the axial, and probably also by the translatory, movement of the earth. No appreciable current, however, will flow continuously in the wire unless the electricity is allowed to leak out into the air. Its escape is greatly facilitated by providing at the elevated end of the wire a conducting terminal of great surface, with many sharp edges or points. We are thus enabled to get a continuous supply of electrical energy by merely supporting a wire at a height, but, unfortunately, the amount of electricity which can be so obtained is small.

The second fact which I have ascertained is that the upper air strata are permanently charged with electricity opposite to that of the earth. So, at least, I have interpreted my observations, from which it appears that the earth, with its adjacent insulating and outer conducting envelop, constitutes a highly charged electrical condenser containing, in all probability, a great amount of electrical energy which might be turned to the uses of man, if it were possible to reach with a wire to great altitudes.

It is possible, and even probable, that there will be, in time, other resources of energy opened up; of which we have no knowledge now. We may even find ways of applying forces such as magnetism or gravity for driving machinery without using any other means. Such realizations, though highly improbable, are not impossible. An example will best convey an idea of what we can hope to attain and what we can never attain. Imagine a disk of some homogeneous material turned perfectly true and arranged to turn in frictionless bearings on a horizontal shaft above the ground. This disk, being under the above conditions perfectly balanced, would rest in any position. Now, it is possible that we may learn how to make such a disk rotate continuously and perform work by the force of gravity without any further effort on our part; but it is perfectly impossible for the disk to turn and to do work without any force from the outside. If it could do so, it would be what is designated scientifically as a "perpetuum mobile," a machine creating its own motive power. To make the disk rotate by the force of gravity we have only to invent a screen against this force. By such a screen we could prevent this force from acting on one half of the disk, and the rotation of the latter would follow. At least, we cannot deny such a possibility until we know exactly the nature of the force of gravity. Suppose that this force were due to a movement comparable to that of a stream of air passing from above toward the center of the earth. The effect of such a stream upon both halves of the disk would be equal, and the latter-would not rotate ordinarily; but if one half should be guarded by a plate arresting the movement, then it would turn.

A Departure from Known Methods — Possibility of a "Self-acting" Engine or Machine, Inanimate, Yet Capable, like a Living Being, of Deriving Energy from the Medium — the Ideal Way of Obtaining Motive Power

When I began the investigation of the subject under consideration, and when the preceding or similar ideas presented themselves :to me for the first time, though I was then unacquainted with a number of the .facts mentioned, a survey of the various ways of utilizing the energy of the medium convinced me, nevertheless, that to arrive at a thoroughly satisfactory practical solution a radical departure from the methods then known had to be made. The windmill, the solar engine, the engine driven by terrestrial heat, had their limitations in the amount of power obtainable. Some new way had to be discovered which would enable us to get more energy. There was enough heat-energy in the medium, but only a small part of it was available for the operation of an engine in the ways then known. Besides, the energy was obtainable only at a very slow rate. Clearly, then, the problem was to discover some new method which would make it possible both to utilize more of the heat-energy of the medium and also to draw it away from the camp at a more rapid rate.

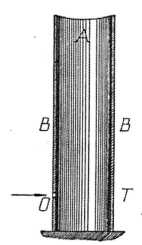

Diagram b. **Obtaining Energy from the Ambient Medium.** A, medium with little energy; B, B, ambient medium with much energy; O, path of the energy.

I was vainly endeavoring to form an idea of how this might be accomplished, when I read some statements from Cannot and Lord Kelvin (then Sir William Thomson) which meant virtually that it is impossible for an inanimate mechanism of self-acting machine to cool a portion of the medium below the temperature of the surrounding, and operate by the heat abstracted. These statements interested the intensely. Evidently a living being could do this very thing, and since the experiences of my early life which I have related had convinced me that a living being is only an automaton, or, otherwise stated, a "self-acting engine," I came to the conclusion that it was possible to construct a machine which would do the same. As the first step toward this realization I conceived the following mechanism. Imagine a thermopile consisting of a number of bars of metal extending from the earth to the outer space beyond the atmosphere. The heat from below, conducted upward along these metal bars, would cool the earth or the sea or the air, according to the location of the lower parts of the bars, and the result, as is well known, would be an electric current circulating in these bars. The two terminals of the thermopile could now be joined through an electric motor, and, theoretically, this motor would run on and on, until the media below would be cooled down to the temperature of the outer space. This would be an inanimate engine which, to all evidence, would be cooling a portion of the medium below the temperature of the surrounding, and operating by the heat abstracted.

But was it not possible to realize a similar condition without necessarily going to a height? Conceive, for the sake of illustration, an inclosure T, as illustrated in diagram b, such that energy could not be transferred across it except through a channel or path O, and that, by some means or other, in this inclosure a medium were maintained which would have little energy, and that on the outer side of the same there would be the ordinary ambient medium with much energy. Under these assumptions the energy would flow through the path O, as indicated by the arrow, and might then be converted on its passage into some other form of energy. The question was, Could such a condition be attained? Could we produce artificially such a "sink" for the energy of the ambient medium to flow in? Suppose that an extremely low temperature could be maintained by some process in a given space; the surrounding medium would then be compelled to give off heat, which could be converted into mechanical or other form of energy, and utilized. By realizing such a plan, we should be enabled to get at any point of the globe a continuous supply of energy, day and night. More than this, reasoning in the abstract, it would seem possible to cause a quick circulation of the medium; and thus draw the energy at a very rapid rate.

Here, then, was an idea which, if realizable, afforded a happy solution of the problem of getting energy from the medium. But was it realizable? I convinced myself that it was so in a number of ways, of which one is the following. As regards heat, we are at a high level, which may be represented by the surface of a mountain lake considerably above the sea, the level of which may mark the absolute zero of temperature existing in the interstellar space. Heat, like water, flows from high to low level, and, consequently, just as we can let the water of the lake run down to the sea, so we are able to let heat from the earth's surface travel up into the cold region above. Heat, like water, can perform work in flowing down, and if we had any doubt as to whether we could derive energy from the medium by means of a thermopile, as before described, it would be dispelled by this analogue. But can we produce cold in a given portion of the space and cause the heat to flow in continually? To create such a "sank," or "cold hole," as we might say, in the medium, would be equivalent to producing in the lake a space either empty or filled with something much lighter than water. This we could do by placing in the lake .a tank, and pumping all the water out of the latter. We know, then, that the water, if allowed to flow back into the tank, would, theoretically, be able to perform exactly the same amount of work which was used in pumping it out, but not a bit more. Consequently nothing could be gained in this double operation of first raising the water and then letting it fall down. This would mean that it is impossible to create such a sink in the medium. But let us reflect a moment. Heat, though following certain general laws of mechanics, like a fluid, is not such; it is energy which may be converted into other forms of energy as it passes from a high to a low level. To make our mechanical analogy complete and true, we must, therefore, assume that the water, in its passage into the tank, is converted into something else, which may be taken out of it without using any, or by using very little, power. For example, if heat be represented in this analogue by the water of the lake, the oxygen and hydrogen composing the water may illustrate other forms of energy into which the heat is transformed in passing from hot to cold. If the process: of heat-transformation were absolutely perfect, no heat at all would arrive at the low level, since all of it would be converted into other forms of energy. Corresponding to this ideal case, all the water flowing into the tank would be decomposed into oxygen and hydrogen before reaching the bottom, and the result would be that water would continually flow in, and yet the tank would remain entirely empty, the gases formed escaping. We would thus produce, by expending initially a certain amount of work to create a sink for the heat or, respectively, the water to flow in, a condition enabling us to get any amount of energy without further effort. This would be an ideal way of obtaining motive power. We do not know of any such absolutely perfect process of heat-conversion, and consequently some heat will generally reach the low level, which means to say, in our mechanical analogue, that some water will arrive at the bottom of the tank, and a gradual and slow filling of the latter will take place, necessitating continuous pumping out. But evidently there will be less to pump out than flows, in, or, in other words, less energy will be needed to maintain the initial condition than is developed by the fall, and this is to say that some energy will be gained from the medium. What is not converted in flowing down can just be raised up with its own energy, and what is converted is clear gain. Thus the virtue of the principle I have discovered resides wholly in the conversion of the energy on the downward flow.

First Efforts to Produce the Self-acting Engine — the Mechanical Oscillator — Work of Dewar and Linde — Liquid Air

Having recognized this truth, I began to devise means for carrying out my idea, and, after long thought, I finally conceived a combination of apparatus which should make possible the obtaining of power from the medium by a process of continuous cooling of atmospheric air. This apparatus, by continually transforming heat into mechanical work, tended to become colder and colder, and if it only were practicable to reach a very low temperature in this manner, then a sink for the heat could be produced, and energy could be derived from the medium. This seemed to be contrary to the statements of Cannot and Lord Kelvin before referred to, but I concluded from the theory of the process that such a result could be attained. This conclusion I reached, I think, in the latter part of 1883, ,hen I was in Paris, and it was at a time when my mind was being more and more dominated by an invention which I had evolved during the preceding year, and which has since become known under the name of the "rotating magnetic field." During the few years which followed I elaborated further the plan I had imagined, and studied the working conditions, but made little headway. The commercial introduction in this country of the invention before referred to required most of my energies until 1889, when I again took up the idea of the self-acting machine. A closer investigation of the principles involved, and calculation, now showed that the result I aimed at could not be reached in a practical manner by ordinary machinery, as I had in the beginning expected. This led me, as a next step, to the study of a type of engine generally designated as "turbine," which at first seemed to offer better chances for a realization of the idea. Soon I found, however, that the turbine, too, was unsuitable. But my conclusions showed that if an engine of a peculiar kind could be brought to a high degree of perfection, the plan I had conceived was realizable, and I resolved to proceed with the development of such an engine, the primary object of which was to secure the greatest economy of transformation of heat into mechanical energy. A characteristic feature of the engine was that .the work-performing piston was not connected with anything else, but was perfectly free to vibrate at an enormous rate. The mechanical difficulties encountered in the construction of this engine were greater than I had anticipated, and I made slow progress. This work was continued until early in 1892, when I went to London, where I saw Professor Dewar's admirable experiments with liquefied gases. Others had liquefied gases before, and notably Ozlewski and Pictet had performed creditable early experiments in this line, but there was such a vigor about the work of Dewar that even the old appeared new. His experiments showed, though in a way different from that I had imagined, that it was possible to reach a very low temperature by transforming heat into mechanical work, and I returned, deeply impressed with what I had seen, and more than ever convinced that my plan was practicable. The work temporarily interrupted was taken up anew, and soon I had in a fair state of perfection the engine which I have named "the mechanical oscillator." In this machine I succeeded in doing away with all packings, valves, and lubrication, and in producing so rapid a vibration of the piston that shafts of tough steel, fastened to the same and vibrated longitudinally, were torn asunder. By combining this engine with a dynamo of special design I produced a highly efficient electrical generator, invaluable in measurements and determinations of physical quantities on account of the unvarying rate of oscillation obtainable by its means. I exhibited several types of this machine, named "mechanical and electrical oscillator," before the Electrical Congress at the World's Pair in Chicago during the summer of 1893, in a lecture which, on account of other pressing work, I was unable to prepare for publication. On that occasion I exposed the principles of the mechanical oscillator, but the original purpose of this machine is explained here for the first time.

In the process, as I had primarily conceived it, for the utilization of the energy of the ambient medium, there were five essential elements in combination, and each of these had to be newly designed and perfected, as no such machines existed. The mechanical oscillator was the first element of this combination, and having perfected this, I turned to the next, which was an air-compressor of a design in certain respects resembling that of the mechanical oscillator. Similar difficulties in the construction were again encountered, but the work was pushed vigorously, and at the close of 1894 I had completed these two elements of the combination, and thus produced an apparatus for compressing air, virtually to any desired pressure, incomparably simpler, smaller, and more efficient than the ordinary. I was just beginning work on the third element, which together with the first two would give a refrigerating machine of exceptional efficiency and simplicity, when a misfortune befell me in the burning *of my* laboratory, which crippled my labors and delayed me. Shortly afterwards Dr. Carl Linde announced the liquefaction of air by a self-cooling process, demonstrating that it was practicable to proceed with the cooling until liquefaction of the air took place. This was the only experimental proof which I was still wanting that energy was obtainable from the medium in the manner contemplated by me.

The liquefaction of air by a self-cooling process was not, as popularly believed, an accidental discovery, but a scientific result which could not have been delayed much longer, and which, in all probability, could not have escaped Dewar. This fascinating advance, I believe, is largely due to the powerful work of this great Scotchman. Nevertheless, Linde's is an immortal achievement. The manufacture of liquid air has been carried on for four years in Germany, on a scale much larger than in any. other country, and this strange product has been applied for a variety of purposes. Much was expected of it in the beginning, but so far it has been an industrial ignis fatuus. By the use of such machinery as I am perfecting, its cost will probably be greatly lessened, but even then its commercial success will be questionable. When used as a refrigerant it is uneconomical, as its temperature is unnecessarily low. It is as expensive to maintain a body at a very low temperature as it is to keep it very hot; it takes coal to keep air cold. In oxygen manufacture it cannot yet compete with the electrolytic method. For use as an explosive it 'is unsuitable, .because its low temperature again condemns it to a small efficiency, and for motive-power purposes its cost is still by far too high. It is of interest to note, however, that in driving an engine by liquid air a certain amount of energy may be gained from the engine, or, stated otherwise, from the ambient medium which keeps the engine warm, each two hundred pounds of ironcasting of the latter .contributing energy at the rate of about one effective horse-power during one hour. But this gain of the consumer is offset by an equal loss of the producer.

Much of this task on which I have labored so long remains to be done. A number of mechanical details are still to be perfected and some difficulties of a different nature to be mastered, and I cannot hope to produce a self-acting machine deriving energy from the ambient medium for a long time yet, even if all my expectations should materialize. Many circumstances have occurred which have retarded my work of late, but for several reasons the delay was beneficial.

One of these reasons was that I had ample time to consider what the ultimate possibilities of this. development might be. I worked for a long time fully convinced that the practical realization of this method of obtaining energy from the sun would be of incalculable industrial value, but :the continued study of the subject revealed the fact that while it will be commercially profitable if my expectations are well founded, it will not be so to an extraordinary degree.

Discovery of Unexpected Properties of The. Atmosphere — Strange Experiments — Transmission of Electrical Energy Through One Wire Without Return — Transmission Through the Earth Without Any Wire

Another of these reasons was that I was led to recognize, the transmission of electrical energy to any distance through the media as by far the best solution of the great problem of harnessing the sun's energy for the uses of man. For a long time I was convinced that such a transmission on an industrial scale could never be realized, but a discovery which I made changed my view. I observed that under certain conditions the atmosphere, which is normally a high insulator, assumes conducting properties, and so becomes capable of conveying any amount of electrical energy. But the difficulties in the way of a practical utilization of this discovery for the purpose of transmitting electrical energy without wires were seemingly insuperable. Electrical pressures of many millions of volts had to be produced and handled; generating apparatus of a novel kind, capable of withstanding the immense

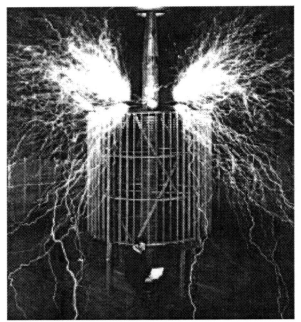

Fig. 3. **Experiment to Illustrate the Supplying of Electrical Energy Through a Single Wire Without Return.** An ordinary incandescent lamp connected with one or both of its terminals to the wire forming the upper free end of the coil shown in the photograph, is lighted by electrical vibrations conveyed to it through the coil from an electrical oscillator, which is worked only to one fifth of one per cent of its full capacity.

electrical stresses, had to be invented and perfected, and a complete safety against the dangers of the high-tension currents had to be attained in the system before its practical introduction could be even thought of. All this could not be done in a few weeks or months, or even years. The work required patience and constant application, but the improvements came, though slowly. Other valuable results were, however, arrived at in the course of this long-continued work, of which I shall endeavor to give a brief account, enumerating the chief advances as they were successively effected.

The discovery of the conducting properties of the air, though unexpected, was only a natural result of experiments in a special field which I had carried on for some years before. It was, I believe, during 1889 that certain possibilities offered by extremely rapid electrical oscillations determined me to design a. number of special machines adapted for their investigation. Owing to the peculiar requirements, the construction of these machines was very difficult, and consumed much time and effort; but my work on them was generously rewarded, for I reached by their means several novel and important results. One of the earliest observations I made with these new machines was that electrical oscillations of an extremely high rate act in an extraordinary manner upon the human organism. Thus, for instance, I demonstrated that powerful electrical discharges of several hundred thousand volts, which at that time were considered absolutely deadly, could be passed through the body without inconvenience or hurtful consequences. These oscillations produced other specific physiological effects, which, upon my announcement, were eagerly taken up by skilled physicians and further investigated. Ibis new field has proved itself fruitful beyond expectation, in the few years which have passed since, it has been developed to such an extent that it now forms a legitimate and important department of medical science. Many results, thought impossible at that time, are now readily obtainable with these oscillations, and many experiments undreamed of then can now be readily performed by their means. I still remember with pleasure how, nine years ago, I passed the discharge of a powerful' induction-coil through my body to demonstrate before a scientific society the comparative harmlessness of very rapidly vibrating electric currents, and I can still recall the astonishment of my audience. I would now undertake,. with much less apprehension than I had in that experiment, to transmit through my body with such currents the entire electrical energy of the dynamos now working at Niagara — forty or fifty thousand horse-power. I have produced electrical oscillations which were of such intensity that when circulating through my arms and chest they have melted wires which joined my hands, and still I felt no inconvenience. I have energized with such oscillations a loop of heavy copper wire so powerfully that masses of metal, and even objects of an electrical resistance specifically greater than that of human tissue, brought close to or placed within the loop, were heated to a high temperature and melted, often with the violence of an explosion, and yet into this very space in which this terribly destructive turmoil was going on I have repeatedly thrust my head without feeling anything or experiencing injurious after-effects.

Fig. 4. Experiment to Illustrate the Transmission of Electrical Energy Through the Earth Without Wire. The coil shown in the photograph has its lower end or terminal connected to the ground, and is exactly attuned to the vibrations of a distant electrical oscillator. The lamp lighted is in an independent wire loop, energized by induction from the coil excited by the .electrical vibrations transmitted to it through the ground from the oscillator, which is worked only to five per cent. of its full capacity.

Another observation was that by means of such oscillations light could be produced in a novel and more economical manner, which promised to lead to an ideal system of electric illumination by vacuum-tubes, dispensing with the necessity of renewal of lamps or incandescent filaments. and possibly also with the use of wires in the interior of buildings. The efficiency of this light increases in proportion to the rate of the oscillations, and its commercial success is, therefore, dependent on the economical production of electrical vibrations of transcending rates. In this direction I have met with gratifying success of late, and the practical introduction of this new system of illumination is not far off.

The investigations led to many other valuable observations and results, one of the more important of which was the demonstration of the practicability of supplying electrical energy through one wire without return. At first I was able to transmit in this novel manner only very small amounts of electrical energy, but in this line also my efforts have been rewarded with similar success.

The photograph shown in Fig. 3 illustrates, as its title explains, an actual transmission of this kind effected with apparatus used in other experiments here described. To what a degree the appliances have been perfected since my first demonstrations early in 1891 before a scientific society, when my apparatus was barely capable of lighting one lamp (which result was considered wonderful), will appear when I state that I have now no difficulty in lighting in this manner four or five hundred lamps, and could light many more. In fact, there is no limit to the amount of energy which may in this way be supplied to operate any kind of electrical device.

After demonstrating the practicability of this method of transmission, the thought naturally occurred to me to use the earth as a conductor, thus dispensing with all wires. Whatever electricity may be, it is a fact that it behaves like an incompressible fluid, and the earth may be looked upon as an immense reservoir of electricity, which, I thought, could be disturbed effectively by a properly designed electrical machine. Accordingly, my next efforts were directed toward perfecting a special apparatus which would be highly effective in creating a disturbance of electricity in the earth. The progress in this new direction was necessarily very slow and the work discouraging, until I finally succeeded in perfecting a novel kind of transformer or induction-coil, particularly suited for this special purpose. That it is practicable, in this manner, not only to transmit minute amounts of electrical energy for operating delicate electrical devices, as I contemplated at first, but also electrical energy in appreciable quantities, will appear from an inspection of Fig. 4, which illustrates an actual experiment of this kind performed with the same apparatus. The result obtained was all the more remarkable as the top end of the coil was not connected to a wire or plate for magnifying the effect.

"Wireless" Telegraphy — the Secret of Tuning — Errors in the Hertzian Investigations — a Receiver of Wonderful Sensitiveness

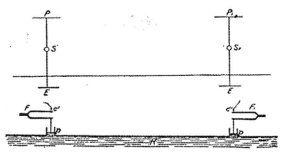

Diagram c. []Wireless[] Telegraphy Mechanically Illustrated.

As the first valuable result of my experiments in this latter line a system of telegraphy without wires resulted, which I described in two scientific lectures in February and March, 1893. It is mechanically illustrated in diagram c, the upper part of which shows the electrical arrangement as I described it then, while the lower part illustrates its mechanical analogue. The system is extremely simple in principle. Imagine two tuning-forks F, F_1, one at the sending and the other at the receiving-station respectively, each having attached to its lower prong a minute piston p, fitting in a cylinder. Both the cylinders communicate with a large reservoir R, with elastic walls, which is supposed to be closed and filled with a light and incompressible fluid. By striking repeatedly one of the prongs of the tuning-fork F, the small piston p below would be vibrated, and its vibrations, transmitted through the fluid, would reach the distant fork F_1, which is "tuned" to the fork F, or stated otherwise, of exactly the same note as the latter. The fork F_1 would now be set vibrating, and its vibration would be intensified by. the continued action of the distant fork F until its upper prong, swinging far out, would make an electrical connection with a stationary contact c, "starting in this manner some electrical or other appliances which may be used for recording the signals. In this simple way messages could be exchanged between the two stations, a similar contact c' being provided for this purpose, close to the upper prong of the fork F, so that the apparatus at each station could be employed in turn as receiver and transmitter.

The electrical system illustrated in the upper figure of diagram c is exactly the same in principle, the two wires or circuits ESP and $E_1S_1P_1$, which extend vertically to a height, representing the two tuning-forks with the pistons attached to them. These circuits are connected with the ground by plates E, E_1, and to two elevated metal sheets P, P_1, which store electricity and thus magnify considerably the effect. The closed reservoir R, with elastic walls, is in this case replaced by the earth, and the fluid by electricity. Both of these circuits are "tuned" and operate just like the two tuning-forks. Instead of striking the fork F at the sending-station, electrical oscillations are produced in the vertical sending- or transmitting-wire ESP, as by the action of a source S, included in this wire, which spread through the ground and reach the distant vertical receiving-wire $E_1S_1P_1$, exciting corresponding electrical oscillations in the same. In the latter wire or circuit is included a sensitive device or receiver S_1, which is thus set in action and made to operate a relay or other appliance. Each station is, of course, provided both with a source of electrical oscillations S and a sensitive receiver S_1, and a simple provision is made for using each of the two wires alternately to send and to receive the Messages.

Fig. 5. Photographic View of Coils Responding to Electrical Oscillations. The picture shows a number of coils, differently attuned and responding to the vibrations transmitted to them through the earth from an electrical oscillator. The large coil on the right, discharging strongly, is tuned to the fundamental vibration, which is fifty thousand per second; the two larger vertical coils to twice that number; the smaller white wire coil to four times that number, and the remaining small coils to higher tones. The vibrations produced by the oscillator were so intense that they affected perceptibly a small coil tuned to the twenty-sixth higher tone.

The exact attunement of the two circuits secures great advantages, and, in fact, it is essential in the practical use of the system. In this respect many popular errors exist, and as a rule, in the technical reports on this subject circuits and appliances are described as affording these advantages when from their very nature it is evident that this is impossible. In order to attain the best results it is essential that the length of each wire or circuit, from the ground connection to the top, should be equal to one quarter of the wavelength of the electrical vibration in the wire, or else equal to that length multiplied by an odd number. Without the observation of this rule it is virtually impossible to prevent the interference and insure the privacy of messages. Therein' lies the secret of tuning. To obtain the most' satisfactory results it is, however, necessary to resort to electrical vibrations of low pitch. The Hertzian spark apparatus, used generally by experimenters which produces oscillations of a very high rate, permits no effective tuning, and slight disturbances are sufficient to render an exchange of messages impracticable. But scientifically designed, efficient appliances allow nearly perfect adjustment. An experiment performed with the improved apparatus repeatedly referred to, and intended to convey an idea of this feature, is illustrated in Fig. 5, which is sufficiently explained by its note.

Since I described these simple principles of telegraphy without wires I have had frequent occasion to note that the identical features and elements have been used, in the evident belief that the signals are being transmitted to considerable distances by "Hertzian" radiations. This is only one of many misapprehensions to which the investigations of the lamented physicist have given rise. About thirty-three years ago Maxwell, following up a suggestive experiment made by Faraday in 1845, evolved an ideally simple theory which intimately connected light, radiant heat, and electrical phenomena, interpreting them as being all due to vibrations of a hypothetical fluid of inconceivable tenuity, called the ether. No experimental verification was arrived at until Hertz, at the suggestion of Helmholtz, undertook a series of experiments to this effect. Hertz proceeded with extraordinary ingenuity and insight, but devoted little energy to the perfection of his old-fashioned apparatus. The consequence was that he failed to observe the important function which the air played in his experiments, and which I subsequently discovered. Repeating his experiments and reaching different results, I ventured to point out this oversight. The strength of the proofs brought forward by Hertz in support of Maxwell's theory resided in the correct estimate of the rates of vibration of the circuits he used. But I ascertained that he could not have obtained the rates he thought he was getting. The vibrations with identical apparatus he employed are, as a rule, much slower, this being due to the presence of air, which produces a dampening effect upon a rapidly vibrating electric circuit of high pressure, as a fluid does upon a vibrating tuning-fork. I have, however, discovered since that time other causes of error, and I have long ago ceased to look upon his results as being an experimental verification of the poetical conceptions of Maxwell. The work of the great German physicist has acted as an immense stimulus to contemporary electrical research; but it has likewise, in a measure, by its fascination, paralyzed the scientific mind, and thus hampered independent inquiry. Every new phenomenon which was discovered was made to fit the theory, and so very often the truth has been unconsciously distorted.

Fig. 6. Photographic View of the Essential Parts of the Electrical Oscillator Used in the Experiments Described.

When I advanced this system of telegraphy, my mind was dominated by the idea of effecting communication to any distance through the earth or environing medium, the practical consummation of which I considered of transcendent importance, chiefly on account of the moral effect which it could not fail to produce universally. As the first effort to this end I proposed, at that time, to employ relay-stations with tuned circuits, in the hope of making thus practicable signaling over vast distances, even with apparatus of very moderate power then at my command. I was confident, however, that with properly designed machinery signals could be transmitted to any point of the globe, no matter what the distance, without the necessity of using such intermediate stations. I gained this conviction through the discovery of a singular electrical phenomenon, which I described early in 1892, in' lectures delivered before some scientific societies abroad, and which I have called a "rotating brush." This is a bundle of light which is formed, under certain conditions, in a vacuum-bulb, and which is of a sensitiveness to magnetic and electric influences bordering, so to speak, on the supernatural. This light-bundle is rapidly rotated by the earth's magnetism as many as twenty thousand times per second, the rotation in these parts being opposite to what it would be in the southern hemisphere, while in the region of the magnetic equator :t should not rotate at all. In its most sensitive state, which is difficult to attain, it is responsive to electric or magnetic influences to an incredible degree. The mere stiffening of the muscles of the arm and consequent slight electrical change in the body of an observer standing at some distance from it, will perceptibly affect it. When in this highly sensitive state it is capable of indicating the slightest magnetic and electric changes taking place in the earth. The observation of this wonderful phenomenon impressed me strongly that communication at any distance could be easily effected by its means, provided that apparatus could be perfected capable of producing an electric or magnetic change of state, however small, in the terrestrial globe or environing medium.

Development of a New Principle — the Electrical Oscillator — Production of Immense Electrical Movements — the Earth Responds to Man — Interplanetary Communication Now Probable

Fig. 7. Experiment to Illustrate an Inductive Effect of an Electrical Oscillator of Great Power. The photograph shows three ordinary incandescent lamps lighted to full candle-power by currents induced in a local loop consisting of a single wire forming a square of fifty feet each side, which includes the lamps, and which is at a distance of one hundred feet from the primary circuit energized by the oscillator. The loop likewise, includes an electrical condenser, and is exactly attuned to the vibrations of the oscillator, which is worked at less than five percent of its total capacity.

I resolved to concentrate my efforts upon this venturesome task, though it involved great sacrifice, for the difficulties to be mastered where such that I could hope to consummate it only after years of labor. It meant delay of other work to which I would have preferred to devote myself, but I gained the conviction that my energies could not be more usefully employed; for I recognized that an efficient apparatus for the production of powerful electrical oscillations, as was needed for that specific purpose, was the key to the solution of other most important electrical and; in fact, human problems. Not only was communication, to any distance, without wires possible by its means, but, likewise, the transmission of energy in great amounts, the burning of the atmospheric nitrogen, the production of an efficient illuminant, and many other results of inestimable scientific and industrial value. Finally, however, I had the satisfaction of accomplishing the task undertaken by the use of a new principle, the virtue of which is based on the marvelous properties of the electrical condenser. One of these is that it can discharge or explode its stored energy in an inconceivably short time. Owing to this it is unequaled in explosive violence. The explosion of dynamite is only the breath of a consumptive compared with its discharge. It is the means of producing the strongest current, the highest electrical pressure, the greatest commotion in the medium. Another of its properties, equally valuable, is that its discharge may vibrate at any rate desired up to many millions per second.

I had arrived at the limit of rates obtainable in other ways when the happy idea presented itself to me to resort to the condenser. I arranged such an instrument so as to be charged and discharged alternately in rapid succession through a coil, with a few turns of stout wire, forming the primary of a transformer of induction-coil. Each time the condenser was discharged the current would quiver in the primary wire and induce corresponding oscillations in the secondary. Thus a transformer or induction-coil on new principles was evolved, which I have called "the electrical oscillator," partaking of those unique qualities which characterize the condenser, and enabling results to be attained impossible by other means. Electrical effects of any desired character and of intensities undreamed of before are now easily producible by perfected apparatus of this kind, to which frequent reference has been made, and the essential parts of which are shown in Fig. 6. For certain purposes a strong inductive effect is required; for others the greatest possible suddenness; for others again, an exceptionally high rate of vibration or extreme pressure; while for . certain other objects immense electrical movements are necessary. The photographs in Figs. 7, 8, 9, and 10, of experiments performed with such an oscillator, may serve to illustrate some of these features and convey an idea of the magnitude of the effects actually produced. The completeness of the titles of the figures referred to makes a further description of them unnecessary.

However extraordinary the results shown may appear, they are but trifling compared with those which are attainable by apparatus designed on these same principles. I have produced electrical discharges the actual path of which, from end to end, was probably more than one hundred feet long; but it would not be difficult to reach lengths one hundred times as great. I have produced electrical movements occurring at the rate of approximately one hundred thousand horse-power, but rates of one, five, or ten million horse-power are easily practicable. In these experiments effects were developed incomparably greater than any ever produced by human agencies, and yet these results are but an embryo of what is to be.

Fig. 8. The coil, partly shown in the photograph, creates an alternative movement of electricity from the earth into a large reservoir and back at the rate of one hundred thousand alternations per second. The adjustments are such that the reservoir is filled full and bursts at each alternation just at the moment when the electrical pressure reaches the maximum. The discharge escapes with a deafening noise, striking an unconnected coil twenty-two feet away, and creating. such a commotion of electricity in the earth that sparks an inch long can be drawn from a watermain at a distance of three hundred feet from the laboratory.

Fig. 9. Experiment to Illustrate the Capacity of the Oscillator for Creating a Great Electrical Movement. The ball shown in the photograph, covered with a polished metallic coating of twenty square feet of surface, represents a large reservoir of electricity, and the inverted tin pan underneath, with a sharp rim, a big opening through which the electricity can escape before filling the reservoir. The quantity of electricity set in movement is so great that, although most of it escapes through the rim of the pan or opening provided, the ball or reservoir is nevertheless alternately emptied and filled to overflowing (as is evident from the discharge escaping on the top of the ball) one hundred and fifty thousand times per second.

That communication without wires to any point of the globe is practicable with such apparatus would need no demonstration, but through a discovery which I made I obtained absolute certitude. Popularly explained, it is exactly this: When we raise the voice and hear an echo in reply, we know that the sound of the voice must have reached a distant wall, or boundary, and must have been reflected from the same. Exactly as the sound, so an electrical wave is reflected, and the same evidence which is afforded by an echo is offered by an electrical phenomenon known as a "stationary" wave — that is, a wave with fixed nodal and ventral regions. Instead of sending sound-vibrations toward a distant wall, I have sent electrical vibrations toward the remote boundaries of the earth, and instead of the wall the earth has replied. In place of an echo I have obtained a stationary electrical wave, a wave reflected from afar.

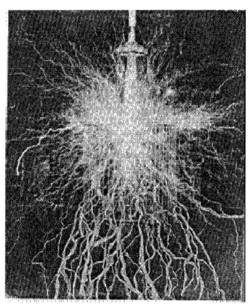

Fig. 10. Photographic View of an Experiment to Illustrate an Effect of an Electrical Oscillator Delivering Energy at a Rate of Seventy-five Thousand Horse-power. The discharge, creating a strong draft owing to the heating of the air, is carried upward through the open roof of the building. The greatest width across is nearly seventy feet. The pressure is over twelve million volts, and the current alternates one hundred and thirty thousand times per second.

Stationary waves in the earth mean something more than mere telegraphy without wires to any distance. They will enable us to attain many important specific results impossible otherwise. For instance, by their use we may produce at will, from a sending-station, an electrical effect in any particular region of the globe; we may determine the relative position or course of a moving object, such as a vessel at sea, the distance traversed by the same, or its speed; or we may send over the earth a wave of electricity traveling at any rate we desire, from the pace of a turtle up to lightening speed.

With these developments we have every reason to anticipate that in a time not very distant most telegraphic messages across the oceans will be transmitted without cables. For short distances we need a "wireless" telephone, which requires no expert operators. The greater the spaces to be bridged, the more rational becomes communication without wires. The cable is not only an easily damaged and costly instrument, but it limits us in the speed of transmission by reason of a certain electrical property inseparable from its construction. A properly designed plant for effecting communication without wires ought to have many times the working capacity of a cable, while it will involve incomparably less expense. Not a long time will pass, I believe, before communication by cable will become obsolete, for not only will signaling by this new method be quicker and cheaper, but also much safer. By using some new means for isolating the messages which I have contrived, an almost perfect privacy can be secured.

I have observed the above effects so far only up to a limited distance of about six hundred miles, but inasmuch as there is virtually no limit to the power of the vibrations producible with such an oscillator, I feel quite confident of the success of such a plant for effecting transoceanic communication. Nor is this all. My measurements and calculations have shown that it is perfectly practicable to produce on our globe, by the use of these principles, an electrical movement of such magnitude that, without the slightest doubt, its effect will be perceptible on some of our nearer planets, as Venus and Mars. Thus from mere possibility interplanetary communication has entered the stage of probability. In fact, that we can produce a distinct effect on one of these planets in this novel manner, namely, by disturbing the electrical condition of the earth, is beyond any doubt. This way of effecting such communication is, however, essentially different from all others which have so far been proposed by scientific men. In all the previous instances only a minute fraction of the total energy reaching the planet — as much as it would be possible to concentrate in a reflector — could be utilized by the supposed observer in his instrument. But by the means I have developed he would be enabled to concentrate the larger portion of the entire energy transmitted to the planet in his instrument, and the chances of affecting the latter are thereby increased many millionfold.

Besides machinery for producing vibrations of the required power, we must have delicate means capable of revealing, the effects of feeble influences exerted upon the earth. For such purposes, too, I have perfected new methods. By their ,use we shall likewise be able, among other things, to detect at considerable distance the presence of an iceberg or other object at sea. By their use, also, I have discovered some terrestrial phenomena still unexplained. That we can send a message to a planet is certain, that we can get an answer is probable: man is not the only being in the Infinite gifted with a mind.

Transmission of Electrical Energy to Any Distance Without Wires — Now Practicable — the Best Means of Increasing the Force Accelerating the Human Mass

The most valuable observation made in the course of these investigations was the extraordinary behavior of the atmosphere toward electric impulses of excessive electromotive force. The experiments showed that the air at the ordinary pressure became distinctly conducting, and this opened up the wonderful prospect of transmitting large amounts of electrical energy for industrial purposes to great distances without wires, a possibility which, up to that time, was thought of only as a scientific dream. Further investigation revealed the important fact that the conductivity imparted to the air by these electrical impulses of many millions of volts increased very rapidly with the degree of rarefaction, so that air strata at very moderate altitudes, which are easily accessible, offer, to all experimental evidence, a perfect conducting path, better than a copper wire, for currents of this character.

Thus the discovery of these new properties of the atmosphere not only opened up the possibility of transmitting, without wires; energy in large amounts, but, what was still more significant, it afforded the certitude that energy could be transmitted in this manner economically. In this new system it matters little — in fact, almost nothing — whether the transmission is effected at a distance of a few miles or of a few thousand miles.

While I have not, as yet, actually effected a transmission of a considerable amount of energy, such as would be of industrial importance, to a great distance by this new method, I have operated several model plants under exactly the same conditions which will exist in a large plant of this kind, and the practicability of the system is thoroughly demonstrated. The experiments have shown conclusively that, with two terminals maintained at an elevation of not more than thirty thousand to thirty-five thousand feet above sea-level, and with an electrical pressure of fifteen to twenty million volts, the energy of thousands of horse-power can be transmitted over distances which may be hundreds and, if necessary, thousands of miles. I am hopeful, however, that I may be able to reduce very considerably the elevation of the terminals now required, and with this object I am following up an idea which promises such a realization. There is, of course, a popular prejudice against using an electrical pressure of millions of volts, which may cause spark's to fly at distances of hundreds of feet, but, paradoxical as it *may* seem, the system, as I have described it in a technical publication, offers greater personal safety than most of the ordinary distribution circuits now used in the cities, This is, in a measure, borne out by the fact that, although I have carried on such experiments for a number of years, no injury has been sustained either by me or any of my assistants.

But to enable a practical introduction of the system, a number of essential requirements are still to be fulfilled. It is not enough to develop appliances by means of which such a transmission can be effected. The machinery must be such as to allow the transformation and transmission of electrical energy under highly economical and practical conditions. Furthermore, an inducement must be offered to those who are engaged in the industrial exploitation of natural sources of power, as waterfalls, by guaranteeing greater returns on the capital invested than they can secure by local development of the property.

From that moment when it was observed that, contrary to the established opinion, low and easily accessible strata of the atmosphere are capable of conducting electricity, the transmission of electrical energy without wires has become a rational task of the engineer, and one surpassing all others in importance. Its practical consummation would mean that energy would be available for the uses of man at any point of the globe, not in small amounts such as might be derived from the ambient medium by suitable machinery, but in quantities virtually unlimited, from waterfalls. Export of power would then become the chief source of income for many happily situated countries, as the United States, Canada, Central and South America, Switzerland, and Sweden. Men could settle down everywhere, fertilize and irrigate the soil with little effort, and convert barren deserts into gardens, and thus the entire globe could be transformed and made a fitter abode for mankind. It is highly probable that if there are intelligent beings on Mars they have long ago realized this very idea, which would explain the changes on its surface noted by astronomers. The atmosphere on that planet, being of considerably smaller density than that of the earth, would make the task much more easy.

It is probable that we shall soon have a self-acting heat-engine capable of deriving moderate amounts of energy from the ambient medium. There is also a possibility — though a small one — that we may obtain electrical energy direct from the sun. This might be the case if the Maxwellian theory is true, according to which electrical vibrations of all rates should emanate from the sun. I am still investigating this subject. Sir William Crookes has shown in his beautiful invention known as the "radiometer" that rays may produce by impact a mechanical effect, and this may lead to some important revelation as to the utilization of the sun's rays in novel ways. Other sources of energy may be opened up, and new methods of deriving energy from the sun discovered, but none of these or similar achievements would equal in importance the transmission of power to any distance through the medium. I can conceive of no technical advance which would tend to unite the various elements of humanity more effectively than this one. or of one which would more add to and more economize human energy. It would be the best means of increasing the force accelerating the human mass. The mere moral influence of such a radical departure would be incalculable. On the other hand, if at any point of the globe energy can be obtained in limited quantities from the ambient medium by means of a self-acting heat-engine or otherwise, the conditions will remain the same as before. Human performance will be increased, but men will remain strangers as they were.

I anticipate that many, unprepared for these results, which, through long familiarity, appear to me simple and obvious, will consider them still far from practical application. Such reserve, and even opposition, of some is as useful a quality and as necessary an element in human progress as the quick receptivity and enthusiasm of others. Thus, a mass which resists the force at first, once set in movement, adds to the energy. The scientific man does not aim at an immediate result. He does not expect that his advanced ideas will be readily taken up. His work is like that of the planter — for the future. His duty is to lay foundation for those who are to come, and point the way. He live and labors and hopes with the poet who says:

Schaff', das Tagwerk meiner Hande,
Hohes Gluck, dass ich's vollende!
Lass, o lass mich niche ermatten!
Nein, es sind nicht leere Traume:
Jetzt nur Slangen, these Baume
Geben eins noch Frucht and Schatten.[1]

Daily work — my hands' employment,
To complete is pure enjoyment!
Let, oh, let me never falter!
No! there is no empty dreaming:
Lo! these trees, but bare poles seeming,
Yet will yield both fruit and shelter!
    —Goethe's "Hope,"

Tesla's New Discovery
New York Sun — Jan. 30, 1901

Capacity of Electrical Conductors is Variable. Not Constant, and Formulas Will Have to Be Rewritten - Capacity Varies With Absolute Height Above Sea Level, Relative Height From Earth and Distance From the Sun.

Nikola Tesla announced yesterday another new discovery in electricity. This time it is a new law and by reason of it, Mr. Tesla asserts, a large part of technical literature will have to be rewritten. Ever since anything has been known about electricity, scientific men have taken for granted that the capacity of an electrical conductor is constant. When Tesla was experimenting in Colorado he found out that this capacity is not constant - but variable. Then he determined to find out the law governing this phenomenon. He did so, and all, this he explained to *The Sun* yesterday. Here is what he said:

"For many years scientific men engaged in the study of physics and electrical research have taken it for granted that certain quantities, entering continuously in their estimates and calculations, are fixed and unalterable. The exact determination of these quantities being of particular importance in electrical vibrations, which are engrossing more and more the attention of experimenters all over the world, it seems to be important to acquaint others with some of my observations, which have finally led me to the results now attracting universal attention. These observations, with which I have long been familiar, show that some of the quantities referred to are variable and that, owing to this, a large portion of the technical literature is defective. I shall endeavor to convey the knowledge of the facts I have discovered in plain language, devoid as much as possible of technicalities.

"It is well known that an electric circuit compacts itself like a spring with a weight attached to it. Such a spring vibrates at a definite rate, which is determined by two quantities, the pliability of the spring and the mass of the weight. Similarly an electric circuit vibrates, and its vibration, too; is dependent on two quantities, designated as electrostatic capacity and inductance. The capacity of the electric circuit corresponds to the pliability of the spring and the inductance to the mass of the weight.

"Exactly as mechanics and engineers have taken it for granted that the pliability of the spring remains the same, no matter how it be placed or used, so electricians and physicists have assumed that the electrostatic capacity of a conducting body, say of a metallic sphere, which is frequently used in experiments, remains a fixed and unalterable quantity, and many scientific results of the greatest importance are dependent on this assumption. Now, I have discovered that this capacity is not fixed and unalterable at all. On the contrary, it is susceptible to great changes, so that under certain conditions it may amount to many times its theoretical value, or may eventually be smaller. Inasmuch as every electrical conductor, besides possessing an inductance, has also a certain amount of capacity, owing to the variations of the latter, the inductance, too, is seemingly modified by the same causes that tend to modify the capacity. These facts I discovered some time before I gave a technical description of my system of energy transmission and telegraphy without wires, which, I believe, became first known through my Belgian and British patents.

"In this system, I then explained, that, in estimating the wave-length of the electrical vibration in the transmitting and receiving circuits, due regard must be had to the velocity with which the vibration is propagated through each of the circuits, this velocity being given by the product of the wave-length and the number of vibrations per second. The rate of vibration being, however, as before stated, dependent on the capacity and inductance in each case, I obtained discordant values. Continuing the investigation of this astonishing phenomenon I observed that the capacity varied with the elevation of the conducting surface above the ground, and I soon ascertained the law of this variation. The capacity increased as the conducting surface was elevated, in open space, from one-half to three-quarters of 1 per cent per foot of elevation. In buildings, however, or near large structures, this increase often amounted to 50 per cent per foot of elevation, and this alone will show to what extent many of the scientific experiments recorded in technical literature are erroneous. In determining the length of the coils or conductors such as I employ in my system of wireless telegraphy, for instance, the rule which I have given is, in view of the above, important to observe.

"Far more interesting, however, for men of science is the fact I observed later, that the capacity undergoes an annual variation with a maximum in summer, and a minimum in winter. In Colorado, where I continued with improved methods of investigations begun in New York, and where I found the rate of increase slightly greater, I furthermore observed that there was a diurnal variation with a maximum during the night. Further, I found that sunlight causes a slight increase in capacity. The moon also produces an effect, but I do not attribute it to its light.

"The importance of these observations will be better appreciated when it is stated that owing to these changes of a quantity supposed to be constant an electrical circuit does not vibrate at a uniform rate, but its rate is modified in accordance with the modifications of the capacity. Thus a circuit vibrates a little slower at an elevation than when at a lower level. An oscillating system, as used in telegraphy without wires, vibrates a little quicker when the ship gets into the harbor than when on open sea. Such a circuit oscillates quicker in the winter than in the summer, though it be at the same temperature, and a trifle quicker at night than in daytime, particularly if the sun is shining.

"Taking together the results of my investigations I find that this variation of the capacity and consequently of the vibration period is evidently dependent, first, on the absolute height above sea level, though in a smaller degree; second, on the relative height of the conducting surface or capacity with respect to the bodies surrounding it; third, on the distance of the earth from the sun, and fourth, on the relative change of the circuit with respect to the sun, caused by the diurnal rotation of the earth. These facts may be of particular interest to meteorologists and astronomers, inasmuch as practical methods of inquiry may result from these observations, which may be useful in their respective fields. It is probable that we shall perfect instruments for indicating the altitude of a place by means of a circuit, properly constructed and arranged, and I have thought of a number of other uses to which this principle may be put.

"It was in the course of investigations of this kind in Colorado that I first noted certain variations in electrical systems arranged in peculiar ways. These variations I first discovered by calculating over the results I had previously noted, and it was only subsequently that I actually perceived them. It will thus be clear that some who have ventured to attribute the phenomena I have observed to ordinary atmospheric disturbances have made a hasty conclusion."

Tesla's Wireless Light
Scientific American — Feb. 2, 1901

Nikola Tesla has given to The New York Sun an authorized statement concerning his new experiments on the production of light without the aid of wires. Mr. Tesla says:

"This light is the result of continuous efforts since my early experimental demonstrations before scientific societies here and abroad. In order to make it suitable for commercial use, I had to overcome great difficulties. One of these was to produce from ordinary currents of supply electrical oscillations of enormous rapidity in a simple and economical manner. This, I am glad to say, I have now accomplished, and the results show that with this new form of light a higher economy is practicable than with the present illuminants. The light offers, besides, many specific advantages, not the least of which is found in its hygienic properties. It is, I believe, the closest approach to daylight which has yet been reached from any artificial source.

"The lamps are glass tubes which may be bent in any ornamental way. I most generally use a rectangular spiral, containing about twenty to twenty-five feet of tubing making some twelve to fourteen convolutions. The total illuminating surface of a lamp is from 300 to 400 square inches. The ends of the spiral tube are covered with a metallic coating, and provided with hooks for hanging the lamp on the terminals of the source of oscillations. The tube contains gases rarefied to a certain degree, determined in the course of long experimentation as being conductive to the best results.

"The process of light production is, according to my views, as follows: The street current is passed through a machine which is an electrical oscillator of peculiar construction and transforms the supply current, be it direct or alternating, into electrical oscillations of very high frequency. These oscillations, coming to the metallically-coated ends of the glass tube, produce in the interior corresponding electrical oscillations, which set the molecules and atoms of the inclosed rarefied gases into violent commotion, causing them to vibrate at enormous rates and emit those radiations which we know as light. The gases are not rendered incandescent in the ordinary sense, for if it were so, they would be hot, like an incandescent filament. As a matter of fact, there is very little heat noticeable, which speaks well for the economy of the light, since all heat would be loss.

"This high economy results chiefly from three causes: First, from the high rate of the electrical oscillations; second, from the fact that the entire light-giving body, being a highly attenuated gas, is exposed and can throw out its radiations unimpeded, and, third, because of the smallness of the particles composing the light-giving body, in consequence of which they can be quickly thrown into a high rate of vibration, so that comparatively little energy is lost in the lower or heat vibrations. An important practical advantage is that the lamps need not be renewed like the ordinary ones, as there is nothing in them to consume. Some of these lamps I have had for years, and they are now in just as good a condition as they ever were. The illuminating power of each of these lamps is, measured by the photometric method, about fifty candle power, but I can make them of any power desired, up to that of several arc lights. It is a remarkable feature of the light that during the day it can scarcely be seen, whereas at night the whole room is brilliantly illuminated.

When the eye becomes used to the light of these tubes, an ordinary incandescent lamp or gas burner produces a violent pain in the eye when it is turned on, showing in a striking manner to what a degree these concentrated sources of light which we now use are detrimental to the eye.

"I have found that in almost all its actions the light produces the same effects as sunlight, and this makes me hopeful that its introduction into dwellings will have the effect of improving, in a measure now impossible to estimate, the hygienic conditions. Since sunlight is a very powerful curative agent, and since this light makes it possible to have sunlight, so to speak, of any desired intensity, day and night in our homes, it stands to reason that the development of germs will be checked and many diseases, as consumption, for instance, successfully combated by continually exposing the patients to the rays of these lamps. I have ascertained unmistakably that the light produces a soothing action on the nerves, which I attribute to the effect which it has upon the retina of the eye. It also improves vision just exactly as the sunlight, and it ozonizes slightly the atmosphere. These effects can be regulated at will. For instance, in hospitals, where such a light is of paramount importance, lamps may be designed which will produce just that quality of ozone which the physician may desire for the purification of the atmosphere, or if necessary, the ozone production can be stopped altogether.

"The lamps are very cheap to manufacture, and by the fact that they need not be exchanged like ordinary lamps or burners they are rendered still less expensive. The chief consideration is, of course, in commercial introduction, the energy consumption. While I am not yet prepared to give exact figures, I can say that, given a certain quantity of electrical energy from the mains, I can produce more light than can be produced by the ordinary methods. In introducing this system of lighting my transformer, or oscillator, will be usually located at some convenient place in the basement, and from there the transformed currents will be led as usual through the building. The lamps can be run with one wire alone, as I have shown in my early demonstrations, and in some cases I can dispense entirely with the wires. I hope that ultimately we shall get to this ideal form of illumination, and that we shall have in our rooms lamps which will be set aglow no matter where they are placed, just as an object is heated by heat rays emanating from a stove. The lamps will then be handled like kerosene lamps, with this difference, however, that the energy will be conveyed through space. The ultimate perfection of apparatus for the production of electrical oscillations will probably bring us to this great realization, and then we shall finally have the light without heat or 'cold' light. I have no difficulty now to illuminate the room with such wireless lamps, but a number of improvements must be made yet before it can be generally introduced."

Talking with the Planets
Collier's Weekly, February 9, 1901

Editor's Note. - *Mr. Nikola Tesla has accomplished some marvelous results in electrical discoveries. Now, with the dawn of the new century, he announces an achievement that wilt amaze the entire universe, and which eclipses the wildest dream of the most visionary scientist. He has received communication, he asserts, from out the great void of space; a call from inhabitants of Mars, or Venus, or some other sister planet! And, furthermore, noted scientists like Sir Norman Lockyer are disposed to agree with Mr. Tesla in his startling deductions.*

Mr. Tesla has not only discovered many important principles, but most of his inventions are in practical use; notably in the harnessing of the Titanic forces of Niagara Falls, and the discovery of a new light by means of a vacuum tube. He has, he declares, solved the problem of telegraphing without wires or artificial conductors of any sort, using the earth as his medium. By means of this principle he expects to be able to send messages under the ocean, and to any distance on the earth's surface. Interplanetary communication has interested him for years, and he sees no reason why we should not soon be within talking distance of Mars or of all worlds in the solar system that many be tenanted by intelligent beings.

At the request of Collier's Weekly *Mr. Tesla present herewith a frank statement of what he expects to accomplish and how he hopes to establish communication with the planets.*

The idea of communicating with the inhabitants of other worlds is an old one. But for ages it has been regarded merely as a poet's dream, forever unrealizable. And yet, with the invention and perfection of the telescope and the ever-widening knowledge of the heavens, its hold upon our imagination has been increased, and the scientific achievements during the latter part of the nineteenth century, together with the development of the tendency toward the nature ideal of Goethe, have intensified it to such a degree that it seems as if it were destined to become the dominating idea of the century that has just begun. The desire to know something of our neighbors in the immense depths of space does not spring from idle curiosity nor from thirst for knowledge, but from a deeper cause, and it is a feeling firmly rooted in the heart of every human being capable of thinking at all.

Whence, then, does it come? Who knows? Who can assign limits to the subtlety of nature's influences? Perhaps, if we could clearly perceive all the intricate mechanism of the glorious spectacle that is continually unfolding before us, and could, also, trace this desire to its distant origin, we might find it in the sorrowful vibrations of the earth which began when it parted from its celestial parent.

But in this age of reason it is not astonishing to find persons who scoff at the very thought of effecting communication with a planet. First of all, the argument is made that there is only a small probability of other planets being inhabited at all. This argument has never appealed to me. In the solar system, there seem to be only two planets--Venus and Mars--capable of sustaining life such as ours: but this does not mean that there might not be on all of them some other forms of life. Chemical processes may be maintained without the aid of oxygen, and it is still a question whether chemical processes are absolutely necessary for the sustenance of organized beings. My idea is that the development of life must lead to forms of existence that will be possible without nourishment and which will not be shackled by consequent limitations. Why should a living being not be able to obtain all the energy it needs for the performance of its life functions from the environment, instead of through consumption of food, and transforming, by a complicated process, the energy of chemical combinations into life-sustaining energy?

If there were such beings on one of the planets we should know next to nothing about them. Nor is it necessary to go so far in our assumptions, for we can readily conceive that, in the same degree as the atmosphere diminishes in density, moisture disappears and the planet freezes up, organic life might also undergo corresponding modifications, leading finally to forms which, according to our present ideas of life, are impossible. I will readily admit, of course, that if there should be a sudden catastrophe of any kind all life processes might be arrested; but if the change, no matter how great, should be gradual, and occupied ages, so that the ultimate results could be intelligently foreseen, I cannot but think that reasoning beings would still find means of existence. They would adapt themselves to their constantly changing environment. So I think it quite possible that in a frozen planet, such as our moon is supposed to be, intelligent beings may still dwell, in its interior, if not on its surface.

Signaling at 100,000,000 Miles!

Then it is contended that it is beyond human power and ingenuity to convey signals to the almost inconceivable distances of fifty million or one hundred million miles. This might have been a valid argument formerly. It is not so now. Most of those who are enthusiastic upon the subject of interplanetary communication have reposed their faith in the light-ray as the best possible medium of such communication. True, waves of light, owing to their immense rapidity of succession, can penetrate space more readily than waves less rapid, but a simple consideration will show that by their means an exchange of signals between this earth and its companions in the solar system is, at least now, impossible. By way of illustration, let us suppose that a square mile of the earth's surface--the smallest area that might possibly be within reach of the best telescopic vision of other worlds--were covered with incandescent lamps, packed closely together so as to form, when illuminated, a continuous sheet of light. It would require not less than one hundred million horse-power to light this area of lamps, and this is many times the amount of motive power now in the service of man throughout the world.

But with the novel means, proposed by myself, I can readily demonstrate that, with an expenditure not exceeding two thousand horse-power, signals can be transmitted to a planet such as Mars with as much exactness and certitude as we now send messages by wire from New

York to Philadelphia. These means are the result of long-continued experiment and gradual improvement.

Some ten years ago, I recognized the fact that to convey electric currents to a distance it was not at all necessary to employ a return wire, but that any amount of energy might be transmitted by using a single wire. I illustrated this principle by numerous experiments, which, at that time, excited considerable attention among scientific men.

This being practically demonstrated, my next step was to use the earth itself as the medium for conducting the currents, thus dispensing with wires and all other artificial conductors. So I was led to the development of a system of energy transmission and of telegraphy without the use of wires, which I described in 1893. The difficulties I encountered at first in the transmission of currents through the earth were very great. At that time I had at hand only ordinary apparatus, which I found to be ineffective, and I concentrated my attention immediately upon perfecting machines for this special purpose. This work consumed a number of years, but I finally vanquished all difficulties and succeeded in producing a machine which, to explain its operation in plain language, resembled a pump in its action, drawing electricity from the earth and driving it back into the same at an enormous rate, thus creating ripples or disturbances which, spreading through the earth as through a wire, could be detected at great distances by carefully attuned receiving circuits. In this manner I was able to transmit to a distance, not only feeble effects for the purposes of signaling, but considerable amounts of energy, and later discoveries I made convinced me that I shall ultimately succeed in conveying power without wires, for industrial purposes, with high economy, and to any distance, however great.

Experiments in Colorado

To develop these inventions further, I went to Colorado in where I continued my investigations along these and other lines, one of which in particular I now consider of even greater importance than the transmission of power without wires. I constructed a laboratory in the neighborhood of Pike's Peak. The conditions in the pure air of the Colorado Mountains proved extremely favorable for my experiments, and the results were most gratifying to me. I found that I could not only accomplish more work, physically and mentally, than I could in New York, but that electrical effects and changes were more readily and distinctly perceived. A few years ago it was virtually impossible to produce electrical sparks twenty or thirty foot long; but I produced some more than one hundred feet in length, and this without difficulty. The rates of electrical movement involved in strong induction apparatus had measured but a few hundred horse-power, and I produced electrical movements of rates of one hundred and ten thousand horse-power. Prior to this, only insignificant electrical pressures were obtained, while I have reached fifty million volts.

The accompanying illustrations, with their descriptive titles, taken from an article I wrote for the "Century Magazine," ["The Problem of Increasing

Human Energy"] may serve to convey an idea of the results I obtained in the directions indicated.

Many persons in my own profession have wondered at them and have asked what I am trying to do. But the time is not far away now when the practical results of my labors will be placed before the world and their influence felt everywhere. One of the immediate consequences will be the transmission of messages without wires, over sea or land, to an immense distance. I have already demonstrated, by crucial tests, the practicability of signaling by my system from one to any other point of the globe, no matter how remote, and I shall soon convert the disbelievers.

I have every reason for congratulating myself that throughout these experiments, many of which were exceedingly delicate and hazardous, neither myself nor any of my assistants received any injury. When working with these powerful electrical oscillations the most extraordinary phenomena take place at times. Owing to some interference of the oscillations, veritable balls of fire are apt to leap out to a great distance, and if any one were within or near their paths, he would be instantly destroyed. A machine such as I have used could easily kill, in an instant, three hundred thousand persons. I observed that the strain upon my assistants was telling, and some of them could not endure the extreme tension of the nerves. But these perils are now entirely overcome, and the operation of such apparatus, however powerful, involves no risk whatever.

As I was improving my machines for the production of intense electrical actions, I was also perfecting the means for observing feeble effects. One of the most interesting results, and also one of great practical importance, was the development of certain contrivances for indicating at a distance of many hundred miles an approaching storm, its direction, speed and distance traveled. These appliances are likely to be valuable in future meteorological observations and surveying, and will lend themselves particularly to many naval uses.

It was in carrying on this work that for the first time I discovered those mysterious effects which have elicited such unusual interest. I had perfected the apparatus referred to so far that from my laboratory in the Colorado mountains I could feel the pulse of the globe, as it were, noting every electrical change that occurred within a radius of eleven hundred miles.

Terrified by Success

I can never forget the first sensations I experienced when it dawned upon me that I had observed something possibly of incalculable consequences to mankind. I felt as though I were present at the birth of a new knowledge or the revelation of a great truth. Even now, at times, I can vividly recall the incident, and see my apparatus as though it were actually before me. My first observations positively terrified me, as there was present in them something mysterious, not to say supernatural, and I was alone in my laboratory at night; but at that time the idea of these disturbances being intelligently controlled signals did not yet present itself to me.

The changes I noted were taking place periodically, and with such a clear suggestion of number and order that they were not traceable to any cause then known to me. I was familiar, of course, with such electrical disturbances as are produced by the sun, Aurora Borealis and earth currents, and I was as sure as I could be of any fact that these variations were due to none of these causes. The nature of my experiments precluded the possibility of the changes being produced by atmospheric disturbances, as has been rashly asserted by some. It was some time afterward when the thought flashed upon my mind that the disturbances I had observed might be due to an intelligent control. Although I could not decipher their meaning, it was impossible for me to think of them as having been entirely accidental. The feeling is constantly growing on me that I had been the first to hear the greeting of one planet to another. A purpose was behind these electrical signals; and it was with this conviction that I announced to the Red Cross Society, when it asked me to indicate one of the great possible achievements of the next hundred years, that it would probably be the confirmation and interpretation of this planetary challenge to us.

Since my return to New York more urgent work has consumed all my attention; but I have never ceased to think of those experiences and of the observations made in Colorado. I am constantly endeavoring to improve and perfect my apparatus, and just as soon as practicable I shall again take up the thread of my investigations at the point where I have been forced to lay it down for a time.

Communicating with the Martians

At the present stage of progress, there would be no insurmountable obstacle in constructing a machine capable of conveying a message to Mars, nor would there be any great difficulty in recording signals transmitted to us by the inhabitants of that planet, if they be skilled electricians. Communication once established, even in the simplest way, as by a mere interchange of numbers, the progress toward more intelligible communication would be rapid. Absolute certitude as to the receipt and interchange of messages would be reached as soon as we could respond with the number "four," say, in reply to the signal "one, two, three." The Martians, or the inhabitants of whatever planet had signaled to us, would understand at once that we had caught their message across the gulf of space and had sent back a response. To convey a knowledge of form by such means is, while very difficult, not impossible, and I have already found a way of doing it.

What a tremendous stir this would make in the world! How soon will it come? For that it will some time be accomplished must be clear to every thoughtful being.

Something, at least, science has gained. But I hope that it will also be demonstrated soon that in my experiments in the West I was not merely beholding a vision, but had caught sight of a great and profound truth.

Wind Power Should Be Used More Now
Philadelphia - North American — May 18

The power of the wind has been overlooked. Some day it will be forcibly brought to the position it deserves through the need of a substitute for the present method of generating power. Given a good breeze, I have estimated that there is as much as half a horse-power to every square foot of area exposed. Imagine what energy is left unused with all this force at hand.

The contrivance that has been at the disposal of mankind from all time, the windmill, is now seen in the rural districts only. The popular mind cannot grasp the power there is in the wind. Many a deluded inventor has spent years of his life in endeavoring to harness the tides, and some have even proposed to compress air by tide or wave power for supplying energy, never understanding the signs of the old windmill on the hill as it sorrowfully waves its arms about and bids them stop.

The fact is that the wave or tide motor would have but small chance of competing commercially with the windmill, which is by far the better machine, allowing a much greater amount of energy to be obtained in a simpler way.

Wind power has been in all times of inestimable value to man, if for nothing else than for enabling him to cross the seas, and it is even now a very important factor in transportation. But there are limitations in this simple method of utilizing the sun's energy. The machines are large for a given output and the power is intermittent, thus necessitating a storage of energy and increasing the cost of the plant. But there is no question as to its usefulness as a substitute for the energy derived from fuel, and the fact that this power is literally as free as air makes it a wonderful factor in the future of the world of industry.

Apart from the views expressed by Lord Kelvin regarding the future, when the coal supply shall have been exhausted, there is need of more attention being paid to it in the present day.

The man who cannot afford to have a furnace in his house may have a windmill on the roof. In this labor-saving age it is astonishing that farmers are the only citizens who call the wind their friend. Dwellers in cities toil up and down stairs hauling and carrying while above them is a good-natured giant who can do all this work for them if they will but force him into service. Why wait for the coal supply of the earth to be exhausted before enlisting the aid of this vast aerial force?

The power to run elevators, pump water to roof tanks, cool houses in the summer and heat them in the winter is above us, at any one's beck and call.

A little ingenuity will enable any householder to harness the wind and leave it to do the work that he has considered part of the curse of Adam.

Electrical oscillator activity ten million Horsepower Burning atmospheric nitrogen by high frequency discharges twelve million volts.
January 1. 1904

I wish to announce that in connection with the commercial introduction of my inventions I shall render professional services in the general capacity of consulting electrician and engineer.

The near future, I expect with confidence, will be a witness of revolutionary departures in the production, transformation and transmission of energy, transportation, lighting, manufacture of chemical compounds, telegraphy, telephony and other arts and industries.

In my opinion, these advances are certain to follow from the universal adoption of high-potential and high-frequency currents and novel regenerative processes of refrigeration to very low temperatures.

Much of the old apparatus will have to be improved, and much of the new developed, and I believe that while furthering my own inventions, I shall be more helpful in this evolution by placing at the disposal of others the knowledge and experience I have gained.

Special attention will be given by me to the solution of problems requiring both expert information and inventive resource — work coming within the sphere of my constant training and predilection.

I shall undertake the experimental investigation and perfection of ideas, methods and appliances, the devising of useful expedients and, in particular, the design and construction of machinery for the attainment of desired results.

Any task submitted to and accepted by me, will be carried out thoroughly and conscientiously.

Nikola Tesla
Laboratory, Long Island, N. Y.
Residence, Waldorf, New York City

The Transmission of Electrical Energy Without Wires
Electrical World and Engineer, March 5, 1904

It is impossible to resist your courteous request extended on an occasion of such moment in the life of your journal. Your letter has vivified the memory of our beginning friendship, of the first imperfect attempts and undeserved successes, of kindnesses and misunderstandings. It has brought painfully to my mind the greatness of early expectations, the quick flight of time, and alas! the smallness of realizations. The following lines which, but for your initiative, might not have been given to the world for a long time yet, are an offering in the friendly spirit of old, and my best wishes for your future success accompany them.

Towards the close of 1898 a systematic research, carried on for a number of years with the object of perfecting a method of transmission of electrical energy through the natural medium, led me to recognize three important necessities: First, to develop a transmitter of great power; second, to perfect means for individualizing and isolating the energy transmitted; and, third, to ascertain the laws of propagation of currents through the earth and the atmosphere. Various reasons, not the least of which was the help proffered by my friend Leonard E. Curtis and the Colorado Springs Electric Company, determined me to select for my experimental investigations the large plateau, two thousand meters above sea-level, in the vicinity of that delightful resort, which I reached late in May, 1899. I had not been there but a few days when I congratulated myself on the happy choice and I began the task, for which I had long trained myself, with a grateful sense and full of inspiring hope. The perfect purity of the air, the unequaled beauty of the sky, the imposing sight of a high mountain range, the quiet and restfulness of the place—all around contributed to make the conditions for scientific observations ideal. To this was added the exhilarating influence of a glorious climate and a singular sharpening of the senses. In those regions the organs undergo perceptible physical changes. The eyes assume an extraordinary limpidity, improving vision; the ears dry out and become more susceptible to sound. Objects can be clearly distinguished there at distances such that I prefer to have them told by someone else, and I have heard—this I can venture to vouch for—the claps of thunder seven and eight hundred kilometers away. I might have done better still, had it not been tedious to wait for the sounds to arrive, in definite intervals, as heralded precisely by an electrical indicating apparatus—nearly an hour before.

In the middle of June, while preparations for other work were going on, I arranged one of my receiving transformers with the view of determining in a novel manner, experimentally, the electric potential of the globe and studying its periodic and casual fluctuations. This formed part of a plan carefully mapped out in advance. A highly sensitive, self-restorative device, controlling a recording instrument, was included in the secondary circuit, while the primary was connected to the ground and an elevated terminal of adjustable capacity. The variations of potential gave rise to electric surgings in the primary; these generated secondary currents, which in turn affected the sensitive device and recorder in proportion to their intensity. The earth was found to be, literally, alive with electrical vibrations, and soon I was deeply absorbed in the interesting investigation. No better opportunities for such observations as I intended to make could be found anywhere. Colorado is a country famous for the natural displays of electric force. In that dry and rarefied atmosphere the sun's rays beat the objects with fierce intensity. I raised steam, to a dangerous pressure, in barrels filled with concentrated salt solution, and the tin-foil coatings of some of my elevated terminals shriveled up in the fiery blaze. An experimental high-tension transformer, carelessly exposed to the rays of the setting sun, had most of its insulating compound melted out and was rendered useless. Aided by the dryness and rarefaction of the air, the water evaporates as in a boiler, and static electricity is developed in abundance. Lightning discharges are, accordingly, very frequent and sometimes of inconceivable violence. On one occasion approximately twelve thousand discharges occurred in two hours, and all in a radius of certainly less than fifty kilometers from the laboratory. Many of them resembled gigantic trees of fire with the trunks up or down. I never saw fire balls, but as compensation for my disappointment I succeeded later in determining the mode of their formation and producing them artificially.

In the latter part of the same month I noticed several times that my instruments were affected stronger by discharges taking place at great distances than by those near by. This puzzled me very much. What was the cause? A number of observations proved that it could not be due to the differences in the intensity of the individual discharges, and I readily ascertained that the phenomenon was not the result of a varying relation between the periods of my receiving circuits and those of the terrestrial disturbances. One night, as I was walking home with an assistant, meditating over these experiences, I was suddenly staggered by a thought. Years ago, when I wrote a chapter of my lecture before the Franklin Institute and the National Electric Light Association, it had presented itself to me, but I dismissed it as absurd and impossible. I banished it again. Nevertheless, my instinct was aroused and somehow I felt that I was nearing a great revelation.

It was on the third of July—the date I shall never forget—when I obtained the first decisive experimental evidence of a truth of overwhelming importance for the advancement of humanity. A dense mass of strongly charged clouds gathered in the west and towards the evening a violent storm broke loose which, after spending much of its fury in the mountains, was driven away with great velocity over the plains. Heavy and long persisting arcs formed almost in regular time intervals. My observations were now greatly facilitated and rendered more accurate by the experiences already gained. I was able to handle my instruments quickly and I was prepared. The recording apparatus being properly adjusted, its indications became fainter and fainter with the increasing distance of the storm, until they ceased altogether. I was watching in eager expectation. Surely enough, in a little while the indications again began, grew stronger and stronger and, after passing through a maximum, gradually decreased and ceased once more. Many times, in regularly recurring intervals, the same actions were repeated until the storm which, as evident from simple computations, was moving with nearly constant speed, had retreated to a distance of about three hundred kilometers. Nor did these strange actions stop then, but continued to manifest themselves with undiminished force. Subsequently, similar observations were also made by my assistant, Mr. Fritz Lowenstein, and shortly afterward several admirable opportunities presented themselves which brought out, still more forcibly, and unmistakably, the true nature of the wonderful phenomenon. No doubt, whatever remained: I was observing stationary waves.

As the source of disturbances moved away the receiving circuit came successively upon their nodes and loops. Impossible as it seemed, this planet, despite its vast extent, behaved like a conductor of limited dimensions. The tremendous significance of this fact in the transmission of energy by my system had already become quite clear to me. Not only was it practicable to send telegraphic messages to any distance without wires, as I recognized long ago, but also to impress upon the entire globe the faint modulations of the human voice, far more still, to transmit power, in unlimited amounts, to any terrestrial distance and almost without loss.

With these stupendous possibilities in sight, and the experimental evidence before me that their realization was henceforth merely a question of expert knowledge, patience and skill, I attacked vigorously the development of my magnifying transmitter, now, however, not so much with the original intention of producing one of great power, as with the object of learning how to construct the best one. This is, essentially, a circuit of very high self-induction and small resistance which in its arrangement, mode of excitation and action, may be said to be the diametrical opposite of a transmitting circuit typical of telegraphy by Hertzian or electromagnetic radiations. It is difficult to form an adequate idea of the marvelous power of this unique appliance, by the aid of which the globe will be transformed. The electromagnetic radiations being reduced to an insignificant quantity, and proper conditions of resonance maintained, the circuit acts like an immense pendulum, storing indefinitely the energy of the primary exciting impulses and impressions upon the earth of the primary exciting impulses and impressions upon the earth and its conducting atmosphere uniform harmonic oscillations of intensities which, as actual tests have shown, may be pushed so far as to surpass those attained in the natural displays of static electricity.

Simultaneously with these endeavors, the means of individualization and isolation were gradually improved. Great importance was attached to this, for it was found that simple tuning was not sufficient to meet the vigorous practical requirements. The fundamental idea of employing a number of distinctive elements, co-operatively associated, for the purpose of isolating energy transmitted, I trace directly to my perusal of Spencer's clear and suggestive exposition of the human nerve mechanism. The influence of this principle on the transmission of intelligence, and electrical energy in general, cannot as yet be estimated, for the art is still in the embryonic stage; but many thousands of simultaneous telegraphic and telephonic messages, through one single conducting channel, natural or artificial, and without serious mutual interference, are certainly practicable, while millions are possible. On the other hand, any desired degree of individualization may be secured by the use of a great number of co-operative elements and arbitrary variation of their distinctive features and order of succession. For obvious reasons, the principle will also be valuable in the extension of the distance of transmission.

Progress though of necessity slow was steady and sure, for the objects aimed at were in a direction of my constant study and exercise. It is, therefore, not astonishing that before the end of 1899 I completed the task undertaken and reached the results which I have announced in my article in the Century Magazine of June, 1900, every word of which was carefully weighed.

Much has already been done towards making my system commercially available, in the transmission of energy in small amounts for specific purposes, as well as on an industrial scale. The results attained by me have made my scheme of intelligence transmission, for which the name of "World Telegraphy" has been suggested, easily realizable. It constitutes, I believe, in its principle of operation, means employed and capacities of application, a radical and fruitful departure from what has been done heretofore. I have no doubt that it will prove very efficient in enlightening the masses, particularly in still uncivilized countries and less accessible regions, and that it will add materially to general safety, comfort and convenience, and maintenance of peaceful relations. It involves the employment of a number of plants, all of which are capable of transmitting individualized signals to the uttermost confines of the earth. Each of them will be preferably located near some important center of civilization and the news it receives through any channel will be flashed to all points of the globe. A cheap and simple device, which might be carried in one's pocket, may then be set up somewhere on sea or land, and it will record the world's news or such special messages as may be intended for it. Thus the entire earth will be converted into a huge brain, as it were, capable of response in every one of its parts. Since a single plant of but one hundred horse-power can operate hundreds of millions of instruments, the system will have a virtually infinite working capacity, and it must needs immensely facilitate and cheapen the transmission of intelligence.

The first of these central plants would have been already completed had it not been for unforeseen delays which, fortunately, have nothing to do with its purely technical features. But this loss of time, while vexatious, may, after all, prove to be a blessing in disguise. The best design of which I know has been adopted, and the transmitter will emit a wave complex of total maximum activity of ten million horse-power, one per cent. of which is amply sufficient to "girdle the globe." This enormous rate of energy delivery. approximately twice that of the combined falls of Niagara, is obtainable only by the use of certain artifices, which I shall make known in due course.

For a large part of the work which I have done so far I am indebted to the noble generosity of Mr. J. Pierpont Morgan, which was all the more welcome and stimulating, as it was extended at a time when those, who have since promised most, were the greatest of doubters. I have also to thank my friend, Stanford White, for much unselfish and valuable assistance. This work is now far advanced, and though the results may be tardy, they are sure to come.

Meanwhile, the transmission of energy on an industrial scale is not being neglected. The Canadian Niagara Power Company have offered me a splendid inducement, and next to achieving success for the sake of the art, it will give me the greatest satisfaction to make their concession financially profitable to them. In this first power plant, which I have been designing for a long time, I propose to distribute ten thousand horse-power under a tension of one hundred million volts, which I am now able to produce and handle with safety.

This energy will be collected all over the globe preferably in small amounts, ranging from a fraction of one to a few horse-power. One its chief uses will be the illumination of isolated homes. I takes very little power to light a dwelling with vacuum tubes operated by high-frequency currents and in each instance a terminal a little above the roof will be sufficient. Another valuable application will be the driving of clocks and other such apparatus. These clocks will be exceedingly simple, will require absolutely no attention and will indicate rigorously correct time. The idea of impressing upon the earth American time is fascinating and very likely to become popular. There are innumerable devices of all kinds which are either now employed or can be supplied, and by operating them in this manner I may be able to offer a great convenience to whole world with a plant of no more than ten thousand horse-power. The introduction of this system will give opportunities for invention and manufacture such as have never presented themselves before.

Knowing the far-reaching importance of this first attempt and its effect upon future development, I shall proceed slowly and carefully. Experience has taught me not to assign a term to enterprises the consummation of which is not wholly dependent on my own abilities and exertions. But I am hopeful that these great realizations are not far off, and I know that when this first work is completed they will follow with mathematical certitude.

When the great truth accidentally revealed and experimentally confirmed is fully recognized, that this planet, with all its appalling immensity, is to electric currents virtually no more than a small metal ball and that by this fact many possibilities, each baffling imagination and of incalculable consequence, are rendered absolutely sure of accomplishment; when the first plant is inaugurated and it is shown that a telegraphic message, almost as secret and non-interferable as a thought, can be transmitted to any terrestrial distance, the sound of the human voice, with all its intonations and inflections, faithfully and instantly reproduced at any other point of the globe, the energy of a waterfall made available for supplying light, heat or motive power, anywhere-on sea, or land, or high in the air-humanity will be like an ant heap stirred up with a stick: See the excitement coming!

Letter from Nikola Tesla
New York Sun — November 27, 1904

My attention has been called to numerous comments on my letter, published in your issue of November 1, and relating to the electrical equipment of the newly opened catacomb in this city. Some of them are based on erroneous assumptions, which it is necessary for me to correct.

When I stated that my system was adopted, I did not mean that I originated every electrical appliance in the subway. For instance, the one which that ill fated electrician was repairing when he was killed, two days after the catacomb was ready for public use, was not invented by me. Nor was that other device on the sidetracked car, which, as will be remembered, caused the burning of two men. I also must deny any connection with that switch or contrivance which was responsible for the premature death of a man immediately afterward, as well as with that other, which cut short the life of his unfortunate successor. None of these funeral devices, I emphatically state, or any of the other which brought on collisions, delays and various troubles and were instrumental in the loss of arms and legs of several victims, are of my invention, nor do they form, in my opinion, necessary appurtenances of an intelligently planned scheme for the propulsion of cars. Referring to these contrivances, it is significant to read in some journals of the 8th inst. that a small firm failed because their bid was too low. This is indicative of keen competition and sharp cutting of prices, and does not seem in keeping with the munificence claimed for the Interborough Company.

I merely intended to say in my letter that my system of power transmission with three-phase generators and synchronous motor converters was adopted in the subway, the same as on the elevated road. I devised it many years ago for the express purpose of meeting the varied wants of a general electrical distribution of light and power. It has been extensively introduced all over the world because of its great flexibility, and under such conditions of use has been found of great value. But the idea of employing in this great city's main artery, in a case presenting such rigid requirements, this flexible system, offering innumerable chances for breakdowns, accidents and injuries to life and property, is altogether too absurd to dignify it with any serious comment. Here only my multiphase system, with induction motors and closed coil armatures - apparatus unfailing in its operation and minimizing the dangers of travel - should have been installed. Nothing, not even ignorance, will prevent its ultimate adoption; and the sooner the change is made the better it will be for all concerned. Personally, I have no financial or other interest in the matter, except that as a long resident of this city I would have been glad to see my inventions properly used to the advantage of the community. Under the circumstances I must forego this gratification.

The consequences of the unpardonable mistake of the Interborough Company are not confined to this first subway or even to this city. We are driven to travel underground. The elevated road is the eighth wonder, as colossal and imposing in the feature of public forbearance as the Pyramid of Cheops in its dimensions. Sooner or later all interurban railways must be transformed into subterranean. This will call for immense investments of capital, and if defective electrical apparatus is generally adopted the damage to life and property will be incalculable, not to speak of inconvenience to the public.

It seems proper to me to acknowledge on this occasion the painstaking suggestions of some friends of mine, mostly unknown to me, both in the large domain of electrical achievement and in the small sphere of my friendship, to again address the American Institute. It is customary with scientific men to present an original subject only once. I have done so and do not desire to depart from this established precedent. A lecture on the defects of the subway offers great opportunities, but would not be original. In view of certain insinuations I may cite a recently published statement of Mr. C. F. Scott, formerly president of the American Institute: 'n a matter of history it is the Tesla principle and the Tesla system which have been the directing factors in modern electrical engineering practice.' There are but a few men whose acknowledgment of my own work I would quote. Mr. Scott is one of them, as the man whose co-operation was most efficient in bringing about the great industrial revolution through these inventions. But the suggestions of my good friends have fallen on fruitful ground, and should it be possible for me to spare time and energy I may ask the city authorities for power to investigate the subway, and make a sworn report to them on all the defects and deficiencies I may discover, in the interest of public welfare.

A few more words in relation to the signs. With all due respect to general opinion, I entertain quite a different view on that subject. Advertising is a useful art, which is being lifted continually to a higher plane, and will soon be quite respectable. It should not be hampered, but rather encouraged. I would give the Interborough Company every facility for exploiting it, restricting it only in so far as the artistic execution is concerned. A commission of capable men comprising a painter, a sculptor, an architect, a literary man, an engineer and an executive business man might be appointed, to pass upon the merits of the signs submitted for acceptance. I do not see why the public should object to them if they were regulated in this manner. They will further business, make travel less tedious, and help many skillful artisans. The subways are bound to become municipal property, and the city will then derive a revenue from them. What is most important for the safety of life and property, quickness and security of travel, should be first considered. All this depends on the electrical equipment. The engineers have built a good tunnel, and proper apparatus should be installed to match it.

Nikola Tesla
New York, Nov. 26

Electric Autos
Manufacturers' Record — Dec. 29, 1904

Mr. Albert Phenis, Special Correspondent Manufacturers' Record, New York:
Dear Sir - Replying to your inquiry of yesterday, the application of electricity to the propulsion of automobiles is certainly a rational idea. I am glad to know that Mr. Lieb has undertaken to put it into practice. His long experience with the General Electric Co. and other concerns must have excellently fitted him for the task.

There is no doubt that a highly-successful machine can be produced on these lines. The field is inexhaustible, and this new type of automobile, introducing electricity between the prime mover and the wheels, has, in my opinion, a great future.

I have myself for many years advocated this principle. Your will find in numerous technical publications statements made by me to this effect. In my article in the Century, June, 1900, I said, in dealing with the subject: 'Steamers and trains are still being propelled by the direct application of steam power to shafts or axles. A much greater percentage of the heat energy of the fuel could be transformed in motive energy by using, in place of the adopted marine engines and locomotives, dynamos driven by specially designed high-pressure steam or gas engines, by utilizing the electricity generated for the propulsion. Again of 50 to 100 percent, in the effective energy derived from the fuel could be secured in this manner. It is difficult to understand why a fact so plain and obvious is not receiving more attention from engineers.

At first glance it may appear that to generate electricity by an engine and then apply the current to turn a wheel, instead of turning it by means of some mechanical connection with the engine, is a complicated and more or less wasteful process. But it is not so; on the contrary, the use of electricity in this manner secures great practical advantages. It is but a question of time when this idea will be extensively applied to railways and also to ocean liners, though in the latter case the conditions are not quite so favorable. How the railroad companies can persist in using the ordinary locomotive is a mystery. By providing an engine generating electricity and operating with the current motors under the cars a train can be propelled with greater speed and more economically. In France this has already been done by Heilman, and although his machinery was not the best, the results he obtained were creditable and encouraging. I have calculated that a notable gain in speed and economy can also be secured in ocean liners, on which the improvement is particularly desirable for many reasons. It is very likely that in the near future oil will be adopted as fuel, and that will make the new method of propulsion all the more commendable. The electric manufacturing companies will scarcely be able to meet this new demand for generators and motors.

In automobiles practically nothing has been done in this direction, and yet it would seem they offer the greatest opportunities for application of this principle. The question, however, is which motor to employ - the direct-current or my induction motor. The former has certain preferences as regards the starting and regulation, but the commutators and brushes are very objectionable on an automobile. In view of this I would advocate the use of the induction motor as an ideally simple machine which can never get out of order. The conditions are excellent, inasmuch as a very low frequency is practicable and more than three phases can be used. The regulation should offer little difficulty, and once an automobile on this novel plan is produced its advantages will be readily appreciated.

Yours very truly,
N. Tesla

The Transmission of Electrical Energy Without Wires as a Means for Furthering Peace
Electrical World and Engineer, January 7, 1905, pp. 21-24

Universal Peace, assuming it to be in the fullest sense realizable, might not require eons for its accomplishment, however probable this may appear, judging from the imperceptibly slow growth of all great reformatory ideas of the past. Man, as a mass in movement, is inseparable from sluggishness and persistence in his life manifestations, but it does not follow from this that any passing phase, or any permanent state of his existence, must necessarily be attained through a *stataclitic* process of development.

Our accepted estimates of the duration of natural metamorphoses, or changes in general, have been thrown in doubt of late. The very foundations of science have been shaken. We can no longer believe in the Maxwellian hypothesis of transversal ether-undulations of electrical vibrations, this most important field of human endeavor, particularly in the advancement of philanthropy and peace, was in no small measure retarded by that fascinating illusion, which I since long hoped to dispel. I have noted with satisfaction the first signs of a change of scientific opinion. The brilliant discovery of the exceptionally "radio-active" substances, radium and polonium, by Mrs. Sklodowska Curie, has likewise afforded me much personal gratification, being an eclatant confirmation of my early experimental demonstrations, of electrified radian streams of primary matter or corpuscular emanations (*Electrical Review*, New York, 1896-1897), which were then received with incredulity. They have awakened us from the poetical dream of an intangible conveyor of energy, weightless, structureless ether, to the plain, palpable reality of a ponderous medium of coarse particles, or bodily carriers of force. They have led us to a radically new interpretation of the changes and transformations we perceive. Enlightened by this recognition, we cannot say the sun is hot, the moon is cold, the star is bright, for all these might be purely electrical phenomena. If this be the case, then even our conceptions of time and space may have to be modified.

So, too, as regards the organic world, a similar revolution of thought is distinctly observable. In biological and zoological research the bold ideas of Haensel have found support in recent discoveries. A heretic belief in such possibilities as the artificial production of simple living material aggregates, the spontaneous natural creation of complex organisms and willful sex control, is gaining ground. We still brush it aside, but not with pedantic disdain as before. The fact is - our faith in the orthodox theory of slow evolution is being destroyed!

Thus a state of human life vaguely defined by the term "Universal Peace," while a result of cumulative effort through centuries past, might come into existence quickly, not unlike a crystal suddenly forms in a solution which has been slowly prepared. But just as no effect can precede its cause, so this state can never be brought on by any pact between nations, however solemn. Experience is made before the law is formulated, both are related like cause and effect. So long as we are clearly conscious of the expectation, that peace is to result from such a parliamentary decision, so long have we a conclusive evidence that we are not fit for peace. Only then when we shall feel that such international meetings are mere formal procedures, unnecessary except in so far as they might serve to give definite expression to a common desire, will peace be assured.

To judge from current events we must be, as yet, very distant from that blissful goal. It is true that we are proceeding towards it rapidly. There are abundant signs of this progress everywhere. The race enmities and prejudices are decidedly waning. A recent act of His Excellency, the President of the United States, is significant in this respect. We begin to think cosmically. Our sympathetic feelers reach out into the dim distance. The bacteria of the "Weltschmerz," are upon us. So far, however, universal harmony has been attained only in a single sphere of international relationship. That is the postal service. Its mechanism is working satisfactorily, but - how remote are we still from that scrupulous respect of the sanctity of the mail bag! And how much farther again is the next milestone on the road to peace - an international judicial service equally reliable as the postal!

The coming meeting at the Hague, now indefinitely postponed, can only consider temporary expedients. General disarmament being for the present entirely out of question, a proportionate reduction might be recommended. The safety of any country and of the world's commerce depending not on the absolute, but relative amount of war material, this would be evidently the first reasonable step to take towards universal economy and peace. But it would be a hopeless task to establish an equitable basis of adjustment. Population, naval strength, force of army, commercial importance, water-power, or any other natural resource, actual or prospective, are equally unsatisfactory standards to consider.

In view of this difficulty a measure suggested by Carnegie might be adopted by a few strong countries to scare all the weaker ones into peace. But while for the time being such a course may seem advisable, the beneficial effects of this homeopathic treatment of the martial disease could hardly be lasting. In the first place, a coalition of the leading powers could not fail to create an organized opposition, which might result in a disaster all the greater as it was long deferred. The ultimate falling out of the virtuous, peace-dictating nations, as certain as the law of gravitation, should be all the more reckoned with, as it would be extremely demoralizing. Again, it is by no means sufficient authority.

To conquer by sheer force is becoming harder and harder every day. Defensive is getting continuously the advantage of offensive, as we progress in the satanic science of destruction. The new art of controlling electrically the movements and operations of individualized automata at a distance without wires, will soon enable any country to render its coasts impregnable against all naval attacks. It is to be regretted, in this connection, that my proposal to the United States Navy four years ago, to introduce this invention, did not receive the least encouragement. Also that my offer to Secretary Long to establish telegraphic communication across the Pacific Ocean by my wireless system was thrown in the naval waste basket in Washington, quite *sans facon*. At that time I had already announced in *The Century Magazine* of June, 1900, my successful "girdling" of the globe with electrical impulses (stationary waves), and my "telautomata" had been publicly exhibited. But that was not the fault of the naval officials, for then these inventions of mine were decried as bald, visionary schemes, loudest indeed by those who have since become

Croesuses of Promise--in "light" storage batteries, "Ocean" telephony and "transatlantic" wireless telegraphy, yet remained to this day—Sisyphuses of Attainment. Had only a few "telautomatic" torpedoes been constructed and adopted by our navy, the mere moral influence of this would have been powerfully and most beneficially felt in the present Eastern Complication. Not to speak of the advantages which might have been secured through the direct and instantaneous transmission of messages to our distant colonies and scenes of the present barbarous conflicts. Since advancing that principle, I have invented a number of improvements, making it possible to direct such a torpedo, submersible at will, from a distance much greater than the range of the largest gun, with unerring precision, upon the object to be destroyed. What is still more surprising, the operator will not need to see the infernal engine or even know its location, and the enemy will be unable to interfere, in the slightest, with its movements by any electrical means. One of these devil-telautomata will soon be constructed, and I shall bring it to the attention of governments. The development of this art must unavoidably arrest the construction of expensive battleships as well as land fortifications, and revolutionize the means and methods of warfare. The distance at which it can strike, and the destructive power of such a quasi-intelligent machine being for all practical purposes unlimited, the gun, the armor of the battleship and the wall of the fortress, lose their import and significance. One can prophesy with a Daniel's confidence that skilled electricians will settle the battles of the near future. But this is the least. In its effect upon war and peace, electricity offers still much greater and more wonderful possibilities. To stop war by the perfection of engines of destruction alone, might consume centuries and centuries. Other means must be employed to hasten the end. What are these to be? Let us consider.

Fights between individuals, as well as governments and nations, invariably result from misunderstandings in the broadest interpretation of this term. Misunderstandings are always caused by the inability of appreciating one another's point of view. This again is due to the ignorance of those concerned, not so much in their own, as in their mutual fields. The peril of a clash is aggravated by a more or less predominant sense of combativeness, posed by every human being. To resist this inherent fighting tendency the best was is to dispel ignorance of the doings of others by a systematic spread of general knowledge. With this object in view, it is most important to aid exchange of thought and intercourse.

Mutual understanding would be immensely facilitated by the use of one universal tongue. But which shall it be, is the great question. At present it looks as if the English might be adopted as such, though it must be admitted that it is not the most suitable. Each language, of course, excels in some feature. The English lends itself to a terse, forceful expression of facts. The French is precise and finely distinctive. The Italian is probably the most melodious and easiest to learn. The Slavic tongues are very rich in sound but extremely difficult to master. The German is unequaled in the facility it offers for coining and combining words. A practical answer to that momentous question must perforce be found in times to come, for it is manifest that by adopting one common language the onward march of man would be prodigiously quickened. I do not believe that an artificial concoction, like Volapuk, will ever find universal acceptance, however time-saving it might be. That would be contrary to human nature. Languages have grown into our hearts. I rather look to the possibility of a reversion to the old Latin or Greek mother tongues, basing myself in this conclusion on the Spencerian law of rhythm [see Spencer's First Principles]. It seems unfortunate that the English-speaking nations, who are now fittest to rule the world, while endowed with extraordinary energy and practical intelligence, are singularly wanting in linguistic talent.

Next to speech we must consider permanent records of all kinds as a means for disseminating general information, or that knowledge of mutual endeavor which is chiefly conducive to harmony. Here the newspapers play by far the most important part. They are undoubtedly more effective than institutions of learning, libraries, museums and individual correspondence, all combined. The knowledge they convey is, on the whole, superficial and sometimes defective, but it is poured out in a mighty stream that reaches far and wide. Disregarding the force of electrical invention, that of journalism is the greatest in urging peace. Our schools are instrumental, mainly, in the furtherance of special thorough knowledge in our own fields, which is destructive of concordance. A world composed of crass specialists only would be perpetually at war. The diffusion of general knowledge through libraries and similar sources of information is very slow. As to individual correspondence, it is principally useful as an indispensable ingredient of the cement of commercial interest, that most powerful binding material between heterogeneous masses of humanity. It would be hard to overestimate the beneficial influence of the marvelous and precise art of photography, nor can that of other arts or means of recording be ignored. But a simple reflection will show that the peace-making force of all permanent, printed, printed or other records, resides not in themselves. It must be sought elsewhere. This is also true of speech.

Our senses enable us to perceive only a minute portion of the outside world. Our hearing extends to a small distance. Our sight is impeded by intervening bodies and shadows. To know each other we must reach beyond the sphere of our sense perceptions. We must transmit our intelligence, travel, transport the materials and transfer the energies necessary for our existence. Following this thought we now realize, forcibly enough to dispense with argument, that of all other conquests of man, without exception, that which is most desirable, which would be most helpful in the establishment of universal peaceful relations is - the complete *Annihilation of Distance*.

To achieve this wonder, electricity is the one and only means. Inestimable good has already been done by the use of this all powerful agent, the nature of which is still a mystery. Our astonishment at what has been accomplished would be uncontrollable were it not held in check by the expectation of greater miracles to come. That one, the greatest of all, can be viewed in three aspects: *Dissemination of intelligence, transportation, and transmission of power*.

Referring to the first, the present systems of telegraphic and telephonic communication are very limited in scope. The conducting channels are costly and of small working capacity. There is serious inductive disturbance, and storms render the service unsafe which, moreover, is too expensive: A vast improvement will be effected by placing the wires underground and insulating them artificially, by refrigeration. Their working capacity also could by indefinitely augmented by resorting to the new principle of "individualization," which I have more recently announced, permitting the simultaneous transmission of thousands of telegraphic and telephonic messages, without interference, over a single wire. The public would be already profiting from these great advances were it not for the stolid indifference of the leading companies engaged in the transmission of intelligence. But new concerns are springing into existence and the near future will witness a great transformation along these two lines of invention. The submarine cables are subject to still greater limitations. Some obstacles to rapid signaling, through them, seem insuperable. The attempts to overcome these have been numerous, but so far all have proved futile. The celebrated mathematician, O. Heaviside, and several able electricians following in his footsteps, have fallen into the singular error that rapid telegraphy and even telephony through ocean cables would be made practicable by the use of induction coils. Inductances might be to some extent helpful on comparatively short lines with thick paper insulation; on long lines insulated with rubber or gutta-percha they would be positively detrimental. Improvements will, undoubtedly, be made, but great electrostatic capacity and unavoidable loss of energy in the insulation and surrounding conductors will always restrict the usefulness of the transmission through artificial conductors is necessarily confined to a small number of stations.

It is therefore evident that the abolishment of all these drawbacks by the conveyance of signals or messages without wires, as I have undertaken in my "world" telegraphy and telephony, will be of the greatest moment in the furtherance of peace. The unifying influence of this advance will be felt all the more, as it will not only completely annihilate distance, but also make it possible to operate from a singe "world" telegraphy plant, an unlimited number of receiving stations distributed all over the globe, and with equal facility, irrespective of location. Within a few years a simple and inexpensive device, readily carried about, will enable one to receive on land or sea the principal news, to hear a speech, a lecture, a song or play of a musical instrument, conveyed from any other region of the globe. The invention will also meet the crying need for cheap transmission to great distances, more especially over the oceans. The small working capacity of the cables and the excessive cost of messages are now fatal impediments in the dissemination of intelligence which can only be removed by transmission without wires.

The deficiencies of Hertzian telegraphy have created in the public mind the impression that exclusive or private messages without the use of artificial channels are impracticable. As a matter of fact, nothing could be more erroneous. Ever since its first appearance in 1891, I have denied the commercial possibilities of the system of signaling by Hertzian or electromagnetic waves, and my forecasts have been fully confirmed. It lends itself little to tuning, still less to the higher artifices of "individualization," and transmission to considerable distances is wholly out of the question. Portentous claims for this method of communication were made three years ago, but they have been unable to stand the hard, cruel test of time. Moreover, I have recently learned through the leading British electrical journal (*Electrician*, London, February 27, 1903), that some experimenters have abandoned all their own and have been "converted" to my methods and appliances, without my approval and officiation. I was both astonished and pained - astonished at the nonchalance and lack of appreciation of these men, pained at the inability exhibited in the construction and use of my apparatus. My high hopes raised by that excellent journal, however, are still to be realized, for I have ascertained that His Majesty the King of England, His excellency the President of the United States, and other persons of exalted positions have, after all, not conferred upon me the imperishable honor of graciously condescending to the use for my coils, transformers and high-potential methods of transmission, but have exchanged their August greetings through the medium of a cable in the old-fashioned way. What has been actually achieved by Hertzian telegraphy can only be conjectured.

Quite different conditions exist in my system in which the electromagnetic waves or radiations are designedly minimized, the connection of one of the terminals of the transmitting circuit to the ground having, itself, the effect of reducing the energy of these radiations to about one-half. Under observance of proper rules and artifices the distance is of little or no consequence, and by skillful application of the principle of "individualization," repeatedly referred to the messages may be rendered both non-interfering and non-interferable. This invention, which I have described in technical publications, attempts to imitate, in a very crude way, the nervous system in the human body. It was the outcome of long-continued tests demonstrating the impossibility of satisfying rigorous commercial requirements by my earlier system, based on simple tuning, in which the selective quality is dependent on a single characteristic feature. In this later improvement the exclusiveness and non-interferability of impulses transmitted through a common channel result from cooperative association for a number of distinctive elements, and can be pushed as far as desired. In actual practice it is found that by combining only two vibrations or tones, a degree of privacy sufficient for most purposes is attained. When three vibrations are combined, it is extremely difficult, even for a skilled expert, to read or to disturb signals not intended for him, with four it is a vain undertaking. The probability of his getting the secret combinations at the right moments and in proper order, is much smaller than that of drawing an ambo, terno or quaterno, respectively, in a lottery. From experimental facts, I conclude that the invention will permit the simultaneous transmission of several millions of separately distinguishable messages through the earth, which, strangely enough, is in this respect much superior to an artificial conductor. This number ought to be sufficient to meet all the pressing necessities of intelligence transmission for at least one century to come. It is important to observe that but on "world" telegraphy plant, such as I am now completing, will have a greater working capacity than all the ocean cables combined. Once these facts are recognized this new art, which I am inaugurating, will sweep the world with the force of a uragan.

In transportation a great change is now going on. The trolley lines are being extended, the steam locomotive is making place for the electric motor. The ocean liners are adopting the turbine. Land travel is being improved by the automobile. The waterfalls are being harnessed and the energy used in the propulsion of cars. The advantages of first generating electricity by a prime mover, are being more and more appreciated. To the majority, this may appear a roundabout way of doing, but in reality it is as direct as the driving of a pulley from another by a belt. The idea is already being applied to railroads, and automobiles of this new type are making their appearance. The ocean vessels are bound to follow. An immense and virgin field will be thus opened up to the manufacturers of electric machinery. Effort towards saving time and money is characteristic of all modern methods of transportation. In many of these new developments, the artificial insulation of the high-tension mains by refrigeration will be very useful. However paradoxical, it is true, that by the use of this invention, power for all industrial purposes can be transmitted to distances of many hundreds of miles, not only without any loss, but with appreciable gain of energy. This is due to the fact that the conductor is much colder than the surrounding medium. The operativeness of this method is restricted to the use of a gaseous refrigerant, no known liquid permitting the attainment of a sufficiently low temperature of the transmission line. Hydrogen is by far the best cooling agent to employ. By its use electric railways can be extended to any desired distance. Owing to the smallness of ohmic loss, the objections to the multiphase system disappear and induction motors with closed coil armatures can be adopted. I find that even transmission through a submarine cable, as from Sweden to England, of great amounts of power is perfectly practicable. But the ideal solution of the problem of transportation will be arrived at only when the complete annihilation of distance in the transmission of power in large amounts shall have become a commercial reality. That day we shall invade the domain of the bird. When the vexing problem of aerial navigation, which has defied his attempts for ages, is solved, man will advance with giant strides.

That electrical energy can be economically transmitted without wires to any terrestrial distance, I have unmistakably established in numerous observations, experiments and measurements, qualitative and quantitative. These have demonstrated that is practicable to distribute power from a central plant in unlimited amounts, with a loss not exceeding a small fraction of one per cent, in the transmission, even to the greatest distance, twelve thousand miles - to the opposite end of the globe. This seemingly impossible feat can now be readily performed by any electrician familiar with the design and construction of my "high-potential magnifying transmitter," the most marvelous electrical apparatus of which I have knowledge, enabling the production of effects of unlimited intensities in the earth and its ambient atmosphere. It is, essentially, a freely vibrating secondary circuit of definite length, very high self-induction and small resistance, which has one of its terminals in intimate direct or inductive connection with the ground and the other with an elevated conductor, and upon which the electrical oscillations of a primary or exciting circuit are impressed under conditions of resonance. To give an idea of the capabilities of this wonderful appliance, I may state that I have obtained, by its means, spark discharges extending through more than one hundred feet and carrying currents of one thousand amperes, electromotive forces approximating twenty million volts, chemically active streamers covering areas of several thousand square feet, and electrical disturbances in the natural media surpassing those caused by lightning, in intensity.

Whatever the future may bring, the universal application of these great principles is fully assured, though it may be long in coming. With the opening of the first power plant, incredulity will give way to wonderment, and this to ingratitude, as ever before. The time is not distant when the energy of falling water will be man's life energy. So far only about million horse-power have been harnessed by my system of alternating-current transmission. This is little, but corresponds, nevertheless to the adding of sixty million indefatigable laborers, working virtually without food and pay, to the world's population. The projects which have come to my own attention, however, contemplate the exploitation of water powers aggregating something like one hundred and fifty million horse-power. Should they be carried out in a quarter of a century, as seems probable from present indications, there will be, on the average two such untiring laborers for every individual. Long before this consummation, coal and oil must cease to be important factors in the sustenance of human life on this planet. It should be borne in mind that electrical energy obtained by harnessing a waterfall is probably fifty times more effective than fuel energy. Since this is the most perfect way of rendering the sun's energy available, the direction of the future material development of man is clearly indicated. He will live on "white coal." Like a babe to the mothers's breast will he cling to his waterfall. "Give us our daily waterfall," will be the prayer of the coming generations. *Deus futurus est deus aquae deiectus!*

But the fact that stationary waves are producible in the earth is of special and, in many ways, still greater significance in the intellectual development of humanity. Popularly explained, such a wave is a phenomenon generically akin to an echo - a result of reflection. It affords a positive and incontrovertible experimental evidence that the electric current, after passing into the earth travels to the diametrically opposite region of the same and rebounding from there, returns to its point of departure with virtually undiminished force. The outgoing and returning currents clash and form nodes and loops similar to those observable on a vibrating cord. To traverse the entire distance of about twenty-five thousand miles, equal to the circumference of the globe, the current requires a certain time interval, which I have approximately ascertained. In yielding this knowledge, nature has revealed one of its most precious secrets, of inestimable consequence to man. So astounding are the facts in this connection, that it would seem as though the Creator, himself, had electrically designed this planet just for the purpose of enabling us to achieve wonders which, before my discovery, could not have been conceived by the wildest imagination. A full account of my discoveries and improvements will be given to the world in a special work which I am preparing. In so far, however, as they relate to industrial and commercial uses, they will be disclosed in patent specifications most carefully drawn.

As stated in a recent article (*Electrical World and Engineer*, March 5, 1904), I have been since some time at work on designs of a power plant which is to transmit ten thousand horse-power without wires. The energy is to be collected al over the earth at many places and in varying amounts. It should not be understood that the practical realization of this undertaking is necessarily far off. The plans could be easily finished this winter, and if some preliminary work on the foundations could be done in the meantime the plant might be ready for operation before the close of next fall. We would then have at our disposal a unique and invaluable machine. Just this one oscillator would advance the world a century. Its civilizing influence would be felt even by the humblest dweller in the wilderness. Millions of instruments of all kinds, for all imaginable purposes, could be operated from that one machine. Universal time could be distributed by simple inexpensive clocks requiring no attention and running with nearly mathematical precision. Stock-tickers, synchronous movements and innumerable devices of this character could be worked in unison all over the earth. Instruments might be provided for indicating the course of a vessel at sea, the distance traversed, the speed, the hour at any particular place, the latitude and longitude. Incalculable commercial advantages could be thus secured and countless accidents and disasters avoided. Here and there a house might be lighted or some other work requiring a few horse-power performed. What is far more important than this, flying machines might be driven in any part of the world. They could be made to travel swiftly because of their small weight and great motive power. My intention would be to utilize this first plant rather as means of enlightenment, to collect its power in very small amounts, and at as many places as possible. The knowledge that there is throbbing through the earth energy readily available everywhere, would exert a strong stimulus on students, mechanics and inventors of all countries. This would be productive of infinite good. Manufacture would receive a fresh and powerful incentive. Conditions, such as never existed before in commerce, would be brought about. Supply would be ever inadequate to demand. The industries of iron, copper, aluminum, insulated wire and many others, could not fail to derive great and lasting benefits from this development.

The economic transmission of power without wires is of all-surpassing importance to man. By its means he will gain complete mastery of the air, the sea and the desert. It will enable him to dispense with the necessity of mining, pumping, transporting and burning fuel, and so do away with innumerable causes of sinful waste. By its means, he will obtain at any place and in any desired amount, the energy of remote waterfalls - to drive his machinery, to construct his canals, tunnels and highways, to manufacture the materials of his want, his clothing and food, to heat and light his home - year in, year out, ever and ever, by day and by night. It will make the living glorious sun his obedient, toiling slave. It will bring peace and harmony on earth.

Over five years have elapsed since that providential lightning storm on the 3d of July, 1899, of which I told in the article before mentioned, and through which I discovered the terrestrial stationary waves; nearly five years since I performed the great experiment which on that unforgettable day, the dark God of Thunder mercifully showed me in his vast, awe-sounding laboratory. I thought then that it would take a year to establish commercially my wireless girdle around the world. Alas! my first "world telegraphy" plant is not yet completed, its construction has progressed but slowly during the past two years. And this machine I am building is but a plaything, an oscillator of a maximum activity of only ten million horse-power, just enough to throw this planet into feeble tremors, by sign and word - to telegraph and to telephone. When shall I see completed that first power plant, that big oscillator which I am designing! From which a current stronger than that of a welding machine, under a tension of one hundred million volts, is to rush through the earth! Which will deliver energy at the rate of one thousand million horse-power - one hundred Falls of Niagara combined in one, striking the universe with blows - blows that will wake from their slumber the sleepiest electricians, if there be any, on Venus or Mars! . . . It is not a dream, *it is a simple feat of scientific electrical engineering*, only expensive - blind, faint-hearted, doubting world! . . . Humanity is not yet sufficiently advanced to be willingly led by the discover's keen searching sense. But who knows? Perhaps it is better in this present world of ours that a revolutionary idea or invention instead of being helped and patted, be hampered and ill-treated in its adolescence - by want of means, by selfish interest, pedantry, stupidity and ignorance; that it be attacked and stifled; that it pass through bitter trials and tribulations, through the heartless strife of commercial existence. So do we get our light. So all that was great in the past was ridiculed, condemned, combated, suppressed - only to emerge all the more powerfully, all the more triumphantly from the struggle.

Tesla on Subway Dangers
New York Sun, June 16, 1905

The flooding of the subway is a calamity apt to repeat itself. As your readers will remember, it did not occur for the first time last Sunday. Water, like fire, will break loose occasionally in spite of precautions. It will never be possible to guard against a casual bursting of a main; for while the conduit can be safely relied upon under normal working conditions, any accidental obstruction to the flow may cause a pressure which no pipe or joint can withstand.

In fact, if we are to place faith in the gloomy forecasts of Commissioner Oakley, who ought to know, such floods may be expected to happen frequently in the future. In view of this it seems timely to call to public attention a danger inherent to the electrical equipment which has been thrust upon the Interborough Company by incompetent advisers.

The subway is bound to be successful, and would be so if the cars were drawn by mules, for it is the ideal means of transportation in crowded cities. But the full measure of success of which it is capable will be attained only when the financiers shall say to the electric companies: "Give us the best, regardless of expense."

It is to be regretted that this important pioneering enterprise, in other respects ably managed and engineered, should have been treated with such gross neglect in its most vital feature. No opportunity was given to myself, the inventor and patentee of the system adopted in the subway and the elevated roads, for offering some useful suggestion, nor was a single electrician or engineer of the General Electric and Westinghouse companies consulted, the very men who should have been thought of first of all.

Once large sums of money are invested in a defective scheme it is difficult to make a change, however desirable it may be. The movement of new capital is largely determined by previous investment. Even the new roads now planned are likely to be equipped with the same claptrap devices, and so the evil will grow. *"Das eben ist der Fluch der boesen Thut, das sie fortzeugend Boeses muss gebaeren."*

The danger to which I refer lies in the possibility of generating an explosive mixture by electrolytic decomposition and thermic dissociation of the water through the direct currents used in the operation of the cars. Such a process might go on for hours and days without being noticed; and with currents of this kind it is scarcely practicable to avoid it altogether.

It will be recalled that an expert found the percentage of free oxygen in the subway appreciably above that which might reasonably have been expected in such a more or less stagnated channel. I have never doubted the correctness of that analysis and have assumed that oxygen is being continuously set free by stray currents passing through the moist ground. The total amperage of the normal working current in the tunnel is very great, and in case of flooding would be sufficient to generate not far from 100 cubic feet of hydrogen per minute. Inasmuch, however, as in railway operation the fuses must be set hard, in order to avoid frequent interruption of the service by their blowing out, in such an emergency the current would be of . much greater volume and hydrogen would be more abundantly liberated.

It is a peculiar property of this gas that it is capable of exploding when mixed with a comparatively large volume of air, and any engineer can convince himself by a simple calculation that, say, 100,000 cubic feet of explosive might be formed before the danger is discovered, reported and preventive measures taken. What the effect of such an explosion might be on life and property is not pleasant to contemplate. True, such a disaster is not probable, but the present electrical equipment makes it possible, and this possibility should be, by all means, removed.

The oppressiveness of the tunnel atmosphere is in a large measure due to the heat supplied by the currents, and to the production of nitrous acid in the arcs, which is enhanced by rarefaction of the air through rapid motion. Some provision for ventilation is imperative. But ventilation will not do away with the danger I have pointed out. It can be completely avoided only by discarding the direct current.

I should say that the city authorities, for this if for no other reason, should forbid its use by a proper act of legislation. Meanwhile, the owners of adjacent property should object to its employment, and the insurance companies should refuse the grant of policies on such property except on terms which it may please them to make.

N. Tesla

Tesla's Reply to Edison
English Mechanic and World of Science — July 14, 1905

As we said last week, Mr. Edison was reported to have said in an interview of the *New York World* that he did not believe with Tesla in being able to talk round the world, but that he thought Marconi would, sooner or later, perfect his system.

Nikola Tesla has replied. He says:
In the course of certain investigations which I carried on for the purpose of studying the effects of lightning discharges upon the electrical condition of the earth I observed that sensitive receiving instruments arranged so as to be capable of responding to electrical disturbances created by the discharges at times failed to respond when they should have done so, and upon inquiring into the causes of this unexpected behavior I discovered it to be due to the character of the electrical waves which were produced in the earth by the lightning discharges, and which had nodal regions following at definite distances the shifting source of the disturbances. From data obtained in a large number of observations of the maxima and minima of these waves I found their length to vary approximately from twenty-five to seventy kilometers, and these results and theoretical deductions led me to the conclusion that waves of this kind may be propagated in all directions over the globe, and that they may be of still more widely differing lengths, the extreme limits being imposed by the physical dimensions and properties of the earth. Recognizing in the existence of these waves an unmistakable evidence that the disturbances created had been conducted from their origin to the most remote portions of the globe, and had been thence reflected, I conceived the idea of producing such waves in the earth by artificial means, with the object of using them for many useful purposes for which they are or might be found applicable.

Beat Lightning Flashes

This problem was rendered extremely difficult, owing to the immense dimensions of the planet, and consequently enormous movement of electricity or rate at which electrical energy had to be delivered in order to approximate, even in a remote degree, movements or rates which are manifestly attained in the displays of electrical forces in nature, and which seemed at first unrealizable by any human agencies; but by gradual and continuous improvements of a generator of electrical oscillations, which I have described in my Patents Nos. 645,576 and 649,621, I finally succeeded in reaching electrical movements or rates of delivery of electrical energy not only approximately, but, as shown in comparative tests and measurements, actually surpassing those of lightning discharges and by means of this apparatus I have found it possible to reproduce, whenever desired, phenomena in the earth the same as or similar to those due to such discharges. With the knowledge of the phenomena discovered by me, and the means at command for accomplishing these results, I am enabled, not only to carry out many operations by the use of known instruments, but also to offer a solution for many important problems involving the operation or control of remote devices which, for want of this knowledge and the absence of these means, have heretofore been entirely impossible. For example, by the use of such a generator of stationary waves and receiving apparatus properly placed and adjusted in any other locality, however remote, it is practicable to transmit intelligible signals, or to control or actuate at will any one or all of such apparatus for many other important and valuable purposes, as for indicating whenever desired the correct time of an observatory, or for ascertaining the relative position of a body or distance of the same with reference to the given point, or for determining the course of a moving object, such as a vessel at sea, the distance traversed by the same or its speed; or for producing many other useful effects at a distance dependent on the intensity, wavelength, direction or velocity of movements, or other feature or property of disturbances of this character.

A Bit of Sarcasm

Permit me to say on this occasion that if there exist to-day no facilities for wireless telegraphic and telephone communication between the most distant countries, it is merely because a series of misfortunes and obstacles have delayed the consummation of my labors, which might have been completed three years ago. In this connection I shall well remember the efforts of some, unwise enough to believe that they can gain an advantage by throwing sand in the eyes of the people and retarding the progress of invention. Should the first messages across the seas prove calamitous to them, it will be a punishment regrettable but fully deserved.

Tesla on the Peary North Pole Expedition
New York Sun — July 16, 1905

To the Editor of the New York Sun:

Everybody must have been pleased to learn that Commodore Peary has finally obtained the financial assistance which will enable him to start without further delay on his important journey. Let us wish the bold navigator the most complete success in his perilous undertaking, in the interest of humanity as well as for his own and his companions' sake and the gratification of the generous donors who have aided him. But, while voicing these sentiments, let us hope that Peary's will be the last attempt to reach the pole in this slow, penible and hazardous way.

We have already sufficiently advanced in the knowledge of electricity and its applications to avail ourselves of better means of transportation, enabling us to reach and to explore without difficulty and in a more perfect manner not only the North, but also the South Pole, and any other still unknown regions of the earth's surface. I refer to the facilities afforded in this respect by the transmission of electrical energy without wires and aerial navigation, which has found in the novel art its ideal solution.

Many of your readers will, no doubt, be under the impression that I am speaking merely of possibilities. As a matter of fact, from the principles involved and the experiments which I have actually performed, not only is the practical success of such distribution of power reduced to a degree of mathematical certitude, but the transmission can be-effected with an economy much greater than possible by the present method involving the use of wires.

It would not take long to build a plant for purposes of aerial navigation and geographical research, nor would it cost as much as might be supposed. Its location would be perfectly immaterial. It might be at the Niagara, or at the Victorian Falls in Africa, without any appreciable difference in the power collected in a flying machine or other apparatus.

A popular error, which I have often opportunity to correct, is to believe that the energy of such a plant would dissipate itself in all directions. This is not so, as I have pointed out in my technical publications. Electricity is displaced by the transmitter in all directions, equally through the earth and the air; that is true, but energy is expended only at the place where it is collected and used to perform some work. To illustrate, a plant of 10,000 hp, such as I have been planning, might be running full blast at Niagara, and there might be but one flying machine, of, say, 50 hp operating in some distant place, the location being of absolutely no consequence. In this case 50 hp would be all the power furnished by the plant to the rest of the universe. Although the electrical oscillations would manifest themselves all over the earth, at the surface as well as high in the air, virtually no power would be consumed. My experiments have shown that the entire electrical movement which keeps the whole globe a-tremble can be maintained with but a few horsepower. Apart from the transmitting and receiving apparatus, the only loss incurred is the energy radiated in the form of Hertzian or electro-magnetic waves, which can be reduced to any entirely insignificant quantity.

I appreciate the difficulty which your non-technical readers must experience in comprehending the working of this system. To gain a rough idea, let them imagine the transmitter and the earth to be two elastic bags, one very small and the other immense, both being connected by a tube and filled with some incompressible fluid. A pump is provided for forcing the fluid from one into the other, alternately and in rapid succession. Now, to produce a great movement of the fluid in a bag of such enormous size as the earth would require a pump so large that it would be a greater task to construct it than to build a thousand Egyptian pyramids. But there is a way of accomplishing this with a pump of very small dimensions. The bag connected to the earth is elastic, and when suddenly struck vibrates at a certain rate. The first artifice consists in so designing and adjusting the parts that the natural vibrations of the bag are in synchronism with the strokes of the pump. Under such conditions the bag is set into violent vibrations, and the fluid is made to rush in and out with terrific force. But the immense bag - the earth, is still comparatively undisturbed. Its size, however, does not exempt it from the laws of nature, and just as the small bag, so too the earth, responds to certain impulses. This fact I discovered in 1899.

The second artifice is to so adjust the transmitter that it will furnish these particular impulses. When all is properly done the large bag is thrown into spasms of vibration, and the effects are bewildering. But no power is yet transmitted, and all this colossal movement requires little energy to maintain. It is like an engine running without load.

Next let your readers imagine that at any place where it may be desired to deliver energy a small elastic bag, not unlike the first, is connected to the large one through a tube. The third artifice consists in so proportioning the parts that the attachment will be responsive to the impulse transmitted, this resulting in a great intensification of the vibration of the bag. Still the pump will not furnish power until these vibrations are made to do work of some kind.

To conduce to an understanding of the fourth artifice, that of "individualization," let your readers follow me a step further, and conceive the flow of energy to any point can be controlled from the place where the pump is located at will, and with equal facility and precision, regardless of distance, and, furthermore, through a device such as the combination lock of a safe, they will then have a crude idea of the processes involved. But only when they realize that all these and many other processes not mentioned, and related to one another like the links of a chain, are completed in a fraction of a second, will your readers be able to appreciate the magical potencies of electrical vibrations and form a conception of the miracles which a skilled electrician can perform by the use of these appliances.

I earnestly hope that in the near future the conditions will be favorable for the construction of a plant such as I have proposed. As soon as this is done it will be possible to adapt electrical motors to flying machines of the type popularized by Santos Dumont. There will be no necessity of carrying a generator or store of motive energy and consequently the machine will be much lighter and smaller. Owing to this and also to the greater power available for propulsion, the speed will be considerably increased. But a few of such machines, properly equipped with photographic and other appliances, will be sufficient to give us in a short time an exact knowledge of the entire earth's surface. It should be borne in mind, however, that for the ordinary uses of a single person a very small machine of not more than one-quarter horse-power, corresponding to the work of two men, would be amply sufficient so that when the first plant of 10,000 hp is installed, the commodity of aerial flight can be offered to a great many individuals all the world over. I can conceive of no improvement which would be more efficient in the furtherance of civilization than this.

N. Tesla

Signaling to Mars —A Problem of Electrical Engineering
Harvard Illustrated — March, 1907

In the early part of 1900, still vividly impressed by certain observations, I had made shortly before, and feeling that the time had come to prepare the world for an experiment which will soon be undertaken, I dwelt on the practicability of interplanetary signaling in an article which appeared in the June number of *Century Magazine* of the same year. In order to correct an erroneous report which gained wide circulation, a statement was published in *Collier's Weekly* of Feb. 9, 1901, defining my position in general terms. Ever since, my thoughts have been centered on the subject, and my original conviction has been strengthened both by reflection and suggestion.

Chief among the stimulating influences was the revelatory work of Percival Lowell, described in a volume with which the observatory, bearing his name, has honored me. No one can look at his globe of Mars without a feeling of profound astonishment, if not awe. These markings, still imperfectly discerned and incomprehensible, but evidently intended for a useful purpose, may they not contain a record of deep meaning left by a superior race, perhaps extinct, to tell its young brethren in other worlds of secrets discovered, of life and struggle, of their own terrible fate? What mighty pathos and love in such a gigantic drama of the universe: But let us hope that the astronomer has seen true, that Mars is not a cold grave, but the abode of happy intelligent creatures, from whom we may learn. In the light of this glorious possibility, signaling to that planet presents itself as a preeminently practical proposition which, to carry out, no human sacrifice could be too great. Can it be done? What chance is there that it will be done?

These questions will be answered definitely the moment all doubt as to the existence of highly developed beings on Mars is dispelled. The straightness of the lines on Lowell's map, their uniform width and other geometrical peculiarities, do not, themselves, appeal to me as strong proofs of artificiality. I should think that a planet large enough not to be frozen stiff in a spasm of volcanic action, like our moon, must, in the course of eons, have all its mountains leveled, the valleys filled, the rocks ground to sand, and ultimately assume the form of a smooth spheroid, with all its rivers flowing in geodetically straight lines. The uniform width of the waterways can be consistently explained, their crossings, however odd and puzzling, might be accidental. But I quite agree with Professor Morse, that this whole wonderful map produces the absolute and irresistible conviction, that these "canals" owe their existence to a guiding intelligence. Their great size is not a valid argument to the contrary. It would merely imply that the Martians have harnessed the energy of waterfalls. We know of no other source of power competent to explain such tremendous feats of engineering. They could not be accomplished by capturing the sun's rays or abstracting heat imparted to the atmosphere, for this, according to our best knowledge, would require clumsy and inefficient machinery. Large falls could be obtained near the polar caps by extensive dams. While much less effective than our own, they could well furnish several billions of horse-power. It should be borne in mind that many Martian tasks in mechanical engineering are much easier than the terrestrial, on account of the smaller mass of the planet and lesser density, which, in the superficial layers, may be considerably below the mean. To a still greater degree this is true of electrical engineering. Taking into account the space encompassed by Mars, a system of wireless transmission of energy, such as I have perfected, would be there much more advantageously applied, for, under similar conditions, a receiving circuit would collect sixteen times as much energy as on the earth.

The astonishing evidences furnished by Lowell are not only indicative of organic life, but they make it appear very probable that Mars is still populated; and furthermore, that its inhabitants are highly developed intelligent beings. Is there any other proof of such existence? I answer, emphatically, yes, prompted both by an instinct which has never yet deceived me, and observation. I refer to the strange electrical disturbances, the discovery of which I announced six years ago. At that time I was only certain that they were of planetary origin. Now, after mature thought and study, I have come to the positive conclusion that they must emanate from Mars.

Life, as a great philosopher has said, is but a continuous adjustment to the environment. Similar conditions must bring forth similar automata. We can have no idea what a Martian might be like, but he certainly has sensitive organs, much as our own, responsive to external stimuli. The indications of these instruments must be real and true. A straight line, a geometrical figure, a number, must convey to his mind a clear and definite conception. He ought to think and reason like ourselves. If he breathes, eats and drinks, he is moved by motives and desires not very different from our own. Such colossal transformation as is observable on the face of Mars could not have been wrought except by beings ages ahead of us in development. What wonder, then, if they have maps of this, our globe, as perfect as Professor Pickering's photographs of the moon? What wonder if they are signaling to us? We are sufficiently advanced in electrical science to know that their task is much easier than ours. The question is, can we transmit electrical energy to that immense distance? This I think myself competent to answer in the affirmative.

N. Tesla

Tuned Lightning
English Mechanic and World of Science, March 8, 1907

I read with interest an article in the Sunday World of Jan. 20 on "Tuned Lightning," described as a mysterious new energy, which is to turn every wheel on earth, and is supposed to have been recently discovered by the Danish inventors Waldemar Poulsen and P. O. Pederson.

From other reports I have gathered that these gentlemen have so far confined themselves to the peaceful production of miniature bolts not many inches long, and I am wondering what an account of their prospective achievements would read like if they had succeeded in obtaining, like myself, electrical discharges of 100 ft., far surpassing lightning in some features of intensity and power.

In view of their limited Jovian experience, the program outlined by the Danish engineers is rather extensive, Lord Armstrong's vast resources notwithstanding. Naturally enough, I shall look with interest to their telephoning across the Atlantic, supplying light and propelling airships without wires. *Anch in suito pittore.* (I, too, am a painter.) In the mean time it may not be amiss to state here incidentally that all the essential processes of and appliances for the generation, transmission, transformation, distribution, storage, regulation, control, and economic utilization of "tuned lightning" have been patented by me, and that I have long since undertaken, and am sparing no effort to render these advances instrumental in insuring the welfare, comfort, and convenience, primarily, of my fellow citizens.

There is nothing remarkable in the demonstration reported to have been made before Sir William Preece and Prof. Sylvanus P. Thompson, nor is there any novelty in the electrical devices employed. The lighting of arc lamps through the human body, the fusing of a piece of copper in mid-air, as described, are simple experiments which by the use of my high-frequency transformers any student of electricity can readily perform. They teach nothing new, and have no bearing on wireless transmission, for the actions virtually cease at a distance of a few feet from the source of vibratory energy. Years ago I gave exhibitions of similar and other much more striking experiments with the same kind of apparatus, many of which have been illustrated and explained in technical journals. The published records are open to inspection.

Regardless of all that, the Danish inventors have not as yet offered the slightest proof that their expectations are realizable, and before advancing seriously the claim that an efficient wireless distribution of light and power to great distances is possible, they should, at least, repeat those of my experiments which have furnished this evidence.

A scientific audience cannot help being impressed by a display of interesting phenomena, but the originality and significance of a demonstration such as that referred to can only be judged by an expert possessed of full knowledge and capable of drawing correct conclusions. A novel effect, spectacular and surprising, might be quite unimportant, while another, seemingly trifling, is of the greatest consequence.

To illustrate, let me mention here two widely different experiments of mine. In one the body of a person was subjected to the rapidly-alternating pressure of an electrical oscillator of two and a half million volts; in the other a small incandescent lamp was lighted by means of a resonant circuit grounded on one end, all the energy being drawn through the earth electrified from a distant transmitter.

The first presents a sight marvelous and unforgettable. One sees the experimenter standing on a big sheet of fierce, blinding flame, his whole body enveloped in a mass of phosphorescent wriggling streamers like the tentacles of an octopus. Bundles of light stick out from his spine. As he stretches out the arms, thus forcing the electric fluid outwardly, roaring tongues of fire leap from his fingertips. Objects in his vicinity bristle with rays, emit musical notes, glow, grow hot. He is the center of still more curious actions, which are invisible. At each throb of the electric force myriads of minute projectiles are shot off from him with such velocities as to pass through the adjoining walls. He is in turn being violently bombarded by the surrounding air and dust. He experiences sensations which are indescribable.

A layman, after witnessing this stupendous and incredible spectacle, will think little of the second modest exhibit. But the expert will not be deceived. He realizes at once that the second experiment is ever so much more difficult to perform and immensely more consequential. He knows that to make the little filament glow, the entire surface of the planet, two hundred million square miles, must be strongly electrified. This calls for peculiar electrical activities, hundreds of times greater than those involved in the lighting of an arc lamp through the human body. What impresses him most, however, is the knowledge that the little lamp will spring into the same brilliancy anywhere on the globe, there being no appreciable diminution of the effect with the increase of distance from the transmitter.

This is a fact of overwhelming importance, pointing with certitude to the final and lasting solution of all the great social, industrial, financial, philanthropic, international, and other problems confronting humanity, a solution of which will be brought about by the complete annihilation of distance in the conveyance of intelligence, transport of bodies and materials, and the transmission of the energy necessary to man's existence. More light has been thrown on this scientific truth lately through Prof. Slaby's splendid and path-breaking experiment in establishing perfect wireless telephone connection between Naum and Berlin, Germany, a distance of twenty miles. With apparatus properly organised such telephonic communication can be effected with the same facility and precision at the greatest terrestrial distance.

The discovery of the stationary terrestrial waves, showing that, despite its vast extent, the entire planet can be thrown into resonant vibration like a little tuning fork; that electrical oscillations suited to its physical properties and dimensions pass through it unimpeded, in strict obedience to a simple mathematical law, has proved beyond the shadow of a doubt that the earth, considered as a channel for conveying electrical energy, even in such delicate and complex transmissions as human speech or musical composition, is infinitely superior to a wire or cable, however well designed.

Very soon it will be possible to talk across an ocean as clearly and distinctly as across a table. The first practical success, already forecast by Slaby's convincing demonstration, will be the signal for revolutionary improvements which will take the world by storm.

However great the success of the telephone, it is just beginning its evidence of usefulness. Wireless transmission of speech will not only provide new but also enormously extend existing facilities. This will be merely the forerunner of ever so much more important development, which will proceed at a furious pace until, by the application of these same great principles, the power of waterfalls can be focused whenever desired; until the air is conquered, the soil fructified and embellished; until, in all departments of human life distance has lost its meaning, and even the immense gulf separating us from other worlds is bridged.

Tesla's Wireless Torpedo
New York Times — March 20, 1907

To the Editor of The *New York Times*:

A report in the *Times* of this Morning says that I have attained no practical results with my dirigible wireless torpedo. This statement should be qualified. I have constructed such machines, and shown them in operation on frequent occasions. They have worked perfectly, and everybody who saw them was amazed at their performance.

It is true that my efforts to have this novel means for attack and defense adopted by our Government have been unsuccessful, but this is no discredit to my invention. I have spent years in fruitless endeavor before the world recognized the value of my rotating field discoveries which are now universally applied. The time is not yet ripe for the telautomatic art. If its possibilities were appreciated the nations would not be building large battleships. Such a floating fortress may be safe against an ordinary torpedo, but would be helpless in a battle with a machine which carries twenty tons of explosive, moves swiftly under water, and is controlled with precision by an operator beyond the range of the largest gun.

As to projecting wave-energy to any particular region of the globe, I have given a clear description of the means in technical publications. Not only can this be done by the use of my devices, but the spot at which the desired effect is to be produced can be calculated very closely, assuming the accepted terrestrial measurements to be correct. This, of course, is not the case. Up to this day we do not know a diameter of the globe within one thousand feet. My wireless plant will enable me to determine it within fifty feet or less, when it will be possible to rectify many geodetical data and make such calculations as those referred to with greater accuracy.

Nikola Tesla
New York, March 19, 1907

Wireless on Railroads
New York Times — March 26, 1907

To the Editor of The New York Times:
No argument is needed to show that the railroads offer opportunities for advantageous use of a practical wireless system. Without question, its widest field of application is the conveyance to the trains of such general information as is indispensable for keeping the traveler in touch with the world. In the near future a telegraphic printer of news, a stock ticker, a telephone, and other kindred appliances will form parts of the regular wireless equipment of a railroad train. Success in this sphere is all the more certain, as the new is not antagonistic, but, on the contrary, very helpful to the old. The technical difficulties are minimized by the employment of a transmitter the effectiveness of which is unimpaired by distance.
In view of the great losses of life and property, improved safety devices on the cars are urgently needed. But upon careful investigation it will be found that the outlook in this direction is not very promising for the wireless art. In the first place the railroads are rapidly changing to electric motive power, and in all such cases the lines become available for the operation of all sorts of signaling apparatus, of which she telephone is by far the most important. This valuable improvement is due to Prof. J. Paley, who introduced it in Germany eight years ago. By enabling the engineer or conductor of any train to call up any other train or station along the track and obtain full and unmistakable information, the liability of collisions and other accidents will be greatly reduced. Public opinion should compel the immediate adoption of this invention.
Those roads which do not contemplate this transformation might avail themselves of wireless transmission for similar purposes, but inasmuch as every train will require in addition to a complete outfit an expert operator, many roads may prefer to use a wire, unless a wireless telephone can be offered to them.
Nikola Tesla
New York, March 25, 1907

Nikola Tesla Objects
New York Times — May 2, 1907

To the Editor of The New York Times:
I have been much surprised to read in The Times of Sunday, April 21, that Admiral H. N. Manney, U.S.N. attributes a well-known invention of mine, a process for the production of continuous electrical oscillations by means of the electric arc and condenser, to Valentine Poulsen, the Danish engineer. This improvement has been embodied by me in numerous forms of apparatus identified with my name; and I have described it minutely in patents and scientific articles. To quote but one of many references, I may mention my experimental lecture on "Light and Other High-Frequency Phenomena," published under the auspices of the Franklin Association, for which both of these societies have distinguished me.

I share with Admiral Manney in the gratification that we are in the lead, and particularly that wireless messages have been transmitted from Pensacola to Point Lorne. Inasmuch, however, as this feat could not have been accomplished except by the use of some of my own devices, it would have been a graceful act on his part to bring this feat to the attention of the wireless conference. My theory has always been that military men are superior to civilians in courtesy. I have not been discouraged by the refusal of our Government to adopt my wireless system six years ago, when I offered it, not by the unpleasant prospect of my passing through the experiences described by Mark Twain in his story of the beef contract, but *I* see no reason why I should be deprived of a well-earned honor and satisfaction.

The Times has hurt me grievously; not by accusing me of commercialism, nor by its unkind editorial comments on those letters I wrote, in condemnation of my system of power transmission in the Subway. It is another injury, perhaps, unintentional, which I have felt most keenly.

The editor of The Times may not have known that I am a student of applied mathematics when he permitted a fellow student of mine to insinuate in The Times of March 28 that I avail myself of inventions of others. I cannot permit such ideas to gain ground in this community, and, just to illuminate the situation, I shall quote from the leading electrical paper, *The London Electrician*, referring to some wireless plants of Braun and Marconi: "The spark occurs between balls in the primary circuit of a Tesla coil. The air wire is in series with a Tesla transformer. The generating plant is virtually a Poldhu in miniature. Evidently Braun, like Marconi, has been converted to the high-potential methods introduced by Tesla." Needless to add that this substitution of the old, ineffective Hertzian appliances for my own has not been authorized by me.

My fellow-student can rest assured that I am scrupulously respecting the rights of others. If I were not prompted to do so by a sense of fairness and pride I would be by the power I have of inventing anything I please.

N. Tesla
New York, April 30, 1907

Tesla's Tidal Wave to Make War Impossible
English Mechanic and World of Science — May 3, 1907

Just at this time, when all efforts towards peaceful arbitration notwithstanding, the nations are preparing to expend immense sums in the design and construction of monstrous battleships, it may be useful to bring to the attention of the general public a singular means for naval attack and defense, which the telautomatic art has made possible, and which is likely to become a deciding factor in the near future.

A few remarks on this invention, of which the wireless torpedo is but a special application, are indispensable to the understanding and full*appreciation of the naval principle of destruction.

The telautomatic art is the result of endeavors to produce an automaton capable of moving and acting as if possessed of intelligence and distinct individuality. Disconnected from its higher embodiment, an organism, such as a human being, is a heat - or thermodynamic engine - comprising:- (1) a complete plant for receiving, transforming, and supplying energy; (2) apparatus for locomotion and other mechanical performance; (3) directive organs; and (4) sensitive instruments responsive to external influences, all these parts constituting a whole of marvelous perfection.

The ambient medium is alive with movement and energy, in a state of unceasing agitation which is beyond comprehension. Strangely enough, to most of this terrible turmoil the human machine is insensible. The automaton does not feel the weight of the atmosphere crushing him with a force of 16 tons. He is unaffected by the shower of particles shooting through his body of cloud and the hurricane of finer substance rushing through him with the speed of light. He is unconscious that he is being whisked through space at the fearful rate of 70,000 miles an hour. But when gentle waves of light or sound strike him his eye and ear respond, his resonant nervefibres transmit the vibrations and his muscles contract and relax. Thus, like a float on a turbulent sea, swayed by external influences, he moves and acts. The average person is not aware of this constant dependence on his environment; but a trained observer has no difficulty in locating the primary disturbance which prompts him into action, and continued exercise soon satisfies him that virtually all of his purely mechanical motions are caused by visual impressions, directly or indirectly received.

Using the Principles of Human Action

A machine of such inconceivable complexity as the body of an organised being, capable of an infinite variety of actions, with controlling organs supersensitive, responsive to influences almost immaterial, cannot be manufactured by man; but the mechanical principles involved in the working of the living automaton are also applicable to an inanimate engine, however crude.

An automobile boat was first employed to carry out the idea. Its storage battery and motor furnished the power; the propeller and rudder, respectively, served as locomotive and directive organs, and a very delicate electrical device, actuated by a circuit tuned to a distant transmitter, took the place of the ear. This mechanism followed perfectly the wireless signals or comments of the operator in control of the transmitter, performing every movement and action as if it had been gifted with intelligence.

The next step was to individualize the machine. The attunement of the controlling circuits gave-it a special feature, but this was not sufficiently distinctive. An individuality implies a number of characteristic traits which, though perhaps extant elsewhere, are unique in that particular combination. Here again the animated automaton, with its nerve-signal system, was coarsely imitated. The action of the delicate device - the ear - was made dependent on a number of sensitized receiving circuits, each recognizable by its own free vibrations, and all together by the character of their operative combination. Correspondingly the transmitter was designed to emit a wave-complex exactly matching the combination in the number and pitch of individual vibrations, their groupment and order of succession.

Wonders of the New Telautomaton

That much is done, but more is to come. A mechanism is being perfected which without operator in control, left to itself, will behave as if endowed with intelligence of its own. It will be responsive to the faintest external influences and from these, unaided, determine its subsequent actions as if possessed of selective qualities, logic, and reason. It will perform the duties of an intelligent slave. Many of us will live to see Bulwer's dream realized.

The reader for whom the preceding short explanation of this novel art is intended may think it simple and easy of execution, but it is far from being so. It has taken years of study and experiment to develop the necessary methods and apparatus, and five inventions, all more or less fundamental and difficult to practice, must be employed to operate successfully and individualized telautomaton.

Such a novel engine of war - a vessel of any kind, submarine or aerial - carrying an agent of unlimited potency of destruction, with no soul aboard, yet capable of doing all it is designed for, as if fully equipped with a fearless crew in command of its captain, must needs bring on a revolution in the present means of attack and defense.

Since ages human ingenuity has been bent upon inventing infernal machines. Of these the modern cannon has been so far the most remarkable. A 12 in. gun charged with cordite is said to hurl a projectile of 850 lb. with the initial velocity of nearly 2,900 ft. per second, imparting to it the energy of 110,000,000 ft. lb. Were it not for the resistance of the air such a projectile would travel about fifty miles before striking the ground. It would take 3,300 H. P. more than a minute to accumulate its mechanical energy. Bear in mind, however, that all this energy is imparted to the projectile while it is being urged through the gun-barrel with a mean force of 1,100 tons. If the barrel is 50 ft. long and the average velocity through it 1,500 ft. per second, the whole energy is transferred to a moving mass in 1/30th of a second; hence the rate of performance is 1,800 times the above-that is about 6,000,000 H.P. This seems wonderful indeed, but is nothing as compared with rates obtained by other means. Electricity can be stored in the form of explosive energy of a violence against which the detonation of cordite is but a breath. With a magnifying transmitter as diagrammatically illustrated, rates of 25,000,000

H. P. have already been obtained. A similar and much improved machine, now under construction, will make it possible to attain maximum explosive rates of over 800,000,000 H.P., twenty times the performance of the Dreadnought's broadside of eight 12 in. guns simultaneously fired. These figures are so incredible that astronomers unacquainted with the marvelous appliance have naturally doubted the practicability of signaling to Mars. In reality, by its means the seemingly visionary project has been reduced to a rational engineering problem.

The time is not far distant when all the tremendous wastes of war will be stopped, and then, if there are battles, they will be fought with water-power and electrical waves. That humanity is moving fast towards this realization is evident from many indications.

What is most to be regretted in the present war regime is that the effort of so many exquisite intelligences must be uneconomically applied, since it cannot be entirely governed by the wavering struggle of opposing principles. This feverish striving to meet the instant demand, to create type after type, one to devour the other, to merge into one contrasting element, leads, like a nightmare, from one to another absurdity. Such a monstrosity is the latest creation of the naval constructor - a 20,000 ton battleship. In theory it is condemned by competent authorities.

Everything points to the development of a small vessel with internal combustion engines, extreme speed, and few weapons of great destructiveness. But the new leviathan is admirably adapted to the practical requirements of the day. In attack it could alone annihilate a nation's fleet. It is equally effective in defense. If equipped with proper acoustic and electrical appliances it has little to fear from a submarine, and an ordinary torpedo will scarcely hurt it. That is why the first of these monsters, built in England, has been name Dreadnought. Now, there is a novel means for attacking a fortress of this kind, from shore or on the high seas, against which all its gun-power and armor resistance are of no avail. It is the tidal wave.

What the Tesla Tidal Wave Will do

Such a wave can be produced with twenty or thirty tons of cheap explosive, carried to its destination and ignited by a non-interferable telautomaton.

The tidal disturbance, as here considered, is a peculiar hydrodynamic phenomenon, in many respects different from the commonly occurring, characterized by a rhythmical succession of waves. It consists generally of but a single advancing swell succeeded by a hollow, the water if not otherwise agitated being perfectly calm in front and very nearly so behind. The wave is produced by some sudden explosion or upheaval, and is, as a rule, asymmetrical for a large part of its course. Those who have encountered a tidal wave must have observed that the sea rises rather slowly, but the descent into the trough is steep. This is due to the fact that the water is lifted, possibly very slowly, under the action of a varying force, great at first, but dying out quickly, while the raised mass is urged downward by the constant force of gravity. When produced by natural causes these waves are not very dangerous to ordinary vessels, because the disturbance originated at a great depth.

To give a fairly accurate idea of the efficacy of this novel means of destruction, particularly adapted for the coast defense, it may be assumed that thirty tons of nitro-glycerine compound, as dynamite, be employed to create the tidal disturbance. This material, weighing about twice as much as water, can be stored in a cubical tank 8 ft. each way, or a spherical vessel of 10 ft. diameter. The reader will now understand that this charge is to be entrusted to a non-interferable telautomaton, heavily protected, and partly submerged or submarine, which is under perfect control of a skilled operator far away. At the propitious moment the signal is given, the charge sunk to the proper depth and ignited.

The water is incompressible. The hydrostatic pressure is the same in all directions. The explosion propagates through the compound with a speed of three miles a second. Owing to all this, the whole mass will be converted into gas before the water can give way appreciably, and a spherical bubble 10 ft. in diameter will form. The gaseous pressure against the surrounding water will be 20,000 atmospheres, or 140 tons per square inch. When the great bubble has expanded to twice its original volume it will weigh as much as the water it displaces, and from that moment on, its lower end tapering more and more into a cone, it will be driven up with a rapidly-increasing force tending towards 20,000 tons. Under the terrific impulsion it would shoot up the surface like a bullet were it not for the water resistance, which will limit its maximum speed to 80 ft. per second.

Consider not the quantity and energy of the upheaval. The caloric potential energy of the compound is 2,800 heat units per pound, or, in mechanical equivalent, almost 1,000 ft.-tons. The entire potential energy of the explosive will thus be 66,000,000 ft.-tons. Of course, only a part of this immense store is transformable into mechanical effort. Theoretically, 40 lb. of good smokeless powder would be sufficient to impart to the Dreadnought's 850 lb. projectile the tremendous velocity mentioned above, but it actually takes a charge of 250 lb. The tidal wave generator is a dynamic transformer much superior to the gun, its greatest possible efficiency being as high as 44 per cent. Taking, to be conservative, 38 per cent, instead, there will be the total potential store about 25 million foot-tons obtained in mechanical energy.

How the Enemy Would Be Engulfed

Otherwise stated, 25,000,000 tons - that is, 860,000,000 cu. ft. of water, could be raised 1 ft., or a smaller quantity to a correspondingly greater elevation. The height and length of the wave will be determined by the depth at which the disturbance originated. Opening in the center like a volcano, the great hollows will belch forth a shower of ice. Some sixteen seconds later a valley of 600 ft. depth, counted from normal ocean level, will form, surrounded by a perfectly circular swell, approximately of equal height, which will enlarge in diameter at the rate of about 220 ft. per second.

It is futile to consider the effect of such an eruption on a vessel situated near by, however large. The entire navy of a great country, if massed around, would be destroyed. But it is instructive to inquire what such a wave could do to a battleship of the Dreadnought type at considerable distance from it origin. A simple calculation will show that when the outer circle has expanded to three-quarters of a mile, the swell, about 1,250 ft. long, would still be more than 100 ft. in height, from crest to normal sea level, and when the circle is one and one-quarter mile in diameter the vertical distance from crest to trough will be over 100 ft.

The first impact of the water will produce pressures of three tons per square foot, which all over the exposed surface of, say, 20,000 sq. ft., may amount to 60,000 tons, eight times the force of the recoil of the broadside. That first impact may in itself be fatal. During more than ten seconds the vessel will be entirely submerged and finally dropped into the hollow from a height of about 75 ft., the descent being effected more or less like a free fall. It will then sink far below the surface, never to rise.

Mr. Tesla on the Wireless Transmission of Power
N. Y. World May 19, 1907

To the Editor of The World:

I have enjoyed very much the odd prediction of Sir Hugh Bell, President of the Iron and Steel Institute, with reference to the wireless transmission of power, reported in The World of the 10th inst.

With all the respect due to that great institution I would take the liberty to remark that if its President is a genuine prophet he must have overslept himself a trifle. Sir Hugh would honor me if he would carefully peruse my British patent No. 8,200, in which I have recorded some of my discoveries and experiments, and which may influence him to considerably reduce his conservative estimate of one hundred years for the fulfillment of his prophecy.

Personally, basing myself on the knowledge of this art to which I have devoted my best energies, I do not hesitate to state here for future reference and as a test of accuracy of my scientific forecast that flying machines and ships propelled by electricity transmitted without wire will have ceased to be a wonder in ten years from now. I would say five were it not that there is such a thing as "inertia of human opinion" resisting revolutionary ideas.

It is idle to believe that because man is endowed with higher attributes his material evolution is governed by other than general physical laws. If the genius of invention were to reveal to-morrow the secret of immortality, of eternal beauty and youth, for which all humanity is aching, the same inexorable agents which prevent a mass from changing suddenly its velocity would likewise resist the force of the new knowledge until time gradually modifies human thought.

What has amused me still more, however, is the curious interview with Lewis Nixon, the naval contractor, printed in the World of the 11th inst. Is it possible that the famous designer of the Oregon is not better versed in editorial matters than some of my farming neighbors of Shoreham? One cannot escape that conviction.

We are not in the dark as regards the electrical energy contained in the earth. It is altogether too insignificant for any industrial use. The current circulating through the globe is of enormous volume but of small tension, and could perform but little work. Beside, how does Nixon propose to coax the current from the natural path of low resistance into an artificial channel of high resistance? Surely he knows that water does not flow up hill. It is absurd of him to compare the inexhaustible dynamic energy of wind with the magnetic energy of the earth, which is minute in amount and in a static condition.

The torpedo he proposes to build is not novel. The principle is old. I could refer him to some of my own suggestions of nine years ago. There are many practical difficulties in the carrying out of the idea, and as much better means for destroying a submarine are available it is doubtful that such a torpedo will ever be constructed.

Nixon has failed to grasp that in my wireless system the effect does not diminish with distance. The Hertz waves have nothing to do with it except that some of my apparatus may be used in their production. So too a Kohinoor might be employed to cut window-glass. And yet, the seeming paradox can be easily understood by any man of ordinary intelligence.

Imagine only that the earth were a hollow shell or reservoir in which the transmitter would compress some fluid, as air, for operating machinery in various localities. What difference would it make when this reservoir is tapped to supply the compressed fluid to the motor? None whatever, for the pressure is the same everywhere. This is also true of my electrical system, with all considerations in its favor. In such a mechanical system of power distribution great losses are unavoidable and definite limits in the quality of the energy transmitted exist. Not so in the electrical wireless supply. It would not be difficult to convey to one of our liners, say, 50,000 horsepower from a plant located at Niagara, Victoria or other waterfall, absolutely irrespective of location. In fact, there would not be a difference of more than a small fraction of one per cent, whether the source of energy be in the vicinity of the vessel or 12,000 miles away, at the antipodes.

Nikola Tesla
New York, May 16, 1907.

Can Bridge the Gap to Mars
New York Times — June 23, 1907

To the Editor of the New York Times:
You have called me an "inventor of some useful pieces of electrical apparatus." It is not quite up to my aspirations, but I must resign myself to my prosaic fate. I cannot deny that you are right.

Nearly four million horse power of waterfalls are harnessed by my alternating current system of transmission, which is like saying that one hundred million men untiring, consuming nothing, receiving no pay - are laboring to provide for one hundred million tons of coal annually. In this great city the elevated roads, the subways, the Street railways are operated by my system, and the lamps and other electrical appliances get the current through machinery of my invention. And as in New York so all the world over where electricity is introduced. The telephone and incandescent lamp fill specific and minor demands, electric power meets the many general and sterner necessities of life. Yes, I must admit, however reluctantly, the truth of your unflattering contention.

But the greater commercial importance of this invention of mine is not the only advantage I have over my celebrated predecessors in the realm of the useful, who have given us the telephone and the incandescent lamp. Permit me to remind you that I did not have, like Bell, such powerful help as the Reis telephone, which reproduced music and only needed a deft turn of an adjusting screw to repeat the human voice; or such vigorous assistance as Edison found in the incandescent lamps of King and Starr, which only needed to be made of high resistance. Not at all. I had to cut the path myself, and my hands are still sore. All the army of my opponents and detractors was ever able to drum up against me in a fanatic contest has simmered down to a short article by an Italian - Prof. Ferraris - dealing with an abstract and meaningless idea of a rotating magnetic pole and published years after my discovery, months even after my complete disclosure of the whole practically developed system in all its essential universally adopted features. It is a publication, pessimistic and discouraging, devoid of the discoverer's virility and force, devoid of results, utterly wanting in the faith and devotion of the inventor, a defective and belated record of a good but feeble man whose only response to my whole-soled brother greeting was a plaintive cry of priority - a sad contrast to the strong and equanimous Schallenberger, a true American engineer, who stoically bore the pain that killed him.

A fundamental discovery or original invention is always useful, but it is often more than that. There are physicists and philosophers to whom the marvelous manifestations of my rotating magnetic field, the suggestive phenomena of rotation without visible connection, the ideal beauty of my induction motor with its contactless armature, mean quite as much as the thousands of millions of dollars invested in enterprises of which it is the foundation.

And this is true of all my discoveries, inventions, and scientific results which I have since announced, for I have never invented what immediate necessity suggested, but what I found as most desirable to invent, irrespective of time. Let me tell you only of one - my "magnifying transmitter," a machine with which I have passed a current of one hundred amperes around the globe, with which I can make the whole earth loudly repeat a word spoken in the telephone, with which I can easily bridge the gulf which separates us from Mars. Do you mean to say that my transmitter is nothing more than a "useful piece of electrical apparatus"?

I do not wish to enlarge on this for obvious reasons. To be compelled by taciturn admirers to dwell on my own achievements is hurting my delicate sensibilities, but as I observe your heroic and increasing efforts in praising your paper, while your distinguished confreres maintain on its merits a stolid silence, I feel that there is, at least, one man in New York able to appreciate the incongeniality of the correspondence. Allow me to ask you just one or two questions in regard to a work which I began in 1892, inspired by a high tribute from Lord Rayleigh at the Royal Institution, most difficult labor which I have carried on for years, encouraged by the sympathetic interest and approval of Hemholtz, Lord Kelvin, and my great friends, Sir William Crookes and Sir James Dewar, ridiculed by small men whose names I have seen displayed in vulgar and deceptive advertisements. I refer to my system of wireless transmission of energy.

The principles which it involves are eternal. We are on a conducting body, insulated in space, of definite and unchangeable dimensions and properties. It will never be possible to transmit electrical energy economically through this body and its environment except by essentially the same means and methods which I have discovered, and the system is so perfect now that it admits of but little improvement. Since I have accepted as true your opinion, which I hope will not be shared by posterity, would you mind telling a reason why this advance should not stand worthily beside the discoveries of Copernicus? Will you state why it should not be ever so much more important and valuable to the progress and welfare of man?

We could still believe in the geocentric theory and yet advance virtually as we do. The work of the astronomer would suffer, for some of his deductions would rest on erroneous assumptions. But, after all, we shall never know the intimate nature of things. So long as our perceptions are accurate our logic will be true. No one can estimate to what an extent the great knowledge he conveyed has been instrumental in developing the power of our minds and furthering discovery and invention. Yet, it has left all the pressing material problems confronting us unsolved.

Now my wireless system offers practical solutions for all. The aerial navigation, which now agitates the minds, is only one of its many and obvious applications equally consequential. The waterfalls of this country alone, its greatest wealth, are adequate to satisfy the wants of humanity for thousands of years to come. Their energy can be used with the same facility to dig the Panama Canal as to operate the Siberian Railway or to irrigate and fertilize the Sahara. The Anglo-Saxon race has a great past and present, but its real greatness is in the future, when the water power it owns or controls shall supply the necessities of the entire world.

As to universal peace - if there is nothing in the order of nature which makes war indispensable to the safe and sane progress of man, if that utopian existence is at all possible, it can be only attained through this very means, for all international friction can be traced to but one cause - the immense extension of the planet. My system of wireless transmission completely annihilates distance in all departments of human activity.

If this does not appeal to you sufficiently to recognize in me a discoverer of principles, do me, at least, the justice of calling me an "inventor of some beautiful pieces of electrical apparatus."

Nikola Tesla
New York, June 21, 1907

Sleep From Electricity
New York Times — Oct. 19, 1907

To the Editor of The New York Times:
I have read with interest the reports in The Times of the 13th and 15th inst. referring to Prof. Leduc's discovery of causing sleep by electric means. While it is possible that he has made a distinct advance there is no novelty in the effect itself.

The narcotic influence of certain periodic currents was long ago discovered by me and has been pointed out in some of my technical publications, among which I may mention a paper on "High Frequency Oscillators for Electro Therapeutic and Other Purposes," read before the American Electro Therapeutic Association, Sept. 13, 1898. I have also shown that human tissues offer little resistance to the electric flow and suggested an absolutely painless method of electrocution by passing the currents through the brain. It is very likely that Prof. Leduc has taken advantage of the same general principles though he applies the currents in a different manner.

In one respect, however, my observations are at variance with those reported. From the special dispatch in The Times of the 13th inst. it would appear that sleep is induced the moment the currents are turned on, and that awakening follows as soon as the electrodes are withdrawn. It is, of course, impossible to tell how strong a current was employed, but the resistance of the head might have been, perhaps, 3,000 ohms, so that at thirty volts the current could have been only about 1-100 of an ampere. Now, I have passed a current of at least 5,000 times stronger through my head and did not lose consciousness, but I invariably fell into a lethargic sleep some time after. This fact impresses me with certain arguments of Prof. Barker of Columbia University in your issue of Sept. 15.

I have always been convinced that electric anaesthesia will become practical, but the application of currents to the brain is so delicate and dangerous an operation that the new method will require long ark careful experimentation before it ,can be used with certitude.

Nikola Tesla
New York, Oct. 16, 1907

Possibilities of "Wireless"
New York Times — Oct. 22, 1907

To the Editor of The New York Times:
In your issue of the 19th inst. Edison makes statements which cannot fail to create erroneous impressions.

There is a vast difference between primitive Hertzwave signaling, practicable to but a few miles, and the great art of wireless transmission of energy, which enables an expert to transmit, to any distance, not only signals, but power in unlimited amounts, and of which the experiments across the Atlantic are a crude application. The plants are quite inefficient, unsuitable for finer work, and totally doomed to an effect less than one percent of that I attained in my test in 1899.

Edison thinks that Sir Hiram Maxim is blowing hot air. The fact is my Long Island plant will transmit almost its entire energy to the antipodes, if desired. As to Martin's communication I can only say, that I shall be able to attain a wave activity of 800,000,000 horse power and a simple calculation will show, that the inhabitants of that planet, if there be any, need not have a Lord Raleigh to detect the disturbance.

Referring to your editorial comment of even date, the question of wireless interference is puzzling only because of its novelty. The underlying principle is old, and it has presented itself for consideration in numerous forms. Just now it appears in the novel aspects of aerial navigation and wireless transmission. Every human effort must of necessity create a disturbance. What difference is there in essence, between the commotion produced by any revolutionary idea or improvement and that of a wireless transmitter? The specter of interference has been conjured by Hertzwave or radio telegraphy in which attunement is absolutely impossible, simply because the effect diminishes rapidly with distance. But to my system of energy transmission, based on the use of impulses not sensibly diminishing with distance, perfect attunement and the higher artifice of individualization are practicable. As ever, the ghost will vanish with the wireless dawn.

Nikola Tesla
New York, Oct. 21, 1907.

Tesla on Wireless
New York Daily Tribune — Oct. 25, 1907

To the Editor of The Tribune:
Sir: In so far as wireless art is concerned there is a vast difference between the great inventor Thomas A. Edison and myself, integrally in my favor. Mr. Edison knows little of the theory and practice of electrical vibrations; I have, in this special field, probably more experience than any of my contemporaries. That you are not as yet able to impart your wisdom by wireless telephone to some subscriber in any other part of the world, however remote, and that the presses of your valuable paper are not operated by wireless power is largely due to your own effort and those of some of your distinguished confreres of this city, and to the efficient assistance you have received from my celebrated colleagues, Thomas A. Edison and Michael Pupin, assistant consulting wireless engineers. But it was all welcome to me. Difficulty develops resource.

The transmission across the Atlantic was not made by any device of Mr. Marconi's, but by my system of wireless transmission of energy, and I have already given notice by cable to my friend Sir James DeWar and the Royal Institution of this fact. I shall also request some eminent man of science to take careful note of the whole apparatus, its mode of operation, dimensions, linear and electrical, all constants and qualitative performance, so as to make possible its exact reproduction and repetition of the experiments. This request is entirely impersonal. I am a citizen of the United States, and I know that the time will come when my busy fellow citizens, too absorbed in commercial pursuits to think of posterity, will honor my memory. A measurement of the time interval taken in the passage of the signal necessary to the full and positive demonstration will show that the current crosses the ocean with a mean speed of 625,000 miles a second.

The Marconi plants are inefficient, and do not lend themselves to the practice of two discoveries of mine, the "art of individualization," that makes the message non-interfering and non-interferable, and the "stationary waves," which annihilate distance absolutely and make the whole earth equivalent to a conductor devoid of resistance. Were it not for this deficiency, the number of words per minute could be increased at will by "individualizing."

You have already commented upon this advance in terms which have caused me no small astonishment, in view of your normal attitude. The underlying principle is to combine a number of vibrations, preferably slightly displaced, to reduce further the danger of interference, active and passive, and to make the operation of the receiver dependent on the co-operative effect of a number of attuned elements. Just to illustrate what can be done, suppose that only four vibrations were isolated on each transmitter. let those on one side be respectively a, b, c, and d. Then the following individualized lines would be ab, ac, ad, bc, bd, cd, abc, abd, acd, bcd and abcd. The same article on the other side will give similar combinations, and both together twenty-two lines, which can be simultaneously operated. To transmit one thousand words a minute, only forty-six words on each combination are necessary. If the plants were suitable, not ten years, as Edison thinks, but ten hours would be necessary to put this improvement into practice. To do this Marconi would have to construct the plants, and it will then be observed that the indefatigable Italian has departed from universal engineering customs for the fourth time.
Nikola Tesla
New York, Oct. 24, 1907

My Apparatus, Says Tesla
New York Times — Dec. 20, 1907

To the Editor of the New York Times:
I have read with great interest the report in your issue of to-day that the Danish engineer, Waldemar Poulson, the inventor of the interesting device known as the "telegraphone," has succeeded in transmitting accurately wireless telephonic messages over a distance of 240 miles.

I have looked up the description of the apparatus he has employed in the experiment and find that it comprises:

(1) My grounded resonant transmitting circuit;

(2) my inductive exciter;

(3) the so-called "Tesla transformer";

(4) my inductive coils for raising the tension on the condenser;

(5) my entire apparatus for producing undamped or continuous oscillations;

(6) my concatenated tuned transforming circuits;

(7) my grounded resonant receiving transformer;

(8) my secondary receiving transformer.

I note other improvements of mine, but those mentioned will be sufficient to show that Denmark is a land of easy invention.

The claim that transatlantic wireless telephone service will soon be established by these means is a modest one. To my system distance has absolutely no significance. My own wireless plant will transmit speech across the Pacific with the same precision and accuracy as across the table.

Nikola Tesla
New York, Dec. 19, 1907

The Future of the Wireless Art
Wireless Telegraphy & Telephony — by Walter W. Massie & Charles R. Underhill, 1908

Mr. Nikola Tesla, in a recent interview by the authors, as to the future of the Wireless Art, volunteered the following statement which is herewith produced in his own words.

"A mass in movement resists change of direction. So does the world oppose a new idea. It takes time to make up the minds to its value and importance. Ignorance, prejudice and inertia of the old retard its early progress. It is discredited by insincere exponents and selfish exploiters. It is attacked and condemned by its enemies. Eventually, though, all barriers are thrown down, and it spreads like fire. This will also prove true of the wireless art.

"The practical applications of this revolutionary principle have only begun. So far they have been confined to the use of oscillations which are quickly damped out in their passage through the medium. Still, even this has commanded universal attention. What will be achieved by waves which do not diminish with distance, baffles comprehension.

"It is difficult for a layman to grasp how an electric current can be propagated to distances of thousands of miles without diminution of intention. But it is simple after all. Distance is only a relative conception, a reflection in the mind of physical limitation. A view of electrical phenomena must be free of this delusive impression. However surprising, it is a fact that a sphere of the size of a little marble offers a greater impediment to the passage of a current than the whole earth. Every experiment, then, which can be performed with such a small sphere can likewise be carried out, and much more perfectly, with the immense globe on which we live. This is not merely a theory, but a truth established in numerous and carefully conducted experiments. When the earth is struck mechanically, as is the case in some powerful terrestrial upheaval, it vibrates like a bell, its period being measured in hours. When it is struck electrically, the charge oscillates, approximately, twelve times a second. By impressing upon it current waves of certain lengths, definitely related to its diameter, the globe is thrown into resonant vibration like a wire, stationary waves forming, the nodal and ventral regions of which can be located with mathematical precision. Owing to this fact and the spheroidal shape of the earth, numerous geodetical and other data, very accurate and of the greatest scientific and practical value, can be readily secured. Through the observation of these astonishing phenomena we shall soon be able to determine the exact diameter of the planet, its configuration and volume, the extent of its elevations and depressions, and to measure, with great precision and with nothing more than an electrical device, all terrestrial distances. In the densest fog or darkness of night, without a compass or other instruments of orientation, or a timepiece, it will be possible to guide a vessel along the shortest or orthodromic path, to instantly read the latitude and longitude, the hour, the distance from any point, and the true speed and direction of movement. By proper use of such disturbances a wave may be made to travel over the earth's surface with any velocity desired, and an electrical effect produced at any spot which can be selected at will and the geographical position of which can be closely ascertained from simple rules of trigonometry.

"This mode of conveying electrical energy to a distance is not 'wireless' in the popular sense, but a transmission through a conductor, and one which is incomparably more perfect than any artificial one. All impediments of conduction arise from confinement of the electric and magnetic fluxes to narrow channels. The globe is free of such cramping and hinderment. It is an ideal conductor because of its immensity, isolation in space, and geometrical form. Its singleness is only an apparent limitation, for by impressing upon it numerous non-interfering vibrations, the flow of energy may be directed through any number of paths which, though bodily connected, are yet perfectly distinct and separate like ever so many cables. Any apparatus, then, which can be operated through one or more wires, at distances obviously limited, can likewise be worked without artificial conductors, and with the same facility and precision, at distances without limit other than that imposed by the physical dimensions of the globe.

"It is intended to give practical demonstrations of these principles with the plant illustrated. As soon as completed, it will be possible for a business man in New York to dictate instructions, and have them instantly appear in type at his office in London or elsewhere. He will be able to call up, from his desk, and talk to any telephone subscriber on the globe, without any change whatever in the existing equipment. An inexpensive instrument, not bigger than a watch, will enable its bearer to hear anywhere, on sea or land, music or song, the speech of a political leader, the address of an eminent man of science, or the sermon of an eloquent clergyman, delivered in some other place, however distant. In the same manner any picture, character, drawing, or print can be transferred from one to another place. Millions of such instruments can be operated from but one plant of this kind. More important than all of this, however, will be the transmission of power, without wires, which will be shown on a scale large enough to carry conviction. These few indications will be sufficient to show that the wireless art offers greater. possibilities than any invention or discovery heretofore made, and if the conditions are favorable, we can expect with certitude that in the next few years wonders will be wrought by its application."

Nikola Tesla's Forecast for 1908
N. Y. World — Jan. 5, 1908

To the Editor of The World:

Constant and careful study of the state of things in this particular sphere enables an expert to make a forecast fairly accurate of the next state. The seemingly isolated events are to him but links of a chain. As a rule, the signs he notes are so pronounced that he can predict the changes about to take place with certitude. The performance is a mere banality as compared with the piercing view of the inspired into the distant future. This is a forecast - not a prophecy.

The coming year will be great in thought and result. It will mark the end of a number of erroneous ideas which, by their paralyzing effect on the mind, have throttled independent research and hampered progress and development in various departments of science and engineering.

The first to be dispelled is the illusion of the Hertz or electro-magnetic waves. The expert already realizes that practical wireless telegraphy and telephony are possible only by minimizing this wasteful radiation. The results recently attained in this manner with comparatively crude appliances illustrate strikingly the possibilities of the genuine art. Before the close of the year wireless transmission across the Pacific and trans-Atlantic wireless telephony may be expected with perfect confidence. The use of the wireless telephone in isolated districts will spread like fire.

The year will mark the fall of the illusionary idea that action must diminish with distance. By impressing upon the earth certain vibrations to which it responds resonantly, the whole planet is virtually reduced to the size of a little marble, thus enabling the reproduction of any kind of effect, as human speech, music, picture or character whatever, and even the transmission of power in unlimited amounts with exactly the same facility and economy at any distance, however great.

The next twelve months will witness a similar revolution of ideas regarding radio-activity. That there is no such element as radium, polonium or ronium is becoming more and more evident. These are simply deceptive appearances of a modern phlogiston. As I have stated in my early announcement of these emanations before the discovery of Mme. Curie, they are emitted more or less by all bodies, and are all of the same kind - merely effects of shattered molecules, differentiated not by the nature of substance but by size, speed and electrification.

The coming year will dispel another error which has greatly retarded progress of aerial navigation. The aeronaut will soon satisfy himself that an airplane proportioned according to data obtained by Langley is altogether too heavy to soar, and that such a machine, while it will have some uses, can never fly as fast as a dirigible balloon. Once this is fully recognized the expert will concentrate his efforts on the latter type, and before many months are passed it will be a familiar object in the sky.

There is abundant evidence that distinct improvements will be made in ship propulsion. The numerous theories are giving place to the view that what propels the vessel is a reactive jet; hence the propeller is doomed in efficiency at high speed. A new principle will be introduced.

The World is invited to test the accuracy of this forecast at the close of the year.

Nikola Tesla

Mr. Tesla's Vision
New York Times — April 21, 1908

To the Editor of the New York Times:
From a report in your issue of March 11, which escaped my attention, I notice that some remarks I made on the occasion referred to have been misunderstood. Allow me to make a correction..

When I spoke of future warfare I meant that it should be conducted by direct application of electrical waves without the use of aerial engines or other implements of destruction. This means, as I pointed out, would be ideal, for not only would the energy of war require no effort for the maintenance of its potentiality, but it would be productive in times of peace. This is not a dream. Even now wireless power plants could be constructed by which any region of the globe might be rendered uninhabitable without subjecting the population of other parts to serious danger or inconvenience.

What I said in regard to the greatest achievement of the man of science whose mind is bent upon the mastery of the physical universe, was nothing more than what I stated in one of my unpublished addresses, from which I quote: "According to an adopted theory, every ponderable atom is differentiated from a tenuous fluid, filling all space merely by spinning motion, as a whirl of water in a calm lake. By being set in movement this fluid, the ether, becomes gross matter. Its movement arrested, the primary substance reverts to its normal state. It appears, then, possible for man through harnessed energy of the medium and suitable agencies for starting and stopping ether whirls to cause matter to form and disappear, At his command, almost without effort on his part, old worlds would vanish and new ones would spring into being. He could alter the size of this planet, control its seasons, adjust its distance from the sun, guide it on its eternal journey along any path he might choose, through the depths of the universe. He could make planets collide and produce his suns and stars, his heat and light; he could originate life in all its infinite forms. To cause at will the birth and death of matter would be man's grandest deed, which would give him the mastery of physical creation, make him fulfill his ultimate destiny."

Nothing could be further from my thought than to call wireless telephony around the world "the greatest achievement of humanity" as reported. This is a feat which, however stupefying, can be readily performed by any expert. I have myself constructed a plant for this very purpose. The wireless wonders are only seeming, not results of exceptional skill, as popularly believed. The truth is the electrician has been put in possession of a veritable lamp of Aladdin. All he has to do is to rub it. Now, to rub the lamp of Aladdin is no achievement,

If you are desirous of hastening the accomplishment of still greater and further - reaching wonders you can do no better than by emphatically opposing any measure tending to interfere with the free commercial exploitation of water power and the wireless art. So absolutely does human progress depend on the development of these that the smallest impediment, particularly through the legislative bodies of this country, may set back civilization and the cause of peace for centuries.

Nikola Tesla
New York, April 19, 1908

Little Airplane Progress
New York Times — June 8

To the Editor of the New York Times:

It was not a little amusing to read a short time ago how the "great secret" of the airplane was revealed. By surrounding that old device with an atmosphere of mystery one gives life and interest to the report; but the plain fact is that all forms of aerial apparatus are well known to engineers, and can be designed for any specific duty without previous trials and with a fair degree of accuracy. The flying machine has materialized - not through leaps and bounds of invention, but by progress slow and imperceptible, not through original individual effort, but by a combination of the same forces which brought forth the automobile, and the motorboat. It is due to the enterprise of the steel, oil, electrical, and other concerns, who have been instrumental in the improvement of materials of construction and in the production of high-power fuels, as well as to the untiring labors of the army of skilled but unknown mechanics, who have been for years perfecting the internal combustion engine.

There is no salient difference between the dirigible balloon of Renard and Krebs of thirty years ago and that of Santos Oumont with which the bold Brazilian performed his feats. The Langley and Maxim aerodromes, which did not soar, were in my opinion better pieces of mechanism than their very latest imitations. The powerful gasoline motor which has since come into existence is practically the only radical improvement.

So far, however, only the self-propelled machine or aerial automobile is in sight. While the dirigible balloon is rapidly nearing the commercial stage, nothing practical has as yet been achieved with the heavier-than-air machine. Without exception the apparatus is flimsy and unreliable. The motor, too light for its power, gives out after a few minutes run; the propeller blades fly off; the rudder is broken, and, after a series of such familiar mishaps, there comes the inevitable and general smash-up. In strong contrast with these unnecessarily hazardous trials are the serious and dignified efforts of Count Zeppelin, who is building a real flying machine, safe and reliable, to carry a dozen men and provisions over distances of thousands of miles, and with a speed far in excess of those obtained with airplanes.

The limits of improvement in the flying machine, propelled by its own power, whether light or heavy, are already clearly defined. We know very closely what we may expect from the ultimate perfection of the internal combustion engine, the remittances which are to be overcome, and the limitations of the screw propeller. The margin is not very great. For many reasons the wireless transmission of power is the only perfect and lasting solution to reach very high speeds.

In this respect many experts are mistaken. The popular belief that because the air has only one-hundredth the density of water, enormous velocities should be practicable. But it is not so. It should be borne in mind that the air is one hundred times more viscous than water, and because of this alone the speed of the flying machine could not be much in excess of a properly designed aqueous craft.

The airplanes of the Langley type, such as was used by Forman and others with some success, will hardly ever prove a practical aerial machine, because no provision is made for maintaining it in the air in a downward current. This and the perfect balance independently of the navigator's control is absolutely essential to the success of the heavier-than-air machine. These two improvements I am myself endeavoring to embody in a machine of my own design.

Nikola Tesla
New York, June 6, 1908

Tesla on Airplanes
New York Times — Sept. 15

To the Editor of The New York Times:
The chronicler of current events is only too apt to lose sight of the true perspective and real significance of the phases of progress he records. Naturally enough, his opinions on subjects out of the sphere of his special training are frequently defective, but this is inseparable from the very idea of journalism. If an editor were to project himself into the future and view the happenings of the present or of the past in their proper relations he would make a dismal failure of his paper.

The comments upon the latest performances with airplanes afford interesting examples in this respect. What is there so very different between a man flying half an hour and another, using a more powerful machine, an hour, or two, or three? To be sure, in one instance the supporting planes are larger and the gasoline tank bigger, but there is nothing revolutionary in these departures. No one can deny the merit of the accomplishments. The feats are certainly remarkable and of great educational value.

The majority of human beings are unreceptive to novel ideas. The practical demonstrator comes with forceful arguments which enlighten and convince. But they are nothing more than obvious consequences of what has preceded, steps in advance which, taken singly, are of no particular importance, but which, in their totality, make up the conquest of the world by the new idea. If any one stands out more strongly than the other it is merely because it chances to occur at the psychological moment, when incredulity and doubt are giving way to confidence and expectancy. Such work is often brilliant, never great, as some would make believe. To be great it must be original. Of such feature it is absolutely devoid.

Place any of the later airplanes beside that of Langley, their prototype, and you will not find as much as one decided improvement. There are the same old propellers, the same old inclined planes, rudders, and vanes - not a single notable difference. Some have tried to hide their "discoveries." It is like the hiding of an ostrich who buries his head in the sand. Half a dozen aeronauts have been in turn hailed as conquerors and kings of the air. It would have been much more appropriate to greet John D. Rockefeller as such. But for the abundant supply of high-grade fuel we would still have to wait for an engine capable of supporting not only itself but several times its own weight against gravity.

The capabilities of the Langley aerodrome have been most strikingly illustrated. Notwithstanding this, it is not a practical machine. It has a low efficiency of propulsion, and the starting, balancing, and alighting are attended with difficulties. The chief defect, however, is that it is doomed if it should encounter a downward gust of wind. The helicopter is in these respects much preferable, but is objectionable for other reasons. The successful heavier-than-air flier will be based on principles radically novel and will meet all requirements. It will soon materialize, and when it does it will give an impetus to manufacture and commerce such as was never witnessed before, provided only that the Governments do not resort to the methods of the Spanish Inquisition, which have already proved so disastrous to the wireless art, the ideal means for making man absolute master of the air.

Nikola Tesla
New York, Sept. 13, 1908

How to Signal Mars
New York Times — May 23

To the Editor of the New York Times:
Of all the evidence of narrow mindedness and folly, I know of no greater than the stupid belief that this little planet is singled out to be the seat of life, and that all other heavenly bodies are fiery masses or lumps of ice. Most certainly, some planets are not inhabited, but others are, and among these there must exist II life under all conditions and phases of development.

I In the solar system Venus, the Earth, and Mars represent respectively, youth, full growth, and old age. Venus, with its mountains rising dozens of miles into the atmosphere, is probably as yet unfitted for such existence as ours, but Mars must have passed through all terrestrial states and conditions.

Civilized existence rests on the development of the mechanical arts. The force of gravitation on Mars is only two-thirds of that on earth, hence all mechanical problems must have been much easier of solution. This is even more so of the electrical. The planet being much smaller, the contact between individuals and the mutual exchange of ideas must have been much quicker, and there are many other reasons why intellectual life should have been on that planet, phenomenal in its evolution.

To be sure, we have no absolute proof that Mars is inhabited. The straightness of the canals, which has been held out as a convincing indication to this effect, is not at all such. We can conclude with mathematical certitude that as a planet grows older and the mountains are leveled down, ultimately every river must flow in a geodetically straight line. Such straightening is already noticeable in some rivers of the earth.

But the whole arrangement of the so-called waterways, as pictured by Lowell, would seem to have been designed. Personally I base my faith on the feeble planetary electrical disturbances which I discovered in the summer of 1899, and which, according to my investigations, could not have originated from the sun, the moon, or Venus. Further study since has satisfied me that they must have emanated from Mars. All doubt in this regard will be soon dispelled.

To bring forth arguments why an attempt should be made to establish interplanetary communication would be a useless and ungrateful undertaking. If we had no other reason, it would be justified by the universal interest which it will command, and by the inspiring hopes and expectations to which it would give rise. I shall rather concentrate my efforts upon the examination of the plans proposed and the description of a method by which this seemingly impossible task can be readily accomplished.

The scheme of signaling by rays of light is old, and has been often discussed, perhaps, more by that eloquent and picturesque Frenchman, Camille Flammarion, than anybody else. Quite recently Prof. W. H. Pickering, as stated in several issues of the New York Times, has made a suggestion which deserves careful examination.

The total solar radiation falling on a terrestrial area perpendicular to the rays amounts to eighty-three foot pounds per square foot per second. This activity measured by the adopted standard is a little over fifteen one-thousandth of a horsepower. But only about 10 per cent of this whole is due to waves of light. These, however, are of widely different lengths, making it impossible to use all in the best advantage, and there are specific losses unavoidable in the use of mirrors, so that the power of sunlight reflected from them can scarcely exceed 5.5 foot pounds per square foot per second, or about one one-hundredth of a horse-power.

In view of this small activity, a reflecting surface of at least one-quarter million square feet should be provided for the experiment. This area, of course, should be circular to insure the greatest efficiency, and, with due regard to economy, it should be made up of mirrors rather small, such as to meet best the requirements of cheap manufacture.

The idea has been advanced by some experts that a small reflector would be as efficient as a large one. This is true in a degree, but holds good only in heliographic transmission to small distances when the area covered by the reflected beam is not vastly in excess of that of the mirror. In signaling to Mars, the effect would be exactly proportionate to the aggregate surface of the reflections. With an area of one-quarter million square feet the activity of the reflected sunlight, at the origin would be about 2,500 horse-power.

It scarcely need be stated that these mirrors would have to be ground and polished most carefully. To use ordinary commercial plates, as has been suggested, would be entirely out of the question, for at such an immense distance the imperfections of surface would fatally interfere with efficiency. Furthermore, expensive clock work would have to be employed to rotate the reflectors in the manner of heliostats, and provision would have to be made for protection against destructive atmospheric influence. It is extremely doubtful that so formidable an array of apparatus could be produced for $10,000,000, but this is a consideration of minor importance to this argument.

If the reflected rays were paralled and the heavenly bodies devoid of atmospheres, nothing would be simpler than signaling to Mars, for it is a truth accepted by physicists that a bundle of parallel rays, in vacuum, would illuminated an area with the same intensity, whether it be near or infinitely remote. In other words, there is no sensible loss in the transportation of radiant energy through interplanetary or vacuous space. This being the case, could we but penetrate the prison wall of the' atmosphere, we could clearly perceive the smallest object on the most distant star, so inconceivably tenuous, frictionless, rigid, and elastic is the medium pervading the universe.

The sun's rays are usually considered to be parallel, and are virtually so through a short trajectory, because of the immense distance of the luminary. But the radiations, coming from a distance of 93,000,000 miles, emanate from a sphere 865,000 miles in diameter, and, consequently, most of them will fall on the mirrors at an angle less than 90 degrees, with the result of causing a corresponding divergence of the reflected rays. Owing to the equality of the angles of incidence and reflection, it follows that if Mars were at half the sun's distance, the rays reaching the planet would cover an area of about one-quarter of that of the solar disc, or in round numbers, 147,000,000,000 square miles, which is nearly 16,400,000, 000 times larger than that of the mirrors. This means that the intensity of the radiation received on Mars would be just that many times smaller.

To convey a definite idea, it may be stated that the light we get from the moon is 600,000 times feebler than that of the sun. Accordingly, even under these purely theoretical conditions the Pickering apparatus could do no more than produce an illumination 27,400,000 times feebler. than that of the full moon, or 1,000 times weaker than that of Venus.

The proceeding is based on the assumption that there is nothing in the path of the reflected rays except the tenuous medium filling all space. But the planets have atmospheres which absorb and refract. We see remote objects less distinctly, we perceive stars long after they have fallen below the horizon. This is due to absorption and refraction of the rays passing through the air. While these effects cannot be exactly estimated it is certain that the atmosphere is the chief impediment to the study of the heavens.

By locating our observatories one mile above sea level the quantity of matter which the rays have to traverse on their way to the planet is reduced to one-third. But, as the air becomes less dense, there is comparatively little gain to be derived from greater elevation. What chance would there be that the reflected rays, reduced to an intensity far below that estimated above, would produce a visible signal on Mars? Though I do not deny this possibility, all evidence points to the contrary.

Lowell, a trained and restless observer, who has made the study of Mars his specialty, and is working under ideal conditions, has been so far unable to perceive a light effect of the magnitude such as the proposed signaling apparatus might produce there. Phobos, the smaller of the two satellites of Mars - from seven to 10 miles in diameter - can only be seen at short intervals when the planet is in opposition. The satellite presents to us an area of approximately fifty square miles, reflecting sunlight at least as well as ordinary earth, which has little over one-twelfth of the power of a mirror.

Stated otherwise, an equivalent effect at that distance would be produced by mirrors covering four square miles, which means two square miles of the same reflectors if located on earth, as it receives sunlight of twice the intensity. Now this is an area 222 times larger than that of the ten million dollar reflector, and yet Phobos is hardly perceptible. It is true that the observation of the satellite is rendered difficult by the glare of its mother planet. But this is offset by the fact that it is in vacuum and that its rays suffer little diminution through absorption and refraction of the earth's atmosphere.

What has been stated is thought sufficient to convince the reader that there is little to be expected from the plan under discussion. The idea naturally presents itself that mirrors might be manufactured which will reflect sunlight in parallel beams. For the time being this is a task beyond human power, but no one can set a limit to the future achievement of man.

Still more ineffective would be the attempt of signaling in the manner proposed by Dr. William R. Brooks and others, by artificial light, as the electric arc. In order to obtain a reflected light activity of 2,500 horsepower it would be necessary to install a power plant of not less than 75,000 horsepower, which, with its turbines, dynamos, parabolic reflectors and other paraphernalia, would probably cost more than $10,000,000. While this method would permit operation at favorable times, when the earth is nearer to, and has its dark side turned toward Mars, it has the disadvantage of involving the use of reflected rays necessarily more divergent than those of the sun, it being impossible to construct mirrors of the required perfection and without their use the rays would be scattered to such an extent that the effect would be much smaller.

Reflecting surfaces of great extent can be had readily. Prof. R. W. Wood makes the odd suggestion of using the white alkali desert of the southwest for the purpose. Prof. E. Doolittle advises the employment of large geometric figures. In my opinion none of these suggestions is feasible. The trouble is, that the earth itself is a reflector, not efficient, it is true, but what it lacks in this respect is more than made up by the immensity of its area. To convey a perceptible signal in this manner it might require as much as 100 square miles reflecting surface

But there is one method of putting ourselves in touch with other planets. Though not easy of execution, it is simple in principle. A circuit properly designed and arranged is connected with one of its ends to an insulated terminal at some height and with the other to earth. Inductively linked with it is another circuit in which electrical oscillations of great intensity are set up by means now familiar to electricians. This combination of apparatus is known as my wireless transmitter.

By careful attunement of the circuits the expert can produce a vibration of extraordinary power, but when certain artifices, which I have not yet described are resorted to the oscillation reaches transcending intensity. By this means, as told in my published technical records, I have passed a powerful current around the globe and attained activities of many millions of horsepower. Assuming only a rate of 15,000,000, readily obtainable, it is 6,000 times more than that produced by the Pickering's mirrors.

But, my method has other and still greater advantages. By its employment the electrician on Mars, instead of utilizing the energy received by a few thousand square feet of area, as in a parabolic reflector, is enabled to concentrate in his instrument the energy received by dozens of square miles, thus multiplying the effect many thousands of times. Nor is this all. By proper methods and devices he can magnify the received effect as many times again.

It is evident, then, that in my experiments in 1899 and 1900 I have already produced disturbances on Mars incomparably more powerful than could be attained by any light reflectors, however large.

Electrical science is now so far advanced that our ability of flashing a signal to a planet is experimentally demonstrated. The question is, when will humanity witness that great triumph. This is readily answered. The moment we obtain absolute evidences that an intelligent effort is being made in some other world to this effect, interplanetary transmission of intelligence can be considered an accomplished fact. A primitive understanding can be reached quickly without difficulty. A complete exchange of ideas is a greater problem, but susceptible of solution.

Nikola Tesla

What Science May Achieve this Year: New Mechanical Principle for Conservation of Energy
Denver Rocky Mountain News — Jan. 16, 1910

The spread of civilization may be likened to that of fire: First, a feeble spark, next a flickering flame, then a mighty blaze, ever increasing in speed and power. We are now in this last phase of development.

Human activity has become so widespread and intense that years count as centuries of progress. There is no more groping in the dark or accidentally stumbling upon discoveries. The results follow one another like the links of a chain. 'Such is the force of the accumulated knowledge and the insight into natural laws and phenomena that future events are clearly projected before our vision. To foretell what is coming would be no more than to draw logical conclusions, were it not for the difficulty in accurately fixing the time of accomplishment.

The practical success of an idea, irrespective of its inherent merit, is dependent on the attitude of the contemporaries. If timely it is quickly adopted; if not, it is apt to fare like a sprout lured out of the ground by warm sunshine, only to be injured and retarded in its growth by the succeeding frost. Another determining factor is the amount of change involved in its introduction. To meet with instant success an invention or discovery must come not only as a rational, but a welcome solution. The year 1910 will mark the advent of such an idea. It is a new mechanical principle.

Since the time of Archimedes certain elementary devices were known, which were finally reduced to two, the lever and the inclined plane. Another element is to be added to these, which will give rise to new conceptions and profoundly affect both the practical and theoretical science of mechanics.

This novel principle is capable of embodiment in all kinds of machinery. It will revolutionize the propulsion apparatus on vessels, the locomotive, passenger car and the automobile. It will give us a practical flying machine entirely different from those made heretofore in operation and control, swift, small and compact and so safe that a girl will be able to fly in it to school without the governess. But the greatest value of this improvement will be in its application in a field virtually unexplored and so vast that it will take decades before the ground is broken. It is the field of waste.

We build but to tear down. Most of our work and resource is squandered. Our onward march is marked by devastation. Everywhere there is an appalling loss of time, effort and life. A cheerless view, but true. A single example out of many will suffice for illustration.

The energy necessary to our comfort and safe existence is largely derived from coal. In this country alone nearly one million tons of the life-sustaining material are daily extracted from the bowels of the earth with pain and sacrifice. This is about seven hundred tons per minute, representing a theoretical activity of, say, four hundred and fifty-million horsepower. But only a small percentage of this is usefully applied.

In heating, most of the precious energy escapes through the flue. The chimneys of New York City puff out into the air several million horsepower. In the use of coal for power purposes, we hardly capture more than 10 percent. The exhaust of engines carries off more energy than obtained from live steam.

In many modern plants the power has been actually doubled by obviating this waste, but the machinery employed is cumbersome and expensive. The manufacture of light is in a barbarous state of imperfection, and this may also be said of many industrial processes. Consider just one case, the manufacture of iron and steel.

America produces approximately 30,000,000 tons of pig iron per year. Each ton of iron requires about one and a half tons of coal, hence, in providing the iron market, 70,000,000 tons of coal per annum, or 133 tons per minute, are consumed. In the manufacture of coke a ton of coal yields, roughly, 10,000 cubic feet of gas of a mean heating capacity of 600 heat units per cubic foot.

Bearing in mind that 133 tons are used per minute, the total heat units developed in that time would be 798,000,000, the mechanical equivalent of which is about 19,000,000 horsepower. By the use of the new principle 7,000,000 horsepower might be rendered available. A furnace of 200 tons produces approximately 17,000 cubic feet of gas per minute of heat value of 100 units, corresponding to a theoretical performance of 40,000 horsepower, of which not less than 13,000 might be utilized in the improved apparatus referred to. The power derived by this method from all blast furnaces in the United States would be considerably above 5,000,000 horsepower.

The preceding figures, which are conservative, show that it would be possible to obtain 12,000,000 horsepower merely from the waste gasses in the iron and steel manufacture. The value of this power, fairly estimated, is $180,000,000 per annum, and it must be made worth much more by systematic exploitation.

A part of the power could be advantageously employed for operating the blowers, rollers and other indispensable machinery and supplying electricity for smelting, steel making and other purposes. The bulk might be used in the manufacture of nitrates, aluminum, carbides and ice. The production of nitrates would be particularly valuable from the point of view of national economy. Assuming that 5,000,000 horsepower were apportioned for that purpose, the annual yield would be not less than 10,000,000 tons of concentrated nitric compound, adequate to fertilizing 40, 000,000 acres of land. A great encouragement would be given to agriculture and the condition of the steel and iron workers ameliorated by offering to them a fertilizer at a reduced rate, thus enabling them to cultivate their farms with exceptional profit. Other conveniences and necessities, as light, power, ice and ozonized water could be similarly offered and numerous other improvements, both to the advantage of capital and labor, carried out.

To appreciate the above it should be borne in mind that the iron and steel industry is one of the best regulated in the world. In many other fields the waste is even greater. For example, in the operation of steam railroads, not less than 98 per cent of the total energy of coal burned is lost. An enormous saving could be effected by replacing the present apparatus with new gas turbines and other improved devices for transmitting and storing mechanical energy. A study of this subject will convince that for the time being, at least, there is more opportunity for invention in the utilization of waste than in the opening up of new resources.

N. Tesla

Mr. Tesla on the Future
Modern Electrics — May, 1912

On Tesla Day, at the Northwest Electric Show, held at Minneapolis, Minn., March 16th to 23rd, Mr. Tesla sent, through Archbishop Ireland, the following message to the people of the Twin Cities and the Northwest:

New York, N. Y., March 18, 1912. His Grace, The Most Reverend Archbishop Ireland:

I bespeak your Grace's far-famed eloquence in voicing sentiments and ideas to which I can give but feeble expression. May the exposition prove a success befitting the cities of magical growth, the courage and energy of western enterprise, a credit to its organization, a lasting benefit to the communities and the world through its lessons and stimulating influence as a bewildering, unforgettable record of the triumphant progress of the art. Great as are the past achievements, the future holds out more glorious promise. We are getting an insight into the essence of things; our means and methods are being refined, a new and specialized race is developing with knowledge deep and precise, with greater powers and keener perceptions. Mysterious as ever before, nature yields her precious secrets more readily and the spirit of man asserts its mastery over the physical universe. The day is not distant when the very planet which gave him birth will tremble at the sound of his voice; he will make the sun his slave, harness the inexhaustible and terribly intense energy of microcosmic movement; cause atoms to combine in predetermined forms; he will draw the mighty ocean from its bed, transport it through the air and create lakes and rivers at will; he will command the wild elements; he will push on and on from great to greater deeds until with his intelligence and force he will reach out to spheres beyond the terrestrial.

I am your Grace's most obedient servant.
Nikola Tesla

The Disturbing Influence of Solar Radiation on the Wireless Transmission of Energy
Electrical Review and Western Electrician, July 6, 1912

When Heinrich Hertz announced the results of his famous experiments in confirmation of the Maxwellian electromagnetic theory of light, the scientific mind at once leaped to the conclusion that the newly discovered dark rays might be used as a means for transmitting intelligible messages through space. It was an obvious inference, for heliography, or signaling by beams of light, was a well recognized wireless art. There was no departure in principle, but the actual demonstration of a cherished scientific idea surrounded the novel suggestion with a nimbus of originality and atmosphere of potent achievement. I also caught the fire of enthusiasm but was not long deceived in regard to the practical possibilities of this method of conveying intelligence.

Granted even that all difficulties were successfully overcome, the field of application was manifestly circumscribed. Heliographic signals had been flashed to a distance of 200 miles, but to produce Hertzian rays of such penetrating power as those of light appeared next to impossible, the frequencies obtainable through electrical discharges being necessarily of a much lower order. The rectilinear propagation would limit the action on the receiver to the extent of the horizon and entail interference of obstacles in a straight line joining the stations. The transmission would be subject to the caprices of the air and, chief of all drawbacks, the intensity of disturbances of this character would rapidly diminish with distance.

But a few tests with apparatus, far ahead of the art of that time, satisfied me that the solution lay in a different direction, and after a careful study of the problem I evolved a new plan which was fully described in my addresses before the Franklin Institute and National Electric Light Association in February and March, 1893. It was an extension of the transmission through a single wire without return, the practicability of which I had already demonstrated. If my ideas were rational, distance was of no consequence and energy could be conveyed from one to any point of the globe, and in any desired amount. The task was begun under the inspiration of these great possibilities.

While scientific investigation had laid bare all the essential facts relating to Hertz-wave telegraphy, little knowledge was available bearing on the system proposed by me. The very first requirement, of course, was the production of powerful electrical vibrations. To impart these to the earth in an efficient manner, to construct proper receiving apparatus, and develop other technical details could be confidently undertaken. But the all important question was, how would the planet be affected by the oscillations impressed upon it? Would not the capacity of the terrestrial system, composed of the earth and its conducting envelope, be too great? As to this, the theoretical prospect was for a long time discouraging. I found that currents of high frequency and potential, such as had to be necessarily employed for the purpose, passed freely through air moderately rarefied. Judging from these experiences, the dielectric stratum separating the two conducting spherical surfaces could be scarcely more than 20 kilometers thick and, consequently, the capacity would be over 220,000 microfarads, altogether too great to permit economic transmission of power to distances of commercial importance.

Another observation was that these currents cause considerable loss of energy in the air around the wire. That such waste might also occur in the earth's atmosphere was but a logical inference.

A number of years passed in efforts to improve the apparatus and to study the electrical phenomena produced. Finally my labors were rewarded and the truth was positively established; the globe did not act like a conductor of immense capacity and the loss of energy, due to absorption in the air, was insignificant. The exact mode of propagation of the currents from the source and the laws governing the electrical movement had still to be ascertained. Until this was accomplished the new art could not be placed on the plane of scientific engineering. One could bridge the greatest distance by sheer force, there being virtually no limit to the intensity of the vibrations developed by such a transmitter, but the installment of economic plants and the predetermination of the effects, as required in most practical applications, would be impossible.

Such was the state of things in 1899 when I discovered a new difficulty which I had never thought of before. It was an obstacle which could not be overcome by any improvement devised by man and of such nature as to fill me with apprehension that transmission of power without wires might never be quite practicable. I think it useful, in the present phase of development, to acquaint the profession with my investigations.

It is a well know fact that the action on a wireless receiver is appreciably weaker during the day than at night and this is attributed to the effect of sunlight on the elevated aerials, an explanation naturally suggested through an early observation of Heinrich Hertz. Another theory, ingenious but rather fine-spun, is that some of the energy of the waves is absorbed by ions or electrons, freed in sunlight and caused to move in the direction of propagation. *The Electrical Review and Western Electrician* of June 1, 1912, contains a report of a test, during the recent solar eclipse, between the station of the Royal Dock Yard in Copenhagen and the Blaavandshuk station on the coast of Jutland, in which it was demonstrated that the signals in that region became more distinct and reliable when the sunlight was partially cut off by the moon. The object of this communication is to show that in all the instances reported the weakening of the impulses was due to an entirely different cause.

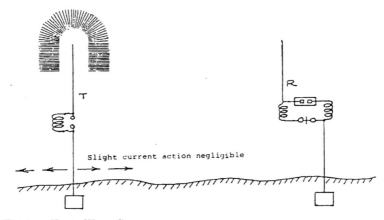

Fig. 1 — Hertz Wave System

It is indispensable to first dispel a few errors under which electricians have labored for years, owing to the tremendous momentum imparted to the scientific mind through the work of Hertz which has hampered independent thought and experiment. To facilitate understanding, attention is called to the annexed diagrams in which Fig. 1 and Fig. 2 represent, respectively, the well known arrangements of circuits in the Hertz-wave system and my own. In the former the transmitting and receiving conductors are separated from the ground through spark gaps, choking coils, and high remittances. This is necessary, as a ground connection greatly reduces the intensity of the radiation by cutting off half of the oscillator and also by increasing the length of the waves from 40 to 100 percent, according to the distribution of capacity and inductance. In the system devised by me a connection to earth, either directly or through a condenser is essential. The receiver, in the first case, is affected only by rays transmitted through the air, conduction being excluded; in the latter instance there is no appreciable radiation and the receiver is energized through the earth while an equivalent electrical displacement occurs in the atmosphere.

Now, an error which should be the focus of investigation for experts is, that in the arrangement shown in Fig. 1 the Hertzian effect has been gradually reduced through the lowering of frequency, so as to be negligible when the usual wavelengths are employed. That the energy is transmitted chiefly, if not wholly, by conduction can be demonstrated in a number of ways. One is to replace the vertical transmitting wire by a horizontal one of the same effective capacity, when it will be found that the action on the receiver is as before. Another evidence is afforded by quantitative measurement which proves that the energy received does not diminish with the square of the distance, as it should, since the Hertzian radiation propagates in a hemisphere. One more experiment in support of this view may be suggested. When transmission through the ground is prevented or impeded, as by severing the connection or otherwise, the receiver fails to respond, at least when the distance is considerable. The plain fact is that the Hertz waves emitted from the aerial are just as much of a loss of power as the short radiations of heat due to frictional waste in the wire. It has been contended that radiation and conduction might both be utilized in actuating the receiver, but this view is untenable in the light of my discovery of the wonderful law governing the movement of electricity through the globe, which may be conveniently expressed by the statement that the projection of the wave-lengths (measured along the surface) on the earth's diameter or axis of symmetry of movement are all equal. Since the surfaces of the zones so defined are the same the law can also be expressed by stating *that the current sweeps in equal times over equal terrestrial areas.* (See among others *"Handbook of Wireless Telegraph,"* by James Erskine-Murray.) Thus the velocity propagation through the superficial layers is variable, dependent on the distance from the transmitter, the mean value being $n/2$ times the velocity of light, while the ideal flow along the axis of propagation takes place with a speed of approximately 300,000 kilometers per second. To illustrate, the current from a transmitter situated at the Atlantic Coast will traverse that ocean—a distance of 4,800 kilometers—in less than 0.006 second with an average speed of 800,000 kilometers. If the signaling were done by Hertz waves the time required would be 0.016 second.

Bearing, then, in mind, that the receiver is operated simply by currents conducted along the earth as through a wire, energy radiated playing no part, it will be at once evident that the weakening of the impulses could not be due to any changes in the air, making it turbid or conductive, but should be traced to an effect interfering with the transmission of the current through the superficial layers of the globe. The solar radiations are the primary cause, that is true, not those of light, but of heat. The loss of energy, I have found, is due to the evaporation of the water on that side of the earth which is turned toward the sun, the conducting particles

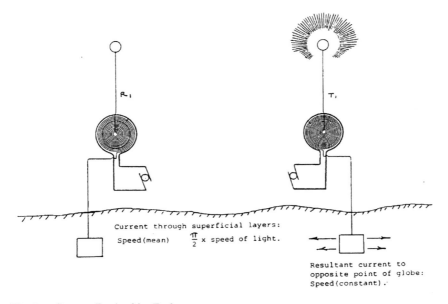

Fig. 2 — System Devised by Tesla

carrying off more or less of the electrical charges imparted to the ground. This subject has been investigated by me for a number of years and on some future occasion I propose to dwell on it more extensively. At present it may be sufficient, for the guidance of experts, to state that the waste of energy is proportional to the product of the square of the electric density induced by the transmitter at the earth's surface and the frequency of the currents. Expressed in this manner it may not appear of very great practical significance. But remembering that the surface density increases with the frequency it may also be stated that the loss is proportional to the cube of the frequency. With waves 300 meters [1 MHZ] in length economic transmission of energy is out of the question, the loss being too great. When using wave-lengths of 6,000 meters [50 kHz] it is still noticeable though not a serious drawback. With wave-lengths of 12,000 meters [25 kHz] it becomes quite insignificant and on this fortunate fact rests the future of wireless transmission of energy.

To assist investigation of this interesting and important subject, Fig. 3 has been added, showing the earth in the position of summer solstice with the transmitter just emerging from the shadow. Observation will bring out the fact that the weakening is not noticeable until the aerials have reached a position, with reference to the sun, in which the evaporation of the water is distinctly more rapid. The maximum will not be exactly when the angle of incidence of the suns rays is greatest, but some time after. It is noteworthy that the experimenters who watched the effect of the recent eclipse, above referred to, have observed the delay.

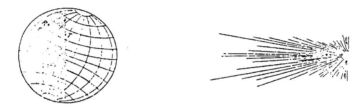

Fig. 3 — Illustrating Disturbing Effect of the Sun on Wireless Transmission.

Nikola Tesla's Plan to Keep "Wireless Thumb" on Ships At Sea
New York Press — Nov. 9

Suggests Transmitters Powerful Enough to Cause the Earth to Vibrate at the Poles and Equator.

He would Determine Vessel's Latitude and Longitude by Measuring the Length of Electric Waves.

Nikola Tesla has come forward to refute the claims of men who recently excited the scientific world with announcements of discovery and invention calculated to crowd the bugbear of scientific warfare back into the primer class, and to safeguard the lives of seafarers. First he takes up and disposes of the announcement of an invention said to enable a receiving ship equipped with wireless to tell the longitude and latitude of a sending ship without the latter vessel offering its own calculations. "It hasn't been done, and it probably will be years before the means for so doing can be applied successfully," he says. As to the power of ultra-violet rays to explode the powder magazine of a warship from a distance, he insists it can't be done through that medium. If charges of powder have been so exploded, he contends, the detonation was accomplished with the familiar waves now utilized by the wireless. But Mr. Tesla admits that in all probability there will come a time when science has so harnessed and developed the means at hand that such results may be obtained. Mr. Tesla sets forth for readers of The Press his views on the two subjects as follows:

The first and incomplete announcements of technical advances should always be taken with a grain of salt. It is true that the newspapers are getting more and more accurate and reliable in putting forth such information, but, nevertheless, the news frequently is misleading.

For instance, not long ago reports widely circulated that powder had been exploded at distance by infra-red or ultra-violet rays, and that a British battleship had been used in a test of this kind, which proved successful. The dispatches gave great opportunity to sensational speculation, but the truth is that there was no novelty whatever in what was done.

A mine or magazine 'may have been blown up, but this was accomplished in a well-known manner through the application of a kind of electrical waves which are now generally adopted in the transmission of signals without wires. Similar experiments were performed in this country many years ago by myself and others, and quite recently John Hays Hammond, Jr., has done credible work in this direction through the application of an art which has been named "Telautomatics," or wireless control of moving mechanism at a distance.

By means of such telautomatic vessels, surface, and submarine or aerial, a perfect system of coast defense can be established. Torpedoes on this plan also can be controlled from battleships, and there is no doubt they sooner or later will be adopted and their introduction will have a revolutionary effect on the methods of warfare.

The results described are, however, not impossible. It is quite practicable to explode by rays of light a mine at a distance, as by acting, on a mixture of chlorine and hydrogen. Certain dark rays also can be employed to produce destructive effects. As far back as 1897, I disclosed before the New York

Academy of Sciences the discovery that Roentgen, or X-rays, projected from certain bulbs have the property of strongly charging an electrical condenser at a distance. The energy so accumulated readily can be discharged and cause the ignition of some explosive compound.

But ultra-violet rays are of very short wave lengths and cannot penetrate steel shells, while the longer and more penetrative waves of the infra-red rays are chemically much less active. There is no doubt in my mind that we soon shall be able to project energy at a distance not only in small, but in large amounts, and what the effect of such an achievement will be on existing conditions, words cannot express.

As regards the determination of latitude and longitude of a vessel at sea by wireless, there is nothing in use as yet which would make such direct observation possible. Some suggestions, however, which I have since many years advocated, have been adopted. They are the flashing of time signals over a wide area and the employment of an instrument known as a wireless compass.

These means enable an expert on a vessel to ascertain the exact hour at any sending station within reach, and also, in an imperfect manner, the direction in which it is situated, and from these data it is possible to get a rough idea of the position of the ship relative to the points of reference.

A perfect means for determining not only such and other data important to the navigator already is available, but it may require years to apply it. I refer to the use of the stationary waves, which were discovered by me fourteen years ago. The subject is too technical to be explained in detail, but the average reader can be made to understand the general principle.

The earth is a conductor of electricity, and as such has its own electrical period of vibration. The time of one complete swing is about one-twelfth of a second. In other words, this is the interval the current requires in passing to, and returning from, the diametrically opposite point of the globe.

Now, the wonderful fact is, that notwithstanding its immense size, the earth responds to a great number of vibrations and can be resonantly excited just like a wire of limited dimensions. When this takes place there are formed on its surface stationary parallel circles of equal electrical activity, which can be revealed by properly attuned instruments.

Imagine that a transmitter capable of exciting the earth were placed at one of the Poles. Then the crests and hollows of the stationary waves would be in parallel circles with their planes at right angles to the axis of the earth, and from readings of a properly graduated instrument the distance of a vessel carrying the same from the Pole could be at once read, giving accurately the geographical latitude.

In like manner, if a transmitter were placed at a point on the Equator, the longitude could be precisely determined by the same means. But the best plan would be to place three transmitters at properly chosen points on the globe so as to establish three non-interferable systems of stationary waves at right angles to one another. If this were done, innumerable results of the greatest practical value could be realized.

From Nikola Tesla: A Tribute to George Westinghouse
Electrical World - N. Y. — March 21

    The first impressions are those to which we cling most-in later life. I like to think of George Westinghouse as he appeared to me in 1888, when I saw him for the first time. The tremendous potential energy of the man had only in part taken kinetic form, but even to a superficial observer the latent force was manifest. A powerful frame, well proportioned, with every joint in working order, an eye as clear as a crystal, a quick and springy step - he presented a rare example of health and strength. Like a lion in a forest, he breathed deep and with delight the smoky air of his factories. Though past forty then, he still had the enthusiasm of youth. Always smiling, affable and polite, he stood in marked contrast to the rough and ready men I met. Not one word which would have been objectionable, not a gesture which might have offended - one could imagine him as moving in the atmosphere of a court, so perfect was his bearing in manner and speech. And yet no fiercer adversary than Westinghouse could have been found when he was aroused. An athlete in ordinary life, he was transformed into a giant when confronted with difficulties which seemed unsurmountable. He enjoyed the struggle and never lost confidence. When others would give up in despair he triumphed. Had he been transferred to another planet with everything against him he would have worked out his salvation. His equipment was such as to make him easily a position of captain among captains, leader among leaders. His was a wonderful career filled with remarkable achievements. He gave to the world a number of valuable inventions and improvements, created new industries, advanced the mechanical and electrical arts and improved in many ways the conditions of modern life. He was a great pioneer and builder whose work was of far-reaching effect on his time and whose name will live long in the memory of men.
    Nikola Tesla

Tesla and Marconi
New York Sun — May 22

To the Editor of The Sun - Sir: The reports contained in The Sun and other journals regarding the issue of a recent wireless patent suit are of a nature to create an erroneous impression. Two of the patents mentioned, namely, Nos. 11,9j3 and 609,154, granted respectively to William Marconi and Sir Oliver Lodge, are of no importance, but another patent of the former expert, dated June 28, 1904, contains arrangements on which I obtained full protection more than three years before and which are essential to the successful practice of the wireless art at any considerable distance.

My patents bear the numbers 645,576 and 649,621 and were secured through Kerr, Page & Cooper, attorneys for the General Electric and Westinghouse companies. The apparatus described by me comprises four circuits peculiarly arranged and carefully attuned so as to secure the greatest possible flow of electrical energy through them. The generator is a transformer of my invention and the oscillations employed are of a kind which are now known in technical literature as the Tesla currents. Every one of these elements, even to the last detail, is contained in the Marconi patent which was involved in the suit, and its use constitutes an infringement of all the fundamental features of my wireless system.

Nikola Tesla
New York, March 21, 1914.

Science and Discovery Are the Great Forces Which Will Lead to the Consummation of the War

The Sun — Dec. 20, 1914

Whatever future ages may have in store for the human race, the development so far would indicate as its probable fate perpetual strife. Civilization alone is evidently insufficient for insuring permanent peace on earth. It but retards the clash to add to its intensity and magnitude, making it all the more dreadful and ruinous.

The present colossal struggle creates an impression apart, a feeling of awe, a sense of solemnity, springing from the knowledge that a terrible calamity, greater than any recorded in the annals of history, has befallen the world. Suddenly awakened from fancied security to the consciousness of suspected and universal danger, the nations stand aghast. It is as if some vast terrestrial upheaval were taking place, as if gigantic forces were unchained, threatening the entire globe.

Never before were such immense armies engaged in battle and such frightfully destructive implements employed; never was so much dependent on a victory of arms. Already the losses incurred amount to tens of billions of dollars; more than three million men have been killed and disabled, and for each of these ten, at least, *have* been turned into nervous wrecks, which will impress their miseries on the succeeding generations and darken their days. All the world over countless sufferers, torn by anxiety, ask themselves how long is this appalling slaughter and sacrilegious waste to continue.

War is essentially a manifestation of energy involving the acceleration and retardation of a mass by a force. In such a. case it is a universally established truth that the time necessary to impart a given velocity and momentum is proportionate to the mass. The same law also applies to the annihilation of velocity and momentum by a resisting force. Translated in popular language this means that the period or duration of an armed conflict is theoretically proportionate to the magnitude of the armies or number of combatants.

It is obviously assumed that the resources are ample and all other conditions equal. Furthermore, in making deductions from previous wars a number of factors have to be taken into consideration and all quantities estimated at their proper value on the basis of statistical and other data. Supposing that, as it appears, 12,000,000 men are engaged in the present struggle, a comparison with some of the past wars gives the following results:

Wars	Number of Combatants	Duration Y.	M.	Remarks
Civil war	4,600.000	4		Protracted by distance, poor communication and ineffective arms.
Present war	12,000.000	10		
Franco-German war	1,700.000		13	Equipment not quite modern.
Present war	12,000.000	7	6	
Russo-Japanese war	2,200.000	1	6	Lengthened by distance, poor communication and nature of campaign.
Present war	12,000.000	8		
First Balkan war	1,200.000		6	In all respects up to date.
Present war	12,000.000	5		
Hypothetical average war	2,425.000	1	9	Various causes affecting duration.
Present war	12,000.000	8	6	

Much more concordant and shorter terms would be obtained in these comparative estimates if the records available were corrected as indicated and due allowances made for the facilities, of transport and communication, increased power and destructiveness of arms and other factors tending to magnify the rate at which energy is delivered, and so to hasten the termination of the clash. The best inference is certainly that drawn from the Balkan war, as the most modern, according to which the term should be five years. Even though this be but a rough approximation, it is sufficient to show that, barring some extraordinary development, this war will be a long, one.

Indeed, it seems on purely scientific grounds that a conflict on such a vast scale can only be ended by exhaustion. The enormous extent of the battle front, owing to numbers and attendant impossibility of striking a decisive blow, is in further support of this theory. It is also highly significant to observe in this connection how the original battle lines, determined in advance by strategy, have been gradually shifted and straightened, contact between the fighting masses being finally established on lines fixed by natural law and brute force of push in defiance of military design. The likelihood of such termination is increased by the fact that the disturbance extends over an immense area, making the supply of necessities to some of the affected regions exceptionally difficult.

Accepting, then, this theory as correct, we are justified in expecting that, conditions remaining normal, the struggle will last more or less according to the form the exhaustion may take. Lack of food, deterioration and shortage of equipment, want of metals, chemicals and ammunition, scarcity of ready capital, failing supply of trained men or sheer giving out of human energy are some of the elements to be reckoned with, any, one of which may compel an early cessation of hostilities. That the war cannot be continued much longer with its present intensity can be easily shown.

The daily cost of operation is more than forty millions of dollars, and , judging from. the casualties recorded to date, twenty-five thousand men, on the average, are killed and disabled in battle every day. At that rate only four more months of active campaign would result in an expenditure of five billions of dollars and a loss of life of three millions of men. This is, manifestly, too great an additional burden to be borne, for even though the fighting material might be available, capital is sure to be lacking. It could be, therefore, concluded with certitude that peace would be restored before next winter, were it not for one possibility, or rather probability, that of a deadlock, which would be the very worst calamity, for, in view of the real cause of ,the trouble and the temper of nations involved, it could not fail to protract the war for years.

Prophesying is an ungrateful occupation, but scientific forecasting is a useful form of 'endeavor and would be much more such if human nature were not so prone to leave advice and lesson unheeded. Having made a careful study of the situation, an expert can predict certain happenings with perfect confidence. There are now only three possible issues of this war: first, collapse of Austria; second, conquest of England by the Germans, and, third, Germany's exhaustion and defeat.

The fall of Austria is inevitable and must occur within the next few months. She may defy German influence and sue independently for peace to save herself, but it is doubtful that she could offer anything acceptable to the Allies. Much more likely it is that the old Emperor, tired of life and recognizing the injustice of Austria's cause, will himself abdicate and recommend partition.

This may not be unwelcome to hard pressed Germany, for it will open up a way of making peace on terms which will not be humiliating and compensate her for the probable loss of Alsace-Lorraine and East Prussia.

The dual monarchy has maintained herself through decades as if by a wonder. It would have been dissolved long ago had it not been for the stubborn adherence of Hungarian magnates to a promise given to Maria Theresa and the extraordinary popularity of the reigning dynasty, largely due to compassion of the subjects of all nationalities aroused by the many strange misfortunes which have befallen the house of Hapsburg. It is well recognized that the unnatural existence of this feudal state has been a constant menace to European peace and is the chief cause of the present upheaval. A division of Austro-Hungarian territory along racial lines will satisfy all warring nations on the European continent. This is sure to come. It is a process natural and unavoidable as the falling of an overripe apple from the tree.

Regarding the second possibility, it is still unsafe to make a prediction and further developments must be awaited before a conclusion can be drawn as to the outcome. There are many indications that Germany is preparing for an attack on England with all energy and speed, and perhaps her operations in the east and west serve- the purpose of masking this move. The tension between the two countries is very great, the causes of the quarrel peculiar and a peaceful solution of the difficulty is next to impossible.

The third of the issues mentioned would mean a very long war. Germany cannot break through the steel wall in France and Belgium; her partial victories in Poland can make no impression on the Russian masses. Gradually she must settle on a defensive. She has the greatest load to carry and must give out first, according to financiers and statisticians.

But with a people so intelligent, industrious, resourceful and solidly united such forecasts are hazardous. The Germans are fully capable of "making two blades of grass grow where one grew before" and it is precisely because of this and their perfect military organization that the danger of a long conflict exists. Such a prospect is enough to cause the gravest apprehensions and the uppermost thought in the minds of seers is how to prevent such paralysis of progress and horrible carnage and waste. Can it be done?

There is a grim determination of all directly concerned to fight the issues to the bitter end on the ground that a premature peace, leaving the vital questions unsettled, would only mean the continuance of the existing pernicious regime and repetition of the evil. A new and irresistible argument must be brought forward to stop the conflict. The case is desperate, but there is a hope. This hope lies in science, discovery and invention. ,

Modern machinery wrought by science is responsible for this calamity; science will also undo the Frankenstein monster it has created. Centuries ago an ingenious contrivance of Archimedes is said to have decided a battle and terminated a great war. Be it a myth or a fact, this story affords an inspiring lesson. What is needed at this psychological moment is some such revelation. A new force, a new agent, a demonstration by any means, old or novel, but of a kind to surprise and suddenly, illuminate, to bring the belligerents to their senses and furnish irrefutable proof of the folly and uselessness of carrying on the brutal fight.

This idea, to which I have myself devoted years of work, has now taken hold of scientific men and experts all the world over. Thousands of inventors, fired on by this unique opportunity, are bent upon developing some process or apparatus for accomplishing the purpose, and there. is feverish. activity in France, Russia, and especially in Germany, among electricians, chemists and engineers. What the genius of nations will bring forth none can tell, but it is not too much to say that the results will be of such character as materially to affect the outcome and duration of the struggle.

It is on this account that importance attaches to vague reports of mysterious experiments with Zeppelins, explosive rays and magic bombs, for though such news items cannot be accepted as true, they reveal just so many startling possibilities. In the production and application of novel means of warfare Germany should be first, no only by reason of superior facilities and excellent training of her experts, but because this has become a dire necessity, a question of life and death in her present trying position.

The uncertain and often conflicting despatches of the daily happenings received from various sources have made it difficult to form a decided opinion as to the actual state of things, but in spite of rigid censorship the main facts have gradually transpired. One of these is that the Germans were the only people ready for war.

Not even the French, who boasted of preparedness; were able to mobilize on time. The invasion of East Prussia was but a daring stroke of the Russians to draw the enemy and relieve the pressure on France, successful but very costly to them. As to the complacent Britons, they were fast asleep. Whatever may be said against Great Britain, her utter unreadiness and the great danger to which she was exposing herself by her ultimatum to Germany would seem to be proof positive that she did not desire to enter the conflict.

Another fact, equally apparent, is that Germany, not content with a partial, even if certain, victory, had determined to defeat all the Allies in quick succession. Her plan of dictating terms of peace first in Paris, then in Petrograd, and finally in London, was not adopted as a military necessity, but as a deliberate program based on the absolute confidence in the overwhelming power of her arms. Nor did she mean to stop at that. Her aim was much higher; she wanted nothing less than to rule all nations.

This is now frankly admitted by many of her leading men. To most of us such an undertaking is dumfounding in its boldness and magnitude, all the more as it is intended to be carried through by force. But it would be a mistake to accuse the Germans of conceit and arrogance. They are convinced of their superiority, and it must be admitted that there is some justification for their attempt.

The question has often been raised as to whether our further development will be in the direction of the artistic and beautiful or the scientific and useful.

The inevitable conclusion is that art must be sacrificed to science. This being so, the rational Germans represent the nearest approach of the humanity of the future. The Slavs, who are in the ascendency and will lead in their turn, will give a fresh impetus to creative and spiritual effort, but they too will have to concentrate on the necessary and practical. A world of bees will be the ultimate result.

Germany has been foiled in her attempt. Though still undefeated, her campaign is a failure. Many statements have been made in explanation of the sudden halt of her victorious armies, as if by a miracle, at the very gates of Paris, but the views expressed are of speculative character and do not deal with the real physical causes. These may be briefly elucidated.

The German war machine is an attempt to substitute for an assemblage of loosely linked temperamental and problematical units a compact and apathetic mass moving at command with clock precision, machinelike, impassive, indifferent to danger and death, in battle the same as on parade. Its conception rests on a deeply scientific foundation. Every human being is swayed by courage and fear, but the former predominates. This is evident, for life or existence itself is a struggle fraught with perils and pains which must be met with determination and fortitude. Fear comes from the consciousness of inimical environment and is accentuated by isolation.

When many men are placed close together the friendly surrounding and sense of connectedness are productive of a distinct psychological mass effect, calming the nerves and subduing the inborn dread and apprehension. On the other hand, frequent and severe drilling kept up for years, besides being conducive to precision and synchrony of movements, is of decided hypnotic influence, still further eliminating individual initiative and incertitude. Thus results a strong and healthy body which moves and acts as a unit, which is without human failings and shortcomings and capable of maximum performance through well directed and simultaneous application .of separate efforts.

Such is the formidable engine Germany has perfected for the protection of her Kultur and conquest of the globe — an unfeeling automaton, a diabolical contrivance for scientific, pitiless, wholesale destruction the: like of which was not dreamed of before. It is believed to express the highest efficiency, but is deceptive in this respect, to none more than the Germans themselves. In reality this modern war machine, considered as a transformer of energy, is barbarously wasteful.

Not only does it call for enormous expenditure of money and effort when idle, but involves a fundamental fallacy which military writers ignore, namely, the conditions determining its performance, and therefore its efficiency, are largely, if not wholly, controlled by the enemy. Indeed, it is lack of appreciation of this truth that is responsible for the Paris failure.

The first of the two chief causes of German unsuccess is found in the admirable defensive tactics of the French, who refused to make a stand for a decisive battle, thus preventing the German machine from developing its full power and compelling it to work at low efficiency. The second, even more important, was the result of undue hurry of the Germans, who drove their engine too fast thereby increasing greatly the losses without adequate gain in useful performance. Had they taken more time, which, as subsequent developments have shown, they could have well afforded, there would have been more energy conserved and the task in all probability successfully accomplished.

The most surprising of the facts which have transpired is that there have been made in diplomatic transaction and conduct of the German campaign a number of grievous mistakes, so patent now that no representations of the press can disguise them. This is a revelation for which the world was least prepared and which shows clearly that German erudition and technical proficiency have been obtained at the expense of intuition, tact and good judgment.

What a blunder was the violation of Belgian neutrality, what an error the expectation that England would tolerate an encroachment so dangerous to her existence, that Italy would sacrifice her fleet and commerce to please the alliance! The Germans had wonderful guns, rendering fortifications useless, and yet in attacking France, instead of the shortest, route, they took a circuitous path through Belgium, thus losing time and conjuring new perils and complications besides. Tens of thousands of men were driven into certain death in vain assaults in mass formation when a few shots from these guns would have been sufficient to level the forts.

Troops were withdrawn from France to less important points at the very moment when their presence meant certain victory. The Germans could have marched on Warsaw and Petrograd before the enemy was ready to put up an effective resistance, yet *they* delayed the invasion until the Russians brought up their millions. They could have taken Dunkirk and Calais without great effort and so avoided the terrible losses which this task, if at all realizable, must now involve. At present they are recklessly venturing far into Russian territory against overwhelming numbers and in a season when snow-storms may cut the communications and put the whole army at the mercy of the enemy.

What explanation can be offered for these and other singular errors of a nation to which economy is religion, which is admittedly the first in achieving successes in the most scientific manner, along lines of least resistance? Only one reason can be given and it is one which has, caused the downfall of many an empire! It is overconfidence and contemptuous disregard of the adversaries.

Germany began the war with a blind faith in an offensive which knows no opposition. She has learned, after a 'frightful and unnecessary sacrifice of life and property, that France can be strong without Napoleon, that the rights of liberty loving nations, as the Belgian and Servian, cannot be trampled upon with impunity, that Russia is no longer the clumsy and helpless beast of the north. She has finally recognized what she should have known from the first, that England is her most dangerous enemy. She might maintain herself against the armies of the Continent, but with Great Britain shutting her off from the sea and strangling her by degrees the task is rendered impossible.

Victory over the Allies in the west, if at all obtainable, would weaken her to the point of danger; in the East the situation is becoming more hopeless every hour. Germany is losing ten thousand men and spending seventy-five millions of marks a day. Her life blood is ebbing fast; in the end she must lose. The only way to win is to crush England. In doing this she frees herself from the deadly grip at her throat and triumphs over all her enemies.

The Fatherland is now aflame with this thought and has started, with energy never shown before, a new campaign which if undertaken four months ago might have terminated the war before it was fairly under way. Germany enters this mortal combat not with the cold deliberation of a military power but with the passionate resolve of a nation animated by that one desire. She depends for success not only on her generals but on her physicists, engineers, inventors, chemists and artisans and on her volunteers who will offer themselves as martyrs for her cause.

She may make raids and demonstrations to trap the enemy, but she does not have the remotest intention to engage the British fleet in open battle. What she proposes to do is to destroy it by hellish means and artifices without losing a single ship of her own. Unless England wakes up immediately to this grave danger and prepares to meet science with science, skill with skill and sacrifice with sacrifice, the next few months may be critical for her reign as mistress of the seas. That the rules set down at The Hague are ineffective in. preventing the use of infernal devices has been already shown. International agreements are of two kinds and may be classified under two captions, which are: "United we stand, divided we fall" and "Circumstances alter cases." The provisions of The Hague are of the latter kind.

Those who would brush aside the above suggestion as highly improbable if not preposterous should bear in mind that a great nation leading in technical achievement is making a fight for its existence and that invention has already provided means by which such destruction can be accomplished, while others are foreshadowed in scientific investigations of recent years., The question that will interest everybody is what methods and contrivances is Germany likely to employ in her cunning undertaking and how can her efforts be met and frustrated?

In her attack upon England four ways are open to Germany: First, forceful invasion in disregard of the British fleet; second, engagement with the fleet in open battle; third, gradual destruction and weakening of the fleet by devices other than guns, and fourth, aerial attacks on land and sea.

History is full of daring conquests. It may be that we are to witness the most remarkable of all. The British Isles have been invaded before, but it was in times' of primitive arms. The means of defense have been brought to great perfection, it is true, but this is largely offset by correspondingly increased ,powers of offensive. The feat is difficult but not impossible.

Strategy, however, can play no important part in its consummation. It is a case of Hannibal crossing the Alps, a problem of overcoming natural barriers. England has a small coast line on which landing can be effected and many places are likely to be well guarded and fortified. If the Germans contemplate invasion it will come like a lightning stroke. They will attempt it in broad daylight and in their favored manner of hacking through the obstacles regardless of loss. Their frantic efforts to get control of the coast would seem to indicate that such is their intention.

Many experts are of the opinion that so long as there is a superior British fleet in existence an undertaking of this nature is wholly out of the question, but this is ,a mistake. It is certainly possible for the Germans to establish an operating zone in the Channel, protected on the sides by impenetrable mine fields and submarines. What is more, the possession of Calais, while it would be of great advantage to them, is not absolutely necessary to their purpose.

Whatever the plan, it will be a piece of engineering worked out in all details with German thoroughness. That is the reason why no credence can be placed in the flimsy proposals which have been described in some papers. No feasible scheme has as yet been disclosed, but I think that I am guessing correctly when I say that the Germans contemplate the use of specially designed floating fortresses, which will be in sections and transportable by rail.

They will be made virtually invulnerable to torpedo and gun attack and will be equipped with guns of great range and destructive power constructed with this very object in view. Under the protection of these fortresses, which will sweep the coast clear, landing of troops and artillery is to be effected while bodies of infantry are transported through the air, this latter operation being performed under cover of darkness. With guns of inferior caliber, and more or less unprepared, it will be hard for the Britons to frustrate the attempt.

There is some foundation to the belief that the Germans may venture a naval engagement on a large scale. They have a smaller number of vessels, but they are mostly of quite modern type and without doubt every unit is in perfect order. All reports agree that their guns are superior to those of the British, both in range and durability. The Germans are masters in the manufacture and treatment of heat resisting materials and many technical branches in other countries are entirely dependent on their product. When we add to this advantage the possibilities offered by mines, torpedoes, submarines, Zeppelins and other means of destruction, skillful manoeuver and surprise the numerical inequality of the fleet assumes secondary importance.

The marvelous exploit of a small German submarine which sank four British cruisers and escaped undamaged is in itself sufficient to justify the conclusion that the impending duel between the two countries will not be decided by guns and armor alone, considered heretofore supreme on the ocean. And yet the full capabilities of this kind of craft remain to be shown.

Germany is apt to go other nations one better. Most inventions originating elsewhere are improved by the Germans. Not only this, but they work for effect, knowing that to surprise is to strike, to strike is to win. It is highly probable that they have developed new things in submarines and may have solved the particular problem confronting them now, which is to destroy battleships in protected harbors.

This might be done by miniature vessels of simplified construction which would be virtually nothing but torpedoes and manned with one or two volunteer operators. The displacement would not need to be more than five tons, so that two or three, if not more, could be lowered from a Zeppelin in convenient localities at night. Such devices controlled by resolute men would be a new terror of the sea hard to guard against.

In general it will be very difficult for the British to combat effectively the submarine peril. An airship or airplane can be fought with a similar machine, but under water this method is impracticable and special craft 'will have to be perfected. Battleships might discourage submarine attacks by small shells filled with explosives of very high velocity so as to produce shocks of great intensity. Minute mines may also be employed, so constructed as to float at a certain depth and to explode on contact. They would do no harm to a large surface vessel but would reveal the presence of and injure a submarine, the delicate apparatus of which is easily deranged.

Next to guns the Zeppelin form of airship is the most valuable war asset of the Germans; at least they think so. Many difficulties had to be overcome in its development. A process of manufacturing cheaply pure hydrogen was perfected, a new alloy of remarkable strength and lightness produced, suitable and highly economical engines constructed and a number of other technical problems successfully solved. While not involving great originality it was a notable advance such as could only be achieved .in Germany. Much has been said, both in exaltation and depreciation, of the Zeppelin, making it necessary to separate the wheat from the chaff `before expressing an opinion as to its merits.

A claim has been advanced that a new non-inflammable gas was recently discovered, by the use of which the carrying capacity of the vessel is increased two and a half times. The only foundation of this persistent report is that according to the periodic hypothesis of elements evolved by the great Russian Mendelejeff, which has proved an unerring guide in chemical research, there should be a gas of an atomic weight 04. In a way its existence is demonstrated in the solar corona — hence the name coronium — and also in the aurora borealis, in which ease it is referred to as terrestrial, or geocoronium.

In order to estimate what Germany might do with her air fleet a correct guess must be made as to its magnitude. Prior to the declaration of war she had thirty-six vessels of various sizes and actual facilities for turning out from eight to ten each month. But under war pressure this rate could be greatly increased.

The machine has passed the experimental stage and it is simply a question of reproduction. In view of the situation it would not. be surprising to find that a hundred or more have been manufactured by this tune. Produced in such numbers the cost of each would not be more than $ 125,000; which means that one hundred could be had for the price of one single dreadnought.

The carrying capacity has heretofore been given on the basis of passenger weight, but for war purposes it could be considerably increased, and in the latest type it might be as much as twenty tons. Such a vessel could transport 200 men with full equipment and a fleet of 100 could land 20,000 men in one operation.

But the possibilities of damage by explosives are much more impressive, especially as it can be inflicted without risk. A Zeppelin fitted out with proper instruments may sail in perfect safety at great height, find the exact point for attack by reference to two wireless plants in absolute darkness, drop many tons of picric compound and do this again and again.

Several experts have expressed themselves in a slighting manner in regard to the destructive effects, but the fact is that the explosion of three tons of dynamite produces an earthquake perceptible at a distance of thirty miles. If ten tons of plastic explosive were dropped into the heart of a large city thousands would be killed and hundreds of millions of property annihilated: Suppose that a fleet of 100 such vessels were to pass over England at night dropping 100,000 bombs of twenty pounds. Who can judge of the damage and demoralization which would ensue?

At the outbreak of the war it was reported that the Germans had devised a shell the poisonous fumes of which were of great destructiveness. Shortly after a marvelous new explosive was said to have been produced in France named turpinite. The first intimation came from military quarters and some weight was attached to the news on this account and also because the discovery was attributed to Eugene Turpin, an ingenious and prolific inventor of chemicals.

The idea of employing poisonous or asphyxiating bombs is old. It is authoritatively stated that some were actually thrown during the second siege of Paris against the army of Versailles, but with the only result of killing the expert who was filling them. There is a natural and deeply rooted prejudice against the employment of poisonous agents in warfare, and many of those who tolerate the present methods of destroying life would shrink from such use. Yet death from many of the toxins known is less painful and disfiguring.

In the absence of demonstrated facts I will endeavor to show in a few words how the effectiveness of such means can be enormously increased. Consider first a large shell which, on striking the ground, liberates a poisonous gas of atmospheric density spreading in half a sphere, and let the effective radius be 1,000 feet. Now imagine that an equivalent charge is subdivided in one million parts, giving that many little shells which can be scattered over a large area. Then, since the gas will be of the same volume as before, the radius of action of each shell will be ten feet and their combined destructive effect will be 100 times greater than that of the large shell; in fact, much more so, for the distribution of the gas will not be uniform. It will be seen the secret lies in the employment of extremely small charges in great numbers.

The same reasoning leads to the conclusion that by using minute projectiles of tungsten dipped in curare or similar poison, paralyzing heart or locomotor function, a means for fighting battles would be provided more humane than the present and incomparably more effective. A complete revolution in methods of attack may be brought on through the use of toxins or asphyxiants heavier than air. This may be illustrated by an example.

Let us suppose that ten tons of such liquefied gas are dropped on a battlefield from an aerial vessel. On evaporating a gas blanket will be formed over the earth's surface, the effective height of which may be assumed to be ten feet. If ten cubic feet of the gas weigh one pound, then ten .tons will give 200,000 cubic feet of gas, which may be more or less diluted, according to its toxic activity. Assume that it is not more poisonous than carbon monoxide, which is fatal when its percentage in the atmosphere is one-half of one .per cent. That means that the gas blanket will contain 40,000,000 cubic feet, and being ten feet high it will cover 4,000,000 square feet, or, roughly, 100 acres.

In a populated city, on account of structures and other objects, the deadly zone would be very much extended.

This is danger enough, but if a gas were employed of lethal power equal to that of prussic acid, aconitine or of the strongest poison known, pseudoaconitine, the destructive area would be a hundred times greater. Evidently then there is a prospect that the chemist, who is largely responsible for the war, may also find the means of compelling its speedy termination.

Telautomatics is a name suggested for the wireless control of the organs and translatory movements of self-propelled automata. Fifteen years ago I showed its first applications and the results were received with an interest such as only few inventions have elicited. My demonstrations were repeated in Germany and other countries, but .on account of the fact that Hertz waves and imperfectly tuned circuits were employed a .general impression was created that such distant control of apparatus was not absolutely reliable.

A further argument was advanced that if it were unfailing, volunteers could always be found ready for sacrifice and more dependable because of intelligence and judgment not possessed *by an* inanimate machine. This view is held by those who are now advocating the use of manned aerial torpedoes, but nothing could be more erroneous. A crewless vessel controlled by proper wireless apparatus is in every way superior as a means of attack.

Large guns are now being manufactured in Germany so expensive and short-lived that a single shot from them costs a small fortune. It would be possible toy produce for less than the price of a shot a telautomatic aerial torpedo of much longer range and greater destructiveness which would hit its mark every time and dispense with the necessity of the gun altogether.

The new principle can also be applied to a submarine, and, particularly in connection with control from great elevation, it will afford the most perfect means for coast defense so far devised. But its full possibilities will only be appreciated when the use of certain electrical waves to which the earth is resonantly responsive becomes general. It will then be practicable to despatch a crewless boat or balloon to distances of hundreds of miles, guide it along any chart at will and release its potential energy at any point desired.

Great many of the present means and methods will then become obsolete. It is very likely that if this war is protracted this invention will prove of importance. Recent reports would indicate that experiments are being made in Germany with telautomatic torpedoes released from balloons.

One good effect of this disastrous disturbance will be a long period of peace. This is a natural consequence of the law that action and reaction are equal. But in the present phase of human development occasional convulsions are in the order of things. A still greater struggle will probably come, that between the united races of the Orient and Occident.

So long as there are different nationalities there will be patriotism. This feeling must be eradicated from our hearts before permanent peace can be established. Its place must be filled by love of nature and scientific ideal. Science and discovery are the great forces which will lead to that consummation.

I have just made known an invention which will show to electricians how to produce immense electrical pressures and activities. By their means many wonderful results will be achieved. The human voice and likeness will be flashed around the globe without wire, energy projected through space, the wastes of the ocean will be made safe to navigation, transport facilitated, rain precipitated at will and, perhaps, the inexhaustible store of atomic energy released.

Advances of this kind will, in times to come, remove the physical causes of war, the chief of which is the vast extent of this planet. The gradual annihilation: of distance will put human beings in closer contact and harmonize their views and aspirations. The harnessing of the forces of nature will banish misery and want and, provide ample. means for safe and comfortable existence.

But one more accomplishment will still be lacking to make the triumph of the mind of man complete. A way must be found to interpret thought and thereby enable the accurate reduction of all forms of human effort. to a common equivalent. The problem is susceptible of solution.

The consequences of such an advance are incalculable. A new epoch in human history would be inaugurated and a colossal revolution in moral, social and other respects accomplished, innumerable causes of trouble would be removed, our lives profoundly modified for the better, and a new and firm foundation laid to all that makes for peace.

How Cosmic Forces Shape Our Destinies
("Did the War Cause the Italian Earthquake")
New York American, February 7, 1915

Every living being is an engine geared to the wheelwork of the universe. Though seemingly affected only by its immediate surrounding, the sphere of external influence extends to infinite distance. There is no constellation or nebula, no sun or planet, in all the depths of limitless space, no passing wanderer of the starry heavens, that does not exercise some control over its destiny—not in the vague and delusive sense of astrology, but in the rigid and positive meaning of physical science.

More than this can be said. There is no thing endowed with life—from man, who is enslaving the elements, to the humblest creature—in all this world that does not sway it in turn. Whenever action is born from force, though it be infinitesimal, the cosmic balance is upset and universal motion result.

Herbert Spencer has interpreted life as a continuous adjustment to the environment, a definition of this inconceivably complex manifestation quite in accord with advanced scientific thought, but, perhaps, not broad enough to express our present views. With each step forward in the investigation of its laws and mysteries our conceptions of nature and its phases have been gaining in depth and breadth.

In the early stages of intellectual development man was conscious of but a small part of the macrocosm. He knew nothing of the wonders of the microscopic world, of the molecules composing it, of the atoms making up the molecules and of the dwindlingly small world of electrons within the atoms. To him life was synonymous with voluntary motion and action. A plant did not suggest to him what it does to us—that it lives and feels, fights for its existence, that it suffers and enjoys. Not only have we found this to be true, but we have ascertained that even matter called inorganic, believed to be dead, responds to irritants and gives unmistakable evidence of the presence of a living principle within.

Thus, everything that exists, organic or inorganic, animated or inert, is susceptible to stimulus from the outside. There is no gap between, no break of continuity, no special and distinguishing vital agent. The same law governs all matter, all the universe is alive. The momentous question of Spencer, "What is it that causes inorganic matter to run into organic forms!" has been answered. It is the sun's heat and light. Wherever they are there is life. Only in the boundless wastes of interstellar space, in the eternal darkness and cold, is animation suspended, and, possibly, at the temperature of absolute zero all matter may die.

Man as a Machine

This realistic aspect of the perceptible universe, as a clockwork wound up and running down, dispensing with the necessity of a hypermechanical vital principle, need not be in discord with our religious and artistic aspirations—those undefinable and beautiful efforts through which the human mind endeavors to free itself from material bonds. On the contrary, the better understanding of nature, the consciousness that our knowledge is true, can only be all the more elevating and inspiring.

It was Descartes, the great French philosopher, who in the seventeenth century, laid the first foundation to the mechanistic theory of life, not a little assisted by Harvey's epochal discovery of blood circulation. He held that animals were simply automata without consciousness and recognized that man, though possessed of a higher and distinctive quality, is incapable of action other than those characteristic of a machine. He also made the first attempt to explain the physical mechanism of memory. But in this time many functions of the human body were not as yet understood, and in this respect some of his assumptions were erroneous.

Great strides have since been made in the art of anatomy, physiology and all branches of science, and the workings of the man-machine are now perfectly clear. Yet the very fewest among us are able to trace their actions to primary external causes. It is indispensable to the arguments I shall advance to keep in mind the main facts which I have myself established in years of close reasoning and observation and which may be summed up as follows:

1. The human being is a self-propelled automaton entirely under the control of external influences. Willful and predetermined though they appear, his actions are governed not from within, but from without. He is like a float tossed about by the waves of a turbulent sea.

2. There is no memory or retentive faculty based on lasting impression. What we designate as memory is but increased responsiveness to repeated stimuli.

3. It is not true, as Descartes taught, that the brain is an accumulator. There is no permanent record in the brain, there is no stored knowledge. Knowledge is something akin to an echo that needs a disturbance to be called into being.

4. All knowledge or form conception is evoked through the medium of the eye, either in response to disturbances directly received on the retina or to their fainter secondary effects and reverberations. Other sense organs can only call forth feelings which have no reality of existence and of which no conception can be formed

5. Contrary to the most important tenet of Cartesian philosophy that the perceptions of the mind are illusionary, the eye transmits to it the true and accurate likeness of external things. This is because light propagates in straight lines and the image cast on the retina is an exact reproduction of the external form and one which, owing to the mechanism of the optic nerve, can not be distorted in the transmission to the brain. What is more, the process must be reversible, that in to say, a form brought to consciousness can, by reflex action, reproduce the original image on the retina just as an echo can reproduce the original disturbance If this view is borne out by experiment an immense revolution in all human relations and departments of activity will be the consequence.

Natural Forces Influence Us

Accepting all this as true let us consider some of the forces and influences which act on such a wonderfully complex automatic engine with organs inconceivably sensitive and delicate, as it is carried by the spinning terrestrial globe in lightning flight through space. For the sake of simplicity we may assume that the earth's axis is perpendicular to the ecliptic and that the human automaton is at the equator. Let his weight be one hundred and sixty pounds then, at the rotational velocity of about 1,520 feet per second with which he is whirled around, the mechanical energy stored in his body will be nearly 5,780,000 foot pounds, which is about the energy of a hundred-pound cannon ball.

This momentum is constant as well as upward centrifugal push, amounting to about fifty-five hundredth of a pound, and both will probably be without marked influence on his life functions. The sun, having a mass 332,000 times that of the earth, but being 23,000 times farther, will attract the automaton with a force of about one-tenth of one pound, alternately increasing and diminishing his normal weight by that amount.

Though not conscious of these periodic changes, he is surely affected by them.

The earth in its rotation around the sun carries him with the prodigious speed of nineteen miles per second and the mechanical energy imparted to him is over 25,160,000,000 foot pounds. The largest gun ever made in Germany hurls a projectile weighing one ton with a muzzle velocity of 3,700 feet per second, the energy being 429,000,000 foot pounds. Hence the momentum of the automaton's body is nearly sixty times greater. It would be sufficient to develop 762,400 horse-power for one minute, and if the motion were suddenly arrested the body would be instantly exploded with a force sufficient to carry a projectile weighing over sixty tons to a distance of twenty-eight miles.

This enormous energy is, however, not constant, but varies with the position of the automaton in relation to the sun. The circumference of the earth has a speed of 1,520 feet per second, which is either added to or subtracted from the translatory velocity of nineteen miles through space. Owing to this the energy will vary from twelve to twelve hours by an amount approximately equal to 1,533,000,000 foot pounds, which means that energy streams in some unknown way into and out of the body of the automaton at the rate of about sixty-four horse-power.

But this is not all. The whole solar system is urged towards the remote constellation Hercules at a speed which some estimate at some twenty miles per second and owing to this there should be similar annual changes in the flux of energy, which may reach the appalling figure of over one hundred billion foot pounds. All these varying and purely mechanical effects are rendered more complex through the inclination of the orbital planes and many other permanent or casual mass actions.

This automaton, is, however subjected to other forces and influences. His body is at the electric potential of two billion volts, which fluctuates violently and incessantly. The whole earth is alive with electrical vibrations in which he takes part. The atmosphere crushes him with a pressure of from sixteen to twenty tons, according, to barometric condition. He receives the energy of the sun's rays in varying intervals at a mean rate of about forty foot pounds per second, and is subjected to periodic bombardment of the sun's particles, which pass through his body as if it were tissue paper. The air is rent with sounds which beat on his eardrums, and he is shaken by the unceasing tremors of the earth's crust. He is exposed to great temperature changes, to rain and wind.

What wonder then that in such a terrible turmoil, in which cast iron existence would seem impossible, this delicate human engine should act in an exceptional manner? If all automata were in every respect alike they would react in exactly the same way, but this is not the case. There is concordance in response to those disturbances only which are most frequently repeated, not to all. It is quite easy to provide two electrical systems which, when subjected to the same influence, will behave in just the opposite way.

So also two human beings, and what is true of individuals also holds good for their large aggregations. We all sleep periodically. This is not an indispensable physiological necessity any more than stoppage at intervals is a requirement for an engine. It is merely a condition gradually imposed upon us by the diurnal revolution of the globe, and this is one of the many evidences of the truth of the mechanistic theory. We note a rhythm or ebb and tide, in ideas and opinions, in financial and political movements, in every department of our intellectual activity.

How Wars Are Started

It only shows that in all this a physical system of mass inertia is involved which affords a further striking proof. If we accept the theory as a fundamental truth and, furthermore, extend the limits of our sense perceptions beyond those within which we become conscious of the external impressions, then all the states in human life, however unusual, can be plausibly explained. A few examples may be given in illustration.

The eye responds only to light vibrations through a certain rather narrow range, but the limits are not sharply defined. It is also affected by vibrations beyond, only in lesser degree. A person may thus become aware of the presence of another in darkness, or through intervening obstacles, and people laboring under illusions ascribe this to telepathy. Such transmission of thought is absurdly impossible.

The trained observer notes without difficulty that these phenomena are due to suggestion or coincidence. The same may be said of oral impressions, to which musical and imitative people are especially susceptible. A person possessing these qualities will often respond to mechanical shocks or vibrations which are inaudible.

To mention another instance of momentary interest reference may be made to dancing, which comprises certain harmonious muscular contractions and contortions of the body in response to a rhythm. How they come to be in vogue just now, can be satisfactorily explained by supposing the existence of some new periodic disturbances in the environment, which are transmitted through the air or the ground and may be of mechanical, electrical or other character.

Exactly so it is with wars, revolutions and similar exceptional states of society.

Though it may seem so, a war can never be caused by arbitrary acts of man.

It is invariably the more or less direct result of cosmic disturbance in which the sun is chiefly concerned.

In many international conflicts of historical record which were precipitated by famine, pestilence or terrestrial catastrophes the direct dependence of the sun is unmistakable. But in most cases the underlying primary causes are numerous and hard to trace.

In the present war it would be particularly difficult to show that the apparently willful acts of a few individuals were not causative. Be it so, the mechanistic theory, being founded on truth demonstrated in everyday experience, absolutely precludes the possibility of such a state being anything but the inevitable consequence of cosmic disturbance.

The question naturally presents itself as to whether there is some intimate relation between wars and terrestrial upheavals. The latter are of decided influence on temperament and disposition, and might at times be instrumental in accelerating the clash but aside from this there seems to be no mutual dependence, though both may be due to the same primary cause.

What can be asserted with perfect confidence is that the earth may be thrown into convulsions through mechanical effects such as are produced in modern warfare. This statement may be startling, but it admits of a simple explanation.

Earthquakes are principally due to two causes—subterranean explosions or structural adjustments. The former are called volcanic, involve immense energy and are hard to start. The latter are named tectonic; their energy is comparatively insignificant and they can be caused by the slightest shock or tremor. The frequent slides in the Culebra are displacements of this kind.

War and the Earthquake

Theoretically, it may be said that one might think of a tectonic earthquake and cause it to occur as a result of the thought, for just preceding the release the mass may be in the most delicate balance. There is a popular error in regard to the energy of such displacements. In a case recently reported as quite extraordinary, extending as it did over a vast territory, the energy was estimated at 65,000,000,000,000 foot tons. Assuming even that the whole work was performed in one minute it would only be equivalent to that of 7,500,000 horse-power during one year, which seems much, but is little for a terrestrial upheaval. The energy of the sun's rays falling on the same area is a thousand times greater.

The explosions of mines, torpedoes, mortars and guns develop reactive forces on the ground which are measured in hundreds or even thousands of tons and make themselves felt all over the globe. Their effect, however, may be enormously magnified by resonance. The earth is a sphere of a rigidity slightly greater than that of steel and vibrates once in about one hour and forty-nine minutes.

If, as might well be possible, the concussions happen to be properly timed their combined action could start tectonic adjustments in any part of the earth, and the Italian calamity may thus have been the result of explosions in France. That man can produce such terrestrial convulsions is beyond any doubt, and the time may be near when it will be done for purposes good or apt.

How Cosmic Forces Shape Our Destinies
("Did the War Cause the Italian Earthquake")
New York American, February 7, 1915

Every living being is an engine geared to the wheelwork of the universe. Though seemingly affected only by its immediate surrounding, the sphere of external influence extends to infinite distance. There is no constellation or nebula, no sun or planet, in all the depths of limitless space, no passing wanderer of the starry heavens, that does not exercise some control over its destiny—not in the vague and delusive sense of astrology, but in the rigid and positive meaning of physical science.

More than this can be said. There is no thing endowed with life—from man, who is enslaving the elements, to the humblest creature—in all this world that does not sway it in turn. Whenever action is born from force, though it be infinitesimal, the cosmic balance is upset and universal motion result.

Herbert Spencer has interpreted life as a continuous adjustment to the environment, a definition of this inconceivably complex manifestation quite in accord with advanced scientific thought, but, perhaps, not broad enough to express our present views. With each step forward in the investigation of its laws and mysteries our conceptions of nature and its phases have been gaining in depth and breadth.

In the early stages of intellectual development man was conscious of but a small part of the macrocosm. He knew nothing of the wonders of the microscopic world, of the molecules composing it, of the atoms making up the molecules and of the dwindlingly small world of electrons within the atoms. To him life was synonymous with voluntary motion and action. A plant did not suggest to him what it does to us—that it lives and feels, fights for its existence, that it suffers and enjoys. Not only have we found this to be true, but we have ascertained that even matter called inorganic, believed to be dead, responds to irritants and gives unmistakable evidence of the presence of a living principle within.

Thus, everything that exists, organic or inorganic, animated or inert, is susceptible to stimulus from the outside. There is no gap between, no break of continuity, no special and distinguishing vital agent. The same law governs all matter, all the universe is alive. The momentous question of Spencer, "What is it that causes inorganic matter to run into organic forms!" has been answered. It is the sun's heat and light. Wherever they are there is life. Only in the boundless wastes of interstellar space, in the eternal darkness and cold, is animation suspended, and, possibly, at the temperature of absolute zero all matter may die.

Man as a Machine

This realistic aspect of the perceptible universe, as a clockwork wound up and running down, dispensing with the necessity of a hypermechanical vital principle, need not be in discord with our religious and artistic aspirations—those undefinable and beautiful efforts through which the human mind endeavors to free itself from material bonds. On the died in crowds like sheep. Contrary to peremptory orders from my father I rushed home and was stricken down. Nine months in bed with scarcely the ability to move seemed to exhaust all my vitality, and I was given up by the physicians. It was an agonizing experience, not so much because of physical suffering as on account of my intense desire to live. On the occasion of one of the fainting spells my father cheered me by a promise to let me study engineering; but it would have remained unfulfilled had it nor been for a marvelous cure brought about by an old lady. There was no force of suggestion or mysterious influence about it. Such means would have had no effect whatever on me, for I was a firm believer in natural laws. The remedy was purely medicinal, heroic if not desperate; but it worked and in one year of mountain climbing and forest life I was fit for the most arduous bodily exertion. My father kept his word, and in 1877 I entered the Joanneum in Gratz, Styria, one of the oldest technical institutions of Europe. I proposed to show results which would repay my parents for their bitter disappointment due to my change of vocation. It was not a passing determination of a light-hearted youth; it was iron resolve. As some young reader of the Scientific American might draw profit from my example I will explain.

Three rotors used with the early induction motor shown below

When I was a boy of seven or eight I read a novel untitled "Abafi"—The Son of Aba—a Servian translation from the Hungarian of Josika, a writer of renown. The lessons it teaches are much like those of "Ben Hur," and in this respect it might be viewed as anticipatory of the work of Wallace. The possibilities of will-power and self-control appealed tremendously to my vivid imagination, and I began to discipline myself. Had I a sweet cake or a juicy apple which I was dying to eat I would give it to another boy and go through the tortures of Tantalus, pained but satisfied. Had I some difficult task before me which was exhausting I would attack it again and again until it was done. So I practiced day by day from morning till night. At first it called for a vigorous mental effort directed against disposition and desire, but as years went by the conflict lessened and finally my will and wish became identical. They are so to-day, and in this lies the secret of whatever success I have achieved. These experiences are as intimately linked with my discovery of the rotating magnetic field as if they formed an essential part of it; but for them I would never have invented the induction motor.

In the first year of my studies at the Joanneum I rose regularly at three o'clock in the morning and worked till eleven at night; no Sundays or holidays excepted. My success was unusual and excited the interest of the professors. Among these was Dr. Allé, who lectured on differential equations and other branches of higher mathematics and whose addresses were unforgettable intellectual treats, and Prof. Poeschl, who held the chair of Physics, theoretical and experimental. These men I always remember with a sense of gratitude. Prof Poeschl was peculiar; it was said of him that he wore the same coat for twenty years. But what he lacked in personal magnetism he made up in the perfection of his exposition. I never saw him miss a word or gesture, and his demonstrations and experiments always went off with clock like precision. Some time in the winter of 1878 a new apparatus was installed in the lecture room. It was a dynamo with a laminated permanent magnet and a Gramme armature. Prof. Poeschi had wound some wire around the field to show the principle of self-excitation, and provided a battery for running the machine as a motor. As he was illustrating this latter feature there was lively sparking at the commutator and brushes, and I ventured to remark that these devices might be eliminated. He said that it was quite impossible and likened my proposal to a perpetual motion scheme, which amused my fellow students and embarrassed me greatly. For a time I hesitated, impressed by his authority, but my conviction grew stronger and I decided to work out the solution. At that time my resolve meant more to me than the most solemn vow.

One of the earliest of induction motors. Although it weighed only a little over 20 pounds, it developed 1/4 horse-power at a speed of 1,800 revolutions, a performance considered remarkable at the time.

I undertook the task with all the fire and boundless confidence of youth. To my mind it was simply a test of will-power. I knew nothing of the technical difficulties. All my remaining term in Gratz was passed in intense but fruitless effort, and I almost convinced myself that the problem was unsolvable. Indeed, I thought, was it possible to transform the steady pull of gravitation into a whirling force! The answer was an emphatic no. And was this not also true of magnetic attraction? The two propositions appeared very much the same.

In 1880 I went to Prague, Bohemia, carrying out my father's wish to complete my school education at a university. The atmosphere of that old and interesting city was favorable to invention. Hungry artists were plentiful and intelligent company could be found everywhere. Here I made the first distinct step in advance, by detaching the commutators from the machines and placing them on distant arbors. Every day I imagined arrangements on this plan without result, but feeling that I was nearing the solution. In the following year there was a sudden change in my views of life. I realized that my parents were making too great sacrifices for me and resolved to relieve them of the burden. The American telephone wave had reached the European continent, and the system was to be installed in Budapest. It appeared an ideal opportunity, and I took the train for that city. By an irony of fate my first employment was as a draughtsman. I hated drawing; it was for me the very worst of annoyances. Fortunately it was not long before I secured the position I sought, that of chief electrician to the telephone company. My duties brought me in contact with a number of young men in whom I became interested. One of these was Mr. Szigety, who was a remarkable specimen of humanity. A big head with an awful lump on one side and a sallow complexion made him distinctly ugly, but from the neck own his body might have served for a statue of Apollo. His strength was phenomenal. At that time I had exhausted myself through hard work and incessant thinking. He impressed me with the necessity of systematic physical development, and I accepted his offer to train me in athletics. We exercised every day and I gained rapidly in strength. My mind also seemed to grow more vigorous and as my thoughts turned to the subject which absorbed me I was surprised at my confidence of success. On one occasion, ever present in my recollection, we were enjoying ourselves in the Varos-liget or City Park. I was reciting poetry, of which I was passionately fond. At that age I knew entire books by heart and could read them from memory word by word. One of these was Faust. It was late in the afternoon, the sun was setting, and I was reminded of the passage:

"Sie rückt und weich, der Tag ist überlebt,
Dort eilt sie bin und fördert neues Leben,
Oh, das kein Flügel mich vom Boden hebt
Ihr nach und immer nach zu streben!

Ach, zu des Geistes Flügeln wird so leicht
Kein körperlicher Flügel sich gesellen!"

"The glow retreats, done is the day of toil
It yonder hastes, new fields of life exploring;
Ah, that no wing can lift me from the soil
Upon its track to follow, follow soaring!

"Alas! the wings that lift the mind no aid
Of wings to lift the body can bequeath me."

As I spoke the last words, plunged in thought and marveling at the power of the poet, the idea came like a lightning flash. In an instant I saw it all, and I drew with a stick on the sand the diagrams which were illustrated in my fundamental patents of May, 1888, and which Szigety understood perfectly.

It is extremely difficult for me to put this experience before the reader in its true light and significance for it is so altogether extraordinary. When an idea presents itself it is, as a rule, crude and imperfect. Birth, growth and development are phases normal and natural. It was different with my invention. In the very moment I became conscious of it. I saw it fully developed and perfected. Then again, a theory, however plausible, must usually be confirmed by experiment. Not so the one I had formulated. It was being daily demonstrated every dynamo and motor was absolute proof of its soundness. The effect on me was indescribable. My imaginings were equivalent to realities. I had carried out what I had undertaken and pictured myself achieving wealth and fame. But more than all this was to me the revelation that I was an inventor. This was the one thing I wanted to be. Archimedes was my ideal. I admired the works of artists, but to my mind, they were only shadows and semblances. The inventor, I thought, gives to the world creations which are palpable, which live and work.

The telephone installation was now completed and in the spring of 1882 an offer was made me to go to Paris, which I accepted eagerly. Here I met a number of Americans whom I befriended and to whom I talked of my invention, and one of them, Mr. D. Cunningham, proposed to form a company for exploitation. This might have been done had not my duties called me to Strasburg, Alsace. It was in this city that I constructed my first motor. I had brought some material from Paris, and a disk of iron with bearings was made for me in a mechanical shop close to the railroad station in which I was installing the light and power plant. It was a crude apparatus, but afforded me the supreme satisfaction of seeing for the first time, rotation affected by alternating currents without commutator. I repeated the experiment with my assistant twice in the summer of 1883. My intercourse with Americans had directed my attention to the practical introduction and I endeavored to secure capital, but was unsuccessful in this attempt and returned to Paris early in 1884. Here, too, I made several ineffectual efforts, and finally resolved to go to America, where I arrived in the summer of 1884. By a previous understanding I entered the Edison Machine Works, where I undertook the design of dynamos and motors. For nine months my regular hours were from 10:30 A. M. till 5 A. M. the next day. All this time I was getting more and more anxious about the invention and was making up my mind to place it before Edison. I still remember an odd incident in this connection. One day in the latter part of 1884 Mr. Bachelor, the manager of the works, took me to Coney Island, where we met Edison in company with his former wife. The moment that I was waiting for was propitious, I was just about to speak when a horrible looking tramp took hold of Edison and drew him away, preventing me from carrying out my intention. Early in 1885 people approached me with a proposition to develop an arc light system and to form a company under my name. I signed the contract, and a year and a half later I was free and in a position to devote myself to the practical development of my discovery. I found financial support, and in April, 1887, a company was organized for the purpose, and what has followed since is well known.

A few words should be said in regard to the various claims for anticipation which were made upon the issuance of my patents in 1888, and in numerous suits conducted subsequently. There were three contestants for the honor, Ferraris, Schallenberger and Cabanellas. All three succumbed to grief. The opponents of my patents advanced the Ferraris claim very strongly, but any one who will peruse his little Italian pamphlet, which appeared in the spring of 1888, and compare it with the patent record filed by me seven months before, and with my paper before the American Institute of Electrical Engineers, will have no difficulty in reaching a conclusion. Irrespective of being behind me in time, Prof. Ferraris's publication concerned only my split-phase motor, and in an application for a patent by him priority was awarded to me. He never suggested any of the essential practical features which constitute my system, and in regard to the split-phase motor he was very decided in his opinion that it was of no value. Both Ferraris and Schallenberger discovered the rotation accidentally while working with a Gullard and Gibbs transformer, and had difficulty in explaining the actions. Neither of them produced a rotating field motor like mine, nor were their theories the same as my own. As to Cabanellas, the only reason for his claim is an abandoned and defective technical document. Some over-zealous friends have interpreted a United States patent granted to Bradley as a contemporary record, but there is no foundation whatever for such a claim. The original application only described a generator with two circuits which were provided for the sole purpose of increasing the output. There was not much novelty in the idea, since a number of such machines existed at that time. To say that these machines were anticipations of my rotary transformer is wholly unjustified. They might have served as one of the elements in my system of transformation, but were nothing more than dynamos with two circuits constructed with other ends in view and in utter ignorance of the new and wonderful phenomena revealed through my discovery.

The Wonder World to Be Created by Electricity
Manufacturer's Record — Sept. 9. 1915

Whoever wishes to get a true appreciation of the greatness of our age should study the history of electrical development. There he will find a story more wonderful than any tale from Arabian Nights. It begins long before the Christian era when Thales, Theophrastus and Pliny tell of the magic properties of electron—the precious substance we call amber—that came from the pure tears of the Heliades, sisters of Phaeton, the unfortunate youth who attempted to run the blazing chariot of Phoebus and nearly burned up the earth. It was but natural for the vivid imagination of the Greeks to ascribe the mysterious manifestations to a hyperphysical cause, to endow the amber with life and with a soul.

Whether this was actual belief or merely poetic interpretation is still a question. When at this very day many of the most enlightened people think that the pearl is alive, that it grows more lustrous and beautiful in the warm contact of the human body. So too, it is the opinion of men of science that a crystal is a living being and this view is being extended to embrace the entire physical universe since Prof. Jagadis Chunder Bose has demonstrated, in a series of remarkable experiments, that inanimate matter responds to stimuli as plant fiber and animal tissue.

The superstitious belief of the ancients, if it existed at all, can therefore not be taken as a reliable proof of their ignorance, but just how much they knew about electricity can only be conjectured. A curious fact is that the ray or torpedo fish, was used by them in electro-therapy. Some old coins show twin stars, or sparks, such as might be produced by a galvanic battery. The records, though scanty, are of a nature to fill us with conviction that a few initiated, at least, had a deeper knowledge of amber-phenomena. To mention one, Moses was undoubtedly a practical and skillful electrician far in advance of his time. The Bible describes precisely and minutely arrangements constituting a machine in which electricity was generated by friction of air against silk curtains and stored in a box constructed like a condenser. It is very plausible to assume that the sons of Aaron were killed by a high tension discharge and that the vestal fires of the Romans were electrical. The belt drive must have been known to engineers of that epoch and it is difficult to see how the abundant evolution of static electricity could have escaped their notice. Under favorable atmospheric conditions a belt may be transformed into a dynamic generator capable of producing many striking actions. I have lighted incandescent lamps, operated motors and performed numerous other equally interesting experiments with electricity drawn from belts and stored in tin cans.

That many facts in regard to the subtle force were known to the philosophers of old can be safely concluded, the wonder is, why two thousand years elapsed before [William] Gilbert in 1600 published his famous work, the first scientific treatise on electricity and magnetism. To an extent this long period of unproductiveness can be explained. Learning was the privilege of a few and all information was jealously guarded. Communication was difficult and slow and a mutual understanding between widely separated investigators hard to reach. Then again, men of those times had no thought of the practical, they lived and fought for abstract principles, creeds, traditions and ideals. Humanity did not change much in Gilbert's time but his clear teachings had a telling effect on the minds of the learned. Friction machines were produced in rapid succession and experiments and observations multiplied. Gradually fear and superstition gave way to scientific in-sight and in 1745 the world was thrilled with the news that Kleist and Leyden had succeeded in imprisoning the uncanny agent in a phial from which it escaped with an angry snap and destructive force. This was the birth of the condenser, perhaps the most marvelous electrical device ever invented.

Two tremendous leaps were made in the succeeding forty years. One was when Franklin demonstrated the identity between the gentle soul of amber and the awe-inspiring belt of Jupiter; the other when Galvany and Volta brought out the contact and chemical battery, from which the magic fluid could be drawn in unlimited quantities. The succeeding forty years bore still greater fruit. Oersted made a significant advance in deflecting a magnetic needle by an electric current, Arago produced the electro-magnet, Seebeck the thermo-pile and in 1831, as the crowning achievement of all, Faraday announced that he had obtained electricity from a magnet, thus discovering the principle of that wonderful engine—the dynamo, and inaugurating a new era both in scientific research and practical application.

From that time on inventions of inestimable value have followed one another at a bewildering rate. The telegraph, telephone, phonograph and incandescent lamp, the induction motor, oscillatory transformer, Roentgen ray, Radium, wireless and numerous other revolutionary advances have been made and all conditions of existence eighty-four years which have since elapsed, the subtle agents dwelling in the living amber and loadstone have been transformed into cyclopean forces turning the wheels of human progress with ever increasing speed. This, in brief, is the fairy tale of electricity from Thales to the present day. The impossible has happened, the wildest dreams have been surpassed and the astounded world is asking: What is coming next?

Electrical Possibilities in Coal and Iron

Many a would-be discoverer, failing in his efforts, has felt the regret to have been born at a time when everything has been already accomplished and nothing is left to be done. This erroneous impression that, as we are advancing, the possibilities of invention are being exhausted, is not uncommon. In reality it is just the opposite. Spenser has conveyed the right idea when he likened civilization to the sphere of light which a lamp throws out in darkness. The brighter the lamp and the larger the sphere the greater is its dark boundary. It is paradoxical, yet true, to say, that the more we know the more ignorant we become in the absolute sense, for it is only through enlightenment that we become conscious of our limitations. Precisely one of the most gratifying results of intellectual evolution is the continuous opening up of new and greater prospects. We are progressing at an amazing pace, but the truth is that, even in fields most successfully exploited, the ground has only been broken. What has been so far done by electricity is nothing compared with what the future has in store. Not only this, but there are now innumerable things done in old-fashioned ways which are much inferior in economy, convenience and many other respects to the new method. So great are the advantages of the latter that whenever an opportunity presents itself the engineer advises his client to "do it electrically."

Consider, in illustration, one of the largest industries, that of coal. From this valuable mineral we chiefly draw the sun's stored energy which is required to meet our industrial and commercial needs. According to statistical records, the output in the United States during the past year was 480,000,000 tons. In perfect engines this fuel would have been sufficient to develop 500,000,000 horse-power steadily for one year, but the squandering is so reckless that we do not get more than 5 per cent of its heating value on the average. There is an appalling waste in mining, handling, transportation, store and use of coal, which could be very much reduced through the adoption of a comprehensive electrical plan in all these operations. The market value of the yearly product would be easily doubled and an immense sum added to the revenues of the country. What is more, inferior grades, billions of tons of which are being thrown away, might be turned to profitable use.

Similar considerations apply to natural gas and mineral oil, the annual loss of which amounts to hundreds of millions of dollars. In the very near future such waste will be looked upon as criminal and the introduction of the new methods will be forced upon the owners of such properties. Here, then, is an immense field for the use of electricity in many ways, vast industries which are bound to be revolutionized through its extensive application.

To give another example, I may refer to the manufacture of iron and steel, which is carried on in this country on a scale truly colossal. During the last year, notwithstanding unfavorable business conditions, 31,000,000 tons of steel have been produced. It would lead too far to dwell on the possibilities of electrical improvements in the manufacturing processes themselves, and I will only indicate what is likely to be accomplished in using the waste gases from the coke ovens and blast furnaces to generate electricity for industrial purposes.

Since in the production of pig-iron for every ton about one ton of coke is employed, the yearly consumption of coke may be put at 31,000,000 tons. The combustion in the blast furnaces yields, per minute, 7,000,000 cubic feet of gas of a heating value of 110 B. T. units per cubic foot. Of this total, without making special provision, 4,000,000 cubic feet may be made available for power purposes. If all the heat energy of this gas could be transformed into mechanical effort, it would develop 10,389,000 horse-power. This result is impossible, but it is perfectly practicable to obtain 2,500,000 horse-power electrical energy at the terminals of the dynamos.

In the manufacture of coke approximately 9400 cubic feet of gas are evolved per ton of coal. This gas is excellent for power purposes, having an average heating value of 600 B. T. units, but very little is now used in engines, largely because of their great cost and other imperfections. A ton of coke requires about 1.32 tons of American coal; hence the total coal consumption per annum on the above basis is nearly 41,000,000 tons, which give, per minute, 733,000 cubic feet of gas. Assuming the yield of surplus or rich gas to be 333,000 cubic feet, the balance of 400,000 cubic feet could be used in gas engines. The heat contents would be, theoretically, sufficient to develop 5,660,000 horse-power, of which 1,500,000 horse-power could be obtained in the form of electric energy.

I have devoted much thought to this industrial proposition, *and find that with new, efficient, extremely cheap and simple thermo-dynamic transformers not less than 4,000,000 horse-power could be developed in electric generators by utilizing the heat of these gases, which, if not entirely wasted, are only in part and inefficiently employed.*

With systematic improvements and refinements much better results could be secured and an annual revenue of $50,000,000 or more derived. The electrical energy could be advantageously used in the fixation of atmospheric nitrogen and production of fertilizers, for which there is an unlimited demand and the manufacture of which is restricted here on account of the high cost of power. I expect confidently the practical realization of the project in the very near future, and look to exceptionally rapid electrical development in this direction.

Hydro-Electric Development

Water-power offers great opportunities for novel electrical applications, particularly in the department of electro-chemistry. The harness of waterfalls is the most economical method known for drawing energy from the sun. This is due to the fact that both water and electricity are incompressible. The net efficiency of the hydro-electric process can be as high as 85 per cent. The initial outlay is generally great, but the cost of maintenance is small and the convenience offered ideal. My alternating system is invariably employed , and so far about 7,000,000 horse-power have been developed. As generally used we do not get more than six-hundredths of a horse-power per ton of coal per year. *This water energy is therefore equivalent to that obtainable from an annual supply of 120,000,000 tons of coal, which is about 25 per cent of the total output in the United States.* The estimate is conservative, and in view of the immense waste of coal 50 per cent may be a closer guess.

We get better appreciation of the tremendous value of this power in our economic development when we remember that, unlike fuel, which demands a terrible sacrifice of human energy and is consumed, it is supplied without effort and destruction of material and equals the mechanical performance of 150,000,000 men—one and one-half times the entire population of this country. These figures are imposing; nevertheless, we have only begun the exploitation of this vast national resource.

There are two chief limitations at present—one in the availability of the energy, the other in its transmission to distance. The theoretical power of the falling water is enormous. If we assume for the rain clouds an average height of 15,000 feet and annual precipitation of 33 inches, the 24 horse-power per square mile is over 4000, and for the whole area in the United States more than 12,000,000,000 horse-power. As a matter of fact, the larger portion of the potential energy is used up in air friction. This, while disappointing to the economist, is a fortunate circumstance, for otherwise the drops would reach the ground with a speed of 800 feet per second—sufficient to raise blisters on our bodies, while hail would be positively deadly. Most of the water, which is available for power purposes comes from a height of about 2000 feet and represents over one and one-half billion horse-power, but we are only able to use an average fall of, say, 100 feet, which means that if all the water-power in this country were harnessed under the existing conditions only 80,000,000 horse-power could be obtained.

The next Great Achievement—Electrical Control of Atmospheric Moisture

But the time is very near when we shall have the precipitation of the moisture of the atmosphere under complete control, and then it will be possible to draw unlimited quantities of water from the oceans, develop any desired amount of energy, and completely transform the globe by irrigation and intensive farming. A greater achievement of man through the medium of electricity can hardly be imagined.

The present limitations in the transmission of power to distance will be overcome in two ways—through the adoption of underground conductors insulated by power, and through the introduction of the wireless art. The first plan I have advanced years ago. The underlying principle is to convey through a tubular conductor hydrogen at a very low temperature, freeze the surrounding material and thus secure a perfect insulation by indirect use of electric energy. In this manner the power derived from falls can be transmitted to distances of hundreds of miles with the highest economy and at a small cost. This innovation is sure to greatly extend the fields of electrical application. As to the wireless method, we have now the means for economic transmission of energy in any desired amount and to distances only limited by the dimensions of this planet. In view of assertions of some misinformed experts to the effect that in the wireless system I have perfected the power of the transmitter is dissipated in all directions, I wish to be emphatic in my statement that such is not the case. The energy goes only to the place where it is needed and to no other.

When these advanced ideas are practically realized we shall get the full benefit of water-power, and it will become our chief dependence in the supply of electricity for domestic, public and other uses in the arts of peace and war.

Economy in Light and Power—Electric Propulsion

In the great departments of electric light and power immense opportunities are offered through the introduction of all kinds of novel devices which can be attached to the circuits at convenient hours for the purpose of equalizing the loads and increasing the revenues from the plants. I have myself knowledge of a number of new appliances of this kind. The most important among them is probably an electrical ice machine which obviates entirely the use of dangerous and otherwise objectionable chemicals. The new machine will also require absolutely no attention and will be extremely economical in operation, so that the refrigeration will be effected very cheaply and conveniently in every household.

An interesting fountain, electrically operated, has been brought out which is likely to be extensively introduced, and will afford an unusual and pleasing sight in squares, parks, hotels and residences.

Cooking devices for all domestic purposed are being provides, and there is great demand for practical designs and suggestions in this field. The same may be stated of electric signs and other attractive means of advertising which can be electrically operated. Some of the effects which it is possible to produce by electric currents are wonderful and lend themselves to exhibitions, and there is no doubt that much can be done in that direction. Theaters, public halls and private dwellings are in need of a great many devices and instruments for convenience and offer ample opportunities to an ingenious and practical inventor.

A vast and absolutely untouched field is the use of electricity for the propulsion of ships. The leading electrical company in this country has just equipped a large vessel with high-speed turbines and electric motors and has achieved a signal success. Applications of this kind will multiply at a rapid rate, for the advantages of the electrical drive are now patent to everybody. In this connection gyroscopic apparatus will probably play an important part, as its general adoption on vessels is sure to come. Very little has yet been done in the introduction of electrical drive in the various branches of industry and manufacture, and the prospects are unlimited.

A Few of the Wonders to Come

Books have already been written on the agricultural uses of electricity, but the fact is that hardly anything has been practically done. The beneficial effects of electricity of high tension have been unmistakably established, and a revolution will be brought about through the extensive adoption of agricultural electrical apparatus. *The safeguarding of forests against fires, the destruction of microbes, insects and rodents will, in due course, be accomplished by electrical means.*

In the near future we shall see a great many new uses of electricity aiming at safety, particularly vessels at sea. *We shall have electrical instruments for preventing collisions, and we shall even be able to disperse fogs by electric force and powerful and penetrative rays. I am hopeful that within the next few years wireless plants will be installed for the purpose of illuminating the oceans.* The project is perfectly feasible, and if carried out will contribute more than any other provision to the safety of property and human lives at sea. The same plant could also produce stationary electrical waves and enable vessels to get at any time accurate bearings and other valuable practical data without resorting to the present means. It could also be used for time signaling and many other purposes of similar nature.

Electrotherapy is another great field in which there are unlimited possibilities for electrical applications. High-frequency currents especially have a great future. The time will come when this form of electrical energy will be available in every private residence. I consider it quite possible that through their surface actions we may do away with the customary bath, as the cleaning of the body can be instantaneously effected simply by connecting it to a source of currents or electric energy of very high potential, which results in the throwing off of dust or any small particles adhering to the skin. Such a dry bath, besides being convenient and time-saving, would also be of beneficial therapeutic influence. New electric devices for use of the deaf and blind are coming and will be a blessing to the afflicted.

In the prevention of crime electrical instruments will soon become an important factor. In court proceedings electric evidence will often be decisive. *In a time not distant it will be possible to flash any image formed in thought on a screen and render it visible at any place desired.* The perfection of this means of reading thought will create a revolution for the better in all our social relations. Unfortunately, it is true, that cunning lawbreakers will avail themselves of such advantages to further their nefarious business.

Telegraphic Photography and Other Advances

Great improvements are still possible in telegraphy and telephony. The use of a new receiving device which will be shortly described, and the sensitiveness of which can be increased almost without limit, will enable telephoning through aerial lines or cables however long by reducing the necessary working current to an infinitesimal value. This invention will dispense with the necessity of resorting to expensive constructions, which, however, are of circumscribed usefulness. It will also enormously extend the wireless transmission of intelligence in all its departments.

The next art to be inaugurated is that of picture transmission by ordinary telegraphic methods and existing apparatus. This idea of telegraphing or telephoning pictures is old, but practical difficulties have hampered commercial realizations. A number of improvements of great promise have been made, and there is every reason to expect that success will soon be achieved.

Another valuable novelty will be a typewriter electrically operated by the human voice. This advance will fill a long-felt want, as it will do away with the operator and save a great deal of labor and time in offices.

A new and extremely simple electric tachometer is being prepared for the market, and it is expected that it will prove useful in power plants and central stations, on boats, locomotives and automobiles.

Many municipal improvements based on the use of electricity are about to be introduced. *We have soon to have everywhere smoke annihilators, dust absorbers, ozonizers, sterilizers of water, air, food and clothing, and accident preventers on streets, elevated roads and in subways. It will become next to impossible to contract disease germs or get hurt in the city, and country folk will got to town to rest and get well.*

Electric Inventions in War

The present international conflict is a powerful stimulus to invention of devices and implements of warfare. An electric gun will soon be brought out. The wonder is that it was not produced long ago. Dirigibles and airplanes will be equipped with small electric generators of high tension, from which the deadly currents will be conveyed through the wires to the ground. Battleships and submarines will be provided with electric and magnetic feelers so delicate that the approach of any body underwater or in darkness will be detected. Torpedoes and floating mines are almost in sight which will direct themselves automatically and without fail get in fatal contact with the object to be destroyed. The art of telautomatics, or wireless control of automatic machines at a distance, will play a very important part in future wars and, possibly, in the next phases of the present one. Such contrivances which act as if endowed with intelligence will be used in innumerable ways for attack as well as defense. They may take the shape of airplanes, balloons, automobiles, surface or under-water boats, or any other form according to the requirement in each special case, and will be of greater range and destructiveness than the implements now employed. *I believe that the telautomatic aerial torpedo will make the large siege gun, on which so much dependence is placed at present, obsolete.*

A volume might be filled with such suggestions without exhausting the possibilities. The advance even under the conditions existing is rapid enough, but when the wireless transmission of energy for general use becomes a practical fact the human progress will assume the character of a hurricane. So all-surpassing is the importance of this marvelous art to the future existence and welfare of the human race that every enlightened person should have a clear idea of the chief factors bearing on its development.

The Power of the Future

We have at our disposal three main sources of life-sustaining energy—fuel, water-power and the heat of the sun's rays. Engineers often speak of harnessing the tides, but the discouraging truth is that the tidewater over one acre of ground will, on the average, develop only one horse-power. Thousands of mechanics and inventors have spent their best efforts in trying to perfect wave motors, not realizing that the power so obtained could never compete with that derived from other sources. The force of wind offers much better chances and is valuable in special instances, but is by far inadequate. Moreover, the tides, waves and winds furnish only periodic and often uncertain power and necessitate the employment of large and expensive storage plants. Of course, there are other possibilities, but they are remote, and we must depend on the first of three resources. If we use fuel to get our power, we are living on our capital and exhausting it rapidly. This method is barbarous and wantonly wasteful, and will have to be stopped in the interest of coming generations. The heat of the sun's rays represents an immense amount of energy vastly in excess of water-power. The earth receives an equivalent of 83 foot-pounds per second for each square foot on which the rays fall perpendicularly. From simple geometrical rules applying to a spherical body it follows that the mean rate per square foot of the earth's surface is one-quarter of that, or 20 3/4 foot-pounds. This is to say over one million horse-power per square mile, or 250 times the water-power for the same area. But that is only true in theory; the practical facts put this in a different aspect. For instance, considering the United States, and taking into account the mean latitude, the daily variation, the diurnal changes, the seasonal variations and casual changes, this power of the sun's rays reduces to about one-tenth, or 100,00 horse-power per square mile, of which we might be able to recover in high-speed low-pressure turbines 10,000 horse-power. To do this would mean the installment of apparatus and storage plants so large and expensive that such a project is beyond the pale of the practical. *The inevitable conclusion is that water-power is by far our most valuable resource. On this humanity must build its hopes for the future.* With its full development and a perfect system of wireless transmission of the energy to any distance man will be able to solve all the problems of material existence. Distance, which is the chief impediment to human progress, will be completely annihilated in thought, word and action. Humanity will be united, wars will be made impossible and peace will reign supreme.

Nikola Tesla Sees A Wireless Vision
New York Times, Oct. 3, 1915

Nikola Tesla announced to The Times last night that he had received a patent on an invention which would not only eliminate static interference, the present bugaboo of wireless telephony, but would enable thousands of persons to talk at once between wireless stations and make it possible for those talking to see one another by wireless, regardless of the distance separating them. He said also that with his wireless station now in the process of construction on Long Island he hoped to make New York one of the central exchanges in a world system of wireless telephony.

Mr. Tesla has been working on wireless problems for many years. Yesterday he exhibited an article published in the Electrical World eleven years ago, in which he predicted not only wireless telephony on a commercial basis but that it would be possible to identify the voice of an acquaintance over any distance. That its operator in Hawaii was able to distinguish the voice of an engineer friend at Arlington, Va., was announced by the American Telephone and Telegraph Company as the most marked triumph of its communication by wireless telephone from the naval radio station at Arlington to Pearl Harbor, Hawaii, a distance of 4,000 miles.

The inventor, who has won fame by his electrical inventions, dictated this statement yesterday.

"The experts carrying out this brilliant experiment are naturally deserving of great credit for the skill they have shown in perfecting the devices. These are of two kinds: First, those serving to control transmission, and, second, those magnifying the received impulse. That the control of transmission is perfect is plain to experts from the fact the Arlington, Mare Island, and Pearl Harbor plants are all inefficient and that the distance of telephonic transmission is equal to that of telegraphic transmission. It is also perfectly apparent that the chief merit of the application lies in the magnification of the microphonic impulse. It must not be imagined that we deal here with new discoveries. The improvement simply concerns the control of the transmitted and the magnification of the received impulse, but the wireless system is the same. This can never be changed.

"That it is practicable to project the human voice not only to a distance of 5,000 miles, but clear across the globe, I demonstrated by experiments in Colorado in 1899. Among my publications I would refer to an article in the Electrical World of March 5, 1904, but describing really tests I made in 1899. The facts which I pointed out in the article were of much greater significance than that of the experiments reported, although this should be taken in a scientific sense, as the experiments were simply scientific demonstrations. I pointed out then that the modulations of the human voice can be reproduced more clearly through the earth than through wire. It is difficult for the layman to understand, but it is an absolute fact that transmission through the earth with the proper apparatus is not more difficult than the sending of a message on a wire strung across a room. This wonderful property of the planet, that, electrically speaking, is through its very bigness, small, is of incalculable significance for the future of mankind.

"These tests made between Washington and Honolulu will act as an immense stimulus to wireless telephony and would be of much more value to the world if the principles of the transmission were understood. But they are not. Even now, fifteen years after the fundamental principles have been demonstrated and the possibilities shown, there are many experts in the dark.

"For instance, it is claimed that static disturbances will fatally interfere with the transmission, while, as a matter of fact, there is no static disturbance possible in properly designed transmission and receiving circuits. Quite recently I have described in a patent, circuits which are absolutely immune to static and other interferences - so much so that when a telephone is attached, there is absolute silence, even lightning in the immediate vicinity not producing a click of the diaphragm, while in the ordinary telephonic conversation there are all kinds of noises. Transmission without static interference has many wonderful properties, besides, first of which is that unlimited amounts of power can be transmitted with very small loss.

"Another contention is that there can be no secrecy in wireless telephone conversation. I say it is absurd to raise this contention when it is positively demonstrated by experiments that the earth is more suitable for transmission than any wire could ever be. A. wireless telephone conversation can be made as secret as thought.

"I have myself erected a plant for the purpose of connecting by wireless telephone the chief centers of the world, and from this plant as many as a hundred will be able to talk absolutely without interference and with absolute secrecy. This plant would simply be connected with the telephone central exchange of New York City, and any subscriber will be able to talk to any other telephone subscriber in the world, and all this without any change in his apparatus. This plan has been called my "world system." By the same means I propose also to transmit pictures and project images, so that the subscriber will not only hear the voice, but see the person to whom he is talking. Pictures transmitted over wires is a perfectly simple art practiced today. Many inventors have labored on it, but the chief credit is due Professor Korn of Munich.

"His apparatus can be attached to a wireless plant and at any other wireless plant can be reproduced. I have undertaken this in the hope of establishing a service which would greatly facilitate the work of the press. A picture could be sent from a battlefield in Europe to New York in five minutes if the proper instruments were available.

"A further advantage would be that transmission is instant and free of the unavoidable delay experienced with the use of wires and cables. As I have already made known, the current passes through the earth, starting from the transmission station with infinite speed, slowing down to the speed of light at a distance of 6,000 miles, then increasing in speed from that region and reaching the receiving station again with infinite velocity.

"It's all a wonderful thing. Wireless is coming to mankind in its full meaning like a hurricane some of these days. Some day there will be, say, six great wireless telephone stations in a world system connecting all the inhabitants on this earth to one another, not only by voice, but by sight. It's surely coming.

Correction by Mr. Tesla
New York Times, *Oct. 4, 1915*

The Times received last night from Nikola Tesla a letter saying the inventor wished to correct a statement in his forecast of the possibilities of wireless, published in the Times of yesterday morning, when he was quoted as saying that the apparatus used by the American Telephone and Telegraph Co. to talk from Arlington to Hawaii was "inefficient." The inventor wrote that he wished to say that the apparatus was "ineffective."

"Although I can guess the character of the apparatus which was employed in projecting the human voice through 4,600 miles of space," Mr. Tesla wrote, "I am unable to judge of its efficiency, but from the technical particulars available I know that the plants are ineffective, inasmuch as they would have furnished currents of much greater volume and tension had they been differently designed. Incidentally, they would then have been immune against static disturbances, unfailing in their operation and adapted to secure secrecy of messages.

"In calling attention to this fact I have meant to give testimony to the excellence of the means of control and magnification resorted to by the experimenters. Had the same devices been used in connection with plants designed for maximum effect the results would have been such as to cause a most profound sensation and to stir great commercial interests all the world over, perhaps to the point of powerfully affecting and hastening the finish of the awful struggle in which nations of the earth are now engaged."

Wonders of the Future
Collier's Weekly, December 2, 1916

Many a would-be discoverer, failing in his efforts, has felt regret at having been born at a time when, as he thinks, everything has been already accomplished and nothing is left to be done. This erroneous impression that, as we are advancing, the possibilities of invention are being exhausted is not uncommon. In reality it is just the opposite. What has been so far done by electricity is nothing as compared with what the future has in store. Not only this, but there are now innumerable things done in old-fashioned ways which are much inferior in economy, convenience, and many other respects to the new method. So great are the advantages of the latter that whenever an opportunity presents itself the engineer advises his client to "do it electrically."

Water power offers great opportunities for novel electrical applications, particularly in the department of electrochemistry. The harnessing of waterfalls is the most economical method known for drawing energy from the sun. This is due to the fact that both water and electricity are incompressible. The net efficiency of the hydroelectric process can be as high as 85 per cent. The initial outlay is generally great, but the cost of maintenance is small and the convenience offered ideal. My alternating system is invariably employed, and so far about 7,000,000 horsepower as been developed. As generally used, we do not get more than six-hundredths of a horsepower per ton of coal per year. This water energy is therefore equivalent to that obtainable from an annual supply of 120,000,000 tons of coal, which is from 25 to 50 per cent of the total output of the United States.

Great possibilities also lie in the use of coal. From this valuable mineral we chiefly draw the sun's stored energy, which is required to meet our industrial and commercial needs. According to statistical records the output in the United States during an average year is 480,000,000 tons. In perfect engines this fuel would be sufficient to develop 500,000,000 horsepower steadily for one year, but the squandering is so reckless that we do not get more than 5 per cent of its heating value on the average. A comprehensive electrical plan for mining, transporting, and using coal could much reduce this appalling waste. What is more, inferior grades, billions of tons of which are being thrown away, might be turned to profitable use.

Similar considerations apply to natural gas and mineral oil, the annual loss of which amounts to hundreds of millions of dollars. In the very near future such waste will be looked upon as criminal and the introduction of the new methods will be forced upon the owners of such properties. Here, then, is an immense field for the use of electricity in many ways. The manufacture of iron and steel offers another large opportunity for the effective application of electricity.

In the production of pig iron about one ton of coke is employed for every ton. Thus 31,000,000 tons of coke are used a year. There are 4,000,000 cubic feet of gases from the blast furnaces which may be used for power purposes. It is practicable to obtain 2,500,000 horsepower electrical energy in this way.

In the manufacture of coke some 41,000,000 tons of coal are employed in this country. From the gases produced in this process some 1,500,000 horsepower could be produced in the form of electrical energy.

I have devoted much thought to this industrial proposition, and find that with new, efficient, extremely cheap, and simple thermodynamic transformers not less than 4,000,000 horsepower could be developed in electric generators by utilizing the heat of these gases, which, if not entirely wasted, are only in part and inefficiently employed.

With systematic improvements and refinements much better results could be secured and an annual revenue of $50,000,000 or more derived. The electrical energy could be advantageously used in the fixation of atmospheric nitrogen and production of fertilizers, for which there is an unlimited demand and the manufacture of which is restricted here on account of the high cost of power. I expect confidently the practical realization of this project in the very near future, and look to exceptionally rapid electrical development in this direction.

But the time is very near when we shall have the precipitation of the moisture of the atmosphere under complete control, and then it will be possible to draw unlimited quantities of water from the oceans, develop any desired amount of energy, and completely transform the globe by irrigation and intensive farming. A greater achievement of man through the medium of electricity can hardly be imagined.

The present limitations in the transmission of power to distance will be overcome in two ways: through the adoption of underground conductors insulated by power, and through the introduction of the wireless art.

When these advanced ideas are practically realized we shall get the full benefit of water power, and it will become our chief dependence in the supply of electricity for domestic, public, and other uses in the arts of peace and war.

A vast and absolutely untouched field is the use of electricity for the propulsion of ships. The leading electrical company in this country equipped a large vessel with high-speed turbines and electric motors. The new equipment was a signal success. Applications of this kind will multiply at a rapid rate, for the advantages of the electrical drive are not patent to everybody. Gyroscopic apparatus will probably play an important part, as its general adoption on vessels is sure to come, Very little has yet been done in the introduction of electrical drive in the various branches of industry and manufacture, but the prospects here are unlimited.

Books have already, been written on the uses of electricity in agriculture, but the fact is that very little has been practically done. The beneficial effects of electricity of high tension have been unmistakably established, so that we are warranted in believing that a revolution will be brought about through the extensive adoption of agricultural electrical apparatus. The safeguarding of forests against fires, the destruction of microbes, insects, and rodents will, in due course, be accomplished by electricity.

In the not far distant future we shall see a great many new uses of electricity that will aim at safety. The safety of vessels at sea will be particularly affected. We shall have electrical instruments which will prevent collisions, and we shall even be able to disperse fogs by electric force and powerful and penetrative rays. I am hopeful that within the next few years wireless plants will be installed for the purpose of illuminating the oceans. The project is perfectly feasible; if carried out it will contribute more than any other provision to the safety of property and human lives at sea. The same plant could also produce stationary electrical waves and enable ships to .get any time accurate bearings and other valuable practical data, thus making the present means unnecessary. It could also be used for time signaling and many other such purposes.

In the great departments of electric light and power great opportunities are offered through the introduction of many kinds of novel devices which can be attached to the circuits at convenient hours to equalize the loads and increase the revenues from the plants. I myself have knowledge of a number of new appliances of this kind. The most important of them is probably an electrical ice machine which obviates entirely the use of dangerous and otherwise objectionable chemicals. The new machine will also require no attention and will be very economical in operation. In this way refrigeration will be effected very cheaply and conveniently in every household.

An interesting fountain, electrically operated, has already been brought out. It will very likely be extensively introduced, and will afford an unusual and pleasing sight in squares, parks, and hotels.

Cooking devices for all domestic purposes are now being made, and there is a large demand for practical designs and suggestions in this field, and for electric signs and other attractive means of advertising which can be electrically operated. Some of the effects which it is possible to produce by electric currents are wonderful and lend themselves to exhibitions. There is no doubt that much can be done in this direction. Theaters, public halls, and private dwellings are in need of a great many devices and instruments for convenience, and offer ample opportunities to ingenious and practical inventors.

Great improvements are also still possible in telegraphy and telephony. The use of a new receiving device, the sensitiveness of which can be increased almost without limit, will enable us to telephone through aerial lines or cables of any length by reducing the necessary working current to an infinitesimal value. This invention will enormously extend the wireless transmission of intelligence in all its departments.

The next art to be introduced is that of picture transmission telegraphically. Existing apparatus will be used. This idea of telegraphing or telephoning pictures was arrived at long ago, but practical difficulties have hampered commercial realization. There have been promising experiments, and there is every reason to believe that success will soon be achieved. Another valuable invention will be a typewriter electrically operated by the human voice. This advance will be of the utmost value, as it will do away with the operator and save a great deal of labor and time in business offices.

Many municipal improvements based on the use of electricity are soon to be introduced. There will be smoke annihilators, dust absorbers, ozonizers, sterilizers of water, air, food, and clothing, and accident preventers on streets, elevated roads, and in subways. It will become next to impossible to contract diseases from germs or get hurt in the city. Country folk will go to town to rest and get well.

Electrotherapy is another great field in which there are unlimited possibilities for the application of electricity. High-frequency currents especially have a great future. The time is bound to come when this form of electrical energy will be on tap in every private residence. It is possible that we may be able to do away with the customary bath. The cleaning of the body can be instantaneously effected simply by connecting it to a source of electric energy of very high potential, which will result in the throwing off of dust or any small particles adhering to the skin. Such a bath, besides being dry and time-saving would also be of beneficial therapeutic influence. New electric devices that will be a blessing to the deaf and blind are coming.

Electrical instruments will soon become an important factor in the prevention of crime. In court proceedings electric evidence can be made decisive. It will, no doubt, be possible before very long to flash any image formed in the mind on a screen and make it visible to a spectator at any place desired. The perfection of this sort of reading thought will create a revolution for the better in all our social relations. It is true that cunning lawbreakers will avail themselves of the same means to further their nefarious business.

The present international conflict is a powerful stimulus to invention of destructive devices and implements. An electric gun will soon be brought out. The wonder is that it was not invented long ago. Dirigibles and airplanes will be furnished with small electric generators of high tension, from which the deadly currents will be conveyed through thin wires to the ground. Battleships and submarines will be provided with electric and magnetic feelers so delicate that the approach of any body under water or in darkness may be easily detected. Torpedoes and floating mines will direct themselves automatically and without fail get in fatal contact with the object to be destroyed - in fact, these are almost in sight. The art of telautomatics, or wireless control of automatic machines at a distance, will play a very important role in future wars and, possibly, in the later phases of the present one. Such contrivances, which act as if endowed with intelligence, may take the shape of airplanes, balloons, automobiles, surface, or underwater boats, or any other form according to the requirement in each special case. They will have far greater ranges and will be much more destructive than the implements now employed. I believe that the telautomatic aerial torpedo will make the large siege gun, on which so much dependence is now placed, utterly obsolete.

Electric Drive for Battle Ships
New York Herald, February 25, 1917

The ideal simplicity of the induction motor, its perfect reversibility and other unique qualities render it eminently suitable for ship propulsion, and ever since I brought my system of power transmission to the attention of the profession through the American Institute of Electrical Engineers I have vigorously insisted on its application for that purpose. During many years the scheme was declared to be impracticable and I was assailed in a manner as vicious as incompetent. In 1900, when an article from me advocating the electric drive appeared in the Century Magazine, Marine Engineering pronounced the plan to be the "climax of asininity," and such was the fury aroused by my proposals that the editor of another technical periodical resigned and severed his connection rather than to allow the publication of some attacks.

A similar reception was accorded to my wireless boat repeatedly described in the Herald of 1898. The patents on these inventions have since expired and they are now common property. Meanwhile insane antagonism and ignorance have been replaced by helpful interest and appreciation of their value. Recently the Navy Department has let contracts aggregating $100,000,000 for the construction of seven war vessels with the induction motor drive, and an equal sum is appropriated to cover the cost of four huge battle cruisers which are to be fitted out in the same way. This latter project is resisted by some shipbuilders, turbine makers, electrical manufacturers and engineers who, in fear of a fatal mistake by the government and under the sway of patriotic motives, urged upon the authorities the employment of the geared turbine.

Controversial Correspondence

Numerous letters of protest have been written to C. A. Swanson, of the Senate Naval Committee, but what has so far come out of this correspondence is purely controversial and of no profit whatever to those who seek information. It is regrettable that the question should have been raised at this critical moment, when speedy preparation against threatening national perils is recognized as imperative, and in view of this no doubt should be permitted to remain in the public mind as to the superiority of the equipment recommended by the naval experts. In the following I shall endeavor to make this clear to the general reader.

The most efficient means of propulsion is a jet of water expelled astern from the body of the vessel. Though the theoretical laws governing its action were precisely expressed fifty years ago by Rankine, a singular and inexplicable prejudice against this device still prevails among engineers and writers of text books on hydraulics. But the far sighted are keenly alive to its possibilities. While our present motive resources do not admit of an advantageous use of the jet, it can be confidently predicted that it will soon be instrumental in a more complete conquest of the deep. I firmly believe that at this writing it is being applied to the submarines devastating the oceans, for their silence alone can explain why they escape so easily detection by microphonic instruments. The sound emitted is the Achilles talon of the undersea boat. Its suppression materially increases the destructiveness of the new weapon.

The Helical Propeller

However, under the existing conditions the best results are obtained in all kinds of surface craft with a helical propeller, which is operated in four different ways. First, straight from the shaft of the prime mover; second, by means of a gear wheel; third, through a hydraulic transformer, and, fourth, by an electric transmitter of power. As the screw in order to save energy must be revolved at a moderate rate, the first mentioned, or "direct drive," lends itself best to a reciprocating or rotary engine. The former is clumsy, the latter impossible, and competition has forced the turbine on the market. But, excessive speed being indispensable to its good performance, it had to be adapted to the propeller. This was, in a measure, accomplished by "staging" - that is, passing the steam through a number of turbines in succession - a plan obviously entailing great drawbacks, financial and otherwise. Need of reducing the bulk and cost of the machinery and insuring better working then compelled the adoption of the second arrangement - "geared turbine drive" - in which a peculiar pinioned wheel, first introduced by De Laval, transmits motion to the screw. Next, efforts to do away with certain limitations of this combination resulted in the third, or "hydraulic drive," the turbine actuating the propeller through a centrifugal pump and water motor. Finally, as the furthest step toward perfection, the last named disposition - "electric drive" - was resorted to. In this case the turbine imparts rotation to a dynamo, which in turn runs a motor carrying the screw on its shaft.

Advantages of Types

Each of these forms has its supporters and champions. In principle the first would be the preferable were it not handicapped in many respects. The second type is cheap, but the gear is a serious objection. Though less economical, the third commends itself by a number of practical and valuable features. As to the last, it is not only very efficient but obtains results impossible with other forms. The law of survival of the fittest is asserting itself, and the struggle for supremacy is now on between the geared turbine and the electric drive.

Through gradual improvement of the cutting tools, scientific design, metallurgical advances and refinement of lubricants, the so-called herringbone gear has been brought to great perfection. De Laval attained an efficiency of ninety-seven per cent and MacAlpine, Melville and Westinghouse ninety-eight and one-half per cent in the transmission from the driving to the driven shaft. On the other hand, ninety-three and three-quarters per cent may be considered as the maximum with electrical apparatus. This means that with the gear the same turbine would impart five per cent more power to the propeller, which should increase the speed of the cruiser from thirty-five to a little more than thirty-five and one-half knots. As it also appears at first sight that the electric drive requires additional space, is heavier and more costly, it is only natural for those who have not made a thorough study of all its phases to decide in favor of the gear.

Some Fatal Mistakes

But a careful inquiry into the subject would induce them to reverse their opinion. In estimating the relative merits of these essentially different propelling means they make two fatal mistakes. The first is to take the power transmitted under abnormal conditions as a criterion; the second to draw a parallel between instalments entirely unlike, one primitive, the other elaborate, the former being incapable of fulfilling important functions of the latter. When premises arc erroneous deductions therefrom must needs be faulty. Thus the opponents of the electric drive have been led to the conclusion that it is less efficient than- the gear, of greater weight, more expensive and uncertain of success. How much truth there is in these contentions will be apparent from an examination of well established facts.

The electric drive is of complex influence on results in ship operation. For the sake of brevity it will be viewed only in the following principal aspects: - (1) turbine performance, (2) power transmitted to the propeller, (3) efficiency of the screw, (4) low power cruising, (5) high power action, (6) fuel consumption by auxiliaries and apparatus for ship use, (7) general economy and (8) promptness and precision of control of all effects, internal and external.

The present turbines are extremely unsuitable for ship propulsion. They offer a striking example of an antiquated invention of small value elevated to a position of extraordinary commercial utility through profound research and astonishing mechanical skill. With hundreds of thousands of thin blades easily destroyed, buckets that through corrosion and erosion soon become wasteful and small clearances between surfaces rotating at terrific rates, they are a cause of constant danger and hazard.

Irreversible Turbines

But their cardinal defect is that they are irreversible, which necessitates the employment of separate turbines for backing. These, besides involving great expense and considerable friction loss, impose narrow limits on the temperature of the working medium. Very high superheat, so desirable in thermo-dynamic conversion, is out of the question in such, perishable structures, but from 200 to 300 degrees F. are permissible.

To that extent, then, the turbine is at an advantage when driving a dynamo. Two hundred degrees superheat will usually effect a saving of about twenty-three per cent of steam and ten per cent of fuel. This, however, is not the only gain. The turbine, freed from all the impediments of the gear drive, is capable of being safely run at a higher peripheral speed with a correspondingly increased efficiency and output. Thus, by moderate superheat and other simple and allowable expedients, it becomes practicable to develop twenty-five per cent more power from the same fuel, and this alone would make the electric drive decidedly superior to its competitor.

A Mechanical Tour De Force

As regards the power transmitted from the turbine to the propeller, it will seem in the light of the preceding that the gear is better by five per cent. That may be the case in exceptional tests, but it is quite different in actual service. To this is to be traced the error of those who are taking results obtained at constant load as standard of comparison. The perfection of the modern high speed gear was a veritable tour de force of the scientific machinist. It is a wonderful device, but it also has its inseparable weaknesses and shortcomings. Since the friction loss in it is sensibly constant through a wide range of performance, a relatively great amount of energy is absorbed at small load. The gear is likewise very sensitive to shocks and vibrations, which break down the capillary oil film, vital to smooth running. In consequence, there is great waste of power when the resisting force is subject to frequent and sudden fluctuations, Measurements I made with turbine gears have shown that while the efficiency with steady and normal effort was ninety-six per cent, not more than ninety per cent was realized with a rapidly varying load. This is what might be expected in practice. Any one who has listened to the tortured engines of a steamship in heavy sea could not have failed to observe how the turning effort varies as the vessel rolls, pitches and ploughs through billows and conflicting undercurrents. A similar state of things is apt to confront a war ship in action, as was evidenced in recent naval engagements, when mountains of water were raised by the exploding shells. Under such circumstances the gear is at a great disadvantage, as the electric drive is susceptible to these drawbacks in a much smaller degree. The idea that the gear transmits more of the primary power to the propeller than the combination of dynamo and motor is, therefore, largely illusionary. There is ample evidence, experimental and inferential, that rather the opposite is true.

Superiority of Electric Drive

Considering the efficiency of the screw as distinct from that attained in the transmission of power, it is admittedly better with the electric drive, this conclusion being entirely based on the superior adaptability and flexibility of the system. But there are deeper causes which should be taken into account. The interposition of electromagnetic means between the turbine and propeller materially reduces the loss due to shocks, vibrations, racing and other disturbances owing to inherent elastic resilience and equalizing tendency. The saving of energy thus effected at high speed and in a heavy sea is considerable.

Economy in cruising is one of the most desirable qualities of a war vessel. This is its ordinary use, for the chance of ever being engaged in battle is remote. The bitterest opponents of the electric drive do not deny that it excels in this feature, upon which the manufacturer chiefly relies in guaranteeing a fuel consumption from 10 to 12 per cent smaller than with the gear. The latter is hopelessly condemned by inability of adjustment to varying speed and wasteful in cruising operations, while the former is readily adaptable and economical under all conditions.

Another quality of the electric drive, which may prove especially valuable in action, is its capacity of carrying great overload without danger, owing to the nature of the connection between the turbine and propeller, as explained. The gear is rigid and unyielding and any increase of effort, particularly if sudden, may cause a breakdown.

Saving of Power

Referring to the auxiliaries proper and other apparatus for ship use, to which are chargeable approximately 20 per cent of the fuel consumed, a very substantial saving of power will be achieved through the introduction of the electrical method.

Quite apart from this central station supply will be operative in reducing other waste, many accessories will be dispensed with and general economy materially increased.

But from the military point of view the quickness, ease and precision of control will perhaps be the most significant of the advantages gained. Everything may be made to respond instantly to the pressure of a button. By reversing the motors the vessel may be brought from full speed to a stop within its length. It will be possible to make it go through all evolutions with extraordinary rapidity and a perfection of manoeuver, undreamed of before, will be attained.

A curious mistake is made by the advocates of the turbine gear in estimating relative weight. It hardly needs be stated that it is unfair, if not absurd, to compare arrangements of widely different character and scope. Only such as are capable of accomplishing the same results should be considered. Now, a gear drive corresponding to the electric would consist of four main turbines with gears, four reversing turbines of the same capacity, and eight smaller driving and reversing turbines for cruising. This agglomeration of complex and not all too rugged machinery, with its network of water, air and oil pipes, valves, pumps and attachments, would by far exceed in weight the proposed electric drive, and would also require better structural protection, not to speak of other defects and shortcomings.

Question of Weight

It should be observed, however, that the weights must be taken in their relation to that of the ship. One equipment may be heavier than another, but if it is more efficient and thereby reduces the weight of fuel and other cargo, it is for all purposes the lighter of the two.

The same is true of the cost. Comparative figures mean nothing. The question is whether the investment of capital is justified by what is to be accomplished. But enough has been said to show that for results in all respects equivalent, assuming them to be possible, the gear type, notwithstanding all claims to the contrary, would be more expensive.

That the electric drive is experimental and uncertain of performance is the least tenable of adverse assertions. In the first place, it has been successfully employed on a number of vessels and a great many more are being built. It also was found to be capable of an efficiency higher than that of any other form. But this is quite immaterial. The confidence that in the present instance all expectations will be realized is not based on a few demonstrations, but on years of experience with power plants ever since my system was commercially inaugurated. Tens of millions of horse power of induction motors are now in use the world over and no failure is recorded.

New Cruisers' Requirements

The new cruisers will require 180,000 horse power each, which, if necessary can be developed by four units of 45,000 horse power. Turbines of that capacity have been constructed and are in operation today. Dynamos of corresponding output have been installed at several places and are supplying light and power to large cities and districts. Induction motors of 15,000 horse power are being turned out by the manufacturers and can be produced in any size that may be desired, for of all kinds of motors this is the simplest and most dependable. Long since the whole system has been worked out and perfected to the least detail. The project is colossal, but it can be readily carried out by any of the few concerns who are possessed of the proper facilities. Not even a new tool has to be made. There is nothing whatever untried or hazardous about the electric drive.

Much stress is laid on reports, still to be verified, that it was rejected by England and Germany. But this is of no consequence. It has been rejected here more than once. Besides, there was war in the air of Europe and the time for radical innovations unpropitious. Moreover, the Diesel engine was looming up with large possibilities and Dr. Föttinger's hydraulic drive was being tried. The beginning must be made somewhere and it would be deplorable indeed if the United States, where the invention was first announced and introduced on a gigantic scale, were the last to recognize it. Such mistakes have happened only too often. The foreign navies are not in the habit of keeping the press informed of their doings and it is safe to predict that if progress in this country is much retarded there will be a repetition of previous disappointments.

It is unnecessary to dwell on other objections which are of minor importance and of no bearing on the principle. Without going into tedious technical discussion, it may be stated that the electric drive, if judiciously designed, will save not less than twenty-five per cent of fuel and, with due regard to this and certain specific and invaluable advantages, will be lighter, cheaper and in every respect more dependable than the gear. In fact, I believe that a scheme can be devised permitting the placing of all vital parts below the water line. In view of this, it is to be hoped that the Secretary of the Navy will not pay attention to the protests of rivals, however patriotic, but will cause the good work to be pushed to completion with all the power at his command.

These statements are to be understood as reflecting the present state of the art. The advent of a reversible turbine will profoundly affect the situation in favor of the gear. Such a machine has been perfected and was described in the Herald of October 15, 1911. It is the lightest prime mover ever produced and can be operated without trouble at red heat, thereby obtaining a very high economy in the transformation of heat energy.. I anticipate its speedy and extensive application to ship propulsion. But although an ideally simple and very inexpensive drive will be thus provided, there will still be weighty reasons for adopting the electric method on war ships. In order to dissipate all doubt created in the mind by diversified engineering opinion, I will make known but One of them, which in itself is sufficiently consequential and convincing to dispense with further arguments.

Disarmament Impossible

It is idle to dream of disarmament and universal peace in the face of the terrible events which are now unfolding. They prove conclusively that no country will be allowed to control all others by any means. Before all peoples can feel secure of their national existence and worldwide harmony established certain obstacles will have to be removed, the chief of which are German militarism, British domination of the seas, the rising tide of Russian millions, the yellow peril and the money power of America. These adjustments will be slow and pennible in conformity with natural laws. International friction and armed conflict will not be banished from the earth for a long period to come. The drag on human progress would not be so great if war energy could be maintained in purely potential form. This can and will be done through the universal introduction of wireless power. Then all destructive energy will be obtained without effort merely by controlling the life sustaining forces of peace.

The maintenance of war ships and other military implements involves an appalling waste. A vessel costing twenty millions of dollars is rendered virtually worthless in the short span of ten years, deteriorating at the mean rate of two million dollars a year, interest not considered. Hardly more than one out of fifty serves its real purpose. To lessen this ruinous loss and exploit certain inventions I elaborated a scheme some years ago. It was recognized as rational bur financially and in other ways difficult of realization. Now, when national economy and preparedness have become burning questions, it assumes special import and significance.

To Use War Ships in Peace

The underlying idea is to make war ships available for purposes of peace in a profitable manner, at the same time improving them in a number of features. I am aware of the proposal lately advanced to employ them as carriers of commerce, but this is not feasible and would be an impediment to further perfection. My project primarily contemplated the installment of the electric drive and the use of the turbo-dynamos for light and power supply and manufacture of various valuable products and articles aboard or on land. This would be a step in the direction of present development meeting the objects of both military and industrial preparedness. I further intended the creation of a type of vessel on radically different principles, which would be a precious asset in peace and ever so much more destructive in war. The new cruisers, if equipped as planned by the Navy Department, will constitute four floating central stations of 180,000 horse power each. The turbines and dynamos are designed for highest efficiency and operate under most favorable conditions. The power they are capable of developing represents a market value of several million dollars a year and could be advantageously utilized at places where fuel can be readily obtained and transport is convenient. The plants would also prove of great value in cases of emergency. They could be quickly sent to any point along the coast of the United States or elsewhere and would enable the government to lend speedy assistance whenever necessary.

But this is not all. There is another and still more potent reason for adopting the electric method. It is founded on the knowledge that at a time not distant the present means and methods of warfare will be revolutionized through novel applications of electric force.

Nikola Tesla Tells of Country's War Problems
New York Herald — April 15, 1917

The conquest of elements, annihilation of distance in the transmission of force and numerous other revolutionary advances have brought us face to face with problems new and unforeseen. To meet these is an imperative necessity rendered especially pressing through the struggle which is now being waged between nations on a stupendous scale unprecedented in history.

This country, finding it impossible to remain an inactive witness of medieval barbarism and disregard of sacred rights, has taken up arms in a spirit broad and impartial and in the interest of humanity and peace. Its participation will be absolutely decisive as regards the final result, but those who expect a speedy termination of the conflict should undeceive themselves.

War, however complex, is essentially a mechanical process, and, in conformity with a universal principle, its duration must be proportionate to the masses set in motion. The truth of this law is borne out by previous records, from which it may be calculated that, barring conditions entirely out of the ordinary, the period should be from five to six years.

Great freedom of institutions, such as we are privileged to enjoy, is not conducive to safety. Militarism is objectionable, but a certain amount of organized discipline is indispensable to a healthy national body. Fortunately, the recognition of this fact has not come too late, for there is no immediate danger, as alarmists would make us believe. The geographical position of this country, its vast resources and wealth, the energy and superior intelligence of its people, make it virtually unconquerable.

There is no nation to attack us that would not be ultimately defeated in the attempt. But events of the last three years have shown that a combination of many inimical powers is possible, and for such an emergency the United States is wholly unprepared. The first efforts must therefore be devoted to the perfection of the best plea for national protection. This idea has taken hold of the minds of people and great results may be expected from its creative imagination fired by this occasion, such as may in a larger measure recompense for the awful wastage of war.

While the chief reliance in this perilous situation must be placed on the army and navy, it is of the greatest importance to provide a big fleet of airplanes and dirigibles for quick movement and observation; also a great number of small high speed craft capable of fulfilling various vital duties as carriers and instruments of defense. These, together with the wireless, will be very effective against the U-boat, of which the cunning and scientific enemy has made a formidable weapon, threatening to paralyze the commerce of the world.

As the first expedient for breaking the submarine blockade, the scheme of employing hundreds of small vessels, advanced by Mr. W. Denman, chairman of the United States Shipping Board, is a most excellent one, which cannot fail to succeed. Another measure which will considerably reduce the toll is to use every possible means for driving the lurking enemy far out into the sea, thus extending the distance at which he must operate and thereby lessening his chances. But a perfect apparatus for revealing his presence is what is most needed at this moment.

A number of devices, magnetic, electric, electro-magnetic or mechanical, more or less known, are available for this purpose. In my own experience it was demonstrated that the small packet boat is capable of affecting a sensitive magnetic indicator at a distance of a few miles. But this effect can be nullified in several ways. With a different form of wireless instrument devised by me some years ago it was found practicable to locate a body of metallic ore below the ground, and it seems that a submarine could be similarly detected.

Sound waves may also be resorted to, but they cannot be depended upon. Another method is that of reflection, which might be rendered practicable, though it is handicapped by experimental difficulties well nigh insuperable. In the present state of the art the wireless principle is the most promising of all, and there is no doubt that it will be applied with telling effect. But we must be prepared for the advent of a large armored submarine of great cruising radius, speed and destructive power which will have to be combated in other ways.

For the time being no effort should be spared to develop aerial machines and motor boats. The effectiveness of these can be largely increased by the use of a turbine, which has been repeatedly referred to in the *Herald* and is ideally suited for such purposes on account of its extreme lightness, reversibility and other mechanical features.

N. Tesla

A Speech Delivered Before the American Institute of Electrical Engineers

The Engineering Societies Building - New York City —May 18, 1917

Mr. President, Ladies and Gentlemen. - I wish to thank you heartily for your kind sympathy and appreciation. I am not deceiving myself in the fact, of which you must be aware, that the speakers have greatly magnified my modest achievements. One should in such a situation be neither diffident nor self-assertive, and in that-sense I will concede that some measure of credit may be due to me for the first steps in certain new directions; but the ideas I advanced have triumphed, the forces and elements have been conquered, and greatness achieved, through the co-operation of many able men some of whom, I am glad to say, are present this evening. Inventors, engineers, designers, manufacturers and financiers have done their share until, as Mr. Behrend said, a gigantic revolution has been wrought in the transmission and transformation of energy. While we are elated over the results achieved we are pressing on, inspired with the hope and conviction that this is just a beginning, a forerunner of further and still greater accomplishments.

On this occasion, you might want me to say something of a personal and more intimate character bearing on my work. One of the speakers suggested: "Tell us something about yourself, about your early struggles." If I am not mistaken in this surmise I will, with your approval, dwell briefly on this rather delicate subject. Some of you who have been impressed by what has been said, and would be disposed to accord me more than I have deserved, might be mystified and wonder how so much as Mr. Terry has outlined could have been done by a man as manifestly young as myself. Permit me to explain this. I do not speak often in public, and wish to address just a few remarks directly to the members of my profession, so that there will be no mistake in the future. In the first place, I come from a very wiry and long-lived race. Some of my ancestors have been centenarians, and one of them lived one hundred and twenty-nine years. I am determined to keep up the record and please myself with prospects of great promise. Then again, nature has given me a vivid imagination which, through incessant exercise and training, study of scientific subjects and verification of theories through experiment, has become very accurate and precise, so that I have been able to dispense, to a large extent, with the slow, laborious, wasteful and expensive process of practical development of the ideas I conceive. It has made it possible for me to explore extended fields with great rapidity and get results with the least expenditure of vital energy. By this means I have it in my power to picture the objects of my desires in forms real and tangible and so rid myself of that morbid craving for perishable possessions to which so many succumb. I may say, also, that I am deeply religious at heart, although not in the orthodox meaning, and that I give myself to the constant enjoyment of believing that the greatest mysteries of our being are still to be fathomed and that, all the evidence of the senses and the teachings of exact and dry sciences to the contrary notwithstanding, death itself may not be the termination of the wonderful metamorphosis we witness. In this way I have managed to maintain an undisturbed peace of mind, to make myself proof against adversity, and to achieve contentment and happiness to a point of extracting some satisfaction even from the darker side of life, the trials and tribulations of existence. I have fame and untold wealth, more than this, and yet - how many articles have been written in which I was declared to be an impractical unsuccessful man, and how many poor, struggling writers, have called me a visionary. Such is the folly and shortsightedness of the world!

Now that I have explained why I have preferred my work to the attainment of worldly rewards, I will touch upon a subject which will lend me to say something of greater importance and enable me to explain how I invent and develop ideas. But first I must say a few words regarding my life which was most extraordinary and wonderful in its varied impressions and incidents. In the first place, it was charmed. You have heard that one of the provisions of the Edison Medal was that the recipient should be alive. Of course the men who have received this medal have fully deserved it, in that respect, because they were alive when it was conferred upon them, but none has deserved it in anything like the measure I do, when it comes to that feature. In my youth my ignorance and lightheartedness brought me into innumerable difficulties, dangers and scrapes, from which I extricated myself as by enchantment. That occasioned my parents great concern more, perhaps, because I was the last male than because I was of their own flesh and blood. You should know that Serbians desperately cling to the preservation of the race. I was nearly drowned a dozen times. I was almost cremated three or four times and just missed being boiled alive. I was buried, abandoned and frozen. I have had narrow escapes from mad dogs, hogs and other wild animals. I have passed through dreadful diseases - have been given up by physicians three or four times in my life for good. I have met with all sorts of odd accidents - I cannot think of anything that did not happen to me, and to realize that I am here this evening, hale and hearty, young in mind and body, with all these fruitful years behind me, is little short of a miracle.

But my life was wonderful in another respect - in my capacity of inventor. Not so much, perhaps, in concentrated mentality, or physical endurance and energy; for these are common enough. If you inquire into the career of successful men in the inventor's profession you will find, as a rule, that they are as remarkable for their physical as for their mental performance. I know that when I worked with Edison, after all of his assistants had been exhausted, he said to me: "I never saw such a thing, you take the cake." That was a characteristic way for him to express what I did. We worked from half past ten in the morning until five o'clock the next morning. I carried this on for nine months without a single day's exception; everybody else gave up. Edison stuck, but he occasionally dozed off on the table. What I wish to say particularly is that my early life was really extraordinary in certain experiences which led to everything I ever did afterwards. It is important that this should be explained to you as otherwise you would not know how I discovered the rotating field. From childhood I was afflicted in a singular way - I would see images of objects and scenes with a strong display of light and of much greater vividness than those I had observed before. They were always images of objects and scenes I had actually seen, never of such as I imagined. I have asked students of psychology, physiology and other experts about it, but none of them has been able to explain the phenomena which seems to have been unique, although I was probably predisposed, because my brother also saw images in the same way. My theory is that they were simply reflex actions from the brain on the retina, superinduced by hyper-excitation of the nerves. You might think that I had hallucinations, That is impossible. They are produced only in diseased and anguished brains. My head was always clear as a bell, and I had no fear. Do you want me to tell of my recollections bearing on this? (Turning to the gentlemen on the platform). This is traditional with me, for I was too young to remember anything of what I said. I had two old aunts, I recall, with wrinkled faces, one of them with two great protruding teeth which she used to bury into my cheek when she kissed me. One day they asked me which of the two was prettier. After looking them over I answered: "This one is not as ugly as the other one." That was evidence of good sense. Now as I told you, I had no fear. They used to ask me, "Are you afraid of robbers?" and I would reply "No." "Of wolves?" "No." Then they would ask, "Are you afraid of crazy Luka?" (A fellow who would tear through the village and nothing could stop him) "No, I am not afraid of Luka." "Are you afraid of the gander?" "Yes, I am," I would reply and cling to my mother. That was because once they put me in the court yard with nothing on, and that beast ran up and grabbed me by the soft part of the stomach tearing off a piece of flesh. I still have the mark.

These images I saw caused me considerable discomfort. I will give you and illustration: Suppose I had witnessed a funeral. In my country the rites are but intensified torture. They smother the dead body with kisses, then they bathe it, expose it for three days, and finally one hears the dull thuds of the earth, when all is over. Same of the pictures as that of the coffin, for instance, would not appear vividly but were sometimes so persistent that when I would stretch my hand out I would see it penetrate the image. As I look at it now these images were simply reflex actions through the optic nerve on the retina, producing on the same an effect identical to that of a projection through the lens, and if my view is correct, then it will be possible, (and certainly my experience has demonstrated that), to project the image of any object one conceives in thought on a screen and make it visible. If this could be done it would revolutionize all human relations. I am convinced that it can and will be accomplished.

In order to free myself of these tormenting appearances, I tried to fix my mind on some other picture or image which I had seen, and in this way I would manage to get some relief; but in order to get this relief I had to let the images come one after the other very fast. Then I found that I soon exhausted all I had at my command, my "reel" was out, as it were. I had seen little of the world, only objects around my own home, and they took me a few times to some neighbors, that was all I knew. When I did so the second or third time, in order to chase the appearance from my vision, I found that this remedy lost all the force: Then I began to make excursions beyond the limits of the little world I knew, and I saw new scenes. These were at first very blurred and indistinct', and would flit away when I tried to concentrate my attention upon them, but by and by I succeeded in fixing them; they gained in force and distinctness and finally assumed the intensity of real things. Soon I observed that my best comfort was attained if I simply went on in my vision farther and farther, getting new impressions all the time, and so I started to travel - of course, in my mind. You know that there have been great discoveries made - when Columbus found America that was one, but when I hit upon the idea of traveling it seemed to me that was the greatest discovery possible to man. Every night (and sometimes during the day), as soon as I was alone I would start on my travels. I would see new places, cities and countries, I would live there, meet people and make friendships and acquaintances, and these were just as dear to me as those in real life and not a bit less intense. That is the way I did until I reached almost manhood. When I turned my thoughts to invention, I found that I could visualize my conceptions with the greatest facility. I did not need any models, drawings or experiments, I could do it all in my mind, and I did. In this way I have unconsciously evolved what I consider a new method of materializing inventive concepts and ideas, which is exactly opposite to the purely experimental of which undoubtedly Edison is the greatest and most successful exponent. The moment you construct a device to carry into practice a crude idea you will find yourself inevitably engrossed with the details and defects of the apparatus. As you go on improving and reconstructing, your force of concentration diminishes and you lose sight of the great underlying principle. You obtain results, but at the sacrifice of quality. My method is different, I do not rush into constructive work. When I get an idea, I start right away to build it up in my mind. I change the structure, I make improvements, I experiment, I run the device in my mind. It is absolutely the same to me whether I operate my turbine in thought or test it actually in my shop. It makes no difference, the results are the same. In this way, you see, I can rapidly develop and perfect an invention, without touching anything. When I have gone so far that I have put into the device every possible improvement I can think of, that I can see no fault anywhere, I then construct this final product of my brain. Every time my device works as I conceive it should and my experiment comes out exactly as I plan it. In twenty years there has not been a single solitary experiment which did not turn out precisely as I thought it would. Why should it not? Engineering, electrical and mechanical, is positive in results. Almost any subject presented can be mathematically treated and the effects calculated; but if it is such that results cannot be had by simple methods of mathematics or short cuts, there is all the experience, and all the data on which to draw and from which to build; - why, then, should one carry out the crude idea? It is not necessary, it is a waste of energy, money and time. Now, that is just the way I produced the rotating field.

If I am to give you in a few words the history of that invention, I must begin with my birthday, and you will see the reason why. I was born exactly at midnight, I have no birthday and I never celebrate it. But something else must have happened on that date. I have learned that my heart beat on the right side and did so for many years after. As I grew up it beat on both sides, and finally settled on the left. I remember that I was surprised, when I developed into a very strong man, to find my heart on the left side. Nobody understands how it happened. I had two or three falls and on one occasion nearly all my chest bones were crushed in. Something that was quite unusual must have occurred at my birth and my parents destined me for the clergy then and there. When I was six years old I managed to have myself imprisoned in a little chapel at an inaccessible mountain, and visited only once a year. It was a place of many bloody encounters and there was a grave yard near by. I was locked in there while looking for some sparrows' nests, and had the most dreadful night I ever passed in my life, in company with the ghosts of the dead. American boys will not understand it, of course, for there are no ghosts in America - the people are too sensible; but my country was full of them, and every one from the small boy to greatest hero, who was plastered all over with medals for courage and bravery, had a fear of ghosts. Finally, as by a wonder, they rescued me, and then my parents said: "Surely he must go to the clergy, he must become a churchman." Whatever happened after that, no matter what it was, simply fortified them in that resolution. One day, to tell you a little story, I fell from the top of one of the farm buildings into a large kettle of milk, which was boiling over a roaring fire. Did I say boiling milk? - It was not boiling - not according to the thermometer - though I would have sworn it was when I fell into it, and they pulled me out. But I only got a blister on the knee where I struck the hot kettle. My parents said again: "Was not that wonderful? Did you ever hear of such a thing? He will surely be a bishop, a metropolitan, perhaps a_ patriarch." In my eighteenth year I came to the cross roads. I had passed through the preliminary schools and had to make up my mind either to embrace the clergy or to run away. I had a profound respect for my parents, and so I resigned myself to take up studies for the clergy. Just then one thing occurred, and if it had not been for that, I would not have had my name connected with the occasion of this evening. A tremendous epidemic of cholera broke out, which decimated the population and, of course, I got immediately. Later it developed into dropsy, pulmonary trouble, and all sorts of diseases until finally my coffin was ordered. In one of the fainting spells when they thought I was dying, my father came to my bedside and cheered me: "You are going to get well." "Perhaps," I replied, "if you will let me study engineering." "Certainly I will," he assured me, "you will go to the best polytechnic school in Europe." I recovered to the amazement of everybody. My father kept his word, and after a year of roaming through the mountains and getting myself in good physical shape, I went to the Polytechnic School at Gratz, Styria, one of the oldest institutions. Something else occurred, however, of which I must tell you as it is vitally linked with this discovery. In the preparatory schools there was no liberty in the choice of subjects, and unless a student was proficient in all of them he could not pass. I found myself in this predicament every year. I could not draw. My faculty for imagining things paralyzed whatever gift I might have had in this respect. I have made some mechanical drawings, of course; practicing so many years one must needs learn to make simple sketches, but if I draw for half an hour I am all exhausted. I never was qualified and passed only through my father's influence. Now, when I went to the polytechnic school I had free choice of subjects and proposed myself to show my parents what I could do. The first year at the polytechnic school was spent in this way - I got up at three o'clock in the morning and worked until eleven o'clock at night, for one whole dear, with a single day's exception. Well, you know when a man with a reasonably healthy brain works that way he must accomplish something. Naturally, I did. I graduated nine times that year and some of the professors were not satisfied with giving me the highest distinction, because they said, that did not express their idea of what I did, and here is where I come to the rotating field. In addition to the regular graduating papers they gave me some certificates which I brought to my father believing that I had achieved a great triumph. He took the certificates and threw them into the waste basket, remarking contemptuously: "I know how these testimonials are obtained." That almost killed my ambition; but later, after my father had died, I was mortified to find a package of letters, from which I could see that there had been considerable correspondence going on between him and the professors who had written to the effect that unless he took me away from school I would kill myself with work. Then I understood why he had slighted my success, which I was told was greater than any previous one at that institution; in fact the best students had only graduated twice. My record in the first year had the result that the professors became very much interested in and attached to me, particularly three of them; Prof. Rogner who was teaching arithmetical subjects and geometry; Prof. Alle, one of the most brilliant and wonderful lecturers I have ever seen, who specialized in differential equations, about which he wrote quite a number of works in German, and Prof. Poeschl, who was my instructor in physics. These three men were simply in love with me and used to give me problems to solve. Prof. Poeschl was a curious man. I never saw such feet in my life. They were about that size. (Indicating) His hands were like paws, but when he performed experiments they were so convincing and the whole went off so beautifully that one never realized how they were done. It was all in the method. He did all with the precision of a clock work, and everything succeeded.

It was in the second year of my studies that we received a Gramme machine from Paris, having a horse-shoe form of laminated magnet, and a wound armature with a commutator. We connected it up and showed various effects of currents. During the time Prof. Poeschl was making demonstrations running the machine as a motor we had some trouble with the brushes. They sparked very badly, and I observed: "Why should not we operate without the brushes?" Prof. Poeschl declared that it could not be done, and in view of my success in the past year he did me ,the honor of delivering a lecture touching on the subject. He remarked: "Mr. Tesla may accomplish great things, but he certainly never will do this," and he reasoned that it would be equivalent to converting a steadily pulling force, like that of gravity, into a rotary effort, a sort of perpetual motion scheme, an impossible idea. But you know that instinct is something which transcends knowledge. We have, undoubtedly, certain finer fibers that enable us to perceive truths when logical deduction, or any other willful effort of the brain, is futile. We cannot reach beyond certain limits in our reasoning, but with instinct we can go to very great lengths. I was convinced that I was right and that it was possible. It was not a perpetual motion idea, it could be done, and I started to work at once.

I will not tire you with an extended account of this undertaking, but will only say that I began in the summer of 1877 and I proceeded as follows: I would picture first of all, a direct-current machine, run it and see how the currents changed in the armature. Then I would imagine an alternator and do the same thing. Next I would visualize systems comprising motors and generators, and so on. Whatever apparatus I imagined, I would put together and operate in my mind, and I continued this practice incessantly until 1882. In that year somehow or other, I began to feel that a revelation was near. I could not yet see just exactly how to do it, but I knew that I was approaching the solution. While on my vacation, in 1882, sure enough, the idea came to me and I will never forget the moment. I was walking with a friend of mine in the city park of Budapest reciting passages from Faust. It was nothing for me to read from memory the contents of an entire book, with every word between the covers, from the first to the last. My sister and brother, however, could do much better than myself. I would like to know whether any of you has that kind of memory. It is curious, entirely visual and retroactive. To be explicit - when I made my examinations, I had always to read the books three or four days if not a week before, because in that time I could reconstruct the images and visualize them; but if I had an examination the next day after reading, images were not clear and the remembrance was not quite complete. As I say, I was reciting Goethe's poem, and just as the sun was setting I felt wonderfully elated, and the idea came to me like a flash. I saw the whole machinery clearly, the generator, the motor, the connections, I saw it work as if it had been real. With a stick I drew on the sand the diagrams which were shown in my paper before the American Institute of Electrical Engineers and illustrated in my patents, as clearly as possible, and from that time on I carried this image in my mind. Had I been a man possessed of the practical gifts of Edison, I would have gone right away to perform an experiment and push the invention along, but I did not have to do this. I could see pictures so vividly, and what I imagined was so real and palpable, that I did not need any experimenting, nor would it have been particularly interesting to me. I went on and improved the plan continuously, inventing new types, and the day I came to America, practically every form, every kind of construction, every arrangement of apparatus I described in my thirty or forty patents was perfected, except just two or three kinds of motors which were the result of later development.

In 1883, I made some tests in Strasburg, as Mr. Terry pointed out, and there at the railroad station obtained the first rotation. The same experiment was repeated twice.

Now I come to an interesting chapter of my life, when I arrived in America. I had made some improvements in dynamos for a French company who were getting their machinery from here. The improved forms were so much better that the manager of the works said to me: "You must go to America, and design the machines for the Edison Company." So, after ineffectual efforts on the other side to get somebody to interest himself in my plans financially, I came to this country. I wish that I could only give you an idea how what I saw here impressed me. You would be very much astonished. You have a? undoubtedly read those charming Arabian Nights tales, in which the genie transports people into wonderful regions, to go through all sorts of delightful adventures. My case was just the opposite. The genie transported me from a world of dreams into one of realities. My world was beautiful, ethereal, as I could imagine it. The one I found here was a machine world; the contact was rough, but I liked it. I realized from the very moment I saw Castle Garden that I was a good American before I landed. Then came another event. I met Edison, and the effect he produced upon me was extraordinary. When I saw this wonderful man, who had had no theoretical training at all, no advantages, who did all himself, getting great results by virtue of his industry and application, I felt mortified that I had squandered my life. I had studied a dozen languages, delved in literature and art and had spent my best years in ruminating through libraries and reading all sorts of stuff that fell into my hands. I thought to myself, what a terrible thing it was to have wasted my life in those useless efforts. If I had only come to America earlier and devoted all of my brain power to inventive work, what might I have done? In later life though, I realized I would not have produced anything without the scientific training I got, and it is a question whether my surmise as to my possible accomplishment was correct. In Edison's works I passed nearly a year of the most strenuous labor, and then certain capitalists approached me with the project to form my own company. I went into the proposition, and developed the arc light. To show you how prejudiced people were against the alternating-current, as the President has indicated, when I told these friends of mine that I had a great invention relating to alternating-current transmission, they said: "No, we want the arc lamp. We do not care for this alternating-current." Finally I perfected my lighting system and the city adopted it. Then I succeeded in organizing another company, in April, 1886, and a laboratory was put up, where I rapidly developed these motors, and eventually the Westinghouse people approached us, and an arrangement was made for their introduction. You know what has happened since then. The invention has swept the world.

I should like to say just a few words regarding the Niagara Falls enterprise. We have a man here to-night to whom belongs really the credit for the early steps and for the first financiering of the project, which was difficult at that time. I refer to Mr. E. D. Adams. When I heard that such authorities as Lord Kelvin and Prof. W. C. Unwin had recommended - one the direct-current system and the other compressed air - for the transmission of power from Niagara Falls to Buffalo, I thought it was dangerous to let the matter go further, and I went to see Mr. Adams. I remember the interview perfectly. Mr. Adams was much impressed with what I told him. We had some correspondence afterwards, and whether it was in consequence of my enlightening him on the situation, or owing to some other influence, my system was adopted. Since that time, of course, new men, new interests have come in, and what has been done I do not know, except that the Niagara Falls enterprise was the real starting impulse in the great movement inaugurated for the -transmission and transformation of energy on a huge scale.

Mr. Terry. has referred to other inventions of mine. I will just make a few remarks relative to these as some of my work has been misunderstood. It seems to me that I ought to tell you a few words about an effort that absorbed my attention later. In 1892 I delivered a lecture at the Royal Institution and Lord Rayleigh surprised me by acknowledging my work in very generous terms, something that is not customary, and among other things he stated that I had really an extraordinary gift for invention. Up to that time, I can assure you, I had hardly realized that I was an inventor. I remembered, for instance, when I was a boy, I could go out into the forest and catch as many crows as I wanted, and nobody else could do it. Once, when I was seven years of age, I repaired a fire engine which the engineers could not make work, and they carried me in triumph through the city. I constructed turbines, clocks and such devices as no other boy in the community. I said to myself: "If I really have a gift for invention, I will bend it to some great purpose or task and not squander my efforts on small things." Then I began to ponder just what was the greatest deed to accomplish. One day as I was walking in the forest a storm gathered and I ran under a tree for shelter. The air was very heavy, and all at once there was a lightning flash, and immediately after a torrent of rain fell. That gave me the first idea. I realized that the sun was lifting the water vapor, and wind swept it over the regions where it accumulated and reached a condition when it was easily condensed and fell to earth again. This life-sustaining stream of water was entirely maintained by sun power, and lightning, or some other agency of this kind, simply came in a trigger-mechanism to release the energy at the proper moment. I started out and attacked the problem of constructing a machine which would enable us to precipitate this water whenever and wherever desired. If this was possible, then we could draw unlimited amounts of water from the ocean, create lakes, rivers and water falls, and indefinitely increase the hydroelectric power, of which there is now a limited supply. That led me to the production of very intense electrical effects. At the same time my wireless work, which I had already begun, was exactly in that direction, and I devoted myself to the perfection of that device, and in 1908, I filed an application describing an apparatus with which I thought the wonder could be achieved. The Patent Office Examiner was from Missouri, he would not believe that it could be done, and my patent was never granted. But in Colorado I had constructed a transmitter. by which I produced effects in some respects at least greater than those of lightning. I do not mean in potential. The highest potential I reached was something like 20,000,000 volts, which is insignificant as compared to that of lightning, but certain effects produced by my apparatus were greater than those of lightning. For instance, I obtained in my antennae currents of from 1,000 to 1,100 amperes. That was in 1899 and you know that in the biggest wireless plants of today only 250 amperes are used. In Colorado I succeeded one day in precipitating a dense fog. There was a mist outside, but when I turned on the current the cloud in the laboratory became so dense that when the hand was held only a few inches from the face it could not be seen. I am positive in my conviction that we can erect a plant of proper design in an arid region, work it according to certain observations and rules, and by its means draw from the ocean unlimited amounts of water for irrigation and power purposes. If I do not live to carry it out, somebody else will, but I feel sure that I am right.

As to the transmission of power through space, that is a project which I considered absolutely certain of success long since. Years ago I was in the position to transmit wireless power to any distance without limit other than that imposed by the physical dimensions of the globe. In my system it makes no difference what the distance is. The efficiency of the transmission can be as high as 96 or 97 per cent, and there are practically no losses except such as are inevitable in the running of the machinery. When there is no receiver there is no energy consumption anywhere. When the receiver is put on, it draws power. That is the exact opposite of the Hertz-wave system. In that case, if you have a plant of 1,000 horsepower, it is radiating all the time whether the energy is received or not; but in my system no power is lost. When there are no receivers the plant consumes only a few horsepower necessary to maintain the electric vibration; it runs idle, as the Edison plant when the lamps and motors are shut off.

I have made advances along this line in later years which will contribute to the practical features of the system. Recently I have obtained a patent on a transmitter with which it is practicable to transfer unlimited amount of energy to any distance. I had a very interesting experience with Mr. Stone, whom I consider, if not the ablest, certainly one of the ablest living experts. I said to Mr. Stone: "Did you see my patent?" He replied: "Yes, I saw it, but I thought you were crazy." When I explained it to Mr. Stone he said, "Now, I see; why, that is great," and he understood how the energy is transmitted.

To conclude, gentlemen, we are coming to great results, but we must be prepared for a condition of paralysis for quite a while. We are facing a crisis such as the world has never seen before, and until the situation clears the best thing we can do is to devise some scheme for overcoming the submarines, and that is what I am doing now. (Applause)

The Effect of Static on Wireless Transmission
Electoral Experimenter—January, 1919

A few statements regarding these phenomena in response to a request of the *Electrical Experimenter* may be useful at the present time in view of the increasing and importance of the subject.

The commercial application of the art has led to the construction of larger transmitters and multiplication of their number, greater distances had to be covered and it became imperative to employ receiving devices of ever increasing sensitiveness. All these and other changes have cooperated in emphasizing the trouble and seriously impairing the reliability and value of the plans. To such a degree has this been the ease that conservative business men and financiers have come to look upon this method of conveying intelligence as one offering but very limited possibilities, and the Government has deemed it advisable to assume control. This unfortunate state of affairs, fatal to enlistment of capital and healthful competitive development, could have been avoided had electricians not remained to this day under the spell of a delusive theory and had the practical exploiters of this advance not permitted enterprise to outrun technical competence.

With the publication of Dr. Heinrich Hertz's classical researches it was an obvious inference that the dark rays investigated by him could be used for signaling purposes, as those of light in heliography, and the first steps in this direction were made with his apparatus which, in 1896, was found capable of actuating receivers at a distance of a few miles. Three years prior to this, however, in lectures before the Franklin Institute and National Electric Light Association, I had described a wireless system radically opposite to the Hertzian in principle inasmuch as it depended on currents conducted through the earth instead of on radiations propagated through the atmosphere, presumably in straight lines.

The apparatus then outlined by me consisted of a transmitter comprising a primary circuit excited from an alternator or equivalent source of electrical energy and a high potential secondary resonant circuit, connected with its terminals to ground and to an elevated capacity, and a similar tuned receiving circuit including the operative device. On that occasion I expressed myself confidently on the feasibility of flashing in this manner not only signals to any terrestrial distance but Transmitting power in unlimited amounts for all sorts of industrial purposes. The discoveries made and experimental results attained I made with a wireless power-plant erected in 1899 some of which were disclosed in the *Century Magazine* of June, 1900, and several U. S. patents subsequently granted to me have, I believe, borne-out strikingly my foresight. In the meantime the Hertzian arrangements were gradually modified, one feature after another being abandoned, so that now not a vestige of them can be found and my system of four tuned circuits has been universally adopted, not only in its fundamentals but in every detail as the "quenched sparks," "ticker," "tone wheel," high frequency and rotating field alternators, forms of discharges and mercury breaks, frequency changers, coils, condensers, regulating methods, and devices, etc. This fact would give me supreme satisfaction were it not that the engineers, misinterpreting the nature of the effects, are making installments so defective in construction and mode of operation as to preclude the possibility of the great realization which might be brought within easy reach by proper application of the underlying principles and one of which—the most desirable at present—is the complete elimination of all static and other interference.

During the past few years several emphatic announcements have been made that a perfect solution of this problem had been discovered, but it was manifest from a casual perusal of these publications that the experts were ignoring certain truths of vital bearing on the question, and so long as this was the case no such claim could be substantiated. I achieved early success by keeping these steadily in mind and applying my efforts from the outset in the right and correct scientific direction.

I may contribute to the clearness of the subject in answering a question which I have been asked by the Editors of the *Electrical Experimenter* with reference to the report contained in the last issue, that signals had been received around the globe. An achievement the practicability of which I have fully demonstrated by experiment eighteen yeas ago.

The question is, how can Hertz waves be conveyed to such a distance in view of the curvature of the earth? A few words will be sufficient to show the absurdity of the prevailing opinion propounded in text books.

We are living on a conducting globe surrounded by a thin layer of insulating air, above which is a rarefied and conducting atmosphere. If the earth is represented by a sphere of 12 ½ inch radius, then the layer which may be considered insulating for high frequency currents of great tension is less than 1/64 of an inch thick. It is odd that the Hertz waves, emanating from a transmitter, get to the distant receiver by successive reflections. The utter impossibility of this will be evident when it is shown by a simple calculation that the amount of energy received , even if it could be collected in its totality, is infinitesimal and would not actuate the most sensitive instrument known were it magnified many million times. The fact is these waves have no perceptible influence on a receiver if situated at a much smaller distance. It should be remembered, moreover, that since the first attempts the wave lengths have been increased until those advocated by me were adopted, in which this form of radiation has been reduced to one-billionth.

When a circuit, connected to ground and to an elevated capacity oscillates two effects separate and distinct arc produced; Hertz waves are radiated in a direction at right angles to the axis of symmetry of the conductor, *and simultaneously a current is passed through the earth*. The former propagates with the speed of light, the latter with a velocity proportionate to the cosecant of an angle which from the origin to the opposite point of the globe varies from zero to 180°. Expressed in words, at the start the speed is infinite and diminishes, first rapidly and then slowly until a quadrant is traversed when the current proceeds with the speed of light. From that region on the velocity gradually increases becoming infinite at the opposite point of the globe. In a patent granted to me in April, 1905, I have summed up this law of propagation in the statement that the projections of all half waves on the axis of symmetry of movement are equal, which means that the successive half waves, though of different length, cover exactly the same area. In the near future many wonderful results will be obtained by taking advantage of this fact.

Tesla's Static Eliminator, Patented and Used by Him over Twenty Years Ago. It Will Be Fully Described in an Early Issue of the Electrical Experimenter

There is a vast difference between these two forms of wave movement in their bearing on the transmission. The Hertz waves represent energy which is radiated and unrecoverable. The current energy on the other hand, is preserved and can be recovered theoretically at least, in its entirety. If the experts will free themselves from that illusions under which they are laboring they will find that to overcome static disturbances all that is needed is the properly constructed transmitter and receiver without any additional devices or preventives. I have, however, devised several forms of apparatus eliminating statics even in the present defective wireless installations in which they are magnified many times. Such a form of instrument which I have used successfully is shown in the annexed photograph. These phenomena have been studied by me for a number of years and I have found that there are nine or ten different causes that tend to intensify them, and in due course I shall give a full description of the various improvements I have made, in the *Electrical Experimenter*. For the present I would only point out that in order to perfectly eliminate the static interference, it is indispensable to redesign the whole wireless apparatus as now employed. The sooner this is understood the better it will be for the further evolution of the Art.

Famous Scientific Illusions
Electrical Experimenter — February, 1919

The human brain, with all its wonderful capabilities and power, is far from being a faultless apparatus. Most of its parts may be in perfect working order, but some are atrophied, undeveloped or missing altogether. Great men of all classes and professions-scientists, inventors, and hard-headed financiers — have placed themselves on record with impossible theories, inoperative devices, and unrealizable schemes. It is doubtful that there could be found a single work of any one individual free of error. There is no such thing as an infallible brain. Invariably, some cells or fibers are wasting or unresponsive, with the result of impairing judgment, sense of proportion, or some other faculty. A man of genius eminently practical, whose name is a household word, has wasted the best years of his life in a visionary undertaking. A celebrated physicist was incapable of tracing the direction of an electric current according to a childishly simple rule. The writer, who was known to recite entire volumes by heart, has never been able to retain in memory and recapitulate in their proper order the words designating the colors of the rainbow, and can only ascertain them after long and laborious thought, strange as it may seem.

Our organs of reception, too, are deficient and deceptive. As a semblance of life is produced by a rapid succession of inanimate pictures, so many of our perceptions are but trickery of the senses, devoid of reality. The greatest triumphs of man were those in which his mind had to free itself from the influence of delusive appearances. Such was the revelation of Buddha that self is an illusion caused by the persistence and continuity of mental images: the discovery of Copernicus that contrary to all observation, this planet rotates around the sun; the recognition of Descartes that the human being is an automaton, governed by external influence and the idea that the earth is spherical, which led Columbus to the finding of this continent. And tho the minds of individuals supplement one another and science and experience are continually eliminating fallacies and misconceptions, much of our present knowledge is still incomplete and unreliable. We have sophisms in mathematics which cannot be disproved. Even in pure reasoning, free of the shortcomings of symbolic processes, we are often arrested by doubt which the strongest intelligences have been unable to dispel. Experimental science itself, most positive of all, is not unfailing.

In the following I shall consider three exceptionally interesting errors in the interpretation and application of physical phenomena which have for years dominated the minds of experts and men of science.

I. The Illusion of the Axial Rotation of the Moon

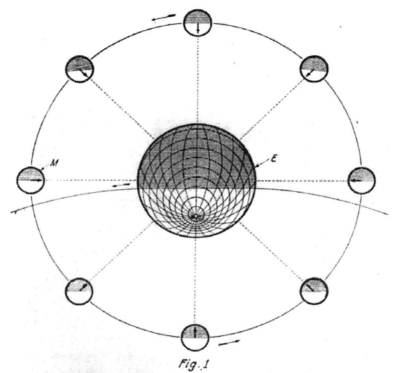

Fig. 1

It Is Well Known That the Moon, M., Always Turns the Same Face Toward the Earth, E, as the Black Arrows Indicate. The Parallel Rays From the Sun Illuminate the Moon in Its Successive Orbital Positions as the Unshaded Semi-circles Indicate. Bearing This in Mind, Do You Believe That the Moon Rotates on Its Own Axis?

It is well known since the discovery of Galileo that the moon, in travelling thru space, always turns the same face towards the earth. This is explained by stating that while passing once around its mother-planet the lunar globe performs just one revolution on its axis. The spinning motion of a heavenly body must necessarily undergo modifications in the course of time, being either retarded by remittances internal or external, or accelerated owing to shrinkage and other causes. An unalterable rotational velocity thru all phases of planetary evolution is manifestly impossible. What wonder, then, that at this very instant of its long existence our satellite should revolve exactly so, and not faster or slower. But many astronomers have accepted as a physical fact that such rotation takes place. It does not, but only appears so; it is an illusion, a most surprising one, too.

I will endeavor to make this clear by reference to Fig. 1, in which E represents the earth and M the moon. The movement thru space is such that the arrow, firmly attached to the latter, always occupies the position indicated with reference to the earth. If one imagines himself as looking down on the orbital plane and follows the motion he will become convinced that the moon *does* turn on its axis as it travels around. But in this very act the observer will have deceived himself. To make the delusion complete let him take a washer similarly marked and supporting it rotatably in the center, carry it around a stationary object, constantly

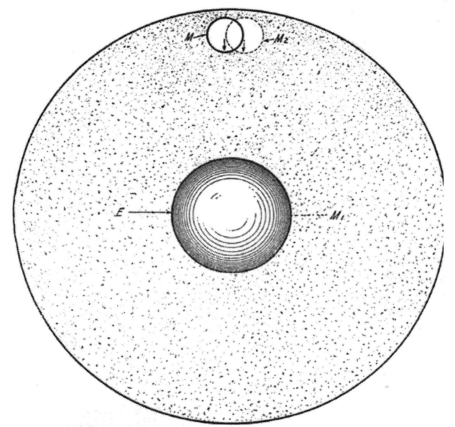

Fig. 2.—Tesla's Conception of the Rotation of the Moon, M, Around the Earth, E; the Moon, In This Demonstration Hypothesis, Being Considered as Embedded in a Solid Mass, M_1. If, As Commonly Believed, the Moon Rotates, This Would Be Equally True For a Portion of the Mass M_2, and the Part Common to Both Bodies Would Turn Simultaneously in "Opposite" Directions.

keeping the arrow pointing towards the latter. Tho to his bodily vision the disk will revolve on its axis, such movement does not exist. He can dispel the illusion at once by holding the washer fixedly while going around. He will now readily see that the supposed axial rotation is only apparent, the impression being produced by successive changes of position in space.

But more convincing proofs can be given that the moon does not, and cannot revolve on its axis. With this object in view attention is called to Fig. 2, in which both the satellite, M, and earth, E, are shown embedded in a solid mass, Mi, (indicated by stippling) and supposed to rotate so as to impact to the moon its normal translatory velocity. Evidently, if the lunar globe could rotate as commonly believed, this would be equally true of any other portion of mass M_1, as the sphere M_1, shown in dotted lines, and then the part common to both bodies would have to turn *simultaneously in opposite directions*. This can be experimentally illustrated in the manner suggested by using instead of one, two overlapping rotatable washers, as may be conveniently represented by circles M and M_2, and carrying them around a center as E, so that the plain and dotted arrows are always pointing towards the same center. No further argument is needed to demonstrate that the two gyrations cannot co-exist or even be pictured in the imagination and reconciled in a purely abstract sense.

The truth is, the so-called "axial rotation" of the moon is a phenomenon deceptive alike to the eye and mind and devoid of physical meaning. It has nothing in common with real mass revolution characterized by effects positive and unmistakable. Volumes have been written on the subject and many erroneous arguments advanced in support of the notion. Thus, it is reasoned, that if the planet did *not* turn on its axis it would expose the whole surface to terrestrial view; as only one-half is visible, it *must* revolve. The first statement is true but the logic of the second is defective, for it admits of only one alternative. The conclusion is not justified as the same appearance can also be produced in another way. The moon does rotate, not on its own, but about an axis passing thru the center of the earth, the true and only one.

The unfailing test of the spinning of a mass is, however, the existence of energy of motion. The moon is not possessed of such *vis viva*. If it were the case then a revolving body as M_1 would contain mechanical energy other than that of which we have experimental evidence. Irrespective of this so exact a coincidence between the axial and orbital periods is, in itself, immensely improbable for this is not the permanent condition towards which the system is tending. Any axial rotation of a mass left to itself retarded by forces external or internal, must cease. Even admitting its perfect control by tides the coincidence would still be miraculous. But when we remember that most of the satellites exhibit this peculiarity, the probability becomes infinitesimal.

Three theories have been advanced for the origin of the moon. According to the oldest suggested by the great German philosopher Kant, and developed by Laplace in his monumental treatise "Mécanique Céleste," the planets have been thrown off from larger central masses by centrifugal force. Nearly forty years ago Prof. George H. Darwin in a masterful essay on tidal friction furnished mathematical proofs, deemed unrefutable, that the moon had separated from the earth. Recently this established theory has been attacked by Prof. T. J. J. See in a remarkable work on the "Evolution of the Stellar Systems," in which he propounds, the view that centrifugal force was altogether inadequate to bring about the separation and that all planets, including the moon, have come from the depths of space and have been captured. Still a third hypothesis of unknown origin exists which has been examined and commented upon by Prof. W. H. Pickering in "Popular Astronomy of 1907," and according to which the moon was torn from the earth when the later was partially solidified, this accounting for the continents which might not have been formed otherwise.

Undoubtedly planets and satellites have originated in both ways and, in my opinion, it is not difficult to ascertain the character of their birth. The following conclusions can be safely drawn:

1. A heavenly body thrown off from a larger one cannot rotate on its axis. The mass, rendered fluid by the combined action of heat and pressure, upon the reduction of the latter immediately stiffens, being at the same time deformed by gravitational pull. The shape becomes permanent upon cooling and solidification and the smaller mass continues to move about the larger one as tho it were rigidly connected to it except for pendular swings or librations due to varying orbital velocity. Such motion precludes the possibility of axial rotation in the strictly physical sense. The moon has never spun around as is well demonstrated by the fact that the most precise measurements have failed to show any measurable flattening in form.

2. If a planetary body in its orbital movement turns the same side towards the central mass this is a positive proof that it has been separated from the latter and is a true satellite.

3. A planet revolving on its axis in its passage around another cannot have been thrown off from the same but must have been captured.

II. The Fallacy of Franklin's Pointed Lightning-Rod

The display of atmospheric electricity has since ages been one of the most marvelous spectacles afforded to the sight of man. Its grandeur and power filled him with fear and For centuries he attributed lightning to agents god like and supernatural and its purpose in the scheme of this universe remained unknown to him. Now we have learned that the waters of the ocean are raised by the sun and maintained in the atmosphere delicately suspended, that they are waited to distant regions of the globe where electric forces assert themselves in upsetting the sensitive balance and causing precipitation, thus sustaining all organic life. There is every reason to hope that man will soon be able to control this life-giving flow of water and thereby solve many pressing problems of his existence.

Atmospheric electricity became of scientific interest in Franklin's time. Faraday had not yet announced his epochal discoveries in magnetic induction but static frictional machines were already generally used in physical laboratories. Franklin's powerful mind and once leaped to the conclusion that frictional and atmosphere electricity were identical. To our present view this inference appears obvious, but in his presence mere thought of it was little short of blasphemy. He investigated the phenomena and argued that if they were of the same nature then the clouds could be drained of their energy exactly as the ball of a static machine, and in 1749 he indicated in a published memoir how this could be done by the use of pointed metal rods.

The earliest trials were made by Dalibrand in France, but Franklin himself was the first to obtain a spark by using a kite, in June, 1752. When these atmospheric discharges manifest themselves today in our wireless station we feel annoyed and wish that they would stop, but to the man who discovered them they brought tears of joy.

The lightning conductor in its classical form was invented by Benjamin Franklin in 1755 and immediately upon its adoption proved a success to a degree. As usual, however, its virtues were often exaggerated. So, for instance, it was seriously claimed that in the city of Piatermaritzburg (capital of Natal, South Africa) no lightning strokes occurred after the

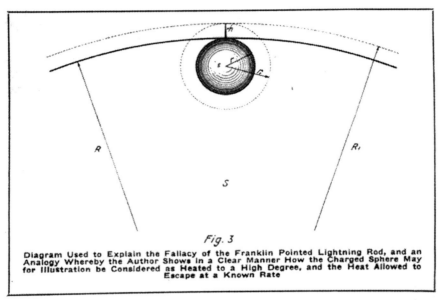

Fig. 3
Diagram Used to Explain the Fallacy of the Franklin Pointed Lightning Rod, and an Analogy Whereby the Author Shows in a Clear Manner How the Charged Sphere May for Illustration be Considered as Heated to a High Degree, and the Heat Allowed to Escape at a Known Rate

pointed rods were installed, altho the storms were as frequent as before. Experience has shown that just the opposite is true. A modern city like New York, presenting innumerable sharp points and projections in good contact with the earth, is struck much more often than equivalent area of land. Statistical records, carefully compiled and published from time to time, demonstrate that the danger from lightning to property and life has been reduced to a small percentage by Franklin's invention, but the damage by fire amounts, nevertheless, to several million dollars annually. It is astonishing that this device, which has been in universal use for more than one century and a half, should be found to involve a gross fallacy in design and construction which impairs its usefulness and may even render its employment hazardous under certain conditions.

For explanation of this curious fact I may first refer to Fig. 3, in which s is a metallic sphere of radius r, such as the capacity terminal of a static machine, provided with a sharply pointed pin of length h, as indicated. It is well known that the latter has the property of quickly dissipating the accumulated charge into the air. To examine this action in the light of present knowledge we may liken electric potential to temperature. Imagine that sphere s is heated to T degrees and that the pin or metal bar is a perfect conductor of heat so that its extreme end is at the same temperature T. Then if another sphere of larger radius, v_1, is drawn about the first and the temperature along this boundary is T_1, it is evident that there will be between the end of the bar and its surrounding a difference of temperature $T - T_1$, which will determine the outflow of heat. Obviously, if the adjacent medium was not affected by the hot sphere this temperature difference would be greater and more heat would be given off. Exactly so in the electric system. Let q be the quantity of the charge, then the sphere — and owing to its great conductivity also the pin — will be at $\frac{q}{r}$ the potential. The medium around the point of the pin will be at the potential $\frac{q}{r_1} = \frac{q}{r+h}$ and, consequently, the difference $\frac{q}{r} - \frac{q}{r+h} = \frac{qh}{r(r+h)}$. Suppose now that a sphere S of much larger radius $R = nr$ is employed containing a charge Q this difference of potential will be, analogously $\frac{Qh}{R(R+h)}$. According to elementary principles of electro-statics the potentials of the two spheres s and S will be equal if $Q = nq$ in which case $\frac{Qh}{R(R+h)} = \frac{nqh}{nr(nr+h)} = \frac{qh}{r(nr+h)}$. Thus the difference of potential between the point of the pin and the medium around the same will be smaller in the ratio $\frac{r+h}{nr+h}$ when the large sphere is used. In many scientific tests and experiments this important observation has been disregarded with the result of causing serious errors. Its significance is that the behavior of the pointed rod entirely depends on the linear dimensions of the electrified body. Its quality to give off the charge may be entirely lost if the latter is very large. For this reason, all points or projections on the surface of a conductor of such vast dimensions as the earth would be quite ineffective were it not for other influences. These will be elucidated with reference to Fig. 4, in which our artist of the Impressionist school has emphasized Franklin's notion that his rod was drawing electricity from the clouds. If the earth were not surrounded by an atmosphere which is generally oppositely charged it would behave, despite all its irregularities of surface, like a polished sphere. But owing to the electrified masses of air and cloud the distribution is greatly modified. Thus in Fig. 4. the positive charge of the cloud induces in the earth an equivalent of opposite charge, the density at the surface of the latter diminishing with the cube of the distance from the static center of the cloud. A brush discharge is then formed at the point of the rod and the action Franklin anticipated takes place. In addition, the surrounding air is ionized and rendered conducting and, eventually, a bolt may hit the building or some other object in the vicinity. The virtue of the pointed end to dissipate the charge, which was uppermost in Franklin's mind is, however, infinitesimal. *Careful measurements show that it would take many years before the electricity stored in a single cloud of moderate site would be drawn off or neutralized thru suck a lightning conductor.* The grounded rod has the quality of rendering harmless most of the strokes it receives, tho occasionally the charge is diverted with damaging results. But, what is very important to note, it invites danger and hazard on account of the fallacy involved in its

Fig. 4. Tesla Explains the Fallacy of the Franklin Pointed Lightning Rod, Here Illustrated, and Shows that Usually Such a Rod Could Not Draw Off the Electricity in a Single Cloud in Many Years. The Density of the Dots Indicates the Intensity of the Charges.

design. The sharp point which was thought advantageous and indispensable to its operation, is really a defect detracting considerably from the practical value of the device. I have produced a much improved form of lightning protector characterized by the employment of a terminal of considerable area and large radius of curvature which makes impossible undue density of the charge and ionization of the air. (*These protectors act as quasi-repellents and so far have never been struck tho exposed a long time.* Their safety is experimentally demonstrated to greatly exceed that invented by Franklin. By their use property worth millions of dollars which is now annually lost, can be saved.

III. The Singular Misconception of the Wireless.

To the popular mind this sensational advance conveys the impression of a single invention but in reality it is an art, the successful practice of which involves the employment of a great many discoveries and improvements. I viewed it as such when I undertook to solve wireless problems and it is due to this fact that my insight into its underlying principles was clear from their very inception.

In the course of development of my induction motors it became desirable to operate them at high speeds and for this purpose I constructed alternators of relatively high frequencies. The striking behavior of the currents soon captivated my attention and in 1889 I started a systematic investigation of their properties and the possibilities of practical application. The first gratifying result of my efforts in this direction was the transmission of electrical energy thru *one wire* without return, of which I gave demonstrations in my lectures and addresses before several scientific bodies here and abroad in 1891 and 1892. During that period, while working with my oscillation transformers and dynamos of frequencies up to 200,000 cycles per second, the idea gradually took hold of me that the earth might be used in place of the wire, thus dispensing with artificial conductors altogether. The immensity of the globe seemed an unsurmountable obstacle but after a prolonged study of the subject I became satisfied that the undertaking was rational, and in my lectures before the Franklin Institute and National Electric Light Association early in 1893 I gave the outline of the system I had conceived. In the latter part of that year, at the Chicago World's Fair, I had the good fortune of meeting Prof. Helmholtz to whom I explained my plan, illustrating it with experiments. On that occasion I asked the celebrated physicist for an expression of opinion on the feasibility of the scheme. He stated unhesitatingly that it was practicable, provided I could perfect apparatus capable of putting it into effect but this, he anticipated, would be extremely difficult to accomplish.

I resumed the work very much encouraged and from that date to 1896 advanced slowly but steadily, making a number of improvements the chief of which was my system of *concatenated tuned circuits* and method of regulation, now universally adopted. In the summer of 1897 Lord Kelvin happened to pass thru New York and honored me by a visit to my laboratory where I entertained him with demonstrations in support of my wireless theory. He was fairly carried away with what he saw but, nevertheless, condemned my protect in emphatic terms, qualifying it as something impossible, "an illusion and a snare." I had expected his approval and was pained and surprised. But the next day he returned and gave me a better opportunity for explanation of the advances I had made and of the true principles underlying the system I had evolved. Suddenly he remarked with evident astonishment: "Then you are not making use of Hertz waves?" "Certainly not," I replied, *"these are radiations."* No energy could be economically transmitted to a distance by any such agency. In my system the process is one of *true conduction* which, theoretically, can be effected at the greatest distance without appreciable loss." I can never forget the magic change that came over the illustrious philosopher the moment he freed himself from that erroneous impression. The skeptic who would not believe was suddenly transformed into the warmest of supporters. He parted from me not only thoroughly convinced of the scientific soundness of the idea but strongly expressed his confidence in its success. In my exposition to him I resorted to the following mechanical analogues of my own and the Hertz wave system.

Imagine the earth to be a bag of rubber filled with water, a small quantity of which is periodically forced in and out of the same by means of a reciprocating pump, as illustrated. If the strokes of the latter are effected in intervals of more than one hour and forty-eight minutes, sufficient for the transmission of the impulse thru the whole mass, the entire bag will expand and contract and corresponding movements will be imparted to pressure gauges or movable pistons with the same intensity, irrespective of distance. By working the pump faster, shorter waves will be produced which, on reaching the opposite end of the bag, may be reflected and give rise to stationary nodes and loops, but in any case, the fluid being incompressible, its inclosure perfectly elastic, and the frequency of oscillations not very high, the energy will be economically transmitted and very little power consumed so long as no work is done in the receivers. This is a crude but correct representation of my wireless system in which, however, I resort to various refinements. Thus, for instance, the pump is made part of a resonant system of great inertia, enormously magnifying the force of the impressed impulses. The receiving devices are similarly conditioned and in this manner the amount of energy collected in them vastly increased.

The Hertz wave system is in many respects the very opposite of this. To explain it by analogy, the piston of the pump is assumed to vibrate to and fro at a terrific rate and the orifice thru which the fluid passes in and out of the cylinder is reduced to a small hole. There is scarcely any movement of the fluid and almost the whole work performed results in the production of radiant heat, of which an infinitesimal part is recovered in a remote locality. However incredible, it is true that the minds of some of the ablest experts have been from the beginning, and still are, obsessed by this monstrous idea, and so it comes that the true wireless art, to which I laid the foundation in 1893, has been retarded in its development

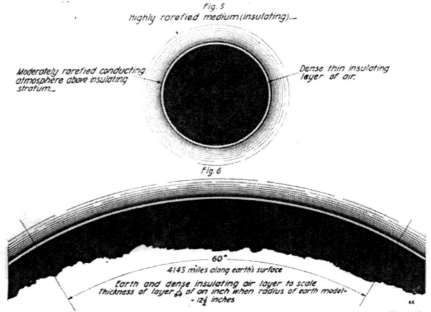

A Section of the Earth and its Atmospheric Envelope Drawn to Scale. It is Obvious That the Hertzian Rays Cannot Traverse So Thin a Crack Between Two Conducting Surfaces For Any Considerable Distance, Without Being Absorbed, Says Dr. Tesla, in Discussing the Ether Space Wave Theory.

for twenty years. This is the reason why the "statics" have proved unconquerable, why the wireless shares are of little value and why the Government has been compelled to interfere.

We are living on a planet of well-nigh inconceivable dimensions, surrounded by a layer of insulating air above which is a rarefied and conducting atmosphere (Fig. 5). This is providential, for if all the air were conducting the transmission of electrical energy thru the natural media would be impossible. My early experiments have shown that currents of high frequency and great tension readily pass thru an atmosphere but moderately rarefied, so that the insulating stratum is reduced to a small thickness as will be evident by inspection of Fig. 6, in which a part of the earth and its gaseous envelope is shown to scale. If the radius of the sphere is 12 ½," then the non-conducting layer is only 1/64" thick and it will be obvious that the Hertzian rays cannot traverse so thin a crack between two conducting surfaces for any considerable distance, without being absorbed. The theory has been seriously advanced that these radiations pass around the globe by *successive reflections*, but to show the absurdity of this suggestion reference is made to Fig. 7 in which this process is diagrammatically indicated. Assuming that there is no

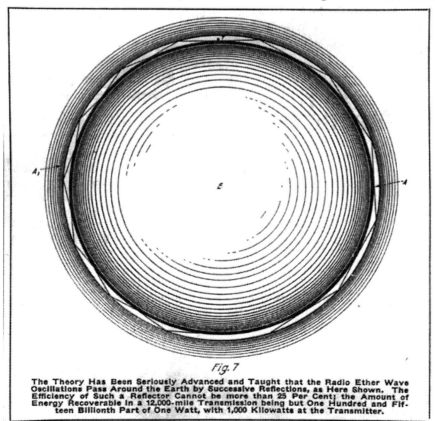

Fig. 7
The Theory Has Been Seriously Advanced and Taught that the Radio Ether Wave Oscillations Pass Around the Earth by Successive Reflections, as Here Shown. The Efficiency of Such a Reflector Cannot be more than 25 Per Cent; the Amount of Energy Recoverable in a 12,000-mile Transmission being but One Hundred and Fifteen Billionth Part of One Watt, with 1,000 Kilowatts at the Transmitter.

refraction, the rays, as shown on the right, would travel along the sides of a polygon drawn around the solid, and inscribed into the conducting gaseous boundary in which case the length of the side would be about 400 miles. As one-half the circumference of the earth is approximately 12,000 miles long there will be, roughly, thirty deviations. The efficiency of such a reflector cannot be more than 25 per cent, so that if none of the energy of the transmitter were lost in other ways, the part recovered would be measured by the fraction (¼)" Let the transmitter radiate Hertz waves at the rate of 1,000 kilowatts. Then about *one hundred and fifteen billionth part of one watt* is all that would be collected in a *perfect* receiver. In truth, the reflections would be much more numerous as shown on the left of the figure, and owing to this and other reasons, on which it is unnecessary to dwell, the amount recovered would be a vanishing quantity.

Consider now the process taking place in the transmission by the instrumentalities and methods of my invention. For this purpose attention is called to Fig. 8, which gives an idea of the mode of propagation of the current waves and is largely self-explanatory. The drawing represents a solar eclipse with the shadow of the moon just touching the surface of the earth at a point where the transmitter is located. As the shadow moves downward it will

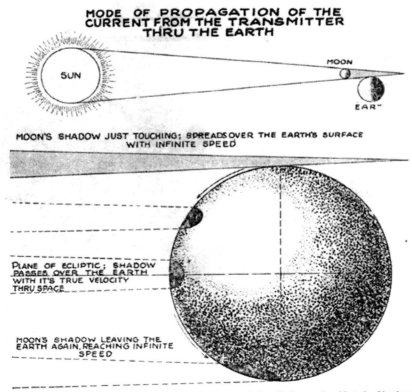

Fig. 8.—This Diagram Illustrates How, During a Solar Eclipse, the Moon's Shadow Passes Over the Earth With Changing Velocity, and Should Be Studied in Connection With Fig. 9. The Shadow Moves Downward With Infinite Velocity at First, Then With Its True Velocity Thru Space, and Finally With Infinite Velocity Again.

spread over the earth's surface, first with infinite and then gradually diminishing velocity until at a distance of about 6,000 miles it will attain its true speed in space. From there on it will proceed with increasing velocity, reaching infinite value at the opposite point of the globe. It hardly need be stated that this is merely an illustration and not an accurate representation in the astronomical sense.

The exact law will be readily understood by reference to Fig. 9, in which a transmitting circuit is shown connected to earth and to an antenna. The transmitter being in action, two effects are produced: Hertz waves pass thru the air, and a current traverses the earth. The former propagate with the speed of light and their energy is *unrecoverable* in the circuit. The latter proceeds with the speed varying as the cosecant of the angle which a radius drawn from any point under consideration forms with the axis of symmetry of the waves. At the origin the speed is infinite but gradually diminishes until a quadrant is traversed, when the velocity is that of light. From there on it again increases, becoming infinite at the antipole. Theoretically the energy of this current is *recoverable* in its entirety, in properly attuned receivers.

Some experts, whom I have credited with better knowledge, have for years contended that my proposals to transmit power without wires are sheer nonsense but I note that they are growing more cautious every day. The latest objection to my system is found in the cheapness of gasoline. These men labor under the impression that the energy flows in all directions and that, therefore, only a minute amount can be recovered in any individual receiver. But this is far from being so. The power is conveyed in only one direction, from the transmitter to the receiver, and none of it is lost elsewhere. It is perfectly practicable to recover at any point of the globe energy enough for driving an airplane, or a pleasure boat or for lighting a dwelling. I am especially sanguine in regard to the lighting of isolated places and believe that a more economical and convenient method can hardly be devised. The future will show whether my foresight is as accurate now as it has proved heretofore.

Tesla's World-Wide Wireless Transmission of Electrical Signals, As Well As Light and Power, is Here Illustrated in Theory, Analogy and Realization. Tesla's Experiments With 100 Foot Discharges At Potentials of Millions of Volts Have Demonstrated That the Hertz Waves Are Infinitesimal In Effect and Unrecoverable; the Recoverable Ground Waves of Tesla Fly "Thru the Earth". Radio Engineers Are Gradually Beginning to See the Light and That the Laws of Propagation Laid Down by Tesla Over a Quarter of a Century Ago Form the Real and True Basis of All Wireless Transmission To-Day.

Tesla Answers Mr. Manierre and Further Explains the Axial Rotation of the Moon
New York Tribune — Feb. 23, 1919

Sirs:

In your article of February 2, Mr. Charles E. Manierre, commenting upon my article in "The Electrical Experimenter" for February, which appeared in The Tribune of January 26, suggests that I give a definition of axial rotation.

I intended to be explicit on this point, as may be judged from the following quotation: "The unfailing test of the spinning of a mass is, however, the existence of energy of motion. The moon is not possessed of such vis viva." By this I meant that "axial rotation" is not simply "rotation upon an axis" as nonchalantly defined in dictionaries, but is circular motion in the true physical sense - that is, one in which half the product of the mass with the square of velocity is a definite and positive quantity.

The moon is a nearly spherical body, of a radius of about 1,081.5 miles, from which I calculate its volume to be approximately 5,300,216,300 cubic miles. Since its mean density is 3.27, one cubic foot of material composing it weighs close to 205 pounds. Accordingly, the total weight of the satellite is about 79,969,000,000, 000,000,000,000 and its mass 2,483,500,000,000,000,000 terrestrial short tons. Assuming that the moon does physically rotate upon its axis, it performs one revolution in 27 days 7 hours 43 minutes and 11 seconds, or 2,360,591 seconds. If, in conformity with mathematical principles, we imagine the entire mass concentrated at a distance from the center equal to two-fifths of the radius, then the calculated rotational velocity is 3.04 feet per second, at which the globe would contain 11,474,000,000,000,000,000 short foot tons of energy, sufficient to run 1,000,000, 000 horsepower for a period of 1,323 years. Now, I say that there is not enough energy in the moon to run a delicate watch.

In astronomical treatises usually the argument is advanced that "if the lunar globe did not turn upon its axis it would expose all parts to terrestrial view. As only a little over one-half is visible it must rotate." But this inference is erroneous, for it admits of one alternative. There are an infinite number of axes besides its own on each of which the moon might turn and still exhibit the same peculiarity.

I have stated in my article that the moon rotates about an axis, passing through the center of the earth, which is not strictly true, but does not vitiate the conclusions I have drawn. It is well known, of course, that the two bodies revolve around a common center of gravity which is at a distance of a little over 2,899 miles from the earth's center.

Another mistake in books on astronomy is made in considering this motion equivalent to that of a weight whirled on a string or in a sling. In the first place, there is an essential difference between these two devices though involving the same mechanical principle. If a metal ball attached to a string is whirled around and the latter breaks an axial rotation of the missile results which is definitely related in magnitude and direction to the motion preceding. By way of illustration: If the ball is whirled on the string clockwise, ten times a second, then when it flies off it will rotate on its axis twenty times a second, likewise in the direction of the clock. Quite different are the conditions when the ball is thrown from a sling. In this case a much more rapid rotation is imparted to it in the opposite sense. There is not true analogy to these in the motion of the moon. If the gravitational string, as it were, would snap, the satellite would go off in a tangent without the slightest swerving or rotation, for there is no momentum about the axis and, consequently, no tendency whatever to spinning motion.

Mr. Manierre is mistaken in his surmise as to what would happen if the earth were suddenly eliminated. Let us suppose that this would occur at the instant when the moon is in opposition. Then it would continue on its elliptical path around the sun, presenting to it steadily the face which was always exposed to the earth. If, on the other hand, the latter would disappear at the moment of conjunction, the moon would gradually swing around through 180 degrees and, after a number of oscillations, revolve again with the same face to the sun. In either case there would be no periodic changes, but eternal day and night, respectively, on the sides turned toward and away from the luminary.

Nikola Tesla

The Moon's Rotation
Electrical Experimenter — April, 1919

Since the appearance of my article entitled the "Famous Scientific Illusions" in your February issue, I have received a number of letters criticizing the views I expressed regarding the moon's "axial rotation." These have been partly answered by my statement to the *New York Tribune* of February 23, which allow me to quote:

In your issue of February 2, Mr. Charles E. Manierre, commenting upon my article in the *Electrical Experimenter* for February which appeared in the *Tribune* of January 26, suggests that I give a definition of axial rotation.

I intended to be explicit on this point as may be judged from the following quotation: "The unfailing test of the spinning of a mass is, however, the existence of *energy of motion*. The moon is not possessed of such vis viva." By this I meant that "axial rotation" is not simply "rotation upon an axis nonchalantly defined in dictionaries, but is a circular motion in the true physical sense—that is, one in which half the product of the mass with the square of velocity is a definite and positive quantity. The moon is a nearly spherical body, of a radius of about 1,087.5 miles, from which I calculate its volume to be approximately 5,300,216,300 cubic miles. Since its mean density is *327*, one cubic foot of material composing it weighs close on 205 lbs. Accordingly, the total weight of the satellite is about 79,969,000,000,000,000,000, and its mass 2,483,500,000,000,000,000 terrestrial short tons. Assuming that the moon does physically rotate upon its axis, it performs one revolution in 27 days, 7 hours, 43 minutes and 11 seconds, or 2,360,591 seconds. If, in conformity with mathematical principles, we imagine the entire mass concentrated at a distance from the center equal to two-fifths of the radius, then the calculated rotational velocity is 3.04 feet per second, at which the globe would contain 11,474,000,000,000,000,000 short foot tons of energy sufficient to run 1,000,000,000 horsepower for a period of 1,323 years. Now, I say, that there is not enough of that energy in the moon to run a delicate watch.

In astronomical treaties usually the argument is advanced that "if the lunar globe did not turn upon its axis it would expose all parts to terrestrial view. As only a little over one-half is visible it *must* rotate." But this inference is erroneous, for it only admits of one alternative. There are an infinite number of axis besides its own in each of which the moon might turn and still exhibit the same peculiarity.

I have stated in my article that the moon rotates about an axis passing thru the center of the earth, which is not strictly true, but it does not vitiate the conclusions I have drawn. It is well known, of course, that the two bodies revolve around a common center of gravity, which is at a distance of a little over 2,899 miles from the earth's center.

Another mistake in books on astronomy is made in considering this motion equivalent to that of a weight whirled on a string or in a sling. In the first place there is an essential difference between these two devices tho involving the same mechanical principle. If a metal ball, attached to a string, is whirled around and the latter breaks, an axial rotation of the missile results which is definitely related in magnitude and direction to the motion preceding. By way of illustration—if the ball is whirled on the string clockwise ten times per second, then when it flies off, it will rotate on its axis ten times per second, likewise in the direction of a clock. Quite different are the conditions when the ball is thrown from a sling. In this case a *much more rapid* rotation is imparted to it in the *opposite sense*. There is no true analogy to these in the motion of the moon. If *the gravitational string, as it were, would snap, the satellite would go off in a tangent without the slightest swerving or rotation, for there is no moment about the axis and, consequently, no tendency whatever to spinning motion.*

Mr. Manierre is mistaken in his surmise as to what would happen if the earth were suddenly eliminated. Let us suppose that this would occur at the instant when the moon is in *opposition*. Then it would continue on its elliptical path around the sun, presenting to it steadily the face which was always exposed to the earth. If, on the other hand, the latter would disappear at the moment of *conjunction,* the moon would gradually swing around thru 180° and, after a number of oscillations, revolve, again with the same face to the sun. In either case there would be no periodic changes but eternal day and night, respectively, on the sides turned towards, and away from, the luminary.

Some of the arguments advanced by the correspondents are ingenious and not a few comical. None, however, are valid.

One of the writers imagines the earth in the center of a circular orbital plate, having fixedly attached to its periperal portion a disk-shaped moon, in frictional or geared engagement with another disk of the same diameter and freely rotatable on a pivot projecting from an arm entirely independent of the planetary system. The arm being held continuously parallel to itself, the pivoted disk, of course, is made to turn on its axis as the orbital plate is rotated. This is a well-known drive, and the rotation of the. pivoted disk is as palpable a fact as that of the orbital plate. But. the moon in this model only revolves about the center of the system *without the slightest angular displacement* on its own axis. The same is true of a cart-wheel to which this writer refers. So long as it advances on the earth's surface it turns on the axle in the true physical sense; when one of its spokes is always kept in a perpendicular position the wheel still *revolves* about the earth's center, *but axial rotation has ceased*. Those who think that it then still exists are laboring under an illusion.

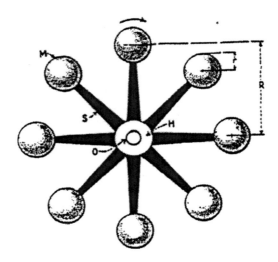

Fig. 1. — If You Still Think That the Moon Rotate on Its Axis, Look at This Diagram and Follow Closely the Successive Positions Taken by One of the Balls M While It is Rotated by a Spoke of the Wheel. Substitute Gravity for the Spoke and the Analogy Solves the Moon Rotation Riddle.

An obvious fallacy is involved in the following abstract reasoning. The orbital plate is assumed to gradually shrink, so that finally the centers of the earth and the satellite coincide when the latter revolves simultaneously about its own and the earth's axis. We may reduce the earth to a mathematical point and the distance between the two planets to the radius of the moon without affecting the system in principle, but a further diminution of the distance is manifestly absurd and of no bearing on the question under consideration.

In all the communications I have received, tho different in the manner of presentation, the successive changes of position in space are mistaken for axial rotation. So, for instance, a positive refutation of my arguments is found in the observation that the moon exposes all sides to other planets! It revolves, to be sure, but none of the evidences is a proof that it turns on its axis. Even the well-known experiment with the Foucault pendulum, altho exhibiting similar phenomena as on our globe, would merely demonstrate a motion of the satellite about *some* axis. The view I have advanced is *Not based on a theory* but on facts *demonstrable by experiment*. It is not a matter of *definition* as some would have it. *A Mass Revolving on its Axis Must Be Possessed of Momentum*. If it has none, there is no axial rotation, all appearances to the contrary notwithstanding.

A few simple reflections based on well established mechanical principles will make this clear. Consider first the case of two equal weights w and w_1, in Fig. 1, whirled about the center O on a string s as shown. Assuming the latter to break at a both weights will fly off on tangents to their circles of gyration, and, being animated with different velocities, they will rotate around their common center of gravity o. If the weights are whirled n times per second then the speed of the outer and the inner one will be, respectively, $V = 2$ ⊛ $+ r) n$ and $V_1 = 2$ p $(R—r) n$, and the difference $V—V_1 = 4$ p r n, will be the length of the circular path of the outer weight. Inasmuch, however, as there will be equalization of the speeds until the mean value is attained, we shall have $\frac{V - V_1}{2} = 2\pi r n = 2\pi r N$, N being the number of revolutions per second of the weights around their center of gravity. Evidently then, the weights continue to rotate at the original rate and in the same direction. I know this to be a fact from actual experiments. It also follows that a ball, as that shown in the figure, will behave in a similar manner for the two half-spherical masses can be concentrated at their centers of gravity and m and m_1, respectively, which will be at a distance from o equal to 3/8 r.

This being understood, imagine a number of balls M carried by as many spokes S radiating from a hub H, as illustrated in Fig. 2, and let this system be rotated n times per second around center O on frictionless bearings. A certain amount of work will be required to bring the structure to this speed, and it will be found that it equals exactly half the product of the masses with the square of the tangential velocity. Now if it be true that the moon rotates in reality on its axis *this must also hold good for each of the balls as it performs the same kind of movement.* Therefore, in imparting to the system a given velocity, energy must have been used up in the axial rotation of the balls. Let M be the mass of one of these and R the radius of gyration, then the rotational energy will be $E = \frac{1}{2}M (2pRn)^2$. Since for one complete turn of the wheel every ball makes one revolution on its axis, according to the prevailing theory, the energy of axial rotation of each ball will be $e = \frac{1}{2}M (2p\, r_1 n)^2$, r_1 being the radius of gyration about the axis and equal to 0.6325 r. We can use as large balls as we like, and so make e a considerable percentage of £ and yet, it is positively established by experiment that each of the rotating balls contain only the energy E, no power whatever being consumed in the supposed axial rotation, which is, consequently, wholly illusionary. Something even more interesting may, however, be stated. As I have shown before, a ball flying off will rotate at the rate of the wheel and in the same direction. But this

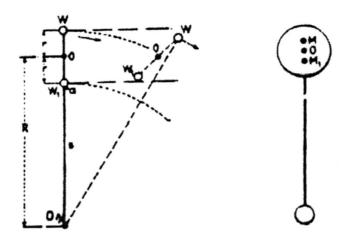

Fig. 2. — Diagram Illustrating the Rotation of Weights Thrown On By Centrifugal Force.

whirling motion, unlike that of a projectile, neither adds to, nor detracts from, the energy of the translatory movement which is exactly equal to the work consumed in giving to the mass the observed velocity.

From the foregoing it will be seen that in order to make one physical revolution on its axis the moon should have twice its present angular velocity, and then it would contain a quantity of stored energy as given in my above letter to the *New York Tribune*, on the assumption that the radius of gyration is 2/5 that of figure. This, of course, is uncertain, as the distribution of density in the interior is unknown. But from the character of motion of the satellite it may be concluded with certitude *that it is devoid of momentum about its axis.* If it be bisected by a plane tangential to the orbit, the masses of the two halves are inversely as the distances of their centers of gravity from the earth's center and, therefore, if the latter were to disappear suddenly, no axial rotation, as in the case of a weight thrown off, would ensue.

We believe the accompanying illustration and its explanation will dispel all doubts as to whether the moon rotates on its axis or not. Each of the balls, as M, depicts a different position of, and rotates exactly like, the moon keeping always the same face turned towards the center O, representing the earth.

But as you study this diagram, can you conceive that any of the balls turn on their axis? Plainly this is rendered physically impossible by the spokes. But if you are still unconvinced, Mr. Tesla's experimental proof will surely satisfy you. A body rotating on its axis must contain rotational energy. Now it is a fact, as Mr. Tesla shows, that no such energy is imparted to the ball as, for instance, to a projectile discharged from a gun.

It is therefore evident that the moon, in which the gravitational attraction is substituted for a spoke, cannot rotate on its axis or, in other words, contain rotational energy. If the earth's attraction would suddenly cease and cause it to fly off in a tangent, the moon would have no other energy except that of translatory movement, and it would not spin like the ball.—Editor.

Tesla on High Frequency Generators
Electrical Experimenter — April, 1919

Editor, Electrical Experimenter:
It is to be regretted that a letter addressed to me by Mr. J. Harris Rogers, in your care, was published in the March number of the *Electrical Experimenter*, altho the concurrence of our views in some wireless features might have made this desirable to so wide-awake and enterprising a periodical as yours.

Mr. Rogers seems to be a very appreciative gentleman and nothing would be farther from my thoughts than to detract anything from his merit, but in a separate contribution, which I expect to prepare for your next issue, I shall express myself on this subject without prejudice and in the interest of truth. However, the article by your Mr. H. Winfield Secor on "America's Greatest War Invention — The Rogers Underground Wireless" contains a reference to "a novel and original high frequency generator" of Mr. Rogers' invention. May I not — to use the President's elegant expression — call attention to the fact that this device was described by me years ago, as will be evident from the following excerpt of a communication which appeared in the *Electrical Review* of March 15, 1899. In speaking of circuit controllers, I said: "I may mention here, based on a different principle, which is incomparably more effective, more efficient, and also simpler on the whole. It comprises a fine stream of conducting fluid which is made to issue, with any desired speed, from an orifice connected with one pole of a generator, thru the primary of the induction coil, against the other terminal of the generator placed at a small distance. This device gives discharges of a remarkable suddenness, and the frequency may be brought within reasonable limits, almost to anything desired. I have used this device for a long time in connection with ordinary coils and in a form of my own coil with results greatly superior in every respect to those obtainable with the form of your letter, make a few statements referring in such make-and-break devices in general, and various forms based on this new principle."

I may add that a great many forms of this apparatus were constructed and employed by me for a long time, proving very convenient and useful. Water does not give particularly good results, being incapable of causing very abrupt changes, but electrolytes have the property of diminishing enormously in resistance when they are heated and the effects are much more intense. Salts of lithium are especially efficient.
Nikola Tesla
New York, February 20, 1919.

The True Wireless
Electrical Experimenter, May 1919

In this remarkable and complete story of his discovery of the "True Wireless" and the principles upon which transmission and reception, even in the present day systems, are based, Dr. Nikola Tesla shows us that he is indeed the "Father of the Wireless." To him the Hertz wave theory is a delusion; it looks sound from certain angles, but the facts tend to prove that it is hollow and empty. He convinces us that the real Hertz waves are blotted out after they have traveled but a short distance from the sender. It follows, therefore, that the measured antenna current is no indication of the effect, because only a small part of it is effective at a distance. The limited activity of pure Hertz wave transmission and reception is here clearly explained, besides showing definitely that in spite of themselves, the radio engineers of today are employing the original Tesla tuned oscillatory system. He shows by examples with different forms of aerials that the signals picked up by the instruments must actually be induced by earth currents—not etheric space waves. Tesla also disproves the "Heaviside layer" theory from his personal observations and tests. —Editor

Ever since the announcement of Maxwell's electro-magnetic theory scientific investigators all the world over had been bent on its experimental verification. They were convinced that it would be done and lived in an atmosphere of eager expectancy, unusually favorable to the reception of any evidence to this end. No wonder then that the publication of Dr. Heinrich Hertz's results caused a thrill as had scarcely ever been experienced before. At that time I was in the midst of pressing work in connection with the commercial introduction of my system of power transmission, but, nevertheless, caught the fire of enthusiasm and fairly burned with desire to behold the miracle with my own eyes. Accordingly, as soon as I had freed myself of these imperative duties and resumed research work in my laboratory on Grand Street, New York, I began, parallel with high frequency alternators, the construction of several forms of apparatus with the object of exploring the field opened up by Dr. Hertz. Recognizing the limitations of the devices he had employed, I concentrated my attention on the production of a powerful induction coil but made no notable progress until a happy inspiration led me to the invention of the oscillation transformer. In the latter part of 1891 I was already so far advanced in the development of this new principle that I had at my disposal means vastly superior to those of the German physicist. All my previous efforts with Rhumkorf coils had left me unconvinced, and in order to settle my doubts I went over the whole ground once more, very carefully, with these improved appliances. Similar phenomena were noted, greatly magnified in intensity, but they were susceptible of a different and more plausible explanation. I considered this so important that in 1892 I went to Bonn, Germany, to confer with Dr. Hertz in regard to my observations. He seemed disappointed to such a degree that I regretted my trip and parted from him sorrowfully. During the succeeding years I made numerous experiments with the same object, but the results were uniformly negative. In 1900, however, after I had evolved a wireless transmitter which enabled me to obtain electro-magnetic activities of many millions of horse-power, I made a last desperate attempt to prove that the disturbances emanating from the oscillator were ether vibrations akin to those of light, but met again with utter failure. For more than eighteen years I have been reading treatises, reports of scientific transactions, and articles on Hertz-wave telegraphy, to keep myself informed, but they have always impressed me like works of fiction.

The history of science shows that theories are perishable. With every new truth that is revealed we get a better understanding of Nature and our conceptions and views are modified. Dr. Hertz did not discover a new principle. He merely gave material support to hypothesis which had been long ago formulated. It was a perfectly well-established fact that a circuit, traversed by a periodic current, emitted some kind of space waves, but we were in ignorance as to their character. He apparently gave an experimental proof that they were transversal vibrations in the ether. Most people look upon this as his great accomplishment. To my mind it seems that his immortal merit was not so much in this as in the focusing of the investigators' attention on the processes taking place in the ambient medium. The Hertz-wave theory, by its fascinating hold on the imagination, has stifled creative effort in the wireless art and retarded it for twenty-five years. But, on the other hand, it is impossible to over-estimate the beneficial effects of the powerful stimulus it has given in many directions.

As regards signaling without wires, the application of these radiations for the purpose was quite obvious. When Dr. Hertz was asked whether such a system would be of practical value, he did not think so, and he was correct in his forecast. The best that might have been expected was a method of communication similar to the heliographic and subject to the same or even greater limitations.

In the spring of 1891 I gave my demonstrations with a high frequency machine before the American Institute of Electrical Engineers at Columbia College, which laid the foundation to a new and far more promising departure. Altho the laws of electrical resonance were well known at that time and my lamented friend, Dr. John Hopkinson, had even indicated their specific application to an alternator in the Proceedings of the Institute of Electrical Engineers, London, Nov.13, 1889, nothing had been done towards the practical use of this knowledge and it is probable that those experiments of mine were the first public exhibition with resonant circuits, more particularly of high frequency. While the spontaneous success of my lecture was due to spectacular features, its chief import was in showing that all kinds of devices could be operated thru a single wire without return. This was the initial step in the evolution of my wireless system. The idea presented itself to me that it might be possible, under observance of proper conditions of resonance, to transmit electric energy thru the earth, thus dispensing with all artificial conductors. Anyone who might wish to examine impartially the merit of that early suggestion must not view it in the light of present day science. I only need to say that as late as 1893, when I had prepared an elaborate chapter on my wireless system, dwelling on its various instrumentalities and future prospects, Mr. Joseph Wetzler and other friends of mine emphatically protested against its publication on the ground that such idle and far-fetched speculations would injure me in the opinion of conservative business men. So it came that only a small part of what I had intended to say was embodied in my address of that year before the Franklin Institute and National Electric Light Association under the chapter "On Electrical Resonance." This little salvage from the wreck has earned me the title of "Father of the Wireless" from many well-disposed fellow workers, rather than the invention of scores of appliances which have brought wireless transmission within the reach of every young amateur and which, in a time not distant, will lead to undertakings overshadowing in magnitude and importance all past achievements of the engineer.

The popular impression is that my wireless work was begun in 1893, but as a matter of fact I spent the two preceding years in investigations, employing forms of apparatus, some of which were almost like those of today. It was clear to me from the very start that the successful consummation could only be brought about by a number of radical improvements. Suitable high frequency generators and electrical oscillators had first to be produced. The energy of these had to be transformed in effective transmitters and collected at a distance in proper receivers. Such a system would be manifestly circumscribed in its usefulness if all extraneous interference were not prevented and exclusiveness secured. In time, however, I recognized that devices of this kind, to be most effective and efficient, should be designed with due regard to the physical properties of this planet and the electrical conditions obtaining on the same. I will briefly touch upon the salient advances as they were made in the gradual development of the system.

The high frequency alternator employed in my first demonstrations is illustrated in Fig. 1. It comprised a field ring, with 384 pole projections and a disc armature with coils wound in one single layer which were connected in various ways according to requirements. It was an excellent machine for experimental purposes, furnishing sinusoidal currents of from 10,000 to 20,000 cycles per second. The output was comparatively large, due to the fact that as much as 30 amperes per square millimeter could be past thru the coils without injury.

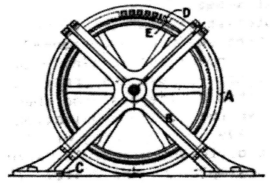

Fig. 1. Alternator of 10.000 Cycles p.s., Capacity 10K.W.. Which Was Employed by Tesla in His First Demonstrations of High Frequency Phenomena Before the American Institute of Electrical Engineers at Columbia College, May 20, 1891.

The diagram in Fig. 2 shows the circuit arrangements as used in my lecture. Resonant conditions were maintained by means of a condenser subdivided into small sections, the finer adjustments being effected by a movable iron core within an inductance coil. Loosely linked with the latter was a high tension secondary which was tuned to the primary.

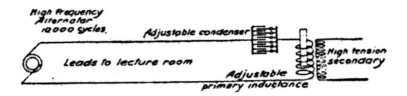

Fig. 2. Diagram Illustrating the Circuit Connections and Tuning Devices Employed by Tesla In His Experimental Demonstrations Before the American Institute of Electrical Engineers With the High Frequency Alternator Shown in Fig. 1.

The operation of devices thru a single wire without return was puzzling at first because of its novelty, but can be readily explained by suitable analogs. For this purpose reference is made to Figs. 3 and 4.

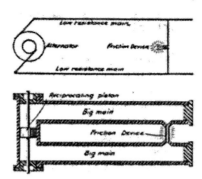

Fig. 3. Electric Transmission Thru Two Wires and Hydraulic Analog.

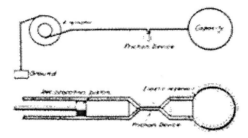

Fig. 4. Electric Transmission Thru a Single Wire Hydraulic Analog.

In the former the low resistance electrical conductors are represented by pipes of large cross section, the alternator by an oscillating piston and the filament of an incandescent lamp by a minute channel connecting the pipes. It will be clear from a glance at the diagram that very slight excursions of the piston would cause the fluid to rush with high velocity thru the small channel and that virtually all the energy of movement would be transformed into heat by friction, similarly to that of the electric current in the lamp filament.

The second diagram will now be self-explanatory. Corresponding to the terminal capacity of the electric system an elastic reservoir is employed which dispenses with the necessity of a return pipe. As the piston oscillates the bag expands and contracts, and the fluid is made to surge thru the restricted passage with great speed, this resulting in the generation of heat as in the incandescent lamp. Theoretically considered, the efficiency of conversion of energy should be the same in both cases.

Granted, then, that an economic system of power transmission thru a single wire is practicable, the question arises how to collect the energy in the receivers. With this object attention is called to Fig. 5, in which a conductor is shown excited by an oscillator joined to it at one end. Evidently, as the periodic impulses pass thru the wire, differences of potential will be created along the same as well as at right angles to it in the surrounding medium and either of these may be usefully applied. Thus at a, a circuit comprising an inductance and capacity is resonantly excited in the transverse, and at b, in the longitudinal sense. At c, energy is collected in a circuit parallel to the conductor but not in contact with it, and again at d, in a circuit which is partly sunk into the conductor and may be, or not, electrically connected to the same. It is important to keep these typical dispositions in mind, for however the distant actions of the oscillator might be modified thru the immense extent of the globe the principles involved are the same.

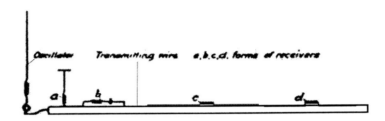

Fig. 5. Illustrating Typical Arrangements for Collecting Energy In a System of Transmission Thru a Single Wire.

Consider now the effect of such a conductor of vast dimensions on a circuit exciting it. The upper diagram of Fig. 6 illustrates a familiar oscillating system comprising a straight rod of self-inductance 2L with small terminal capacities cc and a node in the center. In the lower diagram of the figure a large capacity C is attached to the rod at one end with the result of shifting the node to the right, thru a distance corresponding to self-inductance X. As both parts of the system on either side of the node vibrate at the same rate, we have evidently, (L+X)c = (L-X)C from which X = L(C-c/C+c). When the capacity C becomes commensurate to that of the earth, X approximates L, in other words, the node is close to the ground connection. The exact determination of its position is very important in the calculation of certain terrestrial electrical and geodetic data and I have devised special means with this purpose in view.

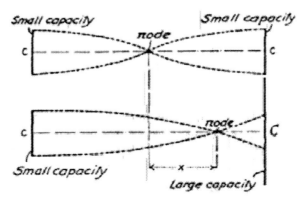

Fig. 6. Diagram Elucidating Effect of Large Capacity on One End.

My original plan of transmitting energy without wires is shown in the upper diagram of Fig. 7, while the lower one illustrates its mechanical analog, first published in my article in the Century Magazine of June, 1900. An alternator, preferably of high tension, has one of its terminals connected to the ground and the other to an elevated capacity and impresses its oscillations upon the earth. At a distant point a receiving circuit, likewise connected to ground and to an elevated capacity, collects some of the energy and actuates a suitable device. I suggested a multiplication of such units in order to intensify the effects, an idea which may yet prove valuable. In the analog two tuning forks are provided, one at the sending and the other at the receiving station, each having attached to its lower prong a piston fitting in a cylinder. The two cylinders communicate with a large elastic reservoir filled with an incompressible fluid. The vibrations transmitted to either of the tuning forks excite them by resonance and, thru electrical contacts or otherwise, bring about the desired result. This, I may say, was not a mere mechanical illustration, but a simple representation of my apparatus for submarine signaling, perfected by me in 1892, but not appreciated at that time, altho more efficient than the instruments now in use.

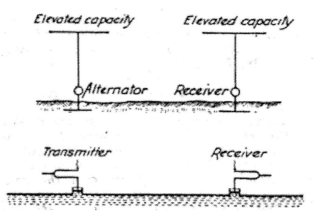

Fig. 7. Transmission of Electrical Energy Thru the Earth as Illustrated in Tesla's Lectures Before the Franklin Institute and Electric Light Association in February and March, 1893, and Mechanical Analog of the Same.

The electric diagram in Fig. 7, which was reproduced from my lecture, was meant only for the exposition of the principle. The arrangement, as I described it in detail, is shown in Fig. 8. In this case an alternator energizes the primary of a transformer, the high tension secondary of which is connected to the ground and an elevated capacity and tuned to the impressed oscillations. The receiving circuit consists of an inductance connected to the ground and to an elevated terminal without break and is resonantly responsive to the transmitted oscillations. A specific form of receiving device was not mentioned, but I had in mind to transform the received currents and thus make their volume and tension suitable for any purpose. This, in substance, is the system of today and I am not aware of a singe authenticated instance of successful transmission at considerable distance by different instrumentalities. It might, perhaps, not be clear to those who have perused my first description of these improvements that, besides making known new and efficient types of apparatus, I gave to the world a wireless system of potentialities far beyond anything before conceived. I made explicit and repeated statements that I contemplated transmission, absolutely unlimited as to terrestrial distance and amount of energy. But, altho I have overcome all obstacles which seemed in the beginning unsurmountable and found elegant solutions of all the problems which confronted me, yet, even at this very day, the majority of experts are still blind to the possibilities which are within easy attainment.

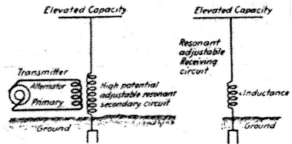

Fig. 8. Tesla's System of Wireless Transmission Thru the Earth as Actually Exposed In His Lectures Before the Franklin Institute and Electric Light Association in February and March, 1893.

My confidence that a signal could be easily flashed around the globe was strengthened thru the discovery of the "rotating brush," a wonderful phenomenon which I have fully described in my address before the Institution of Electrical Engineers, London, in 1892 [Experiments with Alternate Currents of High Potential and High Frequency], and which is illustrated in Fig. 9. This is undoubtedly the most delicate wireless detector known, but for a long time it was hard to produce and to maintain in the sensitive state. These difficulties do not exist now and I am looking to valuable applications of this device, particularly in connection with the high-speed photographic method, which I suggested, in wireless, as well as in wire, transmission.

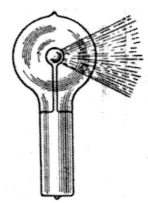

Fig. 9. The Forerunner of the Audion-the Most Sensitive Wireless Detector Known, as Described by Tesla In His Lecture Before the Institution of Electrical Engineers, London, February, 1892.

Possibly the most important advances during the following three or four years were my system of concatenated tuned circuits and methods of regulation, now universally adopted. The intimate bearing of these inventions on the development of the wireless art will appear from Fig. 10, which illustrates an arrangement described in my U.S. Patent No. 568,178 of September 22, 1896, and corresponding dispositions of wireless apparatus. The captions of the individual diagrams are thought sufficiently explicit to dispense with further comment. I will merely remark that in this early record, in addition to indicating how any number of resonant circuits may be linked and regulated, I have shown the advantage of the proper timing of primary impulses and use of harmonics. In a farcical wireless suit in London, some engineers, reckless of their reputation, have claimed that my circuits were not at all attuned; in fact they asserted that I had looked upon resonance as a sort of wild and untamable beast!

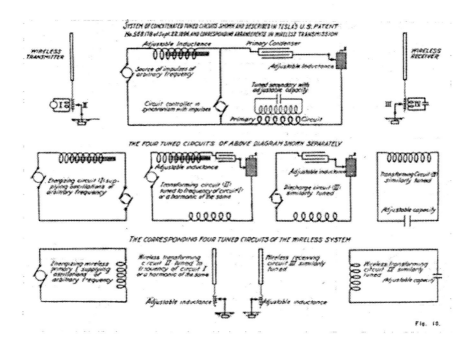

Fig. 10. Tesla's System of Concatenated Tuned Circuits Shown and Described In U. S. Patent No. 568,178 of September 22, 1896, and Corresponding Arrangements in Wireless Transmission.

It will be of interest to compare my system as first described in a Belgian patent of 1897 with the Hertz-wave system of that period. The significant differences between them will be observed at a glance. The first enables us to transmit economically energy to any distance and is of inestimable value; the latter is capable of a radius of only a few miles and is worthless. In the first there are no spark-gaps and the actions are enormously magnified by resonance. In both transmitter and receiver the currents are transformed and rendered more effective and suitable for the operation of any desired device. Properly constructed, my system is safe against static and other interference and the amount of energy which may be transmitted is billions of times greater than with the Hertzian which has none of these virtues, has never been used successfully and of which no trace can be found at present.

A well-advertised expert gave out a statement in 1899 that my apparatus did not work and that it would take 200 years before a message would be flashed across the Atlantic and even accepted stolidly my congratulations on a supposed great feat. But subsequent examination of the records showed that my devices were secretly used all the time and ever since I learned of this I have treated these Borgia-Medici methods with the contempt in which they are held by all fair-minded men. The wholesale appropriation of my inventions was, however, not always without a diverting side. As an example to the point I may mention my oscillation transformer operating with an air gap. This was in turn replaced by a carbon arc, quenched gap, an atmosphere of hydrogen, argon or helium, by a mechanical break with oppositely rotating members, a mercury interrupter or some kind of a vacuum bulb and by such tours de force as many new "systems" have been produced. I refer to this of course, without the slightest ill-feeling, let us advance by all means. But I cannot help thinking how much better it would have been if the ingenious men, who have originated these "systems," had invented something of their own instead of depending on me altogether.

Before 1900 two most valuable improvements were made. One of these was my individualized system with transmitters emitting a wave-complex and receivers comprising separate tuned elements cooperatively associated. The underlying principle can be explained in a few words. Suppose that there are n simple vibrations suitable for use in wireless transmission, the probability that any one tune will be struck by an extraneous disturbance is $1/n$. There will then remain $n-1$ vibrations and the chance that one of these will be excited is 1/n-1 hence the probability that two tunes would be struck at the same time is 1/n(n-1). Similarly, for a combination of three the chance will be 1/n(n-1)(n-2) and so on. It will be readily seen that in this manner any desired degree of safety against the statics or other kind of disturbance can be attained provided the receiving apparatus is so designed that is operation is possible only thru the joint action of all the tuned elements. This was a difficult problem which I have successfully solved so that now any desired number of simultaneous messages is practicable in the transmission thru the earth as well as thru artificial conductors.

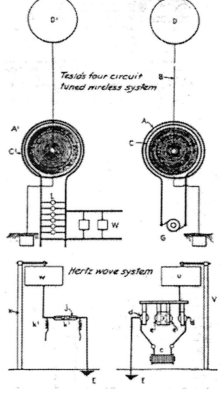

Fig. 11. Tesla's Four Circuit Tuned System Contrasted With the Contemporaneous Hertz-wave System.

The other invention, of still greater importance, is a peculiar oscillator enabling the transmission of energy without wires in any quantity that may ever be required for industrial use, to any distance, and with very high economy. It was the outcome of years of systematic study and investigation and wonders will be achieved by its means.

The prevailing misconception of the mechanism involved in the wireless transmission has been responsible for various unwarranted announcements which have misled the public and worked harm. By keeping steadily in mind that the transmission thru the earth is in every respect identical to that thru a straight wire, one will gain a clear understanding of the phenomena and will be able to judge correctly the merits of a new scheme. Without wishing to detract from the value of any plan that has been put forward I may say that they are devoid of novelty. So for instance in Fig. 12 arrangements of transmitting and receiving circuits are illustrated, which I have described in my U.S. Patent No. 613,809 of November 8, 1898 on a Method of and Apparatus for Controlling Mechanism of Moving Vessels or Vehicles, and which have been recently dished up as original discoveries. In other patents and technical publications I have suggested conductors in the ground as one of the obvious modifications indicated in Fig. 5.

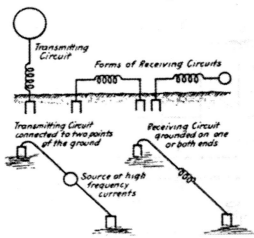

Fig. 12. Arrangements of Directive Circuits Described In Tesla's U. S. Patent No. 613,809 of November 8. 1898, on "Method of and Apparatus for Controlling Mechanism of Moving Vessels or Vehicles."

For the same reason the statics are still the bane of the wireless. There is about as much virtue in the remedies recently proposed as in hair restorers. A small and compact apparatus has been produced which does away entirely with this trouble, at least in plants suitably remodeled.

Nothing is more important in the present phase of development of the wireless art than to dispose of the dominating erroneous ideas. With this object I shall advance a few arguments based on my own observations which prove that Hertz waves have little to do with the results obtained even at small distances.

In Fig. 13 a transmitter is shown radiating space waves of considerable frequency. It is generally believed that these waves pass along the earth's surface and thus affect the receivers. I can hardly think of anything more improbable than this "gliding wave" theory and the conception of the "guided wireless" which are contrary to all laws of action and reaction. Why should these disturbances cling to a conductor where they are counteracted by induced currents, when they can propagate in all other directions unimpeded? The fact is that the radiations of the transmitter passing along the earth's surface are soon extinguished, the height of, the inactive zone indicated in the diagram, being some function of the wave length, the bulk of the waves traversing freely the atmosphere. Terrestrial phenomena which I have noted conclusively show that there is no Heaviside layer, or if it exists, it is of no effect. It certainly would be unfortunate if the human race were thus imprisoned and forever without power to reach out into the depths of space.

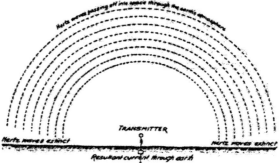

Fig. 13. Diagram Exposing the Fallacy of the Gilding Wave Theory as Propounded In Wireless Text Books.

The actions at a distance cannot be proportionate to the height of the antenna and the current in the same. I shall endeavor to make this clear by reference to diagram in Fig. 14. The elevated terminal charged to a high potential induces an equal and opposite charge in the earth and there are thus Q lines giving an average current I=4Qn which circulates locally and is useless except that it adds to the momentum. A relatively small number of lines q however, go off to great distance and to these corresponds a mean current of ie = 4qn to which is due the action at a distance. The total average current in the antenna is thus Im = 4Qn + 4qn and its intensity is no criterion for the performance. The electric efficiency of the antenna is q/Q+q and this is often a very small fraction.

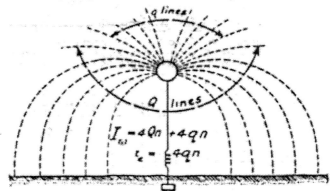

Fig. 14. Diagram Explaining the Relation Between the Effective and the Measured Current In the Antenna.

Dr. L. W. Austin and Mr. J. L. Hogan have made quantitative measurements which are valuable, but far from supporting the Hertz wave theory they are evidences in disproval of the same, as will be easily perceived by taking the above facts into consideration. Dr. Austin's researches are especially useful and instructive and I regret that I cannot agree with him on this subject. I do not think that if his receiver was affected by Hertz waves he could ever establish such relations as he has found, but he would be likely to reach these results if the Hertz waves were in a large part eliminated. At great distance the space waves and the current waves are of equal energy, the former being merely an accompanying manifestation of the latter in accordance with the fundamental teachings of Maxwell.

It occurs to me here to ask the question—why have the Hertz waves, been reduced from the original frequencies to those I have advocated for my system, when in so doing the activity of the transmitting apparatus has been reduced a billion fold? I can invite any expert to perform an experiment such as is illustrated in Fig. 15, which shows the classical Hertz oscillator altho we may have in the Hertz oscillator an activity thousands of times greater, the effect on the receiver is not to be compared to that of the grounded circuit. This shows that in the transmission from an airplane we are merely working thru a condenser, the capacity of which is a function of a logarithmic ratio between the length of the conductor and the distance from the ground. The receiver is affected in exactly the same manner as from an ordinary transmitter, the only difference being that there is a certain modification of the action which can be predetermined from the electrical constants. It is not at all difficult to maintain communication between an airplane and a station on the ground, on the contrary, the feat is very easy.

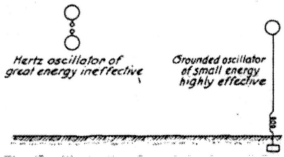

Fig. 15. Illustrating One of the General Evidences Against the Space Wave Transmission.

To mention another experiment in support of my view, I may refer to Fig. 16 in which two grounded circuits are shown excited by oscillations of the Hertzian order. It will be found that the antennas can be put out of parallelism without noticeable change in the action on the receiver, this proving that it is due to currents propagated thru the ground and not to space waves.

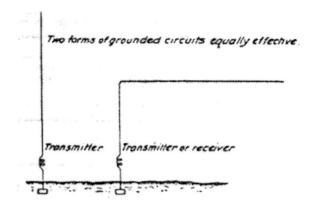

Fig. 16. Showing Unimportance of Relative Position of Transmitting and Receiving Antennae In Disproval of the Hertz-wave Theory.

Particularly significant are the results obtained in cases illustrated in Figures 17 and 18. In the former an obstacle is shown in the path of the waves but unless the receiver is within the effective electrostatic influence of the mountain range, the signals are not appreciably weakened by the presence of the latter, because the currents pass under it and excite the circuit in the same way as if it were attached to an energized wire. If, as in Fig. 18, a second range happens to be beyond the receiver, it could only strengthen the Hertz wave effect by reflection, but as a matter of fact it detracts greatly from the intensity of the received impulses because the electric niveau between the mountains is raised, as I have explained with my lightning protector in the *Experimenter* of February.

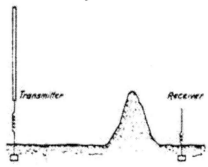

Fig. 17. Illustrating Influence of Obstacle In the Path of Transmission as Evidence Against the Hertz-wave Theory.

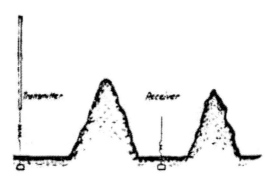

Fig. 18. Showing Effect of Two Hills as Further Proof Against the Hertz-wave Theory.

Again in Fig. 19 two transmitting circuits, one grounded directly and the other thru an air gap are shown. It is a common observation that the former is far more effective, which could not be the case with Hertz radiations. In a like manner if two grounded circuits are observed from day to day the effect is found to increase greatly with the dampness of the ground, and for the same reason also the transmission thru sea-water is more efficient.

Fig. 19. Comparing the Actions of Two Forms of Transmitter as Bearing Out the Fallacy of the Hertz-wave Theory.

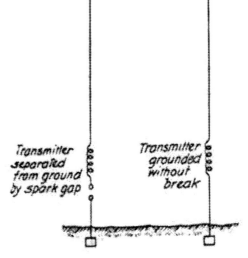

An illuminating experiment is indicated in Fig. 20 in which two grounded transmitters are shown, one with a large and the other with a small terminal capacity. Suppose that the latter be 1/10 of the former but that it is charged to 10 times the potential and let the frequency of the two circuits and therefore the currents in both antennas be exactly the same. The circuit with the smaller capacity will then have 10 times the energy of the other but the effects on the receiver will be in no wise proportionate.

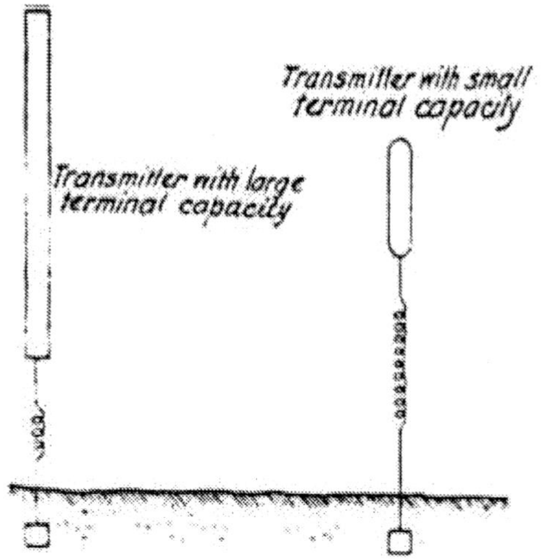

Fig. 20. Disproving the Hertz-wave Theory by Two Transmitters, One of Great and the Other of Small Energy.

Transmitter with large terminal capacity

Transmitter with small terminal capacity

The same conclusions will be reached by transmitting and receiving circuits with wires buried underground. In each case the actions carefully investigated will be found to be due to earth currents. Numerous other proofs might be cited which can be easily verified. So for example oscillations of low frequency are ever so much more effective in the transmission which is inconsistent with the prevailing idea. My observations in 1900 and the recent transmissions of signals to very great distances are another emphatic disproval.

The Hertz wave theory of wireless transmission may be kept up for a while, but I do not hesitate to say that in a short time it will be recognized as one of the most remarkable and inexplicable aberrations of the scientific mind which has ever been recorded in history.

Electrical Oscillators
Electrical Experimenter - July 1919

Few fields have been opened up the exploration of which has proved as fruitful as that of high frequency currents. Their singular properties and the spectacular character of the phenomena they presented immediately commanded universal attention. Scientific men became interested in their investigation, engineers were attracted by their commercial possibilities, and physicians recognized in them a long-sought means for effective treatment of bodily ills. Since the publication of my first researches in 1891, hundreds of volumes have been written on the subject and many invaluable results obtained through the medium of this. new agency. Yet, the art is only in its infancy and the future has incomparably bigger things in store.

From the very beginning I felt the necessity of producing efficient apparatus to meet a rapidly growing demand and during the eight years succeeding my _ original announcements I developed not less than fifty types of these transformers or electrical oscillators, each complete in every detail and refined to such a degree that I could not materially improve any one of them today. Had I been guided by practical considerations 1 might have built up an immense and profitable business, incidentally rendering important services to the world. But the force of circumstances and the ever enlarging vista of greater achievements turned my efforts in other directions. And so it comes that instruments will shortly be placed on the market which, oddly enough, were perfected twenty years ago!

These oscillators are expressly intended to operate on direct and alternating lighting circuits and to generate damped and undamped oscillations' or currents of any frequency, volume and tension within the widest limits. They are compact, self-contained, require no care for long periods of time and will be found very convenient and useful for various purposes as, wireless telegraphy and telephony; conversion of electrical energy; formation of chemical compounds through fusion and combination; synthesis of gases; manufacture of ozone; lighting; welding; municipal, hospital, and domestic sanitation and sterilization, and numerous other applications in scientific laboratories and industrial institutions. While these transformers have never been described before, the general principles underlying them were fully set forth in my published articles and patents, more particularly those of September 22, 1896, and it is thought, therefore, that the appended photographs of a few types, together with a short explanation, will convey all the information that may be desired.

The essential parts of such an oscillator are: a condenser, a self-induction coil for charging the same to a high potential, a circuit controller, and a transformer which is energized by the oscillatory. discharges of the condenser. There are at least three, but usually four, five or six, circuits in tune and the regulation is effected in several ways, most frequently merely by means of an adjusting screw. Under favorable conditions an efficiency as high as 85% is attainable, that is to say, that percentage of the energy supplied can be recovered in the secondary of the transformer. While the chief virtue of this kind of apparatus is obviously due to the wonderful powers of the condenser, special qualities result from concatenation of circuits under observance of accurate harmonic relations, and minimization of frictional and other losses which has been one of the principal objects of the design.

Broadly, the instruments can be divided into two classes: one in which the circuit controller comprises solid contacts, and the other in which the make and break is effected by mercury. Figures 1 to 8, inclusive, belong to the first, and the remaining ones to the second class. The former are capable of an appreciably higher efficiency on account of the fact that the losses involved in the make and break are reduced to the minimum and the resistance component of the damping factor is very small. The latter are preferable for purposes requiring larger output and a great number of breaks per second. The operation of the motor and circuit controller of course consumes a certain amount of energy which, however, is the less significant the larger the capacity of the machine.

In Fig. 1 is shown one of the earliest forms of oscillator constructed for experimental purposes. The condenser is contained in a square box of mahogany upon which is mounted the self-induction or charging coil wound, as will be noted, in two sections connected in multiple or series according to whether the tension of the supply circuit is 110 or 220 volts. From the box protrude four brass columns carrying a plate with the spring contacts and adjusting screws as well as two massive terminals for the reception of the primary of the transformer. Two of the columns serve as condenser connections while the other pair is employed to join the binding posts of the switch in front to the self-inductance and condenser. The primary coil consists of a few turns of topper ribbon to the ends of which are soldered short rods fitting into the terminals referred to. The secondary is made in two parts, wound in a manner to reduce as much as possible the distributed capacity and at the same time enable the coil to withstand a very high pressure between its terminals at the center, which are connected to binding posts on two rubber columns projecting from the primary. The circuit connections may be slightly varied but ordinarily they are as diagrammatically illustrated in the Electrical Experimenter for May on page 89, relating to my oscillation transformer photograph of which appeared on page 16 of the same number. The operation is as follows: When the switch is thrown on, the current from the supply circuit rushes through the self-induction coil, magnetizing the iron core within and separating the contacts of the controller. The high tension induced current then charges the condenser and upon closure of the contacts the accumulated energy is released through the primary, giving rise to a long series of oscillations which excite the tuned secondary circuit.

This device has proved highly serviceable in carrying on laboratory experiments of all kinds. For instance, in studying phenomena of impedance, the transformer was removed and a bent copper bar inserted in the terminals. The latter was often replaced by a large circular loop to exhibit inductive effects at a distance or to excite resonant circuits used in various investigations and measurements. A transformer suitable for any desired performance could be readily improvised and attached to the terminals and in this way much time and labor was saved. Contrary to what might be naturally expected, little trouble was experienced with the contacts, although the currents through them were heavy, namely, proper conditions of resonance existing, the great flow occurs only when the circuit is closed and no destructive arcs can develop. Originally I employed platinum and iridium tips but later replaced them by some of meteorite and finally of tungsten. The last have given the best satisfaction, permitting working for hours and days without interruption.

Fig. 2 illustrates a small oscillator designed for certain specific uses. The underlying idea was to attain great activities during minute intervals of time each succeeded by a comparatively long period of inaction. With this object a large self-induction and a quick-acting break were employed owing to which arrangement the condenser was charged to a very high potential. Sudden secondary currents and sparks of great volume were thus obtained, eminently suitable for welding thin wires, flashing lamp filaments, igniting explosive mixtures and kindred applications. The instrument was also adapted for battery use and in this form was a very effective igniter for gas engines on which a patent bearing number 609,250 was granted to me August 16, 1898.

Fig. 3 represents a large oscillator of the first class intended for wireless experiments, production of Rontgen rays and scientific research in general. It comprises a box containing two condensers of the same capacity on which are supported the charging coil and transformer. The automatic circuit controller, hand switch and connecting posts are mounted on the front plate of the inductance spool as is also one of the contact springs. The condenser box is equipped with three terminals, the two external ones serving merely for connection while the middle one carries a contact bar with a screw for regulating the interval during which the circuit is closed. The vibrating spring, itself, the sole function of which is to cause periodic interruptions, can be adjusted in its strength as well as distance from the iron core in the center of the charging coil by four screws visible on the

top plate so that any desired conditions of mechanical control might be secured. The primary coil of the transformer is of copper sheet and taps are made at suitable points for the purpose of varying at will., the number of turns. As in Fig. 1 the inductance coil is wound in two sections to adapt the instrument both to 110 and 220 volt circuits and several secondaries were provided to suit the various wave lengths of the primary. The output was approximately 500 watt with damped waves of about 50,000 cycles per second. For short periods of time undamped oscillations were produced in screwing the vibrating spring tight against the iron core and separating the contacts by the adjusting screw which also performed the function of a key. With this oscillator I made a number of important observations and it was one of the machines exhibited at a lecture before the New York Academy of Sciences in 1897.

Fig. 4 is a photograph of a type of transformer in every respect similar to the one illustrated in the May, 1919, issue of the Electrical Experimenter to which reference has already been made. It contains the identical essential parts, disposed in like manner, but was specially designed for use on supply circuits of higher tension, from 220 to 500 volts or more. The usual adjustments are made in setting the contact spring and shifting the iron core within the inductance coil up and down by means of two screws. In order to prevent injury through a short-circuit, fuses are inserted in the lines. The instrument was photographed in action, generating undamped oscillations from a 220 volt lighting circuit.

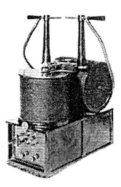

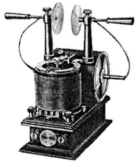

Fig. 5 shows a later form of transformer principally intended to replace Rhumkorf coils. In this instance a primary is employed, having a much greater number of turns and the secondary is closely linked with the same. The currents developed in the latter, having a tension of from 10,000 to 30,000 volts, are used to charge condensers and operate an independent high frequency coil as customary. The controlling mechanism is of somewhat different construction but the core and contact spring are both adjustable as before.

Fig. 6 is a small instrument of this type, particularly intended for ozone production or sterilization. It is remarkably efficient for its size and can be connected either to a 110 or 220 volt circuit, direct or alternating, preferably the former.

In Fig. 7 is shown a photograph of a larger transformer of this kind. The construction and disposition of the parts is as before but there are

two condensers in the box, one of which is connected in the circuit as in the previous cases, while the other is in shunt to the primary coil. In this manner currents of great volume are produced in the latter and the secondary effects are accordingly magnified. The introduction of an additional tuned circuit secures also other advantages but the adjustments are rendered more difficult and for this reason it is desirable to use such an instrument in the production of currents of a definite and unchanging frequency.

Fig. 8 illustrates a transformer with rotary break. There are two condensers of the same capacity in the box which can be connected in series or multiple. The charging inductances are in the form of two long spools upon which are supported the secondary terminals. A small direct current motor, the speed of which can be varied within wide limits, is employed to drive a specially constructed make and break. In other features the oscillator is like the one illustrated in Fig. 3 and its operation will be readily understood from the foregoing. This transformer was used in my wireless experiments and frequently also for lighting the laboratory by my vacuum tubes and was likewise exhibited at my lecture before the New York Academy of Sciences above mentioned.

Coming now to machines of the second class, Fig. 9 shows an oscillatory transformer comprising a condenser and charging inductance enclosed in a box, a transformer and a mercury circuit controller, the latter being of a construction described for the first time in my patent No. 609,251 of August 16, 1898. It consists of a motor driven hollow pulley containing a small quantity of mercury which is thrown outwardly against the walls of the vessel by centrifugal force and entrains a contact wheel which periodically closes and opens the condenser circuit. By means of adjusting screws above the pulley, the depth of immersion of the vanes and consequently, also, the duration of each contact can be varied at desire and thus the intensity of the effects and their character controlled. This form of break has given thorough satisfaction, working continuously with currents of from 20 to 25 amperes. The number of interruptions is usually from 500 to 1,000 per second but higher frequencies are practicable. The space occupied is about 10" X 8" X 10" and the output approximately ½ kW.

In the transformer just described the break is exposed to the atmosphere and a slow oxidation of the mercury takes place. This disadvantage is overcome in the instrument shown in Fig. 10, which consists of a perforated metal box containing the condenser and charging inductance and carrying on the top a motor driving the break, and a transformer. The mercury break is of a kind to be described and operates on the principle of a jet which establishes, intermittently, contact with a rotating wheel in the interior of the pulley. The stationary parts are supported in the vessel on a bar passing through the long hollow shaft of the motor and a mercury seal is employed to effect hermetic closure of the chamber enclosing the circuit controller. The current is led into the interior of the pulley through two sliding rings on the top which are in series with the condenser and primary. The exclusion of the oxygen is a decided improvement, the deterioration of the metal and attendant trouble being eliminated and perfect working.

Fig. 11 is a photograph of a similar oscillator with hermetically inclosed mercury bleak. In this machine the stationary parts of the interrupter in the interior of the pulley were supported on a tube through which was led an insulated wire connecting to one terminal of the break while the other was in contact with the vessel. The sliding rings were, in this manner, avoided and the construction simplified. The instrument was designed for oscillations of lower tension and frequency requiring primary currents of comparatively smaller amperage and was used to excite other resonant circuits.

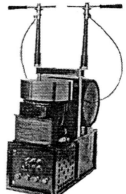

Fig. 12 shows an improved form of oscillator of the kind described in Fig. 10, in which the supporting bar through the hollow motor shaft was done away with, the device pumping the mercury being kept in position by gravity, as will be more fully explained with reference to another figure. Both the capacity of the condenser and primary turns were made variable with the view of producing oscillations of several frequencies.

Fig. 13 is a photographic view of another form of oscillatory transformer with hermetically sealed mercury interrupter, and Fig. 14 diagrams showing the circuit connections and arrangement of parts reproduced from my patent, No. 609,245, of August 16, 1898, describing this particular device. The condenser, inductance, transformer and circuit controller are disposed as before, but the latter is of different construction, which will be clear from an inspection of Fig. 14:

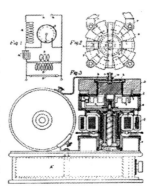

The hollow pulley a is secured to a shaft c which is mounted in a vertical bearing passing through the stationary field magnet d of the motor. In the interior of the vessel is supported, on frictionless bearings, a body h of magnetic material which is surrounded by a dome b in the center of a laminated iron ring, with pole pieces oo wound with energizing coils p. The ring is supported on four columns and, when magnetized, keeps the body h in position while the pulley is rotated, The latter is of steel, but the dome is preferably

made of German silver burnt black by acid or nickeled. The body h carries a short tube k bent, as indicated, to catch the fluid as it is whirled around, and project it against the teeth of a wheel fastened to the pulley. The wheel is insulated and contact from it to the external circuit is established through a mercury cup. As the pulley is rapidly rotated a jet of the fluid is thrown against the wheel, thus peaking and breaking contact about 1,000 times per second. The instrument works silently and, owing to the absence of all deteriorating agents; keeps continually clean and in perfect condition. The number of interruptions per second may be much greater, however, so as to make the currents suitable for wireless telephony and like purposes.

A modified form of oscillator is represented in Figs. 15 and 16, the former being a photographic view and the latter a diagrammatic illustration showing the arrangement of the interior parts of the controller. In this instance the shaft b carrying the vessel a is hollow and supports, in frictionless bearings, a spindle j to which is fastened a weight k. Insulated from the latter, but mechanically fixed to it, is a curved arm L upon which is supported, freely rotatable, a break-wheel with

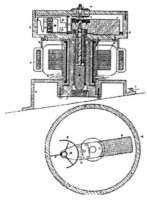

projections QQ. The wheel is in electrical connection with the external circuit through a mercury cup and an insulated plug supported from the top of the pulley. Owing to the inclined position of the motor the weight k keeps the break-wheel in place by the force of gravity and as the pulley is rotated the circuit, including the condenser and primary coil of the transformer, is, rapidly made and broken.

Fig. 17 shows a similar instrument in which, however, the make and break device is a jet of mercury impinging against an insulated toothed wheel carried on an insulated stud in the center of the cover of the pulley as shown. Connection to the condenser circuit is made by brushes bearing on this plug.

Fig. 18 is a photograph of another transformer with a mercury circuit controller of the wheel type, modified in some features on which it is unnecessary to dwell.

These are but a few of the oscillatory transformers I have perfected and constitute only a small part of my high frequency apparatus of which I hope to give a full description, when I shall have freed myself of pressing duties, at some future date.

Nikola Tesla Tells How We May Fly Eight Miles High at 1,000 Miles an Hour

Reconstruction — July, 1919 An interview by Frederick M. Kerby.

As the inventor of the alternating current, the world is indebted to Mr. Tesla for the use of electricity carried Long distances. He now discusses. the probability that airplanes will rise to great heights and travel at speeds that seem incredible. This article is written, in part by Mr. Tesla himself. The rest is written from stenographic notes. It gives, very Likely, a glimpse of the immediate future.

Sitting in his office on the twenty-fifth floor of the Woolworth Tower, Mr. J. Pierpont Jones, American business man, will one day glance at his watch and discover it is 3 o'clock in the afternoon.

"By George," he will say, buzzing for his secretary, "If I don't hurry I'll be late for that dinner engagement at the Savoy!" And as his secretary answers the buzzer:

"Charles, when does the next London bus leave?"

"Three-thirty, sir," says Charles. "You can make it if you hurry. The car is waiting."

And fifteen minutes later Mr. J. Pierpont Jones will emerge from the elevator on the aeronautic landing stage of lower Manhattan, climb into the hermetically sealed steel fuselage of the New York-London Limited, which will rise promptly at 3:30 p. m. At seven that night he will climb out of his compartment on the landing stage on the Thames Embankment, and descend to meet his friend for dinner.

The three-hour airplane trip from New York to London, flying above the storm level at eight miles above the earth's surface is the possibility of the immediate future.

This is not my own prediction. It is the result of sixteen pages of close calculations in higher mathematics made by Nikola Tesla, to test and check up other pages of intricate calculations made by Samuel D. Mott, charter member of the Aero Club of America.

Mr. Mott asserts that the three-hour trip to London from New York is a question of rising into rarefied air where the air pressure is only one-fifth what it is at the earth's surface, at which point the "airplane," as he has named the flying machine of the future, may be expected to fly five times as fast as at the earth's surface. And if the speed of the airplane is increased not five times but only one-fifth, Mr. Mott says the trip will be made anyhow in the rarefied air eight miles above the earth's surface in not more than twelve hours running time.

And Nikola Tesla agrees that taking a plane to such an altitude must result in great increase in speed, although he does not wish, in the absence of exact knowledge of certain factors entering into the problem, to predict exact speeds.

Speaking before the Pan-American Aeronautic Convention at Atlantic City, Mr. Mott asserted that in order to avoid being weather-bound as were the aviators at Newfoundland, it will be necessary to construct planes that will rise above the storm limit.

"I submit," he said, "that waiting indefinitely for ideal weather conditions for long-distance flying over land or sea will not do for the demands of commerce. Therefore I would bring to your attention the possibilities from the airplane or hydroplane, to go into the stillness of nature above the weather.

What The Problem Is

"The problem is evidently one of equipment of our planes to function in rarefied air, and protection of navigators against its tenuity; likewise protection of their body warmth and comfort in extremes of temperature. How high we may go no one may know until tested. Personally I believe it possible to go fifteen or twenty mils aloft, if necessary. It is obviously a matter of equipment plus climbing ability of aircraft designed for the purpose.

"What is the object of high flying? Daily experience shows us that high speed and density are incompatible. We know that we must furnish aircraft with four times the power to go twice as fast, and the marine engineer knows that he must furnish eight times the power to go twice as fast. In other words, from the ultimate height of the air to the earth's core pressure is progressive. Thirty-three feet below the ocean's surface the pressure doubles. For every 1,000 feet ascent the pressure diminishes roughly one-half pound per square inch. The pressure two miles high is 9.8 pounds per square inch; at one mile high, 10.88; at three-quarters of a mile, 12.06; one-half mile, 13.33; one-quarter mile, 14.2, and at sea level, 14.7 pounds, or, in round numbers, 15 pounds per square inch.

"The unknown factor in the high altitude problem is this: Will an airplane in one-fifth density (eight miles high), with equal push, go five times faster or one-fifth faster? The rest is a matter of simple equipment and good construction. In either case the gain is substantial. If the former were true a voyage between New York and London can be made in about three hours by going eight miles high. If the latter is true the same voyage can be made in about twelve hours running time, assuming a surface speed of 200 miles an hour, which is practically a question of power.

"To my mind it is plain that the high altitudes will be determining factors in long distance flying. Greater speed, greater distance, more comfort and less danger because when we double the time to do a risky thing we double the risk incurred; less gasolene, less weight and expense, for if environment permits us to go 100 miles with twice the fuel we formerly used to go twenty-five miles our economic gain is obviously 100 per cent, because we may then go 100 miles with the amount of fuel we formerly consumed to go fifty miles."

That aerial navigation at higher altitudes will Undoubtedly result in great increase of speed is also the opinion of Nikola Tesla, to whom I took Mr. Mott's conclusions in order to get the opinion of this man who has made a life-time study of the air as a medium for the transmission of electrical energy.

"In the propulsion of aerial vessels problems are involved entirely different from those presented in the navigation of the water," said Tesla. "The atmosphere may be likened to a vast ocean, but if one imagines a submarine vessel constructed like an airplane one immediately realizes how inefficient it would be. The energy used in propelling a body through a medium of any kind is wasted in three different ways; first, by skin friction; second, wave making; third, production of eddies. On general principles, however, the resistance can be divided into two parts: one which is due to the friction of the medium and the other to its stickiness, or viscosity, as it is termed. The first is proportionate to the density; the second to this peculiar property of the fluid.

"Everybody will readily understand that the denser the medium the harder it is to push a body through it, but it might not be clear to every person what this other resistance - this viscosity - means. This will be understood if we compare, for instance, water and oil. The latter is lighter, but much more sticky, so that it is a greater obstacle to propulsion than water. Air is a very viscous substance and that part of resistance which is due to this quality is considerable. We must take this latter resistance into account in calculating how fast an airplane could fly in the upper reaches of the air.

"Now, the idea is to fly at a great height where the air is rarefied, and therefore much less power is required to propel the machine through it. If we take the pressure at sea-level at 14.7 pounds and the temperature at 15 degrees centigrade, then, without introducing several corrections that would make for greater accuracy, the pressures at different heights are about as follows: At. 1,000 feet above sea-level, 14.178 lbs.; at one-mile, 12.1457 lbs.; at two miles, 10.035 lbs.; at eight miles, 3.1926 lbs.; at fifteen miles, 0.8392 lbs. and at twenty miles, 0.323 lbs.

Condition Eight Miles Up

"According to these figures that I have worked out, at a height of eight miles the density of the air is 0.2172 or about 22-100th of that at sea level; at fifteen miles it is 0.057, and at twenty miles only 0.0219, or nearly 22-1000th of that at sea-level.

"Let us suppose then that an airplane rises to a height of eight miles where the pressure of the air will be only 3.1926 lbs., or, in other words, the density 0.2172 of that at sea-level. Since, as pointed out, the purely frictional resistance is proportionate to the density of the air, it is obvious that, if there were no other resistance to overcome, only about 22 per cent of power or roughly one-fifth, would be required to propel the vessel at that height, so that extremely high speed, as Mr. Mott points out, would be obtainable.

"And though the other resistance, which is due to the stickiness of the medium, will not be diminished at the same ratio, and therefore the gain will not be strictly in proportion to the decrease of density of the air, nevertheless, the total resistance will be reduced, if not to 22 per cent, perhaps to 30 per cent, so that there will be a great excess of power available for more rapid flight.

"Even allowing for the decreased thrust of the propeller due to the thinness of the air, which cannot be overcome by driving the screw faster, there still will be the very considerable gain and the aircraft will be propelled at a higher speed.

"Of course many incertitudes still exist in the theoretical treatment of a question like this, as there are a number of factors which affect the result and in regard to which we have not yet complete information.

At An Altitude of Twenty Miles

"I doubt that it will be possible to rise as high as fifteen or twenty miles, which is the opinion expressed by Mr. Mott. At the height of twenty miles there is only about 7 per cent of oxygen in the air instead of 21 per cent which is present close to the ground, and there would be great trouble in securing the oxygen supply for the combustion of the fuel, not to speak of other limitations.

"However, at a height of eight miles the decrease of oxygen can be overcome for both engine and aviator. Of course provision would have to be made for supplying the aviator and passengers with oxygen. In all probability they would have to be entirely enclosed just as a diver is enclosed. Our highest mountains are five miles and the rarefication of the air makes climbing them difficult. About five miles provision would certainly have to be made for supplying the aviator. If he were not enclosed the decrease of pressure due to the thinner air would result disastrously. The human mechanism is adjusted to a pressure of nearly 15 pounds per square inch; and if that pressure is reduced to about three pounds, as it would be at an altitude of eight miles, the aviator's ear drums would burst, and even the blood would be forced through the pores and would ooze out of the body.

Tesla explained that the effect would be the same as that of bringing a deep-sea fish, accustomed to live a mile below the surface, to the surface of the water. The fish simply explodes, for lack of the pressure which its body is built to withstand.

With proper protection of the aviator and an artificial supply of oxygen Tesla believes that flights at the eight-mile altitude are quite possible.

"Then there will be great progress with the lighter than air machine and we may soon expect the advent of a dirigible of the Zeppelin type as a common vehicle for travel. Contrary to the general belief, such a vessel can be propelled more rapidly than an airplane and it will be, on the whole, much safer. Furthermore it will give to the passengers the comforts that are necessary in order to make this form of travel popular. Of course in the practical use of these monstrous structures, formidable obstacles will be encountered. They are susceptible to damage by storms, and I believe also from certain danger from lightning, which will not be obviated by the use of helium gas. But I expect to see these difficulties overcome.

The dirigible, supplied with sufficient power, need not fear the storm; it can rise above it, or go around it. The only danger from storm in any case lies in being blown from the course, for while the ship is moving with the storm it is in no danger, since it travels at the same speed as the wind, and the passengers would be in absolutely quiet air, so that a candle might be lighted on deck. Methods of docking and housing the big ships must be devised, but several have been proposed that reduce the danger of landing by making it unnecessary for the ship to come to earth. "

But the revolutionizing influence on aircraft of the future Mr. Tesla believes to lie in the possibility of transmitting power to them through the air.

"For years," he said, " I have advocated my system of wireless transmission of power which is now perfectly practicable and I am looking confidently to its adoption and further development. In the system I have developed, distance is of absolutely no consequence. That is to say, a Zeppelin vessel would receive the same power whether it was 12,000 miles away or immediately above the power plant. The application of wireless power for aerial propulsion will do away with a great deal of complication and waste, and it is difficult to imagine that a more perfect means will ever be found to transport human beings to great distances economically. The power supply is virtually unlimited, as any number of power plants can be operated together, supplying energy to airships just as trains running on tracks are now supplied with electrical energy through rails or wires.

"The transmission of power by wireless will do away with the present necessity for carrying fuel on the airplane or airship. The motors of the plane or airship will be energized by this transmitted power, and there will be no such thing as a limitation on their radius of action, since they can pick up power at any point on the globe.

The advance of science to this point, however, is attended with terrible risks for the world. We are facing a condition that is positively appalling if we ever permit warfare to invade the earth again. For up to the present war the main destructive force was provided by guns which are limited by the size of the projectile and the distance it can be thrown. *In the future nations will fight each other thousands of miles apart. No soldier will see his enemy. In fact future wars will not be conducted by men directly but by the forces which if let loose many well destroy civilization completely.* If war comes again, I look for the extensive use of self-propelled air vehicles carrying enormous charges of explosive which will be sent from any point to another to do their destructive work, with no human being aboard to guide them. The distance to which they can be sent is practically unlimited and the amount of explosive they can carry is likewise practically unlimited. It is practicable to send such an air vessel say to a distance of four or five thousand miles and so control its course either gyroscopically or electrically that it will land at the exact spot where it is intended to have it land, within a few feet, and its cargo of explosive can there be detonated.

"This cannot be done by means of the present wireless plants, but with a proper plant it can be done, and we have here the appalling prospect of a war between nations at a distance of thousands-of miles, with weapons so destructive and demoralizing that the world could not endure them. That is why there must be no more war.

Signals to Mars Based on Hope of Life on Planet
New York Herald — Sunday, Oct. 12, 1919

The idea that other planets are inhabited by intelligent beings might be traced to the very beginnings of civilization. This, in itself, would have little significance, for many of the ancient beliefs had their origin in ignorance, fear or other motives - good or evil, and were nothing more than products of untrained or tortured imagination. But when a conception lives through ages in the minds, growing stronger and stronger with increasing knowledge and intellectual development, it may be safely concluded that there is a solid truth underlying the instinctive perception. The individual is short lived and erring; man, relatively speaking, is imperishable and infallible. Even the positive evidences of the sense and the conclusions of science must be hesitatingly accepted when they are directed against the testimony of the entire body of humanity and the experience of centuries.

Modern investigation has disclosed the fact that there are other worlds, situated much the same as ours, and that organic life is bound to develop wherever there is heat, light and moisture. We know now that such conditions exist on innumerable heavenly bodies. In the solar system, two of these are particularly conspicuous — Venus and Mars. The former is, in many respects like the earth and must undoubtedly be the abode of some kind of life, but as to this we can only conjecture, for the surface is hidden from our view by a dense atmosphere. The latter planet can be readily observed and its periodic changes, which have been exhaustively studied by the late Percival Lowell, are a strong argument in support of the supposition that it is populated by a race vastly superior to ours in the mastery of the forces of nature

If such be the case then all that we can accomplish on this globe is of trifling importance as compared with the perfection of means putting us in possession of the secrets they must have discovered in their struggle against merciless elements. What a tragedy it would be were we to find some day that this wonderful people had finally met its inevitable fate and that all the precious intelligence they might have and, perhaps, had tried to convey to us, was lost. But although scientific research during the last few decades has given substance to the traditional belief, no serious attempt to establish communication could have been made until quite recently for want of proper instrumentalities.

Light Ray Project.

Long ago it was proposed to employ rays of light for this purpose and a number of men of science had devised specific plans which were discussed in the periodicals from time to time. But a careful examination shows that none of them is feasible, even on the assumption that the interplanetary space is devoid of gross matter, being filled only with a homogeneous and inconceivably tenuous medium called the ether. The tails of comets and other phenomena, however, would seem to disprove the theory, so that the successful exchange of signals by that kind of agency is very improbable.

While we can clearly discern the surface of Mars, it does not follow that the reverse is true. In perfect vacuum, of course, a parallel beam of light would be ideally suited for the transmission of energy in any amount for, theoretically, it could pass through infinite distance without any diminution of intensity. Unfortunately, this as well as other forms of radiant energy are rapidly absorbed in traversing the atmosphere.

It is possible that a magnetic force might be produced on the earth sufficient to bridge the gap of 50,000,000 miles and, in fact, it has been suggested to lay a cable around the globe with the object of magnetizing it. But certain electrical observations I made in studying terrestrial disturbances prove conclusively that there can not be much iron or other magnetic bodies in the earth beyond the insignificant quantity in the crust. Everything indicates that it is virtually a ball of glass and it would require many energizing turns to produce perceptible effects at great distance in this manner. Moreover, such an undertaking would be costly and, on account of the low speed of the current through the cable, the signaling would be extremely slow.

The Miracle Performed.

Such was the state of things until twenty years ago when a way was found to perform this miracle. It calls for nothing more than a determined effort and a feat in electrical engineering which, although difficult, is certainly realizable.

In 1899 I undertook to develop a powerful wireless transmitter and to ascertain the mode in which the waves were propagated through the earth. This was indispensable in order to apply my system intelligently for commercial purposes and, after careful study, I selected the high plateau of Colorado (6,000 feet above sea level) for the plant which I erected in the first part of that year. My success in overcoming the technical difficulties was greater than I had expected and in a few months I was able to produce electrical actions comparable to, and in a certain sense surpassing those of lightning. Activities of 18,000,000 horsepower were readily attained and I frequently computed the intensity of the effect in remote localities. During my experiments there, Mars was at a relatively small distance from us and, in that dry and rarefied air, Venus appeared so large and bright that it might have been mistaken for one of those military signaling lights. Its observation prompted me to calculate the energy transmitted by a powerful oscillator at 50,000,000 miles, and I came to the conclusion that it was sufficient to exert a noticeable influence on a delicate receiver of the kind I was, in the meanwhile, perfecting.

My first announcements to this effect were received with incredulity but merely because the potencies of the instrument I had devised were unknown. In the succeeding year, however, I designed a machine for a maximum activity of 1,000,000,000 horsepower which was partly constructed on Long Island in 1902 and would have been put in operation but for reverses and the fact that my project was too far in advance of the time.

It was reported at that period that my tower was intended for signaling to Mars, which was not the case, but it is true that I made a special provision for rendering it suitable to experiments in that direction. For the last few years there has been such a wide application of my wireless transmitter that experts have become, to an extent, familiar with its possibilities, and, if I am not mistaken, there are very few "doubting Thomases" now. But our ability to convey a signal across the gulf separating us from our neighboring planets would be of no avail if they are dead and barren or inhabited by races still undeveloped. Our hope that it might be different rests on what the telescope has revealed, but not on this alone.

Vast Power Found

In the course of my investigations of terrestrial electrical disturbances in Colorado I employed a receiver, the sensitiveness of which is virtually unlimited. It is generally believed that the so-called audion excels all others in this respect and Sir Oliver Lodge is credited with saying that it has been the means of achieving wireless telephony and transforming atomic energy. If the news is correct that scientist must have been victimized by some playful spirits with whom he is communicating. Of course, there is no conversion of atomic energy in such a bulb and many devices are known which can be used in the art with success.

My arrangements enable me to make a number of discoveries, some of which I have already announced in technical periodicals. The conditions under which I operated were very favorable for no other wireless plant of any considerable power existed and the effects t observed were thereafter due to natural causes, terrestrial or cosmic. I gradually learned how to distinguish in my receiver and eliminate certain actions and on one of these occasions my ear barely caught signals coming in regular succession which could not have been produced on the earth, caused by any solar or lunar action or by the influence of Venus, and the possibility that they might have come from Mars flashed upon my mind. In later years I have bitterly regretted that I yielded to the excitement of ideas and pressure of business instead of concentrating all my energies on that investigation.

The time is ripe now to make a systematic study of this transcending problem, the consummation of which may mean untold blessings to the human race. Capital should be liberally provided and a body of competent experts formed -to examine all the plans proposed and to assist in carrying out the best. The mere initiation of such a project in these uncertain and revolutionary times would result in a benefit which cannot be underestimated. In my early proposals I have advocated the application of fundamental mathematical principles for reaching the first elementary understanding. But since that time I have devised a plan akin to picture transmission through which knowledge of form could be conveyed and the barriers to the mutual exchange of ideas largely removed.

Success in Trials

Perfect success cannot be attained in any other way for we know only what we can visualize. Without perception of form there is not precise knowledge. A number of types of apparatus have been already invented with which transmission of pictures has been effected through the medium of wires, and they can be operated with equal facility by the wireless method. Some of these are of primitively simple construction. They are based on the employment of like parts which move in synchronism and transmit in this manner records, however complex. It would not require an extraordinary effort of the minds to hit upon this plan and devise instruments on this or similar principles and by gradual trials finally arrive at a full understanding.

The Herald of Sept. 24 contains a dispatch announcing that Prof. David Todd, of Amhurst College, contemplates an attempt to communicate with the inhabitants of Mars. The idea is to rise in a balloon to a height of about 50,000 feet with the manifest purpose of overcoming the impediments of the dense air stratum. I do not wish to comment adversely upon this undertaking beyond saying that no material advantage will be obtained by this method, for what is gained by height is offset a thousandfold by the inability of using powerful and complex transmitting and receiving apparatus. The physical stress and danger confronting the navigator at such an altitude are very great and he would be likely to lose his life or be permanently injured. In their recent record flights Roelfs and Schroeder have found that at a height of about six miles all their force was virtually exhausted. It would not have taken much more to terminate their careers fatally. If Prof. Todd wants to brave these perils he will have to provide special means of protection and these will be an obstacle to his observations. It is more likely, however, that he merely desires to look at the planet through a telescope in the hope of discerning something new. But it is by no means certain this instrument will be efficient under such conditions.

Developments in Practice and Art of Telephotography
Electrical Review — Dec. 11, 1920.

The recent successful experiments by Edouard Belin of Paris in the transmission of photographs between New York City and St. Louis, a distance of 1000 miles, have naturally aroused new interest in this rather old art. Mr. Belin's apparatus has been examined with a knowledge of previous efforts in that direction, and it must be admitted that the French inventor has achieved a marked improvement. It is true that his apparatus in many of its features is old and well known, but all the details have been worked out skillfully and his photographic reproductions are not only good likenesses of the originals but are expressive in no small degree. In common with other arts the transmission of pictures to a distance has been brought to its present state of perfection by slow and gradual improvements effected in the course of 77 years. The literature on the subject is quite voluminous and difficult to peruse, as the articles are published in various languages and scattered through numerous periodicals. Only one complete and exhaustive work has been published in German by Dr. Arthur Korn of Munich and Dr. Bruno Glatzel.

First Patents Taken out Many Years Ago

The original idea is due to Alexander Bain, a Scotch mechanician, who secured a British patent disclosing the invention in 1843. His plan contemplated the transmission of printed letters, drawings and pictures in the following way: At the sending station a holder with insulated metal points was arranged to glide in the direction of the lines over a frame resting on the printed page to be reproduced at a distance. Within this frame, and at right angles to its plane, a number of short wires were imbedded in sealing wax, their lower ends being in contact with the letters which in turn were all electrically connected. As the holder moved back and forth the insulated metal points would be brought in and out of contact with the upper ends of the short wires, thus controlling the flow of the current through them. Each metal point was joined by a special line to the receiving station where there was a similar holder made to slide over chemically prepared paper laid on a grounded metal plate. When a battery at the transmitting end was connected with one of its poles to the letters and the other to the ground the current impulses traversing the line wires and the chemical paper would cause changes of color in the latter, thus reproducing the characters. A great number of points and line wires were required to attain satisfactory results and, realizing this objection, Bain proposed to avail himself of only one wire, but did not give full information in this regard. Subsequently Bonnelli and other inventors made improvements in his apparatus, reducing the number of the wires to a few. There is no doubt that, despite the manifest crudity of this system, it was quite capable of being used commercially in the transmission of type as well as drawings and pictures and may yet be found valuable.

The first practical success was achieved by an Englishman, Frederick Collier Bakewell, who secured a British patent in 1847 on a process, some features of which have proved to be indispensable in later years. He employed as a transmitter a cylinder on which the characters were written with insulating ink. A metal point bore on the cylinder and advanced slightly with each revolution of the same exactly as in the older form of phonograph. A similar cylinder covered with chemical paper and equipped with a sliding point was provided at the receiving station. The cylinders being grounded and a battery included in the line wire connecting the transmitting and receiving points, the passage of the current resulted in a discoloration of the paper and reproduction of the written characters at the receiving end. Considering the period Bakewell's apparatus was surprisingly perfect, particularly in the feature of maintaining the rotating cylinders in synchronism for which purpose he provided an automatic as well as a hand correction. A controversy was waged between Bakewell and Bain for the honor of priority, but in this respect there can be no mistake. Bain was the originator of the idea while Bakewell was the first to carry it out successfully.

Use of Chemical Paper Considered Impractical

The use of chemical paper was considered objectionable, and in 1851 Hipp eliminated it, producing the impressions at the receiver with a magnet actuated by the transmitted impulses. It is curious, though, to observe that the modern art depends entirely on this very device. In 1855 Casselli modified the Bakewell apparatus by employing carefully synchronized pendulums at the transmitting and receiving stations, thus replacing the rotary motion by a to-and-fro movement as in the Bain arrangement. Casselli seems to have had more enterprise than his predecessors, and the apparatus which he perfected in 1860 was actually used with some success for a short time in service between Paris and several other cities in France. Its abandonment was probably due to the slowness of transmission and lack of demand for this kind of facility. It is singular that many treatises on physics and other text books mention Casselli while ignoring Bain and Bakewell.

Shortly after this Meyer perfected a system which was used with success in France, and may be fairly considered as the first thoroughly practical application of ideas in this field. A curious improvement was made by Gerard who, in 1865, proposed the use of flat disks in place of the cylinders of Bakewell. Ever since one wire was adopted for the transmission it became an imperative necessity to maintain perfect synchronism between the transmitter and receiver, and many inventors devoted their energy to this task. D'Arlincourt resorted to tuning forks, and his idea was subsequently carried out in a more perfect manner by Lacour. At about this time the invention reached America, and in 1870 Sawyer brought his ingenuity to bear on the evolution of a process in which he employed zinc cliches. These were very reliable and constituted a signal advance.

In 1880 Edison devised an apparatus on the principle of that used by Sawyer, except that the impressions were produced on paper in bas-relief. This idea was carried further by Dennison in instruments of the reciprocating type. Through the introduction of the Tesla alternating system of power transmission a novel means was afforded for operating transmitters and receivers. The use of synchronous motors was proposed first in 1893 by Sheehy.

Developments Permit Use of Photographic Films

In all cases without exception it was necessary to provide an actual print, drawing or sketch to be transmitted until Lenoir introduced photographic films into the art, making possible the transmission of any kind of picture. This was a great step forward, but the honor of the first practical success belongs to an American engineer, N. Amstutz, who used photographic sending cliches in relief for the first time and with complete success. Amstutz was a true pioneer, and his improvement is essential in the carrying out of the modern processes. It is true that as early as 1865, a Frenchman, Hubert, had suggested the use of letters written with thick ink, but this was of little value, and Amstutz was undoubtedly the first to "produce and use the cliches on which the up-to-date art vitally depends. Perfectly satisfactory demonstrations were made with his devices in this country more than 20 years ago, when pictures were transmitted over telegraph wires to great distances. Samples of his work have been preserved which clearly show how much be was ahead of his time.

Following Amstutz, Dunlany, Palmer, Mills and other American inventors took up picture transmission with more or less success. By this time the necessity presented itself for increasing the rapidity of the process by greater speed of the devices as well as multiplex transmission. The Belgian inventor, Carbonelle, made an important improvement in this direction when he introduced the telephone diaphragm carrying a stylus for making the impressions.

Of all inventors, however, Dr. Korn was the most successful as well as prolific in the suggestion of improvements, his photographic method of recording carried out in 1903 being the most significant. The general idea of photographic recording had been already advanced by George Little, and a few years later Dillon took out a patent involving the use of sensitized paper and a mirror reflecting a beam of light on the same. But it is obvious that at that time it would have been hardly practicable to use this suggestion, as photography was not sufficiently advanced. In illustration of this it may be mentioned that in 1892 the attention of the scientific world was directed to a wonderfully sensitive receiver, consisting of an electron stream maintained in a delicately balanced condition in a vacuum bulb, by means of which it was proposed to use photography in the transmission of telegraphic and telephonic messages through the

Atlantic cables, and later also by wireless. This proposal was met by unsurmountable objections to the photographic method. Indeed, the Belin process has been rendered possible largely through the great improvements in the sensitive films which have been evolved in response to the urgent demands of the motion picture and also under the stimulus of the recent war.

Selenium Cell and Vacuum Tube Used to Transmit and Receive

In the apparatus invented by Dr. Korn a selenium cell is used at the transmitter to vary the intensity of the sending current, and at the receiving station he employs a vacuum tube of high intensity which throws its light through a fine slot on a sensitive plate. The tube is excited by high-frequency currents supplied from a Tesla transformer and may be flashed up many thousand times per second. The motion of the receiver element is effected either by a wire galvanometer, oscillograph or telephone diaphragm. The Korn system has been used for some years past with success in Germany and other countries. In fact, it has been operated for some time even by wireless. Patents on this mode of transmission have been granted in 1898 and 1899 to Küster and G. Williams, but the arrangements involved the employment of Hertz waves and were impracticable. Later Frederick Braun, Pansa and Knudsen secured patents which, however, are equally defective. Success in this direction has been achieved so far only by Korn, Berjonneau and T. Baker. Invariably the inventors employ a wire galvanometer which is especially suitable for great speed. Telautographic transmission by similar means through wires as well as wireless is now common and is effected by employing a transmitter of two components, the original idea of which is due to an Englishman, Jones, who made that suggestion as early as 1855.

Present Developments Involve Many Old Principles

To this short story of picture transmission Belin has contributed the latest chapter. The process he has finally adopted after many years of persistent effort involves the use of two cylinders rotating in synchronism — one for transmitting and the other for reproducing. The former is of copper and is prepared for operation by having its surface coated with a thin shellac solution, wrapping a carbon print of the photograph about it with its face to the cylinder and immersing the whole in hot water, this causing the gelatine to adhere to the cylinder surface in proportion to the degree of blackness so that a likeness of the print in bas-relief is obtained. On this cylinder bears the stylus of a microphone diaphragm which is slowly moved forward by the revolution of the cylinder as in a phonograph. In this manner the pressure of the carbon contacts is varied in conformity with the changes of the surface, and the microphone currents pass over the transmitting wire to the receiving station where they cause corresponding deflections of a mirror forming part of a highly sensitive dead-beat oscillograph. A strong beam of light reflected from the mirror traverses a screen graduated from full transparency to opacity and is led through a microscopic opening to the sensitive film wrapped around the receiving cylinder. Special provisions are made to keep the cylinders exactly in step as this is indispensable to good performance. The film is, of course, protected against the external light, and when the operation is completed it is developed as usual so that either a positive or negative print is obtained according to the position of the screen. There is nothing in his apparatus which is fundamentally novel; in fact, every feature of the same has been disclosed in the prior art. Even the graduated screen, which is one of the most essential parts, has been employed before by Dr. Korn. But Mr. Belin has displayed considerable ingenuity and skill in all the details, and his reproduced photographs are most excellent. There is every reason to believe that his efforts will be rewarded by an extensive practical application of his devices.

Television to Be next Step in Progress of Transmission

The transmission of photographs constitutes only the first step towards the immeasurably greater achievement of television. By this is meant instantaneous transmission of visual impressions to any distance by wire or wireless. It is a subject to which I have devoted more than 25 years of close study. Two of the impediments which years ago seemed insurmountable have been successfully overcome, but great difficulties are still in the way. These are encountered in the inertia of the sensitive cells and the enormous speed required to make possible the vision of persons, objects or scenes as in life. It is the problem of constructing a transmitter analogous to the lens and retina of the eye, providing a medium of conveyance corresponding to the optic nerve, and a receiver organized similarly to the brain. It is a gigantic task, but I am confident that the world will witness its actual accomplishment in the near future.

Interplanetary Communication
Electrical World — Sept. 24, 1921

To the Editors of the Electrical World:

There are countless worlds such as ours in the universe - planets revolving around their suns in elliptical orbits and spinning on their axes like gigantic tops. They are composed of the same elements and subject to the same forces as the earth. Inevitably at some period in their evolution light, heat and moisture are bound to be present, when inorganic matter will begin to run into organic forms. The first impulse is probably given by heliotropism; then other influences assert themselves, and in the course of ages, through continuous adjustment to the environment, automata of inconceivable complexity of structure result. In the workshop of nature these automatic engines are turned out in all essential respects alike and exposed to the same external influences.

The identity of construction and sameness of environment result in a concordance of action, giving birth to reason; thus intelligence, as the human, is gradually developed. The chief controlling agent in this process must be radiant energy acting upon a sense organ as the eye, which conveys a true conception of form. We may therefore conclude with certitude that, however constructively different may be the automata on other planets, their response to rays of light and their perceptions of the outside world must be similar to a degree so that the difficulties in the way of mutual understanding should not be insuperable.

Irrespective of astronomical and electrical evidences, such as have been obtained by the late Percival Lowell and myself, there is a solid foundation for a systematic attempt to establish communication with one of our heavenly neighbors, as Mars, which through some inventions of mine is reduced to a comparatively simple problem of electrical engineering. Others may scoff at this suggestion or treat it as a practical joke, but I have been in deep earnest about it ever since I made the first observations at my wireless plant in Colorado Springs from 1889 to 1900. Those who are interested in the subject may be referred to my articles in the *Century Magazine* of June, 1900, *Collier's Weekly* of Feb. 9, 1901, the *Harvard Illustrated Magazine* of March, 1907, the *New York Times* of May 23, 1909, and the *New York Herald* of Oct. 12, 1919.

At the time I carried on those investigations there existed no wireless plant on the globe other than mine, at least none that could produce a disturbance perceptible in a radius of more than a few miles. Furthermore, the conditions under which I operated were ideal, and I was well trained for the work. The arrangement of my receiving apparatus and the character of the disturbances recorded precluded the possibility of their being of terrestrial origin, and I also eliminated the influence of the sun, moon and Venus. As I then announced, the signals consisted in a regular repetition of numbers, and subsequent study convinced me that they must have emanated from Mars; this planet having been just then close to the earth.

Since 1900 I have spent a great deal of my time in trying to develop a thoroughly practical apparatus for the purpose and have evolved numerous designs. In one of these I find that an activity of 10,000,000,000 hp in effective wave energy could be attained. Assuming the most unfavorable conditions - namely, half-spherical propagation - then at a distance of 34,000,000 miles the energy rate would be about 1/730,000 hp per square mile, which is far more than necessary to affect a properly designed receiver. In fact, apparatus similar to that used in the transmission of pictures could be operated, and in this manner mathematical, geometrical and other accurate information could be conveyed.

I was naturally very much interested in reports given out about two years ago that similar observations had been made, but soon ascertained that these supposed planetary signals were nothing else than interfering undertones of wireless transmitters, and since I announced this fact other experts have apparently taken the same view. These disturbances I observed for the first time from 1906 to 1907. At that time they occurred rarely, but subsequently they increased in frequency. Every transmitter emits undertones, and these give by interference long beats, the wave length being anything from 50 miles to 300 or 400 miles. In all probability they would have been observed by many other experimenters if it were not so troublesome to prepare receiving circuits suitable for such long waves.

The idea that they would be used in interplanetary signaling by any intelligent beings is too absurd to be seriously commented upon. These waves have no suitable relation to any dimensions, physical constants or succession of events, such as would be naturally and logically considered in an intelligent attempt to communicate with us, and every student familiar with the fundamental theoretical principles will readily see that such waves would be entirely ineffective. The activity being inversely as the cube of the wave length, a short wave would be immensely more efficient as a means for planetary signaling, and we must assume that any beings who had mastered the art would also be possessed of this knowledge. On careful reflection I find, however, that the disturbances as reported, if they have been actually noted, cannot be anything else but forced vibrations of a transmitter and in all likelihood beats of undertones.

While I am not prepared to discuss the various aspects of this subject at length, I may say that a skillful experimenter who is in the position to expend considerable money and time will undoubtedly detect waves of about 25,470,000 m.

Nikola Tesla
New York City.

Nikola Tesla on Electric Transmission
New York Evening Post — Sept. 26

To the Editor of the New York Evening Post:

Sir: Your issue of the 14th inst. contains a report relative to an experimental demonstration at the Pittsfield plant of the General Electric Company in which a pressure of one million volts was used for transmission of power by alternating currents. This is said to be the result of more than thirty years' work and to constitute a dramatic advance in electrical development so much, indeed, that it was deemed proper to record the time of its consummation with greater precision than that when Joshua commanded the sun to stand still over Gibeon and the moon in the Valley of Ajalon. But the prosaic fact is that I have long ago perfected and patented the invention instrumental in this achievement and applied it successfully in the production of pressures amounting to many millions of volts. It may not be amiss to state furthermore that a license was offered to the General Electric Company under my basic patent which bears the No. 1,119,732 and was granted December 1, 1914, the original application having been filed January 16, 1902.

The economic transmission of electrical energy at great distances necessitates the employment of very high pressures and at the outset two serious difficulties were encountered in their application. One was the breaking down of the insulation under the excessive stress. Upon careful investigation of the causes, I found that this was due to the presence of air or gas bubbles which were heated by the action of the currents and impaired the resisting quality of the dielectric. The trouble was done away with entirely by a process of manufacture developed by me which has been universally adopted. But the second obstacle was much harder to overcome. It was met in the apparent impossibility of confining the high tension flow to the conductors. In my early experiments I covered them with the best insulating material, several inches thick, but it was of no avail. Finally my efforts were rewarded and I found a simple and perfect remedy.

An idea of the underlying principle of the invention and its practical significance may be conveyed by an analogue. Alternating currents transmitted through a wire can be likened to pulses of some liquid, as water, forced through a woven hose So long as the pressure is moderate the fluid entering one end will be integrally discharged at the other, but if the pressure is increased beyond a certain critical value the hose will leak and a large portion of the fluid may thus be wasted. Similarly in electric transmission, when the voltage becomes excessive the prison walls of the dielectric yield and the charge escapes. The loss of energy occasioned thereby, although emphasized by engineers, is not a fatal drawback; the real harm lies in the limitation thus imposed to the attainment of many results of immense value. Now, what I did was equivalent to making the hose capable of withstanding any desired pressure, however great. This was accomplished by so constructing the transmitting conductor that its outer surface has itself a large radius of curvature or is composed of separate parts which, irrespective of their own curvature, are arranged in proximity to one another and on an ideal enveloping symmetrical surface of large radius. These parts may be in the shape of shells, hoods, discs, cylinders, or strands, according to the requirement in each special case, but it is always essential that the aggregate outer conducting area be considerable.

I believe that many arts and industries will be revolutionized through the application of the enormous electric pressures which are easily producible by this means, but perhaps the purely scientific results will be more important than the commercial.

N. Tesla
New York, September 23

My Inventions

My Early Life

The progressive development of man is vitally dependent on invention. It is the most important product of his creative brain. Its ultimate purpose is the complete mastery of mind over the material world, the harnessing of the forces of nature to human needs. This is the difficult task of the inventor who is often misunderstood and unrewarded. But he finds ample compensation in the pleasing exercises of his powers and in the knowledge of being one of that exceptionally privileged class without whom the race would have long ago perished in the bitter struggle against pitiless elements.

Speaking for myself, I have already had more than my full measure of this exquisite enjoyment, so much that for many years my life was little short of continuous rapture. I am credited with being one of the hardest workers and perhaps I am, if thought is the equivalent of labor, for I have devoted to it almost all of my waking hours. But if work is interpreted to be a definite performance in a specified time according to a rigid rule, then I may be the worst of idlers. Every effort under compulsion demands a sacrifice of life-energy. I never paid such a price. On the contrary, I have thrived on my thoughts.

In attempting to give a connected and faithful account of my activities in this series of articles which will be presented with the assistance of the Editors of the *Electrical Experimenter* and are chiefly addressed to our young men readers, I must dwell, however reluctantly, on the impressions of my youth and the circumstances and events which have been instrumental in determining my career.

Our first endeavors are purely instinctive, promptings of an imagination vivid and undisciplined. As we grow older reason asserts itself and we become more and more systematic and designing. But those early impulses, tho not immediately productive, are of the greatest moment and may shape our very destinies. Indeed, I feel now that had I understood and cultivated instead of suppressing them, I would have added substantial value to my bequest to the world. But not until I had attained manhood did I realize that I was an inventor.

This was due to a number of causes. In the first place I had a brother who was gifted to an extraordinary degree--one of those rare phenomena of mentality which biological investigation has failed to explain. His premature death left my parents disconsolate. We owned a horse which had been presented to us by a dear friend. It was a magnificent animal of Arabian breed, possessed of almost human intelligence, and was cared for and petted by the whole family, having on one occasion saved my father's life under remarkable circumstances. My father had been called one winter night to perform an urgent duty and while crossing the mountains, infested by wolves, the horse became frightened and ran away, throwing him violently to the ground. It arrived home bleeding and exhausted, but after the alarm was sounded immediately dashed off again, returning to the spot, and before the searching party were far on the way they were met by my father, who had recovered consciousness and remounted, not realizing that he had been lying in the snow for several hours. This horse was responsible for my brother's injuries from which he died. I witnessed the tragic scene and altho fifty-six years have elapsed since, my visual impression of it has lost none of its force. The recollection of his attainments made every effort of mine seem dull in comparison.

Anything I did that was creditable merely caused my parents to feel their loss more keenly. So I grew up with little confidence in myself. But I was far from being considered a stupid boy, if I am to judge from an incident of which I have still a strong remembrance. One day the Aldermen were passing thru a street where I was at play with other boys. The oldest of these venerable gentlemen—a wealthy citizen—paused to give a silver piece to each of us. Coming to me he suddenly stopped and commanded, "Look in my eyes." I met his gaze, my hand outstretched to receive the much valued coin, when, to my dismay, he said, "No, not much, you can get nothing from me, you are too smart." They used to tell a funny story about me. I had two old aunts with wrinkled faces, one of them having two teeth protruding like the tusks of an elephant which she buried in my cheek every time she kissed me. Nothing would scare me more than the prospect of being hugged by these as affectionate as unattractive relatives. It happened that while being carried in my mother's arms they asked me who was the prettier of the two. After examining their faces intently, I answered thoughtfully, pointing to one of them, "This here is not as ugly as the other."

Then again, I was intended from my very birth for the clerical profession and this thought constantly oppressed me. I longed to be an engineer but my father was inflexible. He was the son of an officer who served in the army of the Great Napoleon and, in common with his brother, professor of mathematics in a prominent institution, had received a military education but, singularly enough, later embraced the clergy in which vocation he achieved eminence. He was a very erudite man, a veritable natural philosopher, poet and writer and his sermons were said to be as eloquent as those of Abraham a Sancta-Clara. He had a prodigious memory and frequently recited at length from works in several languages. He often remarked playfully that if some of the classics were lost he could restore them. His style of writing was much admired. He penned sentences short and terse and was full of wit and satire. The humorous remarks he made were always peculiar and characteristic. Just to illustrate, I may mention one or two instances. Among the help there was a cross-eyed man called Mane, employed to do work around the farm. He was chopping wood one day. As he swung the axe my father, who stood nearby and felt very uncomfortable, cautioned him, "For God's sake, Mane, do not strike at what you are looking but at what you intend to hit." On another occasion he was taking out for a drive a friend who carelessly permitted his costly fur coat to rub on the carriage wheel. My father reminded him of it saying, "Pull in your coat, you are ruining my tire." He had the odd habit of talking to himself and would often carry on an animated conversation and indulge in heated argument, changing the tone of his voice. A casual listener might have sworn that several people were in the room.

Altho I must trace to my mother's influence whatever inventiveness I possess, the training he gave me must have been helpful. It comprised all sorts of exercises--as, guessing one another's thoughts, discovering the defects of some form or expression, repeating long sentences or performing mental calculations. These daily lessons were intended to strengthen memory and reason and especially to develop the critical sense, and were undoubtedly very beneficial.

My mother descended from one of the oldest families in the country and a line of inventors. Both her father and grandfather originated numerous implements for household, agricultural and other uses. She was a truly great woman, of rare skill, courage and fortitude, who had braved the storms of life and past thru many a trying experience. When she was sixteen a virulent pestilence swept the country. Her father was called away to administer the last sacraments to the dying and during his absence she went alone to the assistance of a neighboring family who were stricken by the dread disease. All of the members, five in number, succumbed in rapid succession. She bathed, clothed and laid out the bodies, decorating them with flowers according to the custom of the country and when her father returned he found everything ready for a Christian burial. My mother was an inventor of the first order and would, I believe, have achieved great things had she not been so remote from modern life and its multifold opportunities. She invented and constructed all kinds of tools and devices and wove the finest designs from thread which was spun by her. She even planted the seeds, raised the plants and separated the fibers herself. She worked indefatigably, from break of day till late at night, and most of the wearing apparel and furnishings of the home was the product of her hands. When she was past sixty, her fingers were still nimble enough to tie three knots in an eyelash.

There was another and still more important reason for my late awakening. In my boyhood I suffered from a peculiar affliction due to the appearance of images, often accompanied by strong flashes of light, which marred the sight of real objects and interfered with my thought and action. They were pictures of things and scenes which I had really seen, never of those I imagined. When a word was spoken to me the image of the object it designated would present itself vividly to my vision and sometimes I was quite unable to distinguish whether what I saw was tangible or not. This caused me great discomfort and anxiety. None of the students of psychology or physiology whom I have consulted could ever explain satisfactorily these phenomena. They seem to have been unique altho I was probably predisposed as I know that my brother experienced a similar trouble. The theory I have formulated is that the images were the result of a reflex action from the brain on the retina under great excitation. They certainly were not hallucinations such as are produced in diseased and anguished minds, for in other respects I was normal and composed. To give an idea of my distress, suppose that I had witnessed a funeral or some such nerve-racking spectacle. Then, inevitably, in the stillness of night, a vivid picture of the scene would thrust itself before my eyes and persist despite all my efforts to banish it. Sometimes it would even remain fixed in space tho I pushed my hand thru it. If my explanation is correct, it should be able to project on a screen the image of any object one conceives and make it visible. Such an advance would revolutionize all human relations. I am convinced that this wonder can and will be accomplished in time to come; I may add that I have devoted much thought to the solution of the problem.

To free myself of these tormenting appearances, I tried to concentrate my mind on something else I had seen, and in this way I would of ten obtain temporary relief; but in order to get it I had to conjure continuously new images. It was not long before I found that I had exhausted all of those at my command; my "reel" had run out, as it were, because I had seen little of the world--only objects in my home and the immediate surroundings. As I performed these mental operations for the second or third time, in order to chase the appearances from my vision, the remedy gradually lost all its force. Then I instinctively commenced to make excursions beyond the limits of the small world of which I had knowledge, and I saw new scenes. These were at first very blurred and indistinct, and would flit away when I tried to concentrate my attention upon them, but by and by I succeeded in fixing them; they gained in strength and distinctness and finally assumed the concreteness of real things. I soon discovered that my best comfort was attained if I simply went on in my vision farther and farther, getting new impressions all the time, and so I began to travel--of course, in my mind. Every night (and sometimes during the day), when alone, I would start on my journeys--see new places, cities and countries--live there, meet people and make friendships and acquaintances and, however unbelievable, it is a fact that they were just as dear to me as those in actual life and not a bit less intense in their manifestations.

This I did constantly until I was about seventeen when my thoughts turned seriously to invention. Then I observed to my delight that I could visualize with the greatest facility. I needed no models, drawings or experiments. I could picture them all as real in my mind. Thus I have been led unconsciously to evolve what I consider a new method of materializing inventive concepts and ideas, which is radically opposite to the purely experimental and is in my opinion ever so much more expeditious and efficient. The moment one constructs a device to carry into practice a crude idea he finds himself unavoidably engrossed with the details and defects of the apparatus. As he goes on improving and reconstructing, his force of concentration diminishes and he loses sight of the great underlying principle. Results may be obtained but always at the sacrifice of quality.

My method is different. I do not rush into actual work. When I get an idea I start at once building it up in my imagination. I change the construction, make improvements and operate the device in my mind. It is absolutely immaterial to me whether I run my turbine in thought or test it in my shop. I even note if it is out of balance. There is no difference whatever, the results are the same. In this way I am able to rapidly develop and perfect a conception without touching anything. When I have gone so far as to embody in the invention every possible improvement I can think of and see no fault anywhere, I put into concrete form this final product of my brain. Invariably my device works as I conceived that it should, and the experiment comes out exactly as I planned it. In twenty years there has not been a single exception. Why should it be otherwise? Engineering, electrical and mechanical, is positive in results. There is scarcely a subject that cannot be mathematically treated and the effects calculated or the results determined beforehand from the available theoretical and practical data. The carrying out into practice of a crude idea as is being generally done is, I hold, nothing but a waste of energy, money and time.

My early affliction had, however, another compensation. The incessant mental exertion developed my powers of observation and enabled me to discover a truth of great importance. I had noted that the appearance of images was always preceded by actual vision of scenes under peculiar and generally very exceptional conditions and I was impelled on each occasion to locate the original impulse. After a while this effort grew to be almost automatic and I gained great facility in connecting cause and effect. Soon I became aware, to my surprise, that every thought I conceived was suggested by an external impression. Not only this but all my actions were prompted in a similar way. In the course of time it became perfectly evident to me that I was merely an automaton endowed with power of movement, responding to the stimuli of the sense organs and thinking and acting accordingly. The practical result of this was the art of telautomatics which has been so far carried out only in an imperfect manner. Its latent possibilities will, however, be eventually shown. I have been since years planning self-controlled automata and believe that mechanisms can be produced which will act as if possessed of reason, to a limited degree, and will create a revolution in many commercial and industrial departments.

I was about twelve years old when I first succeeded in banishing an image from my vision by wilful effort, but I never had any control over the flashes of light to which I have referred. They were, perhaps, my strangest experience and inexplicable. They usually occurred when I found myself in a dangerous or distressing situation, or when I was greatly exhilarated. In some instances I have seen all the air around me filled with tongues of living flame. Their intensity, instead of diminishing, increased with time and seemingly attained a maximum when I was about twenty-five years old. While in Paris, in 1883, a prominent French manufacturer sent me an invitation to a shooting expedition which I accepted. I had been long confined to the factory and the fresh air had a wonderfully invigorating effect on me. On my return to the city that night I felt a positive sensation that my brain had caught fire. I saw a light as tho a small sun was located in it and I past the whole night applying cold compresses to my tortured head. Finally the flashes diminished in frequency and force but it took more than three weeks before they wholly subsided. When a second invitation was extended to me my answer was an emphatic NO!

These luminous phenomena still manifest themselves from time to time, as when a new idea opening up possibilities strikes me, but they are no longer exciting, being of relatively small intensity. When I close my eyes I invariably observe first, a background of very dark and uniform blue, not unlike the sky on a clear but starless night. In a few seconds this field becomes animated with innumerable scintillating flakes of green, arranged in several layers and advancing towards me. Then there appears, to the right, a beautiful pattern of two systems of parallel and closely spaced lines, at right angles to one another, in all sorts of colors with yellow-green and gold predominating. Immediately thereafter the lines grow brighter and the whole is thickly sprinkled with dots of twinkling light. This picture moves slowly across the field of vision and in about ten seconds vanishes to the left, leaving behind a ground of rather unpleasant and inert grey which quickly gives way to a billowy sea of clouds, seemingly trying to mold themselves in living shapes. It is curious that I cannot project a form into this grey until the second phase is reached. Every time, before falling asleep, images of persons or objects flit before my view. When I see them I know that I am about to lose consciousness. If they are absent and refuse to come it means a sleepless night.

To what an extent imagination played a part in my early life I may illustrate by another odd experience. Like most children I was fond of jumping and developed an intense desire to support myself in the air. Occasionally a strong wind richly charged with oxygen blew from the mountains rendering my body as light as cork and then I would leap and float in space for a long time. It was a delightful sensation and my disappointment was keen when later I undeceived myself.

During that period I contracted many strange likes, dislikes and habits, some of which I can trace to external impressions while others are unaccountable. I had a violent aversion against the earrings of women but other ornaments, as bracelets, pleased me more or less according to design. The sight of a pearl would almost give me a fit but I was fascinated with the glitter of crystals or objects with sharp edges and plane surfaces. I would not touch the hair of other people except, perhaps, at the point of a revolver. I would get a fever by looking at a peach and if a piece of camphor was anywhere in the house it caused me the keenest discomfort. Even now I am not insensible to some of these upsetting impulses. When I drop little squares of paper in a dish filled with liquid, I always sense a peculiar and awful taste in my mouth. I counted the steps in my walks and calculated the cubical contents of soup plates, coffee cups and pieces of food--otherwise my meal was unenjoyable. All repeated acts or operations I performed had to be divisible by three and if I mist I felt impelled to do it all over again, even if it took hours.

Up to the age of eight years, my character was weak and vacillating. I had neither courage or strength to form a firm resolve. My feelings came in waves and surges and vibrated unceasingly between extremes. My wishes were of consuming force and like the heads of the hydra, they multiplied. I was oppressed by thoughts of pain in life and death and religious fear. I was swayed by superstitious belief and lived in constant dread of the spirit of evil, of ghosts and ogres and other unholy monsters of the dark. Then, all at once, there came a tremendous change which altered the course of my whole existence. Of all things I liked books the best. My father had a large library and whenever I could manage I tried to satisfy my passion for reading. He did not permit it and would fly into a rage when he caught me in the act. He hid the candles when he found that I was reading in secret. He did not want me to spoil my eyes. But I obtained tallow, made the wicking and cast the sticks into tin forms, and every night I would bush the keyhole and the cracks and read, often till dawn, when all others slept and my mother started on her arduous daily task. On one occasion I came across a novel entitled "Abafi" (the Son of Aba), a Serbian translation of a well known Hungarian writer, Josika. This work somehow awakened my dormant powers of will and I began to practice self-control. At first my resolutions faded like snow in April, but in a little while I conquered my weakness and felt a pleasure I never knew before--that of doing as I willed. In the course of time this vigorous mental exercise became second nature. At the outset my wishes had to be subdued but gradually desire and will grew to be identical. After years of such discipline I gained so complete a mastery over myself that I toyed with passions which have meant destruction to some of the strongest men. At a certain age I contracted a mania for gambling which greatly worried my parents. To sit down to a game of cards was for me the quintessence of pleasure. My father led an exemplary life and could not excuse the senseless waste of time and money in which I indulged. I had a strong resolve but my philosophy was bad. I would say to him, "I can stop whenever I please but is it worth while to give up that which I would purchase with the joys of Paradise?" On frequent occasions he gave vent to his anger and contempt but my mother was different. She understood the character of men and knew that one's salvation could only be brought about thru his own efforts. One afternoon, I remember, when I had lost all my money and was craving for a game, she came to me with a roll of bills and said, "Go and enjoy yourself. The sooner you lose all we possess the better it will be. I know that you will get over it." She was right. I conquered my passion then and there and only regretted that it had not been a hundred times as strong. I not only vanquished but tore it from my heart so as not to leave even a trace of desire. Ever since that time I have been as indifferent to any form of gambling as to picking teeth.

During another period I smoked excessively, threatening to ruin my health. Then my will asserted itself and I not only stopped but destroyed all inclination. Long ago I suffered from heart trouble until I discovered that it was due to the innocent cup of coffee I consumed every morning. I discontinued at once, tho I confess it was not an easy task. In this way I checked and bridled other habits and passions and have not only preserved my life but derived an immense amount of satisfaction from what most men would consider privation and sacrifice.

After finishing the studies at the Polytechnic Institute and University I had a complete nervous breakdown and while the malady lasted I observed many phenomena strange and unbelievable.

My First Efforts At Invention

I shall dwell briefly on these extraordinary experiences, on account of their possible interest to students of psychology and physiology and also because this period of agony was of the greatest consequence on my mental development and subsequent labors. But it is indispensable to first relate the circumstances and conditions which preceded them and in which might be found their partial explanation.

From childhood I was compelled to concentrate attention upon myself. This caused me much suffering but, to my present view, it was a blessing in disguise for it has taught me to appreciate the inestimable value of introspection in the preservation of life, as well as a means of achievement. The pressure of occupation and the incessant stream of impressions pouring into our consciousness thru all the gateways of knowledge make modern existence hazardous in many ways. Most persons are so absorbed in the contemplation of the outside world that they are wholly oblivious to what is passing on within themselves.

The premature death of millions is primarily traceable to this cause. Even among those who exercise care it is a common mistake to avoid imaginary, and ignore the real dangers. And what is true of an individual also applies, more or less, to a people as a whole. Witness, in illustration, the prohibition movement. A drastic, if not unconstitutional, measure is now being put thru in this country to prevent the consumption of alcohol and yet it is a positive fact that coffee, tea, tobacco, chewing gum and other stimulants, which are freely indulged in even at the tender age, are vastly more injurious to the national body, judging from the number of those who succumb. So, for instance, during my student years I gathered from the published necrologues in Vienna, the home of coffee drinkers, that deaths from heart trouble sometimes reached sixty-seven per cent of the total. Similar observations might probably be made in cities where the consumption of tea is excessive. These delicious beverages superexcite and gradually exhaust the fine fibers of the brain. They also interfere seriously with arterial circulation and should be enjoyed all the more sparingly as their deleterious effects are slow and imperceptible. Tobacco, on the other hand, is conducive to easy and pleasant thinking and detracts from the intensity and concentration necessary to all original and vigorous effort of the intellect. Chewing gum is helpful for a short while but soon drains the glandular system and inflicts irreparable damage, not to speak of the revulsion it creates. Alcohol in small quantities is an excellent tonic, but is toxic in its action when absorbed in larger amounts, quite immaterial as to whether it is taken in as whiskey or produced in the stomach from sugar. But it should not be overlooked that all these are great eliminators assisting Nature, as they do, in upholding her stern but just law of the survival of the fittest. Eager reformers should also be mindful of the eternal perversity of mankind which makes the indifferent "laissez-faire" by far preferable to enforced restraint.

The truth about this is that we need stimulants to do our best work under present living conditions, and that we must exercise moderation and control our appetites and inclinations in every direction. That is what I have been doing for many years, in this way maintaining myself young in body and mind. Abstinence was not always to my liking but I find ample reward in the agreeable experiences I am now making. Just in the hope of converting some to my precepts and convictions I will recall one or two.

A short time ago I was returning to my hotel. It was a bitter cold night, the ground slippery, and no taxi to be had. Half a block behind me followed another man, evidently as anxious as myself to get under cover. Suddenly my legs went up in the air. In the same instant there was a flash in my brain, the nerves responded, the muscles contracted, I swung thru 180 degrees and landed on my hands. I resumed my walk as tho nothing had happened when the stranger caught up with me. "How old are you?" he asked, surveying me critically. "Oh, about fifty-nine," I replied. "What of it?" "Well," said he, "I have seen a cat do this but never a man." About a month since I wanted to order new eyeglasses and went to an oculist who put me thru the usual tests. He looked at me incredulously as I read off with ease the smallest print at considerable distance. But when I told him that I was past sixty he gasped in astonishment. Friends of mine often remark that my suits fit me like gloves but they do not know that all my clothing is made to measurements which were taken nearly 35 years ago and never changed. During this same period my weight has not varied one pound.

In this connection I may tell a funny story. One evening, in the winter of 1885, Mr. Edison, Edward H. Johnson, the President of the Edison Illuminating Company, Mr. Batchellor, Manager of the works, and myself entered a little place opposite 65 Fifth Avenue where the offices of the company were located. Someone suggested guessing weights and I was induced to step on a scale. Edison felt me all over and said: "Tesla weighs 152 lbs. to an ounce," and he guest it exactly. Stripped I weighed 142 lbs. and that is still my weight. I whispered to Mr. Johnson: "How is it possible that Edison could guess my weight so closely?" "Well," he said, lowering his voice. "I will tell you, confidentially, but you must not say anything. He was employed for a long time in a Chicago slaughter-house where he weighed thousands of hogs every day! That's why." My friend, the Hon. Chauncey M. Depew, tells of an Englishman on whom he sprung one of his original anecdotes and who listened with a puzzled expression but - a year later - laughed out loud. I will frankly confess it took me longer than that to appreciate Johnson's joke.

Now, my well being is simply the result of a careful and measured mode of living and perhaps the most astonishing thing is that three times in my youth I was rendered by illness a hopeless physical wreck and given up by physicians. More than this, thru ignorance and lightheartedness, I got into all sorts of difficulties, dangers and scrapes from which I extricated myself as by enchantment. I was almost drowned a dozen times; was nearly boiled alive and just mist being cremated. I was entombed, lost and frozen. I had hair-breadth escapes from mad dogs, hogs, and other wild animals. I past thru dreadful diseases and met with all kinds of odd mishaps and that I am hale and hearty today seems like a miracle. But as I recall these incidents to my mind I feel convinced that my preservation was not altogether accidental.

An inventor's endeavor is essentially lifesaving. Whether he harnesses forces, improves devices, or provides new comforts and conveniences, he is adding to the safety of our existence. He is also better qualified than the average individual to protect himself in peril, for he is observant and resourceful. If I had no other evidence that I was, in a measure, possessed of such qualities I would find it in these personal experiences. The reader will be able to judge for himself if I mention one or two instances. On one occasion, when about 14 years old, I wanted to scare some friends who were bathing with me. My plan was to dive under a long floating structure and slip out quietly at the other end. Swimming and diving came to me as naturally as to a duck and I was confident that I could perform the feat. Accordingly I plunged into the water and, when out of view, turned around and proceeded rapidly towards the opposite side. Thinking that I was safely beyond the structure, I rose to the surface but to my dismay struck a beam. Of course, I quickly dived and forged ahead with rapid strokes until my breath was beginning to give out. Rising for the second time, my head came again in contact with a beam. Now I was becoming desperate. However, summoning all my energy, I made a third frantic attempt but the result was the same. The torture of suppressed breathing was getting unendurable, my brain was reeling and I felt myself sinking. At that moment, when my situation seemed absolutely hopeless, I experienced one of those flashes of light and the structure above me appeared before my vision. I either discerned or guest that there was a little space between the surface of the water and the boards resting on the beams and, with consciousness nearly gone, I floated up, pressed my mouth close to the planks and managed to inhale a little air, unfortunately mingled with a spray of water which nearly choked me. Several times I repeated this procedure as in a dream until my heart, which was racing at a terrible rate, quieted down and I gained composure. After that I made a number of unsuccessful dives, having completely lost the sense of direction, but finally succeeded in getting out of the trap when my friends had already given me up and were fishing for my body.

That bathing season was spoiled for me thru recklessness but I soon forgot the lesson and only two years later I fell into a worse predicament. There was a large flour mill with a dam across the river near the city where I was studying at that time. As a rule the height of the water was only two or three inches above the dam and to swim out to it was a sport not very dangerous in which I often indulged. One day I went alone to the river to enjoy myself as usual. When I was a short distance from the masonry, however, I was horrified to observe that the water had risen and was carrying me along swiftly. I tried to get away but it was too late.

Luckily, tho, I saved myself from being swept over by taking hold of the wall with both hands. The pressure against my chest was great and I was barely able to keep my head above the surface. Not a soul was in sight and my voice was lost in the roar of the fall. Slowly and gradually I became exhausted and unable to withstand the strain longer. just as I was about to let go, to be dashed against the rocks below, I saw in a flash of light a familiar diagram illustrating the hydraulic principle that the pressure of a fluid in motion is proportionate to the area exposed, and automatically I turned on my left side. As if by magic the pressure was reduced and I found it comparatively easy in that position to resist the force of the stream. But the danger still confronted me. I knew that sooner or later I would be carried down, as it was not possible for any help to reach me in time, even if I attracted attention. I am ambidextrous now but then I was left-handed and had comparatively little strength in my right arm. For this reason I did not dare to turn on the other side to rest and nothing remained but to slowly push my body along the dam. I had to get away from the mill towards which my face was turned as the current there was much swifter and deeper. It was a long and painful ordeal and I came near to failing at its very end for I was confronted with a depression in the masonry. I managed to get over with the last ounce of my force and fell in a swoon when I reached the bank, where I was found. I had torn virtually all the skin from my left side and it took several weeks before the fever subsided and I was well. These are only two of many instances but they may be sufficient to show that had it not been for the inventor's instinct I would not have lived to tell this tale.

Interested people have often asked me how and when I began to invent. This I can only answer from my present recollection in the light of which the first attempt I recall was rather ambitious for it involved the invention of an apparatus and a method. In the former I was anticipated but the latter was original. It happened in this way. One of my playmates had come into the possession of a hook and fishing-tackle which created quite an excitement in the village, and the next morning all started out to catch frogs. I was left alone and deserted owing to a quarrel with this boy. I had never seen a real hook and pictured it as something wonderful, endowed with peculiar qualities, and was despairing not to be one of the party. Urged by necessity, I somehow got hold of a piece of soft iron wire, hammered the end to a sharp point between two stones, bent it into shape, and fastened it to a strong string. I then cut a rod, gathered some bait, and went down to the brook where there were frogs in abundance. But I could not catch any and was almost discouraged when it occurred to me to dangle the empty hook in front of a frog sitting on a stump. At first he collapsed but by and by his eyes bulged out and became bloodshot, he swelled to twice his normal size and made a vicious snap at the hook.

Immediately I pulled him up. I tried the same thing again and again and the method proved infallible. When my comrades, who in spite of their fine outfit had caught nothing, came to me they were green with envy. For a long time I kept my secret and enjoyed the monopoly but finally yielded to the spirit of Christmas. Every boy could then do the same and the following summer brought disaster to the frogs.

In my next attempt I seem to have acted under the first instinctive impulse which later dominated me - to harness the energies of nature to the service of man. I did this thru the medium of May-bugs - or June-bugs as they are called in America - which were a veritable pest in that country and sometimes broke the branches of trees by the sheer weight of their bodies. The bushes were black with them. I would attach as many as four of them to a crosspiece, rotably arranged on a thin spindle, and transmit the motion of the same to a large disc and so derive considerable "power." These creatures were remarkably efficient, for once they were started they had no sense to stop and continued whirling for hours and hours and the hotter it was the harder they worked. All went well until a strange boy came to the place. He was the son of a retired officer in the Austrian Army. That urchin ate May-bugs alive and enjoyed them as tho they were the finest blue-point oysters. That disgusting sight terminated my endeavors in this promising field and I have never since been able to touch a May-bug or any other insect for that matter.

After that, I believe, I undertook to take apart and assemble the clocks of my grandfather. In the former operation I was always successful but often failed in the latter. So it came that he brought my work to a sudden halt in a manner not too delicate and it took thirty years before I tackled another clockwork again. Shortly there after I went into the manufacture of a kind of pop-gun which comprised a hollow tube, a piston, and two plugs of hemp. When firing the gun, the piston was pressed against the stomach and the tube was pushed back quickly with both hands. The air between the plugs was compressed and raised to high temperature and one of them was expelled with a loud report. The art consisted in selecting a tube of the proper taper from the hollow stalks. I did very well with that gun but my activities interfered with the window panes in our house and met with painful discouragement. If I remember rightly, I then took to carving swords from pieces of furniture which I could conveniently obtain. At that time I was under the sway of the Serbian national poetry and full of admiration for the feats of the heroes. I used to spend hours in mowing down my enemies in the form of corn-stalks which ruined the crops and netted me several spankings from my mother. Moreover these were not of the formal kind but the genuine article.

I had all this and more behind me before I was six years old and had past thru one year of elementary school in the village of Smiljan where I was born. At this juncture we moved to the little city of Gospic nearby. This change of residence was like a calamity to me. It almost broke my heart to part from our pigeons, chickens and sheep, and our magnificent flock of geese which used to rise to the clouds in the morning and return from the feeding grounds at sundown in battle formation, so perfect that it would have put a squadron of the best aviators of the present day to shame. In our new house I was but a prisoner, watching the strange people I saw thru the window blinds. My bashfulness was such that I would rather have faced a roaring lion than one of the city dudes who strolled about. But my hardest trial came on Sunday when I had to dress up and attend the service. There I meet with an accident, the mere thought of which made my blood curdle like sour milk for years afterwards. It was my second adventure in a church. Not long before I was entombed for a night in an old chapel on an inaccessible mountain which was visited only once a year. It was an awful experience, but this one was worse. There was a wealthy lady in town, a good but pompous woman, who used to come to the church gorgeously painted up and attired with an enormous train and attendants. One Sunday I had just finished ringing the bell in the belfry and rushed downstairs when this grand dame was sweeping out and I jumped on her train. It tore off with a ripping noise which sounded like a salvo of musketry fired by raw recruits. My father was livid with rage. He gave me a gentle slap on the cheek, the only corporal punishment he ever administered to me but I almost feel it now. The embarrassment and confusion that followed are indescribable. I was practically ostracized until something else happened which redeemed me in the estimation of the community.

An enterprising young merchant had organized a fire department. A new fire engine was purchased, uniforms provided and the men drilled for service and parade. The engine was, in reality, a pump to be worked by sixteen men and was beautifully painted red and black. One afternoon the official trial was prepared for and the machine was transported to the river. The entire population turned out to witness the great spectacle. When all the speeches and ceremonies were concluded, the command was given to pump, but not a drop of water came from the nozzle. The professors and experts tried in vain to locate the trouble. The fizzle was complete when I arrived at the scene. My knowledge of the mechanism was nil and I knew next to nothing of air pressure, but instinctively I felt for the suction hose in the water and found that it had collapsed. When I waded in the river and opened it up the water rushed forth and not a few Sunday clothes were spoiled. Archimedes running naked thru the streets of Syracuse and shouting Eureka at the top of his voice did not make a greater impression than myself. I was carried on the shoulders and was the hero of the day.

Upon settling in the city I began a four-years' course in the so-called Normal School preparatory to my studies at the College or Real Gymnasium. During this period my boyish efforts and exploits, as well as troubles, continued. Among other things I attained the unique distinction of champion crow catcher in the country. My method of procedure was extremely simple. I would go in the forest, hide in the bushes, and imitate the call of the bird. Usually I would get several answers and in a short while a crow would flutter down into the shrubbery near me. After that all I needed to do was to throw a piece of cardboard to distract its attention, jump up and grab it before it could extricate itself from the undergrowth. In this way I would capture as many as I desired. But on one occasion something occurred which made me respect them. I had caught a fine pair of birds and was returning home with a friend. When we left the forest, thousands of crows had gathered making a frightful racket. In a few minutes they rose in pursuit and soon enveloped us. The fun lasted until all of a sudden I received a blow on the back of my head which knocked me down. Then they attacked me viciously. I was compelled to release the two birds and was glad to join my friend who had taken refuge in a cave.

In the schoolroom there were a few mechanical models which interested me and turned my attention to water turbines. I constructed many of these and found great pleasure in operating them. How extraordinary was my life an incident may illustrate. My uncle had no use for this kind of pastime and more than once rebuked me. I was fascinated by a description of Niagara Falls I had perused, and pictured in my imagination a big wheel run by the Falls. I told my uncle that I would go to America and carry out this scheme. Thirty years later I saw my ideas carried out at Niagara and marveled at the unfathomable mystery of the mind.

I made all kinds of other contrivances and contraptions but among these the arbalists I produced were the best. My arrows, when shot, disappeared from sight and at close range traversed a plank of pine one inch thick. Thru the continuous tightening of the bows I developed skin on my stomach very much like that of a crocodile and I am often wondering whether it is due to this exercise that I am able even now to digest cobble-stones! Nor can I pass in silence my performances with the sling which would have enabled me to give a stunning exhibit at the Hippodrome. And now I will tell of one of my feats with this antique implement of war which will strain to the utmost the credulity of the reader. I was practicing while walking with my uncle along the river. The sun was setting, the trout were playful and from time to time one would shoot up into the air, its glistening body sharply defined against a projecting rock beyond. Of course any boy might have hit a fish under these propitious conditions but I undertook a much more difficult task and I foretold to my uncle, to the minutest detail, what I intended doing. I was to hurl a stone to meet the fish, press its body against the rock, and cut it in two. It was no sooner said than done. My uncle looked at me almost scared out of his wits and exclaimed "*Vade retro Satanas!*" and it was a few days before he spoke to me again. Other records, how ever great, will be eclipsed but I feel that I could peacefully rest on my laurels for a thousand years.

My Later Endeavors

The Discovery of the Rotating Magnetic Field

At the age of ten I entered the Real Gymnasium which was a new and fairly well equipt institution. In the department of physics were various models of classical scientific apparatus, electrical and mechanical. The demonstrations and experiments performed from time to time by the instructors fascinated me and were undoubtedly a powerful incentive to invention. I was also passionately fond of mathematical studies and often won the professor's praise for rapid calculation. This was due to my acquired facility of visualizing the figures and performing the operations, not in the usual intuitive manner, but as in actual life. Up to a certain degree of complexity it was absolutely the same to me whether I wrote the symbols on the board or conjured them before my mental vision. But freehand drawing, to which many hours of the course were devoted, was an annoyance I could not endure. This was rather remarkable as most of the members of the family excelled in it. Perhaps my aversion was simply due to the predilection I found in undisturbed thought. Had it not been for a few exceptionally stupid boys, who could not do anything at all, my record would have been the worst. It was a serious handicap as under the then existing educational regime, drawing being obligatory, this deficiency threatened to spoil my whole career and my father had considerable trouble in railroading me from one class to another.

In the second year at that institution I became obsessed with the idea of producing continuous motion thru steady air pressure. The pump incident, of which I have told, had set afire my youthful imagination and impressed me with the boundless abilities of a vacuum. I grew frantic in my desire to harness this inexhaustible energy but for a long time I was groping in the dark. Finally, however, my endeavors crystallized in an invention which was to enable me to achieve what no other mortal ever attempted.

Imagine a cylinder freely rotatable on two bearings and partly surrounded by a rectangular trough which fits it perfectly. The open side of the trough is closed by a partition so that the cylindrical segment within the enclosure divides the latter into two compartments entirely separated from each other by air-tight sliding joints. One of these compartments being sealed and once for all exhausted, the other remaining open, a perpetual rotation of the cylinder would result, at least, I thought so. A wooden model was constructed and fitted with infinite care and when I applied the pump on one side and actually observed that there was a tendency to turning, I was delirious with joy. Mechanical flight was the one thing I wanted to accomplish altho still under the discouraging recollection of a bad fall I sustained by jumping with an umbrella from the top of a building. Every day I used to transport myself thru the air to distant regions but could not understand just how I managed to do it. Now I had something concrete--a flying machine with nothing more than a rotating shaft, flapping wings, and--a vacuum of unlimited power! From that time on I made my daily aerial excursions in a vehicle of comfort and luxury as might have befitted King Solomon. It took years before I understood that the atmospheric pressure acted at right angles to the surface of the cylinder and that the slight rotary effort I observed was due to a leak. Tho this knowledge came gradually it gave me a painful shock.

I had hardly completed my course at the Real Gymnasium when I was prostrated with a dangerous illness or rather, a score of them, and my condition became so desperate that I was given up by physicians. During this period I was permitted to read constantly, obtaining books from the Public Library which had been neglected and entrusted to me for classification of the works and preparation of the catalogues. One day I was handed a few volumes of new literature unlike anything I had ever read before and so captivating as to make me utterly forget my hopeless state. They were the earlier works of Mark Twain and to them might have been due the miraculous recovery which followed. Twenty-five years later, when I met Mr. Clemens and we formed a friendship between us, I told him of the experience and was amazed to see that great man of laughter burst into tears.

My studies were continued at the higher Real Gymnasium in Carlstadt, Croatia, where one of my aunts resided. She was a distinguished lady, the wife of a Colonel who was an old war-horse having participated in many battles. I never can forget the three years I past at their home. No fortress in time of war was under a more rigid discipline. I was fed like a canary bird. All the meals were of the highest quality and deliciously prepared but short in quantity by a thousand percent. The slices of ham cut by my aunt were like tissue paper. When the Colonel would put something substantial on my plate she would snatch it away and say excitedly to him: "Be careful, Niko is very delicate." I had a voracious appetite and suffered like Tantalus. But I lived in an atmosphere of refinement and artistic taste quite unusual for those times and conditions. The land was low and marshy and malaria fever never left me while there despite of the enormous amounts of quinin I consumed. Occasionally the river would rise and drive an army of rats into the buildings, devouring everything even to the bundles of the fierce paprika. These pests were to me a welcome diversion. I thinned their ranks by all sorts of means, which won me the unenviable distinction of rat-catcher in the community. At last, however, my course was completed, the misery ended, and I obtained the certificate of maturity which brought me to the cross-roads.

During all those years my parents never wavered in their resolve to make me embrace the clergy, the mere thought of which filled me with dread. I had become intensely interested in electricity under the stimulating influence of my Professor of Physics, who was an ingenious man and often demonstrated the principles by apparatus of his own invention. Among these I recall a device in the shape of a freely rotatable bulb, with tinfoil coatings, which was made to spin rapidly when connected to a static machine. It is impossible for me to convey an adequate idea of the intensity of feeling I experienced in witnessing his exhibitions of these mysterious phenomena. Every impression produced a thousand echoes in my mind. I wanted to know more of this wonderful force; I longed for experiment and investigation and resigned myself to the inevitable with aching heart.

Just as I was making ready for the long journey home I received word that my father wished me to go on a shooting expedition. It was a strange request as he had been always strenuously opposed to this kind of sport. But a few days later I learned that the cholera was raging in that district and, taking advantage of an opportunity, I returned to Gospic in disregard of my parents' wishes. It is incredible how absolutely ignorant people were as to the causes of this scourge which visited the country in intervals of from fifteen to twenty years. They thought that the deadly agents were transmitted thru the air and filled it with pungent odors and smoke. In the meantime they drank the infected water and died in heaps. I contracted the awful disease on the very day of my arrival and altho surviving the crisis, I was confined to bed for nine months with scarcely any ability to move. My energy was completely exhausted and for the second time I found myself at death's door. In one of the sinking spells which was thought to be the last, my father rushed into the room. I still see his pallid face as he tried to cheer me in tones belying his assurance. "Perhaps," I said, "I may get well if you will let me study engineering." "You will go to the best technical institution in the world," he solemnly replied, and I knew that he meant it. A heavy weight was lifted from my mind but the relief would have come too late had it not been for a marvelous cure brought about thru a bitter decoction of a peculiar bean. I came to life like another Lazarus to the utter amazement of everybody.

My father insisted that I spend a year in healthful physical outdoor exercises to which I reluctantly consented. For most of this term I roamed in the mountains, loaded with a hunter's outfit and a bundle of books, and this contact with nature made me stronger in body as well as in mind. I thought and planned, and conceived many ideas almost as a rule delusive. The vision was clear enough but the knowledge of principles was very limited. In one of my inventions I proposed to convey letters and packages across the seas, thru a submarine tube, in spherical containers of sufficient strength to resist the hydraulic pressure. The pumping plant, intended to force the water thru the tube, was accurately figured and designed and all other particulars carefully worked out. Only one trifling detail, of no consequence, was lightly dismissed. I assumed an arbitrary velocity of the water and, what is more, took pleasure in making it high, thus arriving at a stupendous performance supported by faultless calculations. Subsequent reflections, however, on the resistance of pipes to fluid flow determined me to make this invention public property.

Another one of my projects was to construct a ring around the equator which would, of course, float freely and could be arrested in its spinning motion by reactionary forces, thus enabling travel at a rate of about one thousand miles an hour, impracticable by rail. The reader will smile. The plan was difficult of execution, I will admit, but not nearly so bad as that of a well-known New York professor, who wanted to pump the air from the torrid to the temperate zones, entirely forgetful of the fact that the Lord had provided a gigantic machine for this very purpose.

Still another scheme, far more important and attractive, was to derive power from the rotational energy of terrestrial bodies. I had discovered that objects on the earth's surface, owing to the diurnal rotation of the globe, are carried by the same alternately in and against the direction of translatory movement. From this results a great change in momentum which could be utilized in the simplest imaginable manner to furnish motive effort in any habitable region of the world. I cannot find words to describe my disappointment when later I realized that I was in the predicament of Archimedes, who vainly sought for a fixed point in the universe.

At the termination of my vacation I was sent to the Polytechnic School in Gratz, Styria, which my father had chosen as one of the oldest and best reputed institutions. That was the moment I had eagerly awaited and I began my studies under good auspices and firmly resolved to succeed. My previous training was above the average, due to my father's teaching and opportunities afforded. I had acquired the knowledge of a number of languages and waded thru the books of several libraries, picking up information more or less useful. Then again, for the first time, I could choose my subjects as I liked, and free-hand drawing was to bother me no more.

I had made up my mind to give my parents a surprise, and during the whole first year I regularly started my work at three o'clock in the morning and continued until eleven at night, no Sundays or holidays excepted. As most of my fellow-students took thinks easily, naturally enough I eclipsed all records. In the course of that year I past thru nine exams and the professors thought I deserved more than the highest qualifications. Armed with their flattering certificates, I went home for a short rest, expecting a triumph, and was mortified when my father made light of these hard won honors. That almost killed my ambition; but later, after he had died, I was pained to find a package of letters which the professors had written him to the effect that unless he took me away from the Institution I would be killed thru overwork.

Thereafter I devoted myself chiefly to physics, mechanics and mathematical studies, spending the hours of leisure in the libraries. I had a veritable mania for finishing whatever I began, which often got me into difficulties. On one occasion I started to read the works of Voltaire when I learned, to my dismay, that there were close on one hundred large volumes in small print which that monster had written while drinking seventy-two cups of black coffee per diem. It had to be done, but when I laid aside the last book I was very glad, and said, "Never more!"

My first year's showing had won me the appreciation and friendship of several professors. Among these were Prof. Rogner, who was teaching arithmetical subjects and geometry; Prof. Poeschl, who held the chair of theoretical and experimental physics, and Dr. Alle, who taught integral calculus and specialized in differential equations. This scientist was the most brilliant lecturer to whom I ever listened. He took a special interest in my progress and would frequently remain for an hour or two in the lecture room, giving me problems to solve, in which I delighted. To him I explained a flying machine I had conceived, not an illusionary invention, but one based on sound, scientific principles, which has become realizable thru my turbine and will soon be given to the world. Both Professors Rogner and Poeschl were curious men. The former had peculiar ways of expressing himself and whenever he did so there was a riot, followed by a long and embarrassing pause. Prof. Poeschl was a methodical and thoroughly grounded German. He had enormous feet and hands like the paws of a bear, but all of his experiments were skillfully performed with lock-like precision and without a miss.

It was in the second year of my studies that we received a Gramme dynamo from Paris, having the horseshoe form of a laminated field magnet, and a wire-wound armature with a commutator. It was connected up and various effects of the currents were shown. While Prof. Poeschl was making demonstrations, running the machine as a motor, the brushes gave trouble, sparking badly, and I observed that it might be possible to operate a motor without these appliances. But he declared that it could not be done and did me the honor of delivering a lecture on the subject, at the conclusion of which he remarked: "Mr. Tesla may accomplish great things, but he certainly never will do this. It would be equivalent to converting a steadily pulling force, like that of gravity, into a rotary effort. It is a perpetual motion scheme, an impossible idea." But instinct is something which transcends knowledge. We have, undoubtedly, certain finer fibers that enable us to perceive truths when logical deduction, or any other willful effort of the brain, is futile. For a time I wavered, impressed by the professor's authority, but soon became convinced I was right and undertook the task with all the fire and boundless confidence of youth.

I started by first picturing in my mind a direct-current machine, running it and following the changing flow of the currents in the armature. Then I would imagine an alternator and investigate the processes taking place in a similar manner. Next I would visualize systems comprising motors and generators and operate them in various ways. The images I saw were to me perfectly real and tangible. All my remaining term in Gratz was passed in intense but fruitless efforts of this kind, and I almost came to the conclusion that the problem was insolvable.

In 1880 I went to Prague, Bohemia, carrying out my father's wish to complete my education at the University there. It was in that city that I made a decided advance, which consisted in detaching the commutator from the machine and studying the phenomena in this new aspect, but still without result. In the year following there was a sudden change in my views of life. I realized that my parents had been making too great sacrifices on my account and resolved to relieve them of the burden. The wave of the American telephone had just reached the European continent and the system was to be installed in Budapest, Hungary. It appeared an ideal opportunity, all the more as a friend of our family was at the head of the enterprise. It was here that I suffered the complete breakdown of the nerves to which I have referred.

What I experienced during the period of that illness surpasses all belief. My sight and hearing were always extraordinary. I could clearly discern objects in the distance when others saw no trace of them. Several times in my boyhood I saved the houses of our neighbors from fire by hearing the faint crackling sounds which did not disturb their sleep, and calling for help.

In 1899, when I was past forty and carrying on my experiments in Colorado, I could hear very distinctly thunderclaps at a distance of 550 miles. The limit of audition for my young assistants was scarcely more than 150 miles. My ear was thus over thirteen times more sensitive. Yet at that time I was, so to speak, stone deaf in comparison with the acuteness of my hearing while under the nervous strain. In Budapest I could hear the ticking of a watch with three rooms between me and the time-piece. A fly alighting on a table in the room would cause a dull thud in my ear. A carriage passing at a distance of a few miles fairly shook my whole body. The whistle of a locomotive twenty or thirty miles away made the bench or chair on which I sat vibrate so strongly that the pain was unbearable. The ground under my feet trembled continuously. I had to support my bed on rubber cushions to get any rest at all. The roaring noises from near and far often produced the effect of spoken words which would have frightened me had I not been able to resolve them into their accidental components. The sun's rays, when periodically intercepted, would cause blows of such force on my brain that they would stun me. I had to summon all my will power to pass under a bridge or other structure as I experienced a crushing pressure on the skull. In the dark I had the sense of a bat and could detect the presence of an object at a distance of twelve feet by a peculiar creepy sensation on the forehead. My pulse varied from a few to two hundred and sixty beats and all the tissues of the body quivered with twitchings and tremors which was perhaps the hardest to bear. A renowned physician who gave me daily large doses of Bromide of Potassium pronounced my malady unique and incurable.

It is my eternal regret that I was not under the observation of experts in physiology and psychology at that time. I clung desperately to life, but never expected to recover. Can anyone believe that so hopeless a physical wreck could ever be transformed into a man of astonishing strength and tenacity, able to work thirty-eight years almost without a day's interruption, and find himself still strong and fresh in body and mind? Such is my case. A powerful desire to live and to continue the work, and the assistance of a devoted friend and athlete accomplished the wonder. My health returned and with it the vigor of mind. In attacking the problem again I almost regretted that the struggle was soon to end. I had so much energy to spare. When I undertook the task it was not with a resolve such as men often make. With me it was a sacred vow, a question of life and death. I knew that I would perish if I failed. Now I felt that the battle was won. Back in the deep recesses of the brain was the solution, but I could not yet give it outward expression. One afternoon, which is ever present in my recollection, I was enjoying a walk with my friend in the City Park and reciting poetry. At that age I knew entire books by heart, word for word. One of these was Goethe's "Faust." The sun was just setting and reminded me of the glorious passage:

"*Sie ruckt und weicht, der Tag ist uberlebt,
Dort eilt sie hin und fordert neues Leben.
Oh, dass kein Flugel mich vom Boden hebt
Ihr nach und immer nach zu streben!
Ein schoner Traum indessen sie entweicht,
Ach, zu des Geistes Flugeln wird so leicht
Kein korperlicher Flugel sich gesellen!*"

[The glow retreats, done is the day of toil;
It yonder hastes, new fields of life exploring;
Ah, that no wing can lift me from the soil
Upon its track to follow, follow soaring!
A glorious dream! though now the glories fade.
Alas! the wings that lift the mind no aid
Of wings to lift the body can bequeath me.]

As I uttered these inspiring words the idea came like a flash of lightning and in an instant the truth was revealed. I drew with a stick on the sand the diagrams shown six years later in my address before the American Institute of Electrical Engineers, and my companion understood them perfectly. The images I saw were wonderfully sharp and clear and had the solidity of metal and stone, so much so that I told him: "See my motor here; watch me reverse it." I cannot begin to describe my emotions. Pygmalion seeing his statue come to life could not have been more deeply moved. A thousand secrets of nature which I might have stumbled upon accidentally I would have given for that one which I had wrested from her against all odds and at the peril of my existence.

The Discovery of the Tesla Coil and Transformer

For a while I gave myself up entirely to the intense enjoyment of picturing machines and devising new forms. It was a mental state of happiness about as complete as I have ever known in life. Ideas came in an uninterrupted stream and the only difficulty I had was to hold them fast. The pieces of apparatus I conceived were to me absolutely real and tangible in every detail, even to the minute marks and signs of wear. I delighted in imagining the motors constantly running, for in this way they presented to mind's eye a more fascinating sight. When natural inclination develops into a passionate desire, one advances towards his goal in seven-league boots. In less than two months I evolved virtually all the types of motors and modifications of the system which are now identified with my name. It was, perhaps, providential that the necessities of existence commanded a temporary halt to this consuming activity of the mind. I came to Budapest prompted by a premature report concerning the telephone enterprise and, as irony of fate willed it, I had to accept a position as draftsman in the Central Telegraph Office of the Hungarian Government at a salary which I deem it my privilege not to disclose! Fortunately, I soon won the interest of the Inspector-in-Chief and was thereafter employed on calculations, designs and estimates in connection with new installations, until the Telephone Exchange was started, when I took charge of the same. The knowledge and practical experience I gained in the course of this work was most valuable and the employment gave me ample opportunities for the exercise of my inventive faculties. I made several improvements in the Central Station apparatus and perfected a telephone repeater or amplifier which was never patented or publicly described but would be creditable to me even today. In recognition of my efficient assistance the organizer of the undertaking, Mr. Puskas, upon disposing of his business in Budapest, offered me a position in Paris which I gladly accepted.

I never can forget the deep impression that magic city produced on my mind. For several days after my arrival I roamed thru the streets in utter bewilderment of the new spectacle. The attractions were many and irresistible, but, alas, the income was spent as soon as received. When Mr. Puskas asked me how I was getting along in the new sphere, I described the situation accurately in the statement that "the last twenty-nine days of the month are the toughest!" I led a rather strenuous life in what would now be termed "Rooseveltian fashion." Every morning, regardless of weather, I would go from the Boulevard St. Marcel, where I resided, to a bathing house on the Seine, plunge into the water, loop the circuit twenty-seven times and then walk an hour to reach Ivry, where the Company's factory was located. There I would have a woodchopper's breakfast at half-past seven o'clock and then eagerly await the lunch hour, in the meanwhile cracking hard nuts for the Manager of the Works, Mr. Charles Batchellor, who was an intimate friend and assistant of Edison. Here I was thrown in contact with a few Americans who fairly fell in love with me because of my proficiency in billiards. To these men I explained my invention and one of them, Mr. D. Cunningham, Foreman of the Mechanical Department, offered to form a stock company. The proposal seemed to me comical in the extreme. I did not have the faintest conception of what that meant except that it was an American way of doing things. Nothing came of it, however, and during the next few months I had to travel from one to another place in France and Germany to cure the ills of the power plants. On my return to Paris I submitted to one of the administrators of the Company, Mr. Rau, a plan for improving their dynamos and was given an opportunity. My success was complete and the delighted directors accorded me the privilege of developing automatic regulators which were much desired. Shortly after there was some trouble with the lighting plant which had been installed at the new railroad station in Strassburg, Alsace. The wiring was defective and on the occasion of the opening ceremonies a large part of a wall was blown out thru a short-circuit right in the presence of old Emperor William I. The German Government refused to take the plant and the French Company was facing a serious loss. On account of my knowledge of the German language and past experience, I was entrusted with the difficult task of straightening out matters and early in 1883 I went to Strassburg on that mission.

Some of the incidents in that city have left an indelible record on my memory. By a curious coincidence, a number of men who subsequently achieved fame, lived there about that time. In later life I used to say, "There were bacteria of greatness in that old town. Others caught the disease but I escaped!" The practical work, correspondence, and conferences with officials kept me preoccupied day and night, but, as soon as I was able to manage I undertook the construction of a simple motor in a mechanical shop opposite the railroad station, having brought with me from Paris some material for that purpose. The consummation of the experiment was, however, delayed until the summer of that year when I finally had the satisfaction of seeing rotation effected by alternating currents of different phase, and without sliding contacts or commutator, as I had conceived a year before. It was an exquisite pleasure but not to compare with the delirium of joy following the first revelation.

Among my new friends was the former Mayor of the city, Mr. Bauzin, whom I had already in a measure acquainted with this and other inventions of mine and whose support I endeavored to enlist. He was sincerely devoted to me and put my project before several wealthy persons but, to my mortification, found no response. He wanted to help me in every possible way and the approach of the first of July, 1919, happens to remind me of a form of "assistance" I received from that charming man, which was not financial but none the less appreciated. In 1870, when the Germans invaded the country, Mr. Bauzin had buried a good sized allotment of St. Estephe of 1801 and he came to the conclusion that he knew no worthier person than myself to consume that precious beverage. This, I may say, is one of the unforgettable incidents to which I have referred. My friend urged me to return to Paris as soon as possible and seek support there. This I was anxious to do but my work and negotiations were protracted owing to all sorts of petty obstacles I encountered so that at times the situation seemed hopeless.

Just to give an idea of German thoroughness and "efficiency," I may mention here a rather funny experience. An incandescent lamp of 16 c.p. was to be placed in a hallway and upon selecting the proper location I ordered the monteur to run the wires. After working for a while he concluded that the engineer had to be consulted and this was done. The latter made several objections but ultimately agreed that the lamp should be placed two inches from the spot I had assigned, whereupon the work proceeded. Then the engineer became worried and told me that Inspector Averdeck should be notified. That important person called, investigated, debated, and decided that the lamp should be shifted back two inches, which was the place I had marked. It was not long, however, before Averdeck got cold feet himself and advised me that he had informed Ober-Inspector Hieronimus of the matter and that I should await his decision. It was several days before the Ober-Inspector was able to free himself of other pressing duties but at last he arrived and a two-hour debate followed, when he decided to move the lamp two inches farther. My hopes that this was the final act were shattered when the Ober-Inspector returned and said to me: "Regierungsrath Funke is so particular that I would not dare to give an order for placing this lamp without his explicit approval." Accordingly arrangements for a visit from that great man were made. We started cleaning up and polishing early in the morning. Everybody brushed up, I put on my gloves and when Funke came with his retinue he was ceremoniously received. After two hours' deliberation he suddenly exclaimed: "I must be going," and pointing to a place on the ceiling, he ordered me to put the lamp there. It was the exact spot which I had originally chosen,

So it went day after day with variations, but I was determined to achieve at whatever cost and in the end my efforts were rewarded. By the spring of 1884 all the differences were adjusted, the plant formally accepted, and I returned to Paris with pleasing anticipations. One of the administrators had promised me a liberal compensation in case I succeeded, as well as a fair consideration of the improvements I had made in their dynamos and I hoped to realize a substantial sum. There were three administrators whom I shall designate as A, B and C for convenience. When I called on A he told me that B had the say. This gentleman thought that only C could decide and the latter was quite sure that A alone had the power to act. After several laps of this circulus vivios it dawned upon me that my reward was a castle in Spain. The utter failure of my attempts to raise capital for development was another disappointment and when Mr. Batchellor pressed me to go to America with a view of redesigning the Edison machines, I determined to try my fortunes in the Land of Golden Promise. But the chance was nearly mist.

I liquefied my modest assets, secured accommodations and found myself at the railroad station as the train was pulling out,. At that moment I discovered that my money and tickets were gone. What to do was the question. Hercules had plenty of time to deliberate but I had to decide while running alongside the train with opposite feelings surging in my brain like condenser oscillations. Resolve, helped by dexterity, won out in the nick of time and upon passing thru the usual experiences, as trivial as unpleasant, I managed to embark for New York with the remnants of my belongings, some poems and articles I had written, and a package of calculations relating to solutions of an unsolvable integral and to my flying machine. During the voyage I sat most of the time at the stern of the ship watching for an opportunity to save somebody from a watery grave, without the slightest thought of danger. Later when I had absorbed some of the practical American sense I shivered at the recollection and marveled at my former folly.

I wish that I could put in words my first impressions of this country. In the Arabian Tales I read how genii transported people into a land of dreams to live thru delightful adventures. My case was just the reverse. The genii had carried me from a world of dreams into one of realities. What I had left was beautiful, artistic and fascinating in every way; what I saw here was machined, rough and unattractive. A burly policeman was twirling his stick which looked to me as big as a log. I approached him politely with the request to direct me. "Six blocks down, then to the left," he said, with murder in his eyes. "Is this America?" I asked myself in painful surprise. "It is a century behind Europe in civilization." When I went abroad in 1889 - five years having elapsed since my arrival here - I became convinced that it was more than one hundred years AHEAD of Europe and nothing has happened to this day to change my opinion.

The meeting with Edison was a memorable event in my life. I was amazed at this wonderful man who, without early advantages and scientific training, had accomplished so much. I had studied a dozen languages, delved in literature and art, and had spent my best years in libraries reading all sorts of stuff that fell into my hands, from Newton's "Principia" to the novels of Paul de Kock, and felt that most of my life had been squandered. But it did not take long before I recognized that it was the best thing I could have done. Within a few weeks I had won Edison's confidence and it came about in this way.

The S.S. Oregon, the fastest passenger steamer at that time, had both of its lighting machines disabled and its sailing was delayed. As the superstructure had been built after their installation it was impossible to remove them from the hold. The predicament was a serious one and Edison was much annoyed. In the evening I took the necessary instruments with me and went aboard the vessel where I stayed for the night. The dynamos were in bad condition, having several short-circuits and breaks, but with the assistance of the crew I succeeded in putting them in good shape. At five o'clock in the morning, when passing along

Fifth Avenue on my way to the shop, I met Edison with Batchellor and a few others as they were returning home to retire. "Here is our Parisian running around at night," he said. When I told him that I was coming from the Oregon and had repaired both machines, he looked at me in silence and walked away without another word. But when he had gone some distance I heard him remark: "Batchellor, this is a d-n good man," and from that time on I had full freedom in directing the work. For nearly a year my regular hours were from 10.30 A.M. until 5 o'clock the next morning without a day's exception. Edison said to me: "I have had many hard-working assistants but you take the cake." During this period I designed twenty-four different types of standard machines with short cores and of uniform pattern which replaced the old ones. The Manager had promised me fifty thousand dollars on the completion of this task but it turned out to be a practical joke. This gave me a painful shock and I resigned my position.

Immediately thereafter some people approached me with the proposal of forming an arc light company under my name, to which I agreed. Here finally was an opportunity to develop the motor, but when I broached the subject to my new associates they said: "No, we want the arc lamp. We don't care for this alternating current of yours." In 1886 my system of arc lighting was perfected and adopted for factory and municipal lighting, and I was free, but with no other possession than a beautifully engraved certificate of stock of hypothetical value. Then followed a period of struggle in the new medium for which I was not fitted, but the reward came in the end and in April, 1887, the Tesla Electric Company was organized, providing a laboratory and facilities. The motors I built there were exactly as I had imagined them. I made no attempt to improve the design, but merely reproduced the pictures as they appeared to my vision and the operation was always as I expected.

In the early part of 1888 an arrangement was made with the Westinghouse Company for the manufacture of the motors on a large scale. But great difficulties had still to be overcome. My system was based on the use of low frequency currents and the Westinghouse experts had adopted 133 cycles with the object of securing advantages in the transformation. They did not want to depart from their standard forms of apparatus and my efforts had to be concentrated upon adapting the motor to these conditions. Another necessity was to produce a motor capable of running efficiently at this frequency on two wires which was not easy of accomplishment.

At the close of 1889, however, my services in Pittsburgh being no longer essential, I returned to New York and resumed experimental work in a laboratory on Grand Street, where I began immediately the design of high frequency machines. The problems of construction in this unexplored field were novel and quite peculiar and I encountered many difficulties. I rejected the inductor type, fearing that it might not yield perfect sine waves which were so important to resonant action. Had it not been for this I could have saved myself a great deal of labor. Another discouraging feature of the high frequency alternator seemed to be the inconstancy of speed which threatened to impose serious limitations to its use. I had already noted in my demonstrations before the American Institution of Electrical Engineers that several times the tune was lost, necessitating readjustment, and did not yet foresee, what I discovered long afterwards, a means of operating a machine of this kind at a speed constant to such a degree as not to vary more than a small fraction of one revolution between the extremes of load.

From many other considerations it appeared desirable to invent a simpler device for the production of electric oscillations. In 1856 Lord Kelvin had exposed the theory of the condenser discharge, but no practical application of that important knowledge was made. I saw the possibilities and undertook the development of induction apparatus on this principle. My progress was so rapid as to enable me to exhibit at my lecture in 1891 a coil giving sparks of five inches. On that occasion I frankly told the engineers of a defect involved in the transformation by the new method, namely, the loss in the spark gap. Subsequent investigation showed that no matter what medium is employed, be it air, hydrogen, mercury vapor, oil or a stream of electrons, the efficiency is the same. It is a law very much like that governing the conversion of mechanical energy. We may drop a weight from a certain height vertically down or carry it to the lower level along any devious path, it is immaterial insofar as the amount of work is concerned. Fortunately however, this drawback is not fatal as by proper proportioning of the resonant circuits an efficiency of 85 per cent is attainable. Since my early announcement of the invention it has come into universal use and wrought a revolution in many departments. But a still greater future awaits it. When in 1900 I obtained powerful discharges of 100 feet and flashed a current around the globe, I was reminded of the first tiny spark I observed in my Grand Street laboratory and was thrilled by sensations akin to those I felt when I discovered the rotating magnetic field.

The Magnifying Transmitter

As I review the events of my past life I realize how subtle are the influences that shape our destinies. An incident of my youth may serve to illustrate. One winter's day I managed to climb a steep mountain, in company with other boys. The snow was quite deep and a warm southerly wind made it just suitable for our purpose. We amused ourselves by throwing balls which would roll down a certain distance, gathering more or less snow, and we tried to outdo one another in this exciting sport. Suddenly a ball was seen to go beyond the limit, swelling to enormous proportions until it became as big as a house and plunged thundering into the valley below with a force that made the ground tremble. I looked on spellbound, incapable of understanding what had happened. For weeks afterward the picture of the avalanche was before my eyes and I wondered how anything so small could grow to such an immense size. Ever since that time the magnification of feeble actions fascinated me, and when, years later, I took up the experimental study of mechanical and electrical resonance, I was keenly interested from the very start. Possibly, had it not been for that early powerful impression, I might not have followed up the little spark I obtained with my coil and never developed my best invention, the true history of which I'll tell here for the first time.

"Lionhunters" have often asked me which of my discoveries I prize most. This depends on the point of view. Not a few technical men, very able in their special departments, but dominated by a pedantic spirit and nearsighted, have asserted that excepting the induction motor I have given to the world little of practical use. This is a grievous mistake. A new idea must not be judged by its immediate results. My alternating system of power transmission came at a psychological moment, as a long-sought answer to pressing industrial questions, and altho considerable resistance had to be overcome and opposing interests reconciled, as usual, the commercial introduction could not be long delayed. Now, compare this situation with that confronting my turbine, for example. One should think that so simple and beautiful an invention, possessing many features of an ideal motor, should be adopted at once and, undoubtedly, it would under similar conditions. But the prospective effect of the rotating field was not to render worthless existing machinery; on the contrary, it was to give it additional value. The system lent itself to new enterprise as well as to improvement of the old. My turbine is an advance of a character entirely different. It is a radical departure in the sense that its success would mean the abandonment of the antiquated types of prime movers on which billions of dollars have been spent. Under such circumstances the progress must needs be slow and perhaps the greatest impediment is encountered in the prejudicial opinions created in the minds of experts by organized opposition.

Only the other day I had a disheartening experience when I met my friend and former assistant, Charles F. Scott, now professor of Electrical Engineering at Yale. I had not seen him for a long time and was glad to have an opportunity for a little chat at my office. Our conversation naturally enough drifted on my turbine and I became heated to a high degree. "Scott," I exclaimed, carried away by the vision of a glorious future, "my turbine will scrap all the heat-engines in the world." Scott stroked his chin and looked away thoughtfully, as though making a mental calculation. "That will make quite a pile of scrap," he said, and left without another word!

These and other inventions of mine, however, were nothing more than steps forward in certain directions. In evolving them I simply followed the inborn sense to improve the present devices without any special thought of our far more imperative necessities. The "Magnifying Transmitter" was the product of labors extending through years, having for their chief object the solution of problems which are infinitely more important to mankind than mere industrial development.

If my memory serves me right, it was in November, 1890, that I performed a laboratory experiment which was one of the most extraordinary and spectacular ever recorded in the annals of science. In investigating the behavior of high frequency currents I had satisfied myself that an electric field of sufficient intensity could be produced in a room to light up electrodeless vacuum tubes. Accordingly, a transformer was built to test the theory and the first trial proved a marvelous success. It is difficult to appreciate what those strange phenomena meant at that time. We crave for new sensations but soon become indifferent to them. The wonders of yesterday are today common occurrences. When my tubes were first publicly exhibited they were viewed with amazement impossible to describe. From all parts of the world I received urgent invitations and numerous honors and other flattering inducements were offered to me, which I declined.

But in 1892 the demands became irresistible and I went to London where I delivered a lecture before the Institution of Electrical Engineers. It had been my intention to leave immediately for Paris in compliance with a similar obligation, but Sir James Dewar insisted on my appearing before the Royal Institution. I was a man of firm resolve but succumbed easily to the forceful arguments of the great Scotsman. He pushed me into a chair and poured out half a glass of a wonderful brown fluid which sparkled in all sorts of iridescent colors and tasted like nectar. "Now," said he. "you are sitting in Faraday's chair and you are enjoying whiskey he used to drink." In both aspects it was an enviable experience. The next evening I gave a demonstration before that Institution, at the termination of which Lord Rayleigh addressed the audience and his generous words gave me the first start in these endeavors. I fled from London and later from Paris to escape favors showered upon me, and journeyed to my home where I passed through a most painful ordeal and illness. Upon regaining my health I began to formulate plans for the resumption of work in America. Up to that time I never realized that I possessed any particular gift of discovery but Lord Rayleigh, whom I always considered as an ideal man of science, had said so and if that was the case I felt that I should concentrate on some big idea.

One day, as I was roaming in the mountains, I sought shelter from an approaching storm. The sky became overhung with heavy clouds but somehow the rain was delayed until, all of a sudden, there was a lightning flash and a few moments after a deluge. This observation set me thinking. It was manifest that the two phenomena were closely related, as cause and effect, and a little reflection led me to the conclusion that the electrical energy involved in the precipitation of the water was inconsiderable, the function of lightning being much like that of a sensitive trigger.

Here was a stupendous possibility of achievement. If we could produce electric effects of the required quality, this whole planet and the conditions of existence on it could be transformed. The sun raises the water of the oceans and winds drive it to distant regions where it remains in a state of most delicate balance. If it were in our power to upset it when and wherever desired, this mighty life-sustaining stream could be at will controlled. We could irrigate arid deserts, create lakes and rivers and provide motive power in unlimited amounts. This would be the most efficient way of harnessing the sun to the uses of man. The consummation depended on our ability to develop electric forces of the order of those in nature. It seemed a hopeless undertaking, but I made up my mind to try it and immediately on my return to the United States, in the Summer of 1892, work was begun which was to me all the more attractive, because a means of the same kind was necessary for the successful transmission of energy without wires.

The first gratifying result was obtained in the spring of the succeeding year when I reached tensions of about 1,000,000 volts with my conical coil. That was not much in the light of the present art, but it was then considered a feat. Steady progress was made until the destruction of my laboratory by fire in 1895, as may be judged from an article by T. C. Martin which appeared in the April number of the Century Magazine. This calamity set me back in many ways and most of that year had to be devoted to planning and reconstruction. However, as soon as circumstances permitted, I returned to the task.

Although I knew that higher electro-motive forces were attainable with apparatus of larger dimensions, I had an instinctive perception that the object could be accomplished by the proper design of a comparatively small and compact transformer. In carrying on tests with a secondary in the form of a flat spiral, as illustrated in my patents, the absence of streamers surprised me, and it was not long before I discovered that this was due to the position of the turns and their mutual action. Profiting from this observation I resorted to the use of a high tension conductor with turns of considerable diameter sufficiently separated to keep down the distributed capacity, while at the same time preventing undue accumulation of the charge at any point. The application of this principle enabled me to produce pressures of 4,000,000 volts, which was about the limit obtainable in my new laboratory at Houston Street, as the discharges extended through a distance of 16 feet. A photograph of this transmitter was published in the Electrical Review of November, 1898.

In order to advance further along this line I had to go into the open, and in the spring of 1899, having completed preparations for the erection of a wireless plant, I went to Colorado where I remained for more than one year. Here I introduced other improvements and refinements which made it possible to generate currents of any tension that may be desired. Those who are interested will find some information in regard to the experiments I conducted there in my article, "The Problem of Increasing Human Energy" in the Century Magazine of June, 1900, to which I have referred on a previous occasion.

I have been asked by the *Electrical Experimenter* to be quite explicit on this subject so that my young friends among the readers of the magazine will clearly understand the construction and operation of my "Magnifying Transmitter" and the purposes for which it is intended. Well, then, in the first place, it is a resonant transformer with a secondary in which the parts, charged to a high potential, are of considerable area and arranged in space along ideal enveloping surfaces of very large radii of curvature, and at proper distances from one another thereby insuring a small electric surface density everywhere so that no leak can occur even if the conductor is bare. It is suitable for any frequency, from a few to many thousands of cycles per second, and can be used in the production of currents of tremendous volume and moderate pressure, or of smaller amperage and immense electromotive force. The maximum electric tension is merely dependent on the curvature of the surfaces on which the charged elements are situated and the area of the latter.

Judging from my past experience, as much as 100,000,000 volts are perfectly practicable. On the other hand currents of many thousands of amperes may be obtained in the antenna. A plant of but very moderate dimensions is required for such performances. Theoretically, a terminal of less than 90 feet in diameter is sufficient to develop an electromotive force of that magnitude while for antenna currents of from 2,000-4,000 amperes at the usual frequencies it need not be larger than 30 feet in diameter.

In a more restricted meaning this wireless transmitter is one in which the Hertz-wave radiation is an entirely negligible quantity as compared with the whole energy, under which condition the damping factor is extremely small and an enormous charge is stored in the elevated capacity. Such a circuit may then be excited with impulses of any kind, even of low frequency and it will yield sinusoidal and continuous oscillations like those of an alternator.

Taken in the narrowest significance of the term, however, it is a resonant transformer which, besides possessing these qualities, is accurately proportioned to fit the globe and its electrical constants and properties, by virtue of which design it becomes highly efficient and effective in the wireless transmission of energy. Distance is then absolutely eliminated, there being no diminution in the intensity of the transmitted impulses. It is even possible to make the actions increase with the distance from the plant according to an exact mathematical law.

This invention was one of a number comprised in my "World-System" of wireless transmission which I undertook to commercialize on my return to New York in 1900. As to the immediate purposes of my enterprise, they were clearly outlined in a technical statement of that period from which I quote:

"The 'World-System' has resulted from a combination of several original discoveries made by the inventor in the course of long continued research and experimentation. It makes possible not only the instantaneous and precise wireless transmission of any kind of signals, messages or characters, to all parts of the world, but also the inter-connection of the existing telegraph, telephone, and other signal stations without any change in their present equipment. By its means, for instance, a telephone subscriber here may call up and talk to any other subscriber on the Globe. An inexpensive receiver, not bigger than a watch, will enable him to listen anywhere, on land or sea, to a speech delivered or music played in some other place, however distant. These examples are cited merely to give an idea of the possibilities of this great scientific advance, which annihilates distance and makes that perfect natural conductor, the Earth, available for all the innumerable purposes which human ingenuity has found for a line-wire. One far-reaching result of this is that any device capable of being operated thru one or more wires (at a distance obviously restricted) can likewise be actuated, without artificial conductors and with the same facility and accuracy, at distances to which there are no limits other than those imposed by the physical dimensions of the Globe. Thus, not only will entirely new fields for commercial exploitation be opened up by this ideal method of transmission but the old ones vastly extended.

The 'World-System' is based on the application of the following important inventions and discoveries:

1. The 'Tesla Transformer.' This apparatus is in the production of electrical vibrations as revolutionary as gunpowder was in warfare. Currents many times stronger than any ever generated in the usual ways, and sparks over one hundred feet long, have been produced by the inventor with an instrument of this kind.

2. The 'Magnifying Transmitter.' This is Tesla's best invention, a peculiar transformer specially adapted to excite the Earth, which is in the transmission of electrical energy what the telescope is in astronomical observation. By the use of this marvelous device he has already set up electrical movements of greater intensity than those of lightning and passed a current, sufficient to light more than two hundred incandescent lamps, around the Globe.

3. The 'Tesla Wireless System.' This system comprises a number of improvements and is the only means known for transmitting economically electrical energy to a distance without wires. Careful tests and measurements in connection with an experimental station of great activity, erected by the inventor in Colorado, have demonstrated that power in any desired amount can be conveyed, clear across the Globe if necessary, with a loss not exceeding a few per cent.

4. The 'Art of Individualization.' This invention of Tesla's is to primitive 'tuning' what refined language is to unarticulated expression. It makes possible the transmission of signals or messages absolutely secret and exclusive both in the active and passive aspect, that is, non-interfering as well as non-interferable. Each signal is like an individual of unmistakable identity and there is virtually no limit to the number of stations or instruments which can be simultaneously operated without the slightest mutual disturbance.

5. 'The Terrestrial Stationary Waves.' This wonderful discovery, popularly explained, means that the Earth is responsive to electrical vibrations of definite pitch just as a tuning fork to certain waves of sound. These particular electrical vibrations, capable of powerfully exciting the Globe, lend themselves to innumerable uses of great importance commercially and in many other respects.

The first 'World-System' power plant can be put in operation in nine months. With this power plant it will be practicable to attain electrical activities up to ten million horsepower and it is designed to serve for as many technical achievements as are possible without due expense. Among these the following may be mentioned:

(1) The inter-connection of the existing telegraph exchanges or offices all over the world;

(2) The establishment of a secret and non-interferable government telegraph service;

(3) The inter-connection of all the present telephone exchanges or offices on the Globe;

(4) The universal distribution of general news, by telegraph or telephone, in connection with the Press;

(5) The establishment of such a 'World-System' of intelligence transmission for exclusive private use;

(6) The inter-connection and operation of all stock tickers of the world;

(7) The establishment of a 'World-System' of musical distribution, etc.;

(8) The universal registration of time by cheap clocks indicating the hour with astronomical precision and requiring no attention whatever;

(9) The world transmission of typed or handwritten characters, letters, checks, etc.;

(10) The establishment of a universal marine service enabling the navigators of all ships to steer perfectly without compass, to determine the exact location, hour and speed, to prevent collisions and disasters, etc.;

(11) The inauguration of a system of world-printing on land and sea;

(12) The world reproduction of photographic pictures and all kinds of drawings or records.

I also proposed to make demonstrations in the wireless transmission of power on a small scale but sufficient to carry conviction. Besides these I referred to other and incomparably more important applications of my discoveries which will be disclosed at some future date.

A plant was built on Long Island with a tower 187 feet high, having a spherical terminal about 68 feet in diameter. These dimensions were adequate for the transmission of virtually any amount of energy. Originally only from 200 to 300 K.W. were provided but I intended to employ later several thousand horsepower. The transmitter was to emit a wave complex of special characteristics and I had devised a unique method of telephonic control of any amount of energy.

The tower was destroyed two years ago but my projects are being developed and another one, improved in some features, will be constructed. On this occasion I would contradict the widely circulated report that the structure was demolished by the Government which owing to war conditions, might have created prejudice in the minds of those who may not know that the papers, which thirty years ago conferred upon me the honor of American citizenship, are always kept in a safe, while my orders, diplomas, degrees, gold medals and other distinctions are packed away in old trunks. If this report had a foundation I would have been refunded a large sum of money which I expended in the construction of the tower. On the contrary it was in the interest of the Government to preserve it, particularly as it would have made possible--to mention just one valuable result--the location of a submarine in any part of the world. My plant, services, and all my improvements have always been at the disposal of the officials and ever since the outbreak of the European conflict I have been working at a sacrifice on several inventions of mine relating to aerial navigation, ship propulsion and wireless transmission which are of the greatest importance to the country. Those who are well informed know that my ideas have revolutionized the industries of the United States and I am not aware that there lives an inventor who has been, in this respect, as fortunate as myself especially as regards the use of his improvements in the war. I have refrained from publicly expressing myself on this subject before as it seemed improper to dwell on personal matters while all the world was in dire trouble.

I would add further, in view of various rumors which have reached me, that Mr. J. Pierpont Morgan did not interest himself with me in a business way but in the same large spirit in which he has assisted many other pioneers. He carried out his generous promise to the letter and it would have been most unreasonable to expect from him anything more. He had the highest regard for my attainments and gave me every evidence of his complete faith in my ability to ultimately achieve what I had set out to do. I am unwilling to accord to some small-minded and jealous individuals the satisfaction of having thwarted my efforts. These men are to me nothing more than microbes of a nasty disease. My project was retarded by laws of nature. The world was not prepared for it. It was too far ahead of time. But the same laws will prevail in the end and make it a triumphal success.

The Art of Telautomatics

No subject to which I have ever devoted myself has called for such concentration of mind and strained to so dangerous a degree the finest fibers of my brain as the system of which the Magnifying Transmitter is the foundation. I put all the intensity and vigor of youth in the development of the rotating field discoveries, but those early labors were of a different character. Although strenuous in the extreme, they did not involve that keen and exhausting discernment which had to be exercised in attacking the many puzzling problems of the wireless. Despite my rare physical endurance at that period the abused nerves finally rebelled and I suffered a complete collapse, just as the consummation of the long and difficult task was almost in sight.

Without doubt I would have paid a greater penalty later, and very likely my career would have been prematurely terminated, had not providence equipt me with a safety device, which has seemed to improve with advancing years and unfailingly comes into play when my forces are at an end. So long as it operates I am safe from danger, due to overwork, which threatens other inventors and, incidentally, I need no vacations which are indispensable to most people. When I am all but used up I simply do as the darkies, who "naturally fall asleep while white folks worry." To venture a theory out of my sphere, the body probably accumulates little by little a definite quantity of some toxic agent and I sink into a nearly lethargic state which lasts half an hour to the minute. Upon awakening I have the sensation as though the events immediately preceding had occurred very long ago, and if I attempt to continue the interrupted train of thought I feel a veritable mental nausea. Involuntarily I then turn to other work and am surprised at the freshness of the mind and ease with which I overcome obstacles that had baffled me before. After weeks or months my passion for the temporarily abandoned invention returns and I invariably find answers to all the vexing questions with scarcely any effort.

In this connection I will tell of an extraordinary experience which may be of interest to students of psychology. I had produced a striking phenomenon with my grounded transmitter and was endeavoring to ascertain its true significance in relation to the currents propagated through the earth. It seemed a hopeless undertaking, and for more than a year I worked unremittingly, but in vain. This profound study so entirely absorbed me that I became forgetful of everything else, even of my undermined health. At last, as I was at the point of breaking down, nature applied the preservative inducing lethal sleep. Regaining my senses I realized with consternation that I was unable to visualize scenes from my life except those of infancy, the very first ones that had entered my consciousness. Curiously enough, these appeared before my vision with startling distinctness and afforded me welcome relief. Night after night, when retiring, I would think of them and more and more of my previous existence was revealed. The image of my mother was always the principal figure in the spectacle that slowly unfolded, and a consuming desire to see her again gradually took possession of me. This feeling grew so strong that I resolved to drop all work and satisfy my longing. But I found it too hard to break away from the laboratory, and several months elapsed during which I had succeeded in reviving all the impressions of my past life up to the spring of 1892. In the next picture that came out of the mist of oblivion, I saw myself at the Hotel de la Paix in Paris just coming to from one of my peculiar sleeping spells, which had been caused by prolonged exertion of the brain. Imagine the pain and distress I felt when it flashed upon my mind that a dispatch was handed to me at that very moment bearing the sad news that my mother was dying. I remembered how I made the long journey home without an hour of rest and how she passed away after weeks of agony! It was especially remarkable that during all this period of partially obliterated memory I was fully alive to everything touching on the subject of my research. I could recall the smallest details and the least significant observations in my experiments and even recite pages of text and complex mathematical formulae.

My belief is firm in a law of compensation. The true rewards are ever in proportion to the labor and sacrifices made. This is one of the reasons why I feel certain that of all my inventions, the Magnifying Transmitter will prove most important and valuable to future generations. I am prompted to this prediction not so much by thoughts of the commercial and industrial revolution which it will surely bring about, but of the humanitarian consequences of the many achievements it makes possible. Considerations of mere utility weigh little in the balance against the higher benefits of civilization. We are confronted with portentous problems which can not be solved just by providing for our material existence, however abundantly. On the contrary, progress in this direction is fraught with hazards and perils not less menacing than those born from want and suffering. If we were to release the energy of atoms or discover some other way of developing cheap and unlimited power at any point of the globe this accomplishment, instead of being a blessing, might bring disaster to mankind in giving rise to dissension and anarchy which would ultimately result in the enthronement of the hated regime of force. The greatest good will comes from technical improvements tending to unification and harmony, and my wireless transmitter is preeminently such. By its means the human voice and likeness will be reproduced everywhere and factories driven thousands of miles from waterfalls furnishing the power; aerial machines will be propelled around the earth without a stop and the sun's energy controlled to create lakes and rivers for motive purposes and transformation of arid deserts into fertile land. Its introduction for telegraphic, telephonic and similar uses will automatically cut out the statics and all other interferences which at present impose narrow limits to the application of the wireless.

This is a timely topic on which a few words might not be amiss. During the past decade a number of people have arrogantly claimed that they had succeeded in doing away with this impediment. I have carefully examined all of the arrangements described and tested most of them long before they were publicly disclosed, but the finding was uniformly negative. A recent official statement from the U.S. Navy may, perhaps, have taught some beguilable news editors how to appraise these announcements at their real worth. As a rule the attempts are based on theories so fallacious that whenever they come to my notice I can not help thinking in a lighter vein. Quite recently a new discovery was heralded, with a deafening flourish of trumpets, but it proved another case of a mountain bringing forth a mouse.

This reminds me of an exciting incident which took place years ago when I was conducting my experiments with currents of high frequency. Steve Brodie had just jumped off the Brooklyn Bridge. The feat has been vulgarized since by imitators, but the first report electrified New York. I was very impressionable then and frequently spoke of the daring printer. On a hot afternoon I felt the necessity of refreshing myself and stepped into one of the popular thirty thousand institutions of this great city where a delicious twelve per cent beverage was served which can now be had only by making a trip to the poor and devastated countries of Europe. The attendance was large and not overdistinguished and a matter was discussed which gave me an admirable opening for the careless remark: "This is what I said when I jumped off the bridge." No sooner had I uttered these words than I felt like the companion of Timotheus in the poem of Schiller. In an instant there was a pandemonium and a dozen voices cried: "It is Brodie!" I threw a quarter on the counter and bolted for the door but the crowd was at my heels with yells: "Stop, Steve!" which must have been misunderstood for many persons tried to hold me up as I ran frantically for my haven of refuge. By darting around corners I fortunately managed - through the medium of a fire-escape - to reach the laboratory where I threw off my coat, camouflaged myself as a hard-working blacksmith, and started the forge. But these precautions proved unnecessary; I had eluded my pursuers. For many years afterward, at night, when imagination turns into specters the trifling troubles of the day, I often thought, as I tossed on the bed, what my fate would have been had that mob caught me and found out that I was not Steve Brodie!

Now the engineer, who lately gave an account before a technical body of a novel remedy against statics based on a "heretofore unknown law of nature," seems to have been as reckless as myself when he contended that these disturbances propagate up and down, while those of a transmitter proceed along the earth. It would mean that a condenser, as this globe, with its gaseous envelope, could be charged and discharged in a manner quite contrary to the fundamental teachings propounded in every elemental text-book of physics. Such a supposition would have been condemned as erroneous, even in Franklin's time, for the facts bearing on this were then well known and the identity between atmospheric electricity and that developed by machines was fully established. Obviously, natural and artificial disturbances propagate through the earth and the air in exactly the same way, and both set up electromotive forces in the horizontal, as well as vertical, sense. Interference can not be overcome by any such methods as were proposed. The truth is this: in the air the potential increases at the rate of about fifty volts per foot of elevation, owing to which there may be a difference of pressure amounting to twenty, or even forty thousand volts between the upper and lower ends of the antenna. The masses of the charged atmosphere are constantly in motion and give up electricity to the conductor, not continuously but rather disruptively, this producing a grinding noise in a sensitive telephonic receiver. The higher the terminal and the greater the space encompassed by the wires, the more pronounced is the effect, but it must be understood that it is purely local and has little to do with the real trouble.

In 1900, while perfecting my wireless system, one form of apparatus comprised four antennae. These were carefully calibrated to the same frequency and connected in multiple with the object of magnifying the action, in receiving from any direction. When I desired to ascertain the origin of the transmitted impulses, each diagonally situated pair was put in series with a primary coil energizing the detector circuit. In the former case the sound was loud in the telephone; in the latter it ceased, as expected, the two antennae neutralizing each other, but the true statics manifested themselves in both instances and I had to devise special preventives embodying different principles.

By employing receivers connected to two points of the ground, as suggested by me long ago, this trouble caused by the charged air, which is very serious in the structures as now built, is nullified and besides, the liability of all kinds of interference is reduced to about one-half, because of the directional character of the circuit. This was perfectly self-evident, but came as a revelation to some simple-minded wireless folks whose experience was confined to forms of apparatus that could have been improved with an axe, and they have been disposing of the bear's skin before killing him. If it were true that strays performed such antics, it would be easy to get rid of them by receiving without aerials. But, as a matter of fact, a wire buried in the ground which, conforming to this view, should be absolutely immune, is more susceptible to certain extraneous impulses than one placed vertically in the air. To state it fairly, a slight progress has been made, but not by virtue of any particular method or device. It was achieved simply by discarding the enormous structures, which are bad enough for transmission but wholly unsuitable for reception, and adopting a more appropriate type of receiver. As I pointed out in a previous article, to dispose of this difficulty for good, a radical change must be made in the system, and the sooner this is done the better.

It would be calamitous, indeed, if at this time when the art is in its infancy and the vast majority, not excepting even experts, have no conception of its ultimate possibilities, a measure would be rushed through the legislature making it a government monopoly. This was proposed a few weeks ago by Secretary Daniels, and no doubt that distinguished official has made his appeal to the Senate and House of Representatives with sincere conviction. But universal evidence unmistakably shows that the best results are always obtained in healthful commercial competition. There are, however, exceptional reasons why wireless should be given the fullest freedom of development. In the first place it offers prospects immeasurably greater and more vital to betterment of human life than any other invention or discovery in the history of man. Then again, it must be understood that this wonderful art has been, in its entirety, evolved here and can be called "American" with more right and propriety than the telephone, the incandescent lamp or the airplane. Enterprising press agents and stock jobbers have been so successful in spreading misinformation that even so excellent a periodical as the Scientific American accords the chief credit to a foreign country. The Germans, of course, gave us the Hertz-waves and the Russian, English, French and Italian experts were quick in using them for signaling purposes. It was an obvious application of the new agent and accomplished with the old classical and unimproved induction coil - scarcely anything more than another kind of heliography. The radius of transmission was very limited, the results attained of little value, and the Hertz oscillations, as a means for conveying intelligence, could have been advantageously replaced by sound-waves, which I advocated in 1891. Moreover, all of these attempts were made three years after the basic principles of the wireless system, which is universally employed to-day, and its potent instrumentalities had been clearly described and developed in America. No trace of those Hertzian appliances and methods remains today. We have proceeded in the very opposite direction and what has been done is the product of the brains and efforts of citizens of this country. The fundamental patents have expired and the opportunities are open to all. The chief argument of the Secretary is based on interference. According to his statement, reported in the New York Herald of July 29th, signals from a powerful station can be intercepted in every village of the world . In view of this fact, which was demonstrated in my experiments of 1900, it would be of little use to impose restrictions in the United States.

As throwing light on this point, I may mention that only recently an odd looking gentleman called on me with the object of enlisting my services in the construction of world transmitters in some distant land. "We have no money," he said, "but carloads of solid gold and we will give you a liberal amount." I told him that I wanted to see first what will be done with my inventions in America, and this ended the interview. But I am satisfied that some dark forces are at work, and as time goes on the maintenance of continuous communication will be rendered more difficult. The only remedy is a system immune against interruption. It has been perfected, it exists, and all that is necessary is to put it in operation.

The terrible conflict is still uppermost in the minds and perhaps the greatest importance will be attached to the Magnifying Transmitter as a machine for attack and defense, more particularly in connection with Telautomatics. This invention is a logical outcome of observations begun in my boyhood and continued throughout my life. When the first results were published the Electrical Review stated editorially that it would become one of the "most potent factors in the advance and civilization of mankind." The time is not distant when this prediction will be fulfilled. In 1898 and 1900 it was offered to the Government and might have been adopted were I one of those who would go to Alexander's shepherd when they want a favor from Alexander. At that time I really thought that it would abolish war, because of its unlimited destructiveness and exclusion of the personal element of combat. But while I have not lost faith in its potentialities, my views have changed since.

War can not be avoided until the physical cause for its recurrence is removed and this, in the last analysis, is the vast extent of the planet on which we live. Only thru annihilation of distance in every respect, as the conveyance of intelligence, transport of passengers and supplies and transmission of energy will conditions be brought about some day, insuring permanency of friendly relations. What we now want most is closer contact and better understanding between individuals and communities all over the earth, and the elimination of that fanatic devotion to exalted ideals of national egoism and pride which is always prone to plunge the world into primeval barbarism and strife. No league or parliamentary act of any kind will ever prevent such a calamity. These are only new devices for putting the weak at the mercy of the strong. I have expressed myself in this regard fourteen years ago, when a combination of a few leading governments - a sort of Holy Alliance - was advocated by the late Andrew Carnegie, who may be fairly considered as the father of this idea, having given to it more publicity and impetus than anybody else prior to the efforts of the President. While it can not be denied that such a pact might be of material advantage to some less fortunate peoples, it can not attain the chief object sought. Peace can only come as a natural consequence of universal enlightenment and merging of races, and we are still far from this blissful realization.

As I view the world of today, in the light of the gigantic struggle we have witnessed, I am filled with conviction that the interests of humanity would be best served if the United States remained true to its traditions and kept out of "entangling alliances." Situated as it is, geographically, remote from the theaters of impending conflicts, without incentive to territorial aggrandizement, with inexhaustible resources and immense population thoroughly imbued with the spirit of liberty and right, this country is placed in a unique and privileged position. It is thus able to exert, independently, its colossal strength and moral force to the benefit of all, more judiciously and effectively, than as member of a league.

In one of these biographical sketches, published in the *Electrical Experimenter*, I have dwelt on the circumstances of my early life and told of an affliction which compelled me to unremitting exercise of imagination and self observation. This mental activity, at first involuntary under the pressure of illness and suffering, gradually became second nature and led me finally to recognize that I was but an automaton devoid of free will in thought and action and merely responsive to the forces of the environment. Our bodies are of such complexity of structure, the motions we perform are so numerous and involved, and the external impressions on our sense organs to such a degree delicate and elusive that it is hard for the average person to grasp this fact. And yet nothing is more convincing to the trained investigator than the mechanistic theory of life which had been, in a measure, understood and propounded by Descartes three hundred years ago. But in his time many important functions of our organism were unknown and, especially with respect to the nature of light and the construction and operation of the eye, philosophers were in the dark.

In recent years the progress of scientific research in these fields has been such as to leave no room for a doubt in regard to this view on which many works have been published. One of its ablest and most eloquent exponents is, perhaps, Felix Le Dantec, formerly assistant of Pasteur. Prof. Jacques Loeb has performed remarkable experiments in heliotropism, clearly establishing the controlling power of light in lower forms of organisms, and his latest book, "Forced Movements," is revelatory. But while men of science accept this theory simply as any other that is recognized, to me it is a truth which I hourly demonstrate by every act and thought of mine. The consciousness of the external impression prompting me to any kind of exertion, physical or mental, is ever present in my mind. Only on very rare occasions, when I was in a state of exceptional concentration, have I found difficulty in locating the original impulses.

The by far greater number of human beings are never aware of what is passing around and within them, and millions fall victims of disease and die prematurely just on this account. The commonest every-day occurrences appear to them mysterious and inexplicable. One may feel a sudden wave of sadness and rake his brain for an explanation when he might have noticed that it was caused by a cloud cutting off the rays of the sun. He may see the image of a friend dear to him under conditions which he construes as very peculiar, when only shortly before he has passed him in the street or seen his photograph somewhere. When he loses a collar button he fusses and swears for an hour, being unable to visualize his previous actions and locate the object directly. Deficient observation is merely a form of ignorance and responsible for the many morbid notions and foolish ideas prevailing. There is not more than one out of every ten persons who does not believe in telepathy and other psychic manifestations, spiritualism and communion with the dead, and who would refuse to listen to willing or unwilling deceivers.

Just to illustrate how deeply rooted this tendency has become even among the clearheaded American population, I may mention a comical incident. Shortly before the war, when the exhibition of my turbines in this city elicited widespread comment in the technical papers, I anticipated that there would. be a scramble among manufacturers to get hold of the invention, and I had particular designs on that man from Detroit who has an uncanny faculty for accumulating millions. So confident was I that he would turn up some day, that I declared this as certain to my secretary and assistants. Sure enough, one fine morning a body of engineers from the Ford Motor Company presented themselves with the request of discussing with me an important project. "Didn't I tell you?" I remarked triumphantly to my employees, and one of them said, "You are amazing, Mr. Tesla; everything comes out exactly as you predict." As soon as these hard-headed men were seated I, of course, immediately began to extol the wonderful features of my turbine, when the spokesmen interrupted me and said, "We know all about this, but we are on a special errand. We have formed a psychological society for the investigation of psychic phenomena and we want you to join us in this undertaking." I suppose those engineers never knew how near they came to being fired out of my office.

Ever since I was told by some of the greatest men of the time, leaders in science whose names are immortal, that I am possessed of an unusual mind, I bent all my thinking faculties on the solution of great problems regardless of sacrifice. For many years I endeavored to solve the enigma of death, and watched eagerly for every kind of spiritual indication. But only once in the course of my existence have I had an experience which momentarily impressed me as supernatural. It was at the time of my mother's death. I had become completely exhausted by pain and long vigilance, and one night was carried to a building about two blocks from our home. As I lay helpless there, I thought that if my mother died while I was away from her bedside she would surely give me a sign. Two or three months before I was in London in company with my late friend, Sir William Crookes, when spiritualism was discussed, and I was under the full sway of these thoughts. I might not have paid attention to other men, but was susceptible to his arguments as it was his epochal work on radiant matter, which I had read as a student, that made me embrace the electrical career. I reflected that the conditions for a look into the beyond were most favorable, for my mother was a woman of genius and particularly excelling in the powers of intuition. During the whole night every fiber in my brain was strained in expectancy, but nothing happened until early in the morning, when I fell in a sleep, or perhaps a swoon, and saw a cloud carrying angelic figures of marvelous beauty, one of whom gazed upon me lovingly and gradually assumed the features of my mother. The appearance slowly floated across the room and vanished, and I was awakened by an indescribably sweet song of many voices. In that instant a certitude, which no words can express, came upon me that my mother had just died. And that was true. I was unable to understand the tremendous weight of the painful knowledge I received in advance, and wrote a letter to Sir William Crookes while still under the domination of these impressions and in poor bodily health. When I recovered I sought for a long time the external cause of this strange manifestation and, to my great relief, I succeeded after many months of fruitless effort. I had seen the painting of a celebrated artist, representing allegorically one of the seasons in the form of a cloud with a group of angels which seemed to actually float in the air, and this had struck me forcefully. It was exactly the same that appeared in my dream, with the exception of my mother's likeness. The music came from the choir in the church nearby at the early mass of Easter morning, explaining everything satisfactorily in conformity with scientific facts.

This occurred long ago, and I have never had the faintest reason since to change my views on psychical and spiritual phenomena, for which there is absolutely no foundation. The belief in these is the natural outgrowth of intellectual development. Religious dogmas are no longer accepted in their orthodox meaning, but every individual clings to faith in a supreme power of some kind. We all must have an ideal to govern our conduct and insure contentment, but it is immaterial whether it be one of creed, art, science or anything else, so long as it fulfills the function of a dematerializing force. It is essential to the peaceful existence of humanity as a whole that one common conception should prevail.

While I have failed to obtain any evidence in support of the contentions of psychologists and spiritualists, I have proved to my complete satisfaction the automatism of life, not only through continuous observations of individual actions, but even more conclusively through certain generalizations. These amount to a discovery which I consider of the greatest moment to human society, and on which I shall briefly dwell. I got the first inkling of this astounding truth when I was still a very young man, but for many years I interpreted what I noted simply as coincidences. Namely, whenever either myself or a person to whom I was attached, or a cause to which I was devoted, was hurt by others in a particular way, which might be best popularly characterized as the most unfair imaginable, I experienced a singular and undefinable pain which, for want of a better term, I have qualified as "cosmic," and shortly thereafter, and invariably, those who had inflicted it came to grief. After many such cases I confided this to a number of friends, who had the opportunity to convince themselves of the truth of the theory which I have gradually formulated and which may be stated in the following few words:

Our bodies are of similar construction and exposed to the same external influences. This results in likeness of response and concordance of the general activities on which all our social and other rules and laws are based. We are automata entirely controlled by the forces of the medium being tossed about like corks on the surface of the water, but mistaking the resultant of the impulses from the outside for free will. The movements and other actions we perform are always life preservative and tho seemingly quite independent from one another, we are connected by invisible links. So long as the organism is in perfect order it responds accurately to the agents that prompt it, but the moment that there is some derangement in any individual, his self-preservative power is impaired. Everybody understands, of course, that if one becomes deaf, has his eyesight weakened, or his limbs injured, the chances for his continued existence are lessened. But this is also true, and perhaps more so, of certain defects in the brain which deprive the automaton, more or less, of that vital quality and cause it to rush into destruction. A very sensitive and observant being, with his highly developed mechanism all intact, and acting with precision in obedience to the changing conditions of the environment, is endowed with a transcending mechanical sense, enabling him to evade perils too subtle to be directly perceived. When he comes in contact with others whose controlling organs are radically faulty, that sense asserts itself and he feels the "cosmic" pain. The truth of this has been borne out in hundreds of instances and I am inviting other students of nature to devote attention to this subject, believing that thru combined and systematic effort results of incalculable value to the world will be attained.

The idea of constructing an automaton, to bear out my theory, presented itself to me early but I did not begin active work until 1893, when I started my wireless investigations. During the succeeding two or three years a number of automatic mechanisms, to be actuated from a distance, were constructed by me and exhibited to visitors in my laboratory. In 1896, however, I designed a complete machine capable of a multitude of operations, but the consummation of my labors was delayed until late in 1897. This machine was illustrated and described in my article in the Century Magazine of June, 1900, and other periodicals of that time and, when first shown in the beginning of 1898, it created a sensation such as no other invention of mine has ever produced. In November, 1898, a basic patent on the novel art was granted to me, but only after the Examiner-in-Chief had come to New York and witnessed the performance, for what I claimed seemed unbelievable. I remember that when later I called on an official in Washington, with a view of offering the invention to the Government, he burst out in laughter upon my telling him what I had accomplished. Nobody thought then that there was the faintest prospect of perfecting such a device. It is unfortunate that in this patent, following the advice of my attorneys, I indicated the control as being effected thru the medium of a single circuit and a well-known form of detector, for the reason that I had not yet secured protection on my methods and apparatus for individualization. As a matter of fact, my boats were controlled thru the joint action of several circuits and interference of every kind was excluded. Most generally I employed receiving circuits in the form of loops, including condensers, because the discharges of my high-tension transmitter ionized the air in the hall so that even a very small aerial would draw electricity from the surrounding atmosphere for hours. Just to give an idea, I found, for instance, that a bulb 12" in diameter, highly exhausted, and with one single terminal to which a short wire was attached, would deliver well on to one thousand successive flashes before all charge of the air in the laboratory was neutralized. The loop form of receiver was not sensitive to such a disturbance and it is curious to note that it is becoming popular at this late date. In reality it collects much less energy than the aerials or a long grounded wire, but it so happens that it does away with a number of defects inherent to the present wireless devices. In demonstrating my invention before audiences, the visitors were requested to ask any questions, however involved, and the automaton would answer them by signs. This was considered magic at that time but was extremely simple, for it was myself who gave the replies by means of the device.

At the same period another larger telautomatic boat was constructed a photograph of which is shown in this number of the *Electrical Experimenter*. It was controlled by loops, having several turns placed in the hull, which was made entirely water-tight and capable of submergence. The apparatus was similar to that used in the first with the exception of certain special features I introduced as, for example, incandescent lamps which afforded a visible evidence of the proper functioning of the machine.

These automata, controlled within the range of vision of the operator, were, however, the first and rather crude steps in the evolution of the Art of Telautomatics as I had conceived it. The next logical improvement was its application to automatic mechanisms beyond the limits of vision and at great distance from the center of control, and I have ever since advocated their employment as instruments of warfare in preference to guns. The importance of this now seems to be recognized, if I am to judge from casual announcements thru the press of achievements which are said to be extraordinary but contain no merit of novelty, whatever. In an imperfect manner it is practicable, with the existing wireless plants, to launch an airplane, have it follow a certain approximate course, and perform some operation at a distance of many hundreds of miles. A machine of this kind can also be mechanically controlled in several ways and I have no doubt that it may prove of some usefulness in war. But there are, to my best knowledge, no instrumentalities in existence today with which such an object could be accomplished in a precise manner. I have devoted years of study to this matter and have evolved means, making such and greater wonders easily realizable.

As stated on a previous occasion, when I was a student at college I conceived a flying machine quite unlike the present ones. The underlying principle was sound but could not be carried into practice for want of a prime-mover of sufficiently great activity. In recent years I have successfully solved this problem and am now planning aerial machines devoid of sustaining planes, ailerons, propellers and other external attachments, which will be capable of immense speeds and are very likely to furnish powerful arguments for peace in the near future. Such a machine, sustained and propelled entirely by reaction, is shown on page 108 and is supposed to be controlled either mechanically or by wireless energy. By installing proper plants it will be practicable to project a missile of this kind into the air and drop it almost on the very spot designated, which may be thousands of miles away. But we are not going to stop at this. Telautomata will be ultimately produced, capable of acting as if possessed of their own intelligence, and their advent will create a revolution. As early as 1898 I proposed to representatives of a large manufacturing concern the construction and public exhibition of an automobile carriage which, left to itself, would perform a great variety of operations involving something akin to judgment. But my proposal was deemed chimerical at that time and nothing came from it.

At present many of the ablest minds are trying to devise expedients for preventing a repetition of the awful conflict which is only theoretically ended and the duration and main issues of which I have correctly predicted in an article printed in the Sun of December 20, 1914. The proposed League is not a remedy but on the contrary, in the opinion of a number of competent men, may bring about results just the opposite. It is particularly regrettable that a punitive policy was adopted in framing the terms of peace, because a few years hence it will be possible for nations to fight without armies, ships or guns, by weapons far more terrible, to the destructive action and range of which there is virtually no limit. A city, at any distance whatsoever from the enemy, can be destroyed by him and no power on earth can stop him from doing so. If we want to avert an impending calamity and a state of things which may transform this globe into an inferno, we should push the development of flying machines and wireless transmission of energy without an instant's delay and with all the power and resources of the nation.